Nonlinear Dynamics and Vibration Control of Flexible Systems

This book is an essential guide to nonlinear dynamics and vibration control, detailing both the theory and the practical industrial applications within all aspects of engineering. Demonstrating how to improve efficiency through reducing unwanted vibration, it will aid both students and engineers in practically and safely improving flexible structures through control methods.

Increasing demand for light-weight robotic systems and space applications has actuated the design and construction of more flexible structures. These flexible structures, involving numerous dynamic systems, experience unwanted vibrations, impacting accuracy, operating speed, safety, and, importantly, efficiency. This book aids engineers in assuaging this issue through vibration control methods, including nonlinear dynamics. It covers topics such as dynamic modeling of nonlinear system, nonlinear oscillators, and modal analyses of multiple-mode system. It also looks at vibration control methods including linear control, nonlinear control, intelligent control, and command smoothers. These control methods are effective and reliable methods to counteract unwanted vibrations. The book is practically minded, using industrial applications throughout, such as bridge cranes, tower cranes, aerial cranes, and liquid sloshing. It also discusses cable-suspension structures, light-weight links, and fluid motions which exhibit flexible-structure dynamics.

The book will be of interest to students and engineers alike, in the field of mechatronics, mechanical systems and signal processing, nonlinear dynamics, vibration, and control engineering.

Nonlinear Dynamics and Vibration Control of Flexible Systems

Jie Huang

CRC Press is an imprint of the
Taylor & Francis Group, an **informa** business

First edition published 2023
by CRC Press
6000 Broken Sound Parkway NW, Suite 300, Boca Raton, FL 33487-2742

and by CRC Press
4 Park Square, Milton Park, Abingdon, Oxon, OX14 4RN

CRC Press is an imprint of Taylor & Francis Group, LLC

ISBN: 978-1-032-16128-0 (hbk)
ISBN: 978-1-032-16137-2 (pbk)
ISBN: 978-1-003-24721-0 (ebk)

DOI: 10.1201/9781003247210

Typeset in Times
by MPS Limited, Dehradun

Contents

Preface

The demands of light-weight robotic systems and space applications have actuated the flexible structures. Numerous dynamic systems with flexible structures suffer from undesirable vibrations. Detrimental effects will cause serious problems for positioning accuracy, operating speed, effectiveness, and safety. Therefore, there is a need to study nonlinear dynamics and vibration control of flexible structures such that unwanted vibrations can be effectively reduced for safe and effective operations.

This book describes theoretical knowledge on nonlinear dynamics and vibration control. Then some industrial applications with flexible structures have been reported by using the theoretical knowledge. Those applications include bridge cranes, tower cranes, aerial cranes, dual cranes, flexible link manipulators, and liquid sloshing.

Theoretical knowledge of nonlinear dynamics includes dynamic modeling of nonlinear system, analyses of nonlinear oscillator, approximate solution of nonlinear oscillator, and modal analyses of multiple-mode system. Meanwhile, theoretical knowledge of vibration control includes linear control, nonlinear control, intelligent control, input shaper, and command smoother. Those control methods are effective and robust for attenuating vibrations.

Cable-suspension structures, light-weight links, and fluid motions in the container exhibit flexible-structure dynamics. Therefore, the dynamic modeling and dynamic analyses are very challenging for academic staff and graduate students. The vibration-reduction solutions are also provided by using the advanced control methods.

The author would like to acknowledge the support of the Beijing Institute of Technology (BIT) and the National Natural Science Foundation of China (NSFC). The authors would also like to thank the anonymous reviewers, Nicola Sharpe, and Nishant Bhagat, for their helpful comments and suggestions, on which the quality of this book has been improved.

Jie Huang
School of Mechanical Engineering,
Beijing Institute of Technology,
Beijing, China

Author Biography

Jie Huang received the Ph.D. degree in mechanical engineering from the Beijing Institute of Technology, China. He then joined the School of Mechanical Engineering, Beijing Institute of Technology. He is an associate professor in mechanical engineering at Beijing Institute of Technology now. He also held visiting appointments at Georgia Institute of Technology, University of Technology Sydney, and University of Southampton. He published 62 journal papers and 3 books. His research interests include dynamics and control of many types of flexible systems including industrial cranes, tower cranes, flexible manipulators, and liquid sloshing.

1 Nonlinear Dynamics

Numerous dynamic systems with flexible structures suffer from undesirable vibrations, which exhibit nonlinear dynamic behavior. Detrimental effects will cause serious problems for positioning accuracy, operating speed, effectiveness, and safety. Therefore, there is a need to study dynamics and control of flexible structures so that unwanted vibrations can be effectively reduced for safe and effective operations.

1.1 DYNAMIC MODELING

There are many analytical methods for deriving equations of motion for flexible systems including Newton-Euler methods, Lagrange's equation, Hamilton's equation, and Kane's equations. The first three methods are traditional, while the last one is modern. In this book, Kane's equations are applied for modeling of industrial cranes in Chapters 3–6. Newton-Euler methods have been used for modeling of flexible link manipulators in Chapter 7 and liquid sloshing in Chapter 8. Hamilton's equation has been applied for modeling of nonlinear sloshing in Chapter 8. While many books described the first three traditional methods, an explanation of Kane's equations will be given later.

The generalized coordinates, $q_1, q_2, \ldots, q_n$, should be established for a system with n degree of freedom. The derivative of the generalized coordinates with respect to time is the generalized speeds, $u_1, u_2, \ldots, u_n$. The flexible multibody system includes particles and rigid bodies. The particle can be considered as a body without size, or point-mass body. The velocity of a particle, P_k, in the Newtonian reference frame is denoted by ${}^N v^{P_k}$, which can be expressed as:

$$ {}^N v^{P_k} = \sum_{i=1}^{n} \left[{}^N v_i^{P_k} u_i + {}^N v_t^{P_k} \right] \tag{1.1} $$

where ${}^N v_i^{P_k}$ represents the ith partial velocity of the particle, and ${}^N v_t^{P_k}$ is corresponding coefficient of the generalized speed. The velocity, ${}^N v^{B_k^*}$, of mass center, B_k^*, of a body, B_k, is given by:

$$ {}^N v^{B_k^*} = \sum_{i=1}^{n} \left[{}^N v_i^{B_k^*} u_i + {}^N v_t^{B_k^*} \right] \tag{1.2} $$

where ${}^N v_i^{B_k^*}$ represents the ith partial velocity of the mass center of the body, and ${}^N v_t^{B_k^*}$ is corresponding coefficient of the generalized speed. Meanwhile, the angular velocity, ${}^N \omega^{B_k}$, of the body, B_k, is given by:

DOI: 10.1201/9781003247210-1

$$ {}^{N}\omega^{B_k} = \sum_{i=1}^{n} \left[{}^{N}\omega_i^{B_k} u_i + {}^{N}\omega_t^{B_k} \right] \tag{1.3} $$

where ${}^{N}\omega_i^{B_k}$ represents the ith partial velocity of the body, B_k, and ${}^{N}\omega_t^{B_k}$ is corresponding coefficient of the generalized speed. The acceleration of the particle can be derived from the derivative (1.1) with respect to time:

$$ {}^{N}a^{P_k} = \frac{{}^{N}d\,{}^{N}v^{P_k}}{dt} \tag{1.4} $$

The acceleration of mass center, B_k^*, of the body, B_k, can be derived from the derivative (1.2) with respect to time:

$$ {}^{N}a^{B_k^*} = \frac{{}^{N}d\,{}^{N}v^{B_k^*}}{dt} \tag{1.5} $$

The angular acceleration of the body, B_k, can be derived from the derivative (1.3) with respect to time:

$$ {}^{N}\alpha^{B_k} = \frac{{}^{N}d\,{}^{N}\omega^{B_k}}{dt} \tag{1.6} $$

The ith partial velocity, ${}^{N}v_i^{P_k}$, of the particle, P_k, derives from the derivative (1.1) with respect to the generalized speed, u_i:

$$ {}^{N}v_i^{P_k} = \frac{\partial\,{}^{N}v^{P_k}}{\partial u_i} \tag{1.7} $$

The ith partial velocity, ${}^{N}v_i^{B_k^*}$, of the mass center, B_k^*, of the body, B_k, derives from the derivative (1.2) with respect to the generalized speed, u_i:

$$ {}^{N}v_i^{B_k^*} = \frac{\partial\,{}^{N}v^{B_k^*}}{\partial u_i} \tag{1.8} $$

The ith partial angular velocity, ${}^{N}\omega_i^{B_k}$, of the body, B_k, derives from the derivative (1.3) with respect to the generalized speed, u_i:

$$ {}^{N}\omega_i^{B_k} = \frac{\partial\,{}^{N}\omega^{B_k}}{\partial u_i} \tag{1.9} $$

The corresponding generalized inertia forces can be written as:

$$F_i^* = -\sum_{k=1}^{NR}\left[m_k\cdot{}^N a^{B_k^*}.{}^N v_i^{B_k^*} + \left(\mathrm{I}^{B_k/B_k^*}.{}^N\alpha^{B_k} + {}^N\omega^{B_k}\times \mathrm{I}^{B_k/B_k^*}.{}^N\omega^{B_k}\right).{}^N\omega_i^{B_k}\right]$$
$$-\sum_{k=1}^{NP}\left[m_k\cdot{}^N a^{P_k}.{}^N v_i^{P_k}\right] \tag{1.10}$$

where $\cdot$ is the dot product, $\times$ is the cross product, NR is the number of bodies, NP is the number of particles, m_k denotes the mass of the bodies or particles, and I^{B_k/B_k^*} denotes inertia dyadic of the body, B_k, about the mass center, B_k^*.

The corresponding generalized active forces can be written as:

$$F_i = \sum_{k=1}^{NR}\left[F^{B_k^*}.{}^N v_i^{B_k^*} + T^{B_k}.{}^N\omega_i^{B_k}\right] + \sum_{k=1}^{NP}\left[F^{P_k}.{}^N v_i^{P_k}\right] \tag{1.11}$$

where F^{P_k} denotes the force acting on the particle, P_k, $F^{B_k^*}$ denotes the force acting on the mass center, B_k^*, of the body, B_k, and T^{B_k} denotes the moment acting on the body, B_k. Forcing the sum of the generalized active forces (1.11) and generalized inertia forces (1.10) to be zero gives rise to dynamic equations of the motion for the flexible system:

$$F_i + F_i^* = 0 \tag{1.12}$$

or

$$\sum_{k=1}^{NR}\left[F^{B_k^*}.{}^N v_i^{B_k^*} + T^{B_k}.{}^N\omega_i^{B_k}\right] + \sum_{k=1}^{NP}\left[F^{P_k}.{}^N v_i^{P_k}\right]$$
$$= \sum_{k=1}^{NP}\left[m_k\cdot{}^N a^{P_k}.{}^N v_i^{P_k}\right] \tag{1.13}$$
$$+\sum_{k=1}^{NR}\left[m_k\cdot{}^N a^{B_k^*}.{}^N v_i^{B_k^*} + \left(\mathrm{I}^{B_k/B_k^*}.{}^N\alpha^{B_k} + {}^N\omega^{B_k}\times \mathrm{I}^{B_k/B_k^*}.{}^N\omega^{B_k}\right).{}^N\omega_i^{B_k}\right]$$

1.2 NONLINEAR OSCILLATORS

1.2.1 Nonlinear Pendulum

Figure 1.1 shows a model of single pendulum. A massless cable supports a point-mass payload. The swing angle denotes θ, and the cable length is l. By using Newton's second law, the equation of motion can be derived:

$$l\ddot{\theta} + g\ \sin\ \theta = 0 \tag{1.14}$$

A column matrix, x, is defined as $x = [\theta,\ \dot{\theta}]^T$. Then, Equation (1.14) can be re-written as:

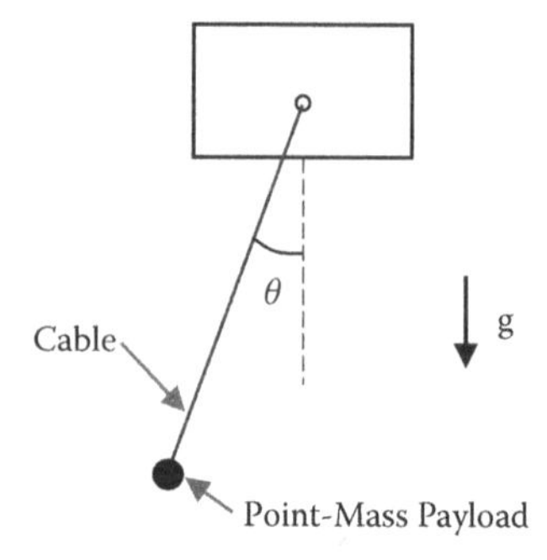

FIGURE 1.1 Model of pendulum.

$$\dot{x} = \begin{bmatrix} \dot{\theta} \\ \ddot{\theta} \end{bmatrix} = \begin{bmatrix} \dot{\theta} \\ \frac{-g \sin \theta}{l} \end{bmatrix} \tag{1.15}$$

Forcing the right side of Equation (1.15) to zero yields the equilibrium point of Equation (1.14). Therefore, the corresponding equilibrium point is zero angular displacement and zero angular velocity of the swing.

The frequency of pendulum (1.14) is described as:

$$\omega^2 = \frac{g \sin \theta}{l\theta} = \frac{g}{l}\left(1 - \frac{\theta^2}{6} + \frac{\theta^4}{120} \cdots\right) \tag{1.16}$$

The frequency of pendulum is dependent on the cable length and vibrational amplitude. Increasing cable length decreases the pendulum frequency. Meanwhile, the pendulum frequency decreases with increasing vibrational amplitude. At the equilibrium point, the vibrational amplitude of the swing is zero, and the pendulum frequency is equal to the linearized frequency, $\sqrt{g/l}$. The system, in which frequency decreases as the vibrational amplitudes increase, corresponds to the softening spring type.

1.2.2 Nonlinear Mass-Spring Oscillators

Figure 1.2 shows a model of mass-spring oscillator. A spring supports a mass block. The mass denotes m. The displacement of the mass block is y. The linear stiffness of the spring is k_1, and the nonlinear stiffness denotes k_3. By using Newton's second law, the equation of motion can be derived:

$$m\ddot{y} + k_1 y + k_3 y^3 = 0 \tag{1.17}$$

A column matrix, x, is defined as $x = [y, \dot{y}]^T$. Then, Equation (1.17) can be rewritten as:

$$\dot{x} = \begin{bmatrix} \dot{y} \\ \ddot{y} \end{bmatrix} = \begin{bmatrix} \dot{y} \\ \frac{-k_1 y - k_3 y^3}{m} \end{bmatrix} \tag{1.18}$$

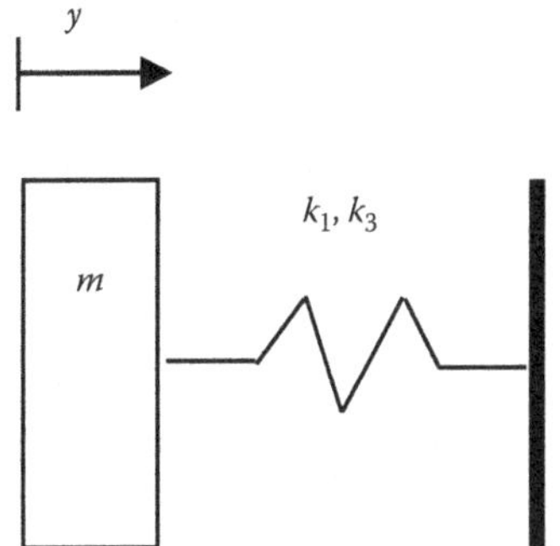

FIGURE 1.2 Model of mass-spring oscillators.

Forcing the right side of Equation (1.18) to zero yields the equilibrium point of Equation (1.17). Therefore, the corresponding equilibrium point is zero displacement and zero velocity of the mass block.

The frequency of mass-spring oscillator (1.17) is described as:

$$\omega^2 = \frac{k_1}{m} + \frac{k_3 y^2}{m} \tag{1.19}$$

The frequency of mass-spring oscillator is dependent on the system parameters and vibrational amplitude. System parameters include mass and stiffness.

The frequency increases with increasing vibrational amplitude. At the equilibrium point, the vibrational amplitude of the mass-spring oscillator is zero, and the frequency is equal to the linearized frequency, $\sqrt{k_1/m}$. The system, in which frequency increases as the vibrational amplitudes increase, corresponds to the hardening spring type.

1.3 APPROXIMATE SOLUTION

Exact analytical solutions or closed-form solutions are challenging to derive for nonlinear systems. Of course, numerical solutions can be derived for nonlinear systems. Some methods have been presented to get approximate solutions for nonlinear systems, including harmonic balance method, averaging method, normal form transformations, perturbation method, and multiple scales method. The last two methods will be described.

1.3.1 Perturbation Method

Consider a one-DOF unforced nonlinear system:

$$\begin{cases} \ddot{x} + \omega_0^2 x = \varepsilon p(x, \dot{x}) \\ x(0) = A \\ \dot{x}(0) = 0 \end{cases} \tag{1.20}$$

where ε denotes the small nonlinear coefficient, $\varepsilon p(x, \dot{x})$ is the small nonlinear term, and A is the initial displacement. The solutions of Equation (1.20) and corresponding nonlinear frequency, ω, can be assumed to be:

$$\begin{cases} x = x_0 + \varepsilon x_1 + \varepsilon^2 x_2 + \ldots + \varepsilon^k x_k + \ldots \\ \omega^2 = \omega_0^2 + \varepsilon b_1 + \varepsilon^2 b_2 + \ldots + \varepsilon^k b_k + \ldots \end{cases} \tag{1.21}$$

Substituting Equation (1.21) into (1.20) gives rise to:

$$\begin{cases} (\ddot{x}_0 + \varepsilon\ddot{x}_1 + \varepsilon^2\ddot{x}_2 + \ldots) \\ \quad + (\omega^2 - \varepsilon b_1 - \varepsilon^2 b_2 + \ldots)(x_0 + \varepsilon x_1 + \varepsilon^2 x_2 + \ldots) \\ \quad = \varepsilon p(x_0 + \varepsilon x_1 + \varepsilon^2 x_2 + \ldots, \dot{x}_0 + \varepsilon\dot{x}_1 + \varepsilon^2\dot{x}_2 + \ldots) \\ x_0(0) + \varepsilon x_1(0) + \varepsilon^2 x_2(0) + \ldots = A \\ \dot{x}_0(0) + \varepsilon\dot{x}_1(0) + \varepsilon^2\dot{x}_2(0) + \ldots = 0 \end{cases} \tag{1.22}$$

The assumptions can be balanced for all powers of ε. The ε^0, ε^1, and ε^2 terms are balanced:

$$\varepsilon^0: \begin{cases} \ddot{x}_0 + \omega^2 x_0 = 0 \\ x_0(0) = A \\ \dot{x}_0(0) = 0 \end{cases} \tag{1.23}$$

$$\varepsilon^1: \begin{cases} \ddot{x}_1 + \omega^2 x_1 = p(x_0, \dot{x}_0) + b_1 x_0 \\ x_1(0) = 0 \\ \dot{x}_1(0) = 0 \end{cases} \tag{1.24}$$

$$\varepsilon^2: \begin{cases} \ddot{x}_2 + \omega^2 x_2 = p_1(x_0, \dot{x}_0)x_1 + p_2(x_0, \dot{x}_0)\dot{x}_1 + b_2 x_0 + b_1 x_1 \\ x_2(0) = 0 \\ \dot{x}_2(0) = 0 \end{cases} \tag{1.25}$$

The high-order terms are also balanced for high exact expression. Solutions of Equations (1.23), (1.24), and (1.25) can be substituted into (1.21) to produce the approximate solutions for nonlinear systems. This method is named after Lindstedt-Poincaré perturbation method.

Example 1.1: Approximate solutions of nonlinear pendulum.

A Taylor series expansion has been applied to the nonlinear pendulum (1.14):

$$\ddot{\theta} + \left(\frac{g}{l}\theta - \frac{g\theta^3}{6l} + \ldots\right) = 0 \tag{1.26}$$

The Taylor series expansion is approximated by the first two terms:

$$\ddot{\theta} + \frac{g}{l}\theta - \frac{g\theta^3}{6l} = 0 \tag{1.27}$$

The simplified system (1.27) can be rewritten as:

$$\begin{cases} \ddot{\theta} + \omega_0^2\theta + \varepsilon\mu\theta^3 = 0 \\ \theta(0) = 0; \\ \dot{\theta}(0) = V \end{cases} \tag{1.28}$$

where

$$\omega_0 = \sqrt{\frac{g}{l}}; \quad \varepsilon = -\frac{1}{6}; \quad \mu = \frac{g}{l} \tag{1.29}$$

The solution and corresponding nonlinear frequency can be assumed to be:

$$\begin{cases} \theta = \theta_0 + \varepsilon\theta_1 \\ \omega^2 = \omega_0^2 + \varepsilon b_1 \end{cases} \tag{1.30}$$

Substituting Equation (1.30) into (1.28) yields:

$$\begin{cases} (\ddot{\theta}_0 + \varepsilon\ddot{\theta}_1) + (\omega^2 - \varepsilon b_1)\cdot(\theta_0 + \varepsilon\theta_1) + \varepsilon\mu\cdot(\theta_0 + \varepsilon\theta_1)^3 = 0 \\ \theta_0(0) + \varepsilon\theta_1(0) = 0 \\ \dot{\theta}_0(0) + \varepsilon\dot{\theta}_1(0) = V \end{cases} \tag{1.31}$$

Equation (1.31) can be balanced for ε^0, and ε^1, terms:

$$\varepsilon^0: \begin{cases} \ddot{\theta}_0 + \omega^2\theta_0 = 0 \\ \theta_0(0) = 0 \\ \dot{\theta}_0(0) = V \end{cases} \tag{1.32}$$

$$\varepsilon^1: \begin{cases} \ddot{\theta}_1 + \omega^2\theta_1 = b_1\theta_0 - \mu\theta_0^3 \\ \theta_1(0) = 0; \\ \dot{\theta}_1(0) = 0 \end{cases} \tag{1.33}$$

The solution of Equation (1.32) is given by:

$$\theta_0 = \frac{V}{\omega}\sin(\omega t) \tag{1.34}$$

Substituting Equation (1.34) into (1.33) gives rise to:

$$\begin{cases} \ddot{\theta}_1 + \omega^2\theta_1 = \left(\frac{V}{\omega}b_1 - \frac{3\mu V^3}{4\omega^3}\right)\sin(\omega t) + \frac{\mu V^3}{4\omega^3}\sin(3\omega t) \\ \theta_1(0) = 0 \\ \dot{\theta}_1(0) = 0 \end{cases} \tag{1.35}$$

The first term of the right side of the first equation of (1.35) is secular term. This term must be kept to zero for bounded solutions:

$$\frac{V}{\omega}b_1 - \frac{3\mu V^3}{4\omega^3} = 0 \tag{1.36}$$

The solution of Equation (1.36) is given by:

$$b_1 = \frac{3\mu V^2}{4\omega^2} \tag{1.37}$$

Substituting Equations (1.36) and (1.37) into (1.35) gives rise to:

$$\begin{cases} \ddot{\theta}_1 + \omega^2\theta_1 = \frac{\mu V^3}{4\omega^3}\sin(3\omega t) \\ \theta_1(0) = 0 \\ \dot{\theta}_1(0) = 0 \end{cases} \tag{1.38}$$

The solution of Equation (1.38) is:

$$\theta_1 = \frac{3\mu V^3}{8\omega^5}\sin(\omega t) - \frac{\mu V^3}{8\omega^5}\sin(3\omega t) \tag{1.39}$$

Substituting Equations (1.34) and (1.39) into (1.30) gives rise to the approximate solutions of nonlinear pendulum:

$$\begin{cases} \theta = \frac{V}{\omega}\sin(\omega t) + \varepsilon\left[\frac{3\mu V^3}{8\omega^5}\sin(\omega t) - \frac{\mu V^3}{8\omega^5}\sin(3\omega t)\right] \\ \omega^2 = \omega_0^2 + \varepsilon\frac{3\mu V^2}{4\omega^2} \end{cases} \tag{1.40}$$

Example 1.2: Approximate solutions of a nonlinear damped system.

$$\begin{cases} \ddot{x} + 2\zeta\omega_0\dot{x} + \omega_0^2 x + \varepsilon\mu x^3 = 0 \\ x(0) = A; \quad \dot{x}(0) = 0 \end{cases} \tag{1.41}$$

where ε denotes the small nonlinear coefficient, ζ is the damping ratio, ω_0 is the linear frequency, $\varepsilon\mu x^3$ is nonlinear term, and A is the initial displacement. The solution and nonlinear frequency can be assumed to be:

$$\begin{cases} x = x_0 + \varepsilon x_1 \\ \omega = \omega_0 + \varepsilon b_1 \\ \zeta = \varepsilon\zeta_1 \end{cases} \tag{1.42}$$

Substituting Equation (1.42) into (1.41) gives rise to:

$$\begin{cases} (\ddot{x}_0 + \varepsilon\ddot{x}_1) + 2\varepsilon\zeta_1(\omega - \varepsilon b_1)(\dot{x}_0 + \varepsilon\dot{x}_1) + (\omega - \varepsilon b_1)^2(x_0 + \varepsilon x_1) + \varepsilon\mu(x_0 + \varepsilon x_1)^3 \\ \quad = 0 \\ x_0(0) + \varepsilon x_1(0) = A; \quad \dot{x}_0(0) + \varepsilon\dot{x}_1(0) = 0 \end{cases} \tag{1.43}$$

Equation (1.43) can be balanced for ε^0, and ε^1, terms:

$$\begin{cases} \ddot{x}_0 + \omega^2 x_0 = 0 \\ x_0(0) = A; \quad \dot{x}_0(0) = 0 \end{cases} \tag{1.44}$$

$$\begin{cases} \ddot{x}_1 + 2\zeta_1\omega\dot{x}_0 + \omega^2 x_1 - 2b_1 x_0 + \mu x_0{}^3 = 0 \\ x_1(0) = 0; \quad \dot{x}_1(0) = 0 \end{cases} \tag{1.45}$$

The solution of Equation (1.44) is:

$$x_0 = A\cdot\cos(\omega t) \tag{1.46}$$

The solution of Equation (1.45) is:

$$\begin{cases} \ddot{x}_1 + \omega^2 x_1 = -2\zeta_1\omega^2 A\sin(\omega t) + \left(2b_1 A - \frac{3\mu A^3}{4}\right)\cos(\omega t) - \frac{\mu A^3}{4}\cos(3\omega t) \\ x_1(0) = 0; \quad \dot{x}_1(0) = 0 \end{cases} \tag{1.47}$$

The first two terms of the right side of the first equation of (1.47) are secular terms. Those terms must be kept to zero for bounded solutions. Then, Equation (1.47) can be rewritten as:

$$\begin{cases} \ddot{x}_1 + \omega^2 x_1 = -\frac{\mu A^3}{4}\cos(3\omega t) \\ x_1(0) = 0; \quad \dot{x}_1(0) = 0 \end{cases} \tag{1.48}$$

The solution of Equation (1.48) is:

$$x_1 = -\frac{\mu A^3}{32\omega^2}\cos(3\omega t) \tag{1.49}$$

Substituting Equations (1.46) and (1.49) into (1.42) gives rise to the approximate solutions of nonlinear damped system.

1.3.2 Multiple Scales Method

A time scale is defined as:

$$T_k = \varepsilon^k t \tag{1.50}$$

where ε denotes the small nonlinear coefficient and t is the time. Then the derivative can be written as a function of the time scales:

$$\begin{cases} \dfrac{d}{dt} = \dfrac{dT_0}{dt}\dfrac{\partial}{\partial T_0} + \dfrac{dT_1}{dt}\dfrac{\partial}{\partial T_1} + \ldots = \dfrac{\partial}{\partial T_0} + \varepsilon\dfrac{\partial}{\partial T_1} + \ldots = D_0 + \varepsilon D_1 + \ldots \\ \dfrac{d^2}{d^2 t} = \dfrac{dT_0}{dt}\dfrac{\partial}{\partial T_0}\left[\dfrac{dT_0}{dt}\dfrac{\partial}{\partial T_1}\right] + \dfrac{dT_1}{dt}\dfrac{\partial}{\partial T_1}\left[\dfrac{dT_0}{dt}\dfrac{\partial}{\partial T_1}\right] + \ldots \\ \quad = \dfrac{\partial^2}{\partial T_0^2} + 2\varepsilon\dfrac{\partial^2}{\partial T_0 \partial T_1} + \varepsilon^2\dfrac{\partial^2}{\partial T_1^2} + \ldots = D_0^2 + 2\varepsilon D_0 D_1 + \varepsilon^2 D_1^2 + \ldots \end{cases} \tag{1.51}$$

where

$$T_0 = t, \quad T_1 = \varepsilon t \tag{1.52}$$

$$D_0 = \frac{\partial}{\partial T_0}, \quad D_1 = \frac{\partial}{\partial T_1} \tag{1.53}$$

Consider an unforced nonlinear system:

$$\begin{cases} \ddot{x} + \omega_0^2 x = \varepsilon p(x, \dot{x}) \\ x(0) = A \\ \dot{x}(0) = 0 \end{cases} \tag{1.54}$$

The solution can be assumed to be:

$$x = x_0(T_0, T_1, \ldots) + \varepsilon x_1(T_0, T_1, \ldots) + \varepsilon^2 x_2(T_0, T_1, \ldots) + \ldots \tag{1.55}$$

Substituting Equation (1.55) into (1.54), and then balanced for ε^0, and ε^1, terms:

$$\begin{cases} D_0^2 x_0 + \omega_0^2 x_0 = 0 \\ D_0^2 x_1 + \omega_0^2 x_1 = -2D_0 D_1 x_0 + p(x_0, D_0 x_0) \\ D_0^2 x_2 + \omega_0^2 x_2 = -(D_1^2 + 2D_0 D_2)x_0 - 2D_0 D_1 x_1 \\ \qquad + p_1(x_0, D_0 x_0)x_1 + p_2(x_0, D_0 x_0)(D_1 x_0 + D_0 x_1) \\ \ldots \end{cases} \tag{1.56}$$

The solutions of Equation (1.56) will result in the approximate solutions of Equation (1.54).

Example 1.3: Approximate solutions of nonlinear pendulum.

A Taylor series expansion has been applied to the nonlinear pendulum. The Taylor series expansion is approximated by the first two terms. The simplified system can be rewritten as:

$$\begin{cases} \ddot{\theta} + \omega_0^2\theta + \varepsilon\omega_0^2\theta^3 = 0 \\ \theta(0) = M \\ \dot{\theta}(0) = 0 \end{cases} \tag{1.57}$$

where

$$\omega_0 = \sqrt{\frac{g}{l}}, \quad \varepsilon = -\frac{1}{6} \tag{1.58}$$

The solution can be assumed to be:

$$\theta = \theta_0(T_0, T_1) + \varepsilon\theta_1(T_0, T_1) \tag{1.59}$$

Substituting Equation (1.59) into (1.57) gives rise to:

$$\begin{cases} \left[\frac{\partial^2\theta_0}{\partial T_0^2} + 2\varepsilon\frac{\partial^2\theta_0}{\partial T_0\partial T_1} + \varepsilon^2\frac{\partial^2\theta_0}{\partial T_1^2}\right] + \varepsilon\left[\frac{\partial^2\theta_1}{\partial T_0^2} + 2\varepsilon\frac{\partial^2\theta_1}{\partial T_0\partial T_1} + \varepsilon^2\frac{\partial^2\theta_1}{\partial T_1^2}\right] \\ \quad + \omega_0^2[\theta_0 + \varepsilon\theta_1] + \varepsilon\omega_0^2[\theta_0 + \varepsilon\theta_1]^3 = 0 \\ \theta_0(0) + \varepsilon\theta_1(0) = M \\ \dot{\theta}_0(0) + \varepsilon\dot{\theta}_1(0) = 0 \end{cases} \tag{1.60}$$

Equation (1.60) can be balanced for ε^0, and ε^1, terms:

$$\frac{\partial^2\theta_0}{\partial T_0^2} + \omega_0^2\theta_0 = 0 \tag{1.61}$$

$$\frac{\partial^2\theta_1}{\partial T_0^2} + \omega_0^2\theta_1 + 2\frac{\partial^2\theta_0}{\partial T_0\partial T_1} + \omega_0^2\theta_0^3 = 0 \tag{1.62}$$

The solution of Equation (1.61) is:

$$\theta_0 = C(T_1)e^{j\omega_0 T_0} + \overline{C(T_1)}e^{-j\omega_0 T_0} \tag{1.63}$$

Substituting Equation (1.63) into (1.62) is:

$$\begin{aligned} \frac{\partial^2\theta_1}{\partial T_0^2} + \omega_0^2\theta_1 = &\left[-2j\omega_0\frac{\partial[C(T_1)]}{\partial T_1} - 3\omega_0^2 C^2(T_1)\overline{C(T_1)}\right]e^{j\omega_0 T_0} \\ &+ \left[2j\omega_0\frac{\partial[\overline{C(T_1)}]}{\partial T_1} - 3\omega_0^2\overline{C^2(T_1)}C(T_1)\right]e^{-j\omega_0 T_0} \\ &- \omega_0^2 C^3(T_1)e^{j3\omega_0 T_0} - \omega_0^2\overline{C^3(T_1)}e^{-3j\omega_0 T_0} \end{aligned} \tag{1.64}$$

The first two terms of the right side of Equation (1.64) are secular terms. Those terms must be kept to zero for bounded solutions:

$$\begin{cases} -2j\omega_0 \frac{\partial[C(T_1)]}{\partial T_1} - 3\omega_0^2 C^2(T_1)\overline{C(T_1)} = 0 \\ 2j\omega_0 \frac{\partial[\overline{C(T_1)}]}{\partial T_1} - 3\omega_0^2 \overline{C^2(T_1)}C(T_1) = 0 \end{cases} \tag{1.65}$$

A trial solution of Equation (1.65) can be chosen as:

$$C(T_1) = Me^{jA} \tag{1.66}$$

Substituting Equation (1.66) into (1.51) gives rise to:

$$\begin{cases} \frac{dC}{dt} = \frac{\partial C}{\partial T_0} + \varepsilon \frac{\partial C}{\partial T_1} = \varepsilon \frac{\partial C}{\partial T_1} = \varepsilon \frac{3\omega_0^2 C^2(T_1)\overline{C(T_1)}}{-2j\omega_0} \\ \frac{d\overline{C}}{dt} = \frac{\partial \overline{C}}{\partial T_0} + \varepsilon \frac{\partial \overline{C}}{\partial T_1} = \varepsilon \frac{\partial \overline{C}}{\partial T_1} = \varepsilon \frac{3\omega_0^2 C^2(T_1)\overline{C(T_1)}}{2j\omega_0} \end{cases} \tag{1.67}$$

Substituting Equation (1.67) into (1.65) gives rise to:

$$\begin{cases} C = 0.5Me^{j\left[\left(\frac{3}{8}\varepsilon\omega_0 M^2\right)+B\right]} \\ \overline{C} = 0.5Me^{-j\left[\left(\frac{3}{8}\varepsilon\omega_0 M^2\right)+B\right]} \end{cases} \tag{1.68}$$

where B is the coefficient and depends on the initial condition. Then by ignoring the secular terms, Equation (1.64) can be rewritten as:

$$\frac{\partial^2 \theta_1}{\partial T_0^2} + \omega_0^2 \theta_1 = -\omega_0^2 C^3(T_1)e^{j3\omega_0 T_0} - \omega_0^2 \overline{C^3(T_1)}e^{-3j\omega_0 T_0} \tag{1.69}$$

The solution of Equation (1.69) is:

$$\theta_1 = \frac{\omega_0^2 C^3}{8} e^{j3\omega_0 T_0} + \frac{\omega_0^2 \overline{C}^3}{8} e^{-j3\omega_0 T_0} \tag{1.70}$$

Substituting Equations (1.63) and (1.70) into (1.59) gives rise to approximate solutions of nonlinear pendulum:

$$\theta = C(T_1)e^{j\omega_0 T_0} + \overline{C(T_1)}e^{-j\omega_0 T_0} + \varepsilon \left[\frac{\omega_0^2 C^3}{8} e^{j3\omega_0 T_0} + \frac{\omega_0^2 \overline{C}^3}{8} e^{-j3\omega_0 T_0} \right] \tag{1.71}$$

Substituting Equation (1.68) into (1.71) gives rise to:

$$\theta = M\cos\left(\omega_0 t + \frac{3}{8}\varepsilon\omega_0 M^2 t + B\right) + \frac{\varepsilon\omega_0^2 M^3}{32}\cos\left(3\omega_0 t + \frac{9}{8}\varepsilon\omega_0 M^2 t + B\right) \tag{1.72}$$

Resulting from Equation (1.72), the nonlinear frequency is:

$$\omega = \omega_0 + \frac{3}{8}\varepsilon\omega_0 M^2 \tag{1.73}$$

1.3.3 Forced Oscillators

1.3.3.1 Approximate Solutions

Consider a forced nonlinear oscillator:

$$\ddot{x} + \omega_0^2 x + \varepsilon\omega_0^2 x^3 = F\sin(\omega t) \tag{1.74}$$

where ε denotes the small nonlinear coefficient, ω_0 is the linear frequency, F is the magnitude of the input, and ω is the frequency of the input. The magnitude and frequency of the input can be assumed to be:

$$\begin{cases} F = \varepsilon f \\ \omega = \omega_0 + \varepsilon\sigma \end{cases} \tag{1.75}$$

where f and σ are coefficients. Substituting Equation (1.75) into (1.74) yields:

$$\ddot{x} + \omega_0^2 x = -\varepsilon\omega_0^2 x^3 + \varepsilon f\sin[(\omega_0 + \varepsilon\sigma)t] \tag{1.76}$$

The solution of Equation (1.76) can be assumed to be:

$$x = x_0(T_0, T_1) + \varepsilon x_1(T_0, T_1) \tag{1.77}$$

Substituting Equation (1.76) into (1.76), and balanced for ε^0, and ε^1, terms give rise to:

$$\begin{cases} D_0^2 x_0 + \omega_0^2 x_0 = 0 \\ D_0^2 x_1 + \omega_0^2 x_1 = -2D_0 D_1 x_0 - \omega_0^2 x_0^3 + f\cos(\omega_0 T_0 + \sigma T_1) \end{cases} \tag{1.78}$$

The solution of the first equation of (1.78) is:

$$x_0 = C(T_1)e^{j\omega_0 T_0} + \overline{C(T_1)}e^{-j\omega_0 T_0} \tag{1.79}$$

Substituting Equation (1.79) into the second equation of (1.78) gives rise to:

$$D_0^2 x_1 + \omega_0^2 x_1 = \left[-2j\omega_0 D_1 C - 3\omega_0^2 C^2\overline{C} + \frac{f}{2}e^{j\sigma T_1}\right]e^{j\omega_0 T_0} - \omega_0^2 C^3 e^{j3\omega_0 T_0} + cc \tag{1.80}$$

The first term of the right side of Equation (1.80) is secular term. The term must be kept to zero for bounded solutions:

$$-2j\omega_0\frac{\partial C}{\partial T_1} - 3\omega_0^2 C^2\overline{C} + \frac{f}{2}e^{j\sigma T_1} = 0 \tag{1.81}$$

The trial solution of Equation (1.81) is:

$$C = \frac{a(T_1)}{2}e^{j\beta(T_1)} \tag{1.82}$$

Substituting Equation (1.82) into (1.81) gives rise to:

$$\begin{cases} \dfrac{\partial a}{\partial T_1} = \dfrac{f}{2\omega_0}\sin[\sigma T_1 - \beta] \\ \dfrac{\partial \beta}{\partial T_1} = \dfrac{3}{8}\omega_0 a^2 - \dfrac{f}{2\omega_0 a}\cos[\sigma T_1 - \beta] \end{cases} \tag{1.83}$$

The coefficients $a(T_1)$ and $\beta(T_1)$ of Equation (1.82) can be derived from Equation (1.83). By ignoring the secular term, Equation (1.80) can be rewritten as:

$$D_0^2 x_1 + \omega_0^2 x_1 = -\omega_0^2 C^3 e^{j3\omega_0 T_0} + cc \tag{1.84}$$

The solution of Equation (1.84) is:

$$x_1 = \frac{\omega_0^2 C^3}{8}e^{j3\omega_0 T_0} + cc \tag{1.85}$$

Substituting Equations (1.79) and (1.85) into (1.77) results in the approximate solution:

$$x = \frac{a(T_1)}{2}\cdot e^{j[\omega_0 T_0 + \beta(T_1)]} + \frac{\varepsilon\omega_0^2 a^3(T_1)}{16}e^{j[3\omega_0 T_0 + 3\beta(T_1)]} + cc \tag{1.86}$$

1.3.3.2 Amplitude-Frequency Response

A coefficient is defined as:

$$\varphi = \sigma T_1 - \beta \tag{1.87}$$

Substituting Equation (1.87) into (1.83) results in:

$$\begin{cases} \dfrac{\partial a}{\partial T_1} = \dfrac{f}{2\omega_0}\sin\varphi \\ \dfrac{\partial \varphi}{\partial T_1} = \sigma - \dfrac{3}{8}\omega_0 a^2 + \dfrac{f}{2\omega_0 a}\cos\varphi \end{cases} \tag{1.88}$$

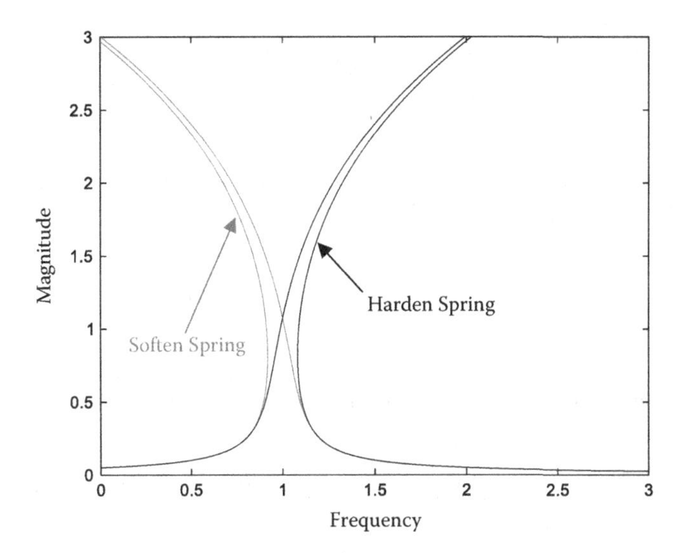

FIGURE 1.3 Amplitude-frequency response.

In order to get the amplitude of steady-state response, the derivative of Equation (1.88) with respect to T_1 should be zero:

$$\begin{cases} \frac{f}{2\omega_0} \sin \overline{\varphi} = 0 \\ \sigma - \frac{3}{8}\omega_0 \overline{a}^2 + \frac{f}{2\omega_0 \overline{a}} \cos \overline{\varphi} = 0 \end{cases} \tag{1.89}$$

where $\overline{a}$ is the amplitude of steady-state response, and $\overline{\varphi}$ is the phase of steady-state response. When the phase, $\overline{\varphi}$, of steady-state response is zero, the amplitude, $\overline{a}$, of steady-state response can be obtained from Equation (1.89):

$$\left| \sigma \overline{a} - \frac{3}{8}\omega_0 \overline{a}^3 \right| = \left| \frac{f}{2\omega_0} \right| \tag{1.90}$$

Then the nonlinear frequency can be derived from:

$$\omega = \omega_0 + \frac{3}{8}\varepsilon\omega_0 \overline{a}^2 \pm \left| \frac{F}{2\omega_0 \overline{a}} \right| \tag{1.91}$$

The amplitude-frequency response shown in Figure 1.3 can be derived from Equation (1.91). Figure 1.3 exhibits the jump phenomena of nonlinear systems.

1.3.4 Coupled Oscillators

A coupled Duffing oscillator with two DOF is given by:

$$\begin{cases} \ddot{q}_1 + \omega_1^2 q_1 + \varepsilon\left[\eta_{11} q_1^3 + \eta_{12} q_1^2 q_2 + \eta_{13} q_1 q_2^2 + \eta_{14} q_2^3\right] = 0 \\ \ddot{q}_2 + \omega_2^2 q_2 + \varepsilon\left[\eta_{21} q_1^3 + \eta_{22} q_1^2 q_2 + \eta_{23} q_1 q_2^2 + \eta_{24} q_2^3\right] = 0 \end{cases} \tag{1.92}$$

where q_i is the displacement of the ith mode, ω_i is the linear frequency of the ith mode, and η_{ij} is the coefficient. The linear frequency satisfies $\omega_1 > 5\omega_2$. The solution of Equation (1.92) can be assumed to be:

$$\begin{cases} q_1 = q_{10}(T_0, T_1) + \varepsilon q_{11}(T_0, T_1) \\ q_2 = q_{20}(T_0, T_1) + \varepsilon q_{21}(T_0, T_1) \end{cases} \tag{1.93}$$

where T_i is the time scale. Substituting Equation (1.93) into (1.92) gives rise to:

$$\begin{aligned} &\left[\frac{\partial^2 q_{10}}{\partial T_0^2} + 2\varepsilon\frac{\partial^2 q_{10}}{\partial T_0 T_1} + \varepsilon^2\frac{\partial^2 q_{10}}{\partial T_1^2}\right] + \varepsilon\left[\frac{\partial^2 q_{11}}{\partial T_0^2} + 2\varepsilon\frac{\partial^2 q_{11}}{\partial T_0 T_1} + \varepsilon^2\frac{\partial^2 q_{11}}{\partial T_1^2}\right] + \omega_1^2(q_{10} + \varepsilon q_{11}) \\ &= \varepsilon\left[-\eta_{11}(q_{10} + \varepsilon q_{11})^3 - \eta_{12}(q_{10} + \varepsilon q_{11})(q_{20} + \varepsilon q_{21})^2 - \eta_{13}(q_{10} + \varepsilon q_{11})^2(q_{20} + \varepsilon q_{21}) \right. \\ &\quad \left. - \eta_{14}(q_{20} + \varepsilon q_{21})^3\right] \end{aligned} \tag{1.94}$$

$$\begin{aligned} &\left[\frac{\partial^2 q_{20}}{\partial T_0^2} + 2\varepsilon\frac{\partial^2 q_{20}}{\partial T_0 T_1} + \varepsilon^2\frac{\partial^2 q_{20}}{\partial T_1^2}\right] + \varepsilon\left[\frac{\partial^2 q_{21}}{\partial T_0^2} + 2\varepsilon\frac{\partial^2 q_{21}}{\partial T_0 T_1} + \varepsilon^2\frac{\partial^2 q_{21}}{\partial T_1^2}\right] + \omega_2^2(q_{20} + \varepsilon q_{21}) \\ &= \varepsilon\left[-\eta_{21}(q_{10} + \varepsilon q_{11})^3 - \eta_{22}(q_{10} + \varepsilon q_{11})(q_{20} + \varepsilon q_{21})^2 - \eta_{23}(q_{10} + \varepsilon q_{11})^2(q_{20} + \varepsilon q_{21}) \right. \\ &\quad \left. - \eta_{24}(q_{20} + \varepsilon q_{21})^3\right] \end{aligned} \tag{1.95}$$

Equations (1.94) and (1.95) can be balanced for ε^0 term:

$$\begin{cases} \dfrac{\partial^2 q_{10}}{\partial T_0^2} + \omega_1^2 q_{10} = 0 \\ \dfrac{\partial^2 q_{20}}{\partial T_0^2} + \omega_2^2 q_{20} = 0 \end{cases} \tag{1.96}$$

The solution of Equation (1.96) is:

$$\begin{cases} q_{10} = C_1(T_1)\cdot e^{jw_1 T_0} + \overline{C_1}(T_1)\cdot e^{-jw_1 T_0} \\ q_{20} = C_2(T_1)\cdot e^{jw_2 T_0} + \overline{C_2}(T_1)\cdot e^{-jw_2 T_0} \end{cases} \tag{1.97}$$

The solution of Equation (1.97) is:

$$\begin{cases} C_1(T_1) = A_1(t) e^{j\varphi_1(t)} \\ C_2(T_1) = A_2(t) e^{j\varphi_2(t)} \end{cases} \tag{1.98}$$

Equations (1.94) and (1.95) can be balanced for ε^1 term:

$$\begin{cases} \frac{\partial^2 q_{11}}{\partial T_0^2} + \omega_1^2 q_{11} = -2\frac{\partial^2 q_{10}}{\partial T_0 \partial T_1} - \eta_{11} q_{10}^3 - \eta_{12} q_{10} q_{20}^2 - \eta_{13} q_{10}^2 q_{20} - \eta_{14} q_{20}^3 \\ \frac{\partial^2 q_{21}}{\partial T_0^2} + \omega_2^2 q_{21} = -2\frac{\partial^2 q_{20}}{\partial T_0 \partial T_1} - \eta_{21} q_{10}^3 - \eta_{22} q_{10} q_{20}^2 - \eta_{23} q_{10}^2 q_{20} - \eta_{24} q_{20}^3 \end{cases} \tag{1.99}$$

Substituting Equation (1.97) into (1.99) gives rise to:

$$\begin{aligned} \frac{\partial^2 q_{11}}{\partial T_0^2} + \omega_1^2 q_{11} = & \left[-j2\omega_1 \frac{\partial C_1(T_1)}{\partial T_1} - 3\eta_{11} C_1^2(T_1)\overline{C_1}(T_1) - 2\eta_{12} C_2(T_1)\overline{C_2}(T_1)C_1(T_1) \right] e^{j\omega_1 T_0} \\ & + \left[j2\omega_1 \frac{\partial \overline{C_1}(T_1)}{\partial T_1} - 3\eta_{11} C_1(T_1)\overline{C_1}^2(T_1) - 2\eta_{12} C_2(T_1)\overline{C_2}(T_1)\overline{C_1}(T_1) \right] e^{-j\omega_1 T_0} \\ & + \left[-2\eta_{13} C_1(T_1)\overline{C_1}(T_1)C_2(T_1) - 3\eta_{14} C_2^2(T_1)\overline{C_2}(T_1) \right] e^{j\omega_2 T_0} \\ & + \left[-2\eta_{13} C_1(T_1)\overline{C_1}(T_1)\overline{C_2}(T_1) - 3\eta_{14} \overline{C_2}^2(T_1)\overline{C_2}(T_1) \right] e^{-j\omega_2 T_0} \\ & - \eta_{11} \left[C_1^3(T_1) e^{j3\omega_1 T_0} + \overline{C_1}^3(T_1) e^{-j3\omega_1 T_0} \right] \\ & - \eta_{12} \left[C_2^2(T_1)C_1(T_1) e^{j(2\omega_2 T_0 + \omega_1 T_0)} + \overline{C_2}^2(T_1)\overline{C_1}(T_1) e^{-j(2\omega_2 T_0 + \omega_1 T_0)} \right] \\ & - \eta_{12} \left[C_2^2(T_1)\overline{C_1}(T_1) e^{j(2\omega_2 T_0 - \omega_1 T_0)} + \overline{C_2}^2(T_1)C_1(T_1) e^{-j(2\omega_2 T_0 - \omega_1 T_0)} \right] \\ & - \eta_{13} \left[C_1^2(T_1)C_2(T_1) e^{j(2\omega_1 T_0 + \omega_2 T_0)} + \overline{C_1}^2(T_1)\overline{C_2}(T_1) e^{-j(2\omega_1 T_0 + \omega_2 T_0)} \right] \\ & - \eta_{13} \left[C_1^2(T_1)\overline{C_2}(T_1) e^{j(2\omega_1 T_0 - \omega_2 T_0)} + \overline{C_1}^2(T_1)C_2(T_1) e^{-j(2\omega_1 T_0 - \omega_2 T_0)} \right] \\ & - \eta_{14} \left[C_2^3(T_1) e^{j3\omega_2 T_0} + \overline{C_2}^3(T_1) e^{-j3\omega_2 T_0} \right] \end{aligned} \tag{1.100}$$

$$\begin{aligned} \frac{\partial^2 q_{21}}{\partial T_0^2} + \omega_2^2 q_{21} = & \left[-j2\omega_2 \frac{\partial C_2(T_1)}{\partial T_1} - 2\eta_{23} C_1(T_1)\overline{C_1}(T_1)C_2(T_1) - 3\eta_{24} C_2^2(T_1)\overline{C_2}(T_1) \right] e^{j\omega_2 T_0} \\ & + \left[j2\omega_2 \frac{\partial \overline{C_2}(T_1)}{\partial T_1} - 2\eta_{23} C_1(T_1)\overline{C_1}(T_1)\overline{C_2}(T_1) - 3\eta_{24} \overline{C_2}^2(T_1)\overline{C_2}(T_1) \right] e^{-j\omega_2 T_0} \\ & + \left[-3\eta_{21} C_1^2(T_1)\overline{C_1}(T_1) - 2\eta_{22} C_2(T_1)\overline{C_2}(T_1)C_1(T_1) \right] e^{j\omega_1 T_0} \\ & + \left[-3\eta_{21} C_1(T_1)\overline{C_1}^2(T_1) - 2\eta_{22} C_2(T_1)\overline{C_2}(T_1)\overline{C_1}(T_1) \right] e^{-j\omega_1 T_0} \\ & - \eta_{21} \left[C_1^3(T_1) e^{j3\omega_1 T_0} + \overline{C_1}^3(T_1) e^{-j3\omega_1 T_0} \right] - \eta_{22} \left[C_2^2(T_1)C_1(T_1) e^{j(2\omega_2 T_0 + \omega_1 T_0)} \right. \\ & \left. + \overline{C_2}^2(T_1)\overline{C_1}(T_1) e^{-j(2\omega_2 T_0 + \omega_1 T_0)} \right] - \eta_{22} \left[C_2^2(T_1)\overline{C_1}(T_1) e^{j(2\omega_2 T_0 - \omega_1 T_0)} \right. \\ & \left. + \overline{C_2}^2(T_1)C_1(T_1) e^{-j(2\omega_2 T_0 - \omega_1 T_0)} \right] - \eta_{23} \left[C_1^2(T_1)C_2(T_1) e^{j(2\omega_1 T_0 + \omega_2 T_0)} \right. \\ & \left. + \overline{C_1}^2(T_1)\overline{C_2}(T_1) e^{-j(2\omega_1 T_0 + \omega_2 T_0)} \right] - \eta_{23} \left[C_1^2(T_1)\overline{C_2}(T_1) e^{j(2\omega_1 T_0 - \omega_2 T_0)} \right. \\ & \left. + \overline{C_1}^2(T_1)C_2(T_1) e^{-j(2\omega_1 T_0 - \omega_2 T_0)} \right] - \eta_{24} \left[C_2^3(T_1) e^{j3\omega_2 T_0} + \overline{C_2}^3(T_1) e^{-j3\omega_2 T_0} \right] \end{aligned} \tag{1.101}$$

The first two terms of the right side of Equation (1.100) are secular terms. Meanwhile, the first two terms of the right side of Equation (1.101) are secular terms. Those secular terms must be kept to zero for bounded solutions:

$$\begin{cases} -j2\omega_1\frac{\partial C_1(T_1)}{\partial T_1}-3\eta_{11}C_1^2(T_1)\overline{C_1}(T_1)-2\eta_{12}C_2(T_1)\overline{C_2}(T_1)C_1(T_1)=0 \\ j2\omega_1\frac{\partial \overline{C_1}(T_1)}{\partial T_1}-3\eta_{11}C_1(T_1)\overline{C_1}^2(T_1)-2\eta_{12}C_2(T_1)\overline{C_2}(T_1)\overline{C_1}(T_1)=0 \\ -j2\omega_2\frac{\partial C_2(T_1)}{\partial T_1}-2\eta_{23}C_1(T_1)\overline{C_1}(T_1)C_2(T_1)-3\eta_{24}C_2^2(T_1)\overline{C_2}(T_1)=0 \\ j2\omega_2\frac{\partial \overline{C_2}(T_1)}{\partial T_1}-2\eta_{23}C_1(T_1)\overline{C_1}(T_1)\overline{C_2}(T_1)-3\eta_{24}\overline{C_2}^2(T_1)\overline{C_2}(T_1)=0 \end{cases} \tag{1.102}$$

Substituting Equation (1.98) into (1.102) yields:

$$\begin{cases} A_1 = const \\ A_2 = const \\ \varphi_1(t) = \frac{\varepsilon}{2\omega_1}\left[3\eta_{11}A_1^2(t)+2\eta_{12}A_2^2(t)\right]t+\varphi_{10} \\ \varphi_2(t) = \frac{\varepsilon}{2\omega_2}\left[3\eta_{23}A_1^2(t)+3\eta_{24}A_2^2(t)\right]t+\varphi_{20} \end{cases} \tag{1.103}$$

Ignoring the secular terms, solutions of Equations (1.100) and (1.101) can be described as:

$$\begin{aligned} q_{11} = & \left[\frac{-2\eta_{13}C_1(T_1)\overline{C_1}(T_1)C_2(T_1)-3\eta_{14}C_2^2(T_1)\overline{C_2}(T_1)}{\omega_1^2-\omega_2^2}\right]e^{j\omega_2T_0} \\ & + \left[\frac{-2\eta_{13}C_1(T_1)\overline{C_1}(T_1)\overline{C_2}(T_1)-3\eta_{14}\overline{C_2}^2(T_1)\overline{C_2}(T_1)}{\omega_1^2-\omega_2^2}\right]e^{-j\omega_2T_0} \\ & - \eta_{11}\left[\frac{C_1^3(T_1)e^{j3\omega_1T_0}}{-8\omega_1^2}+\frac{\overline{C_1}^3(T_1)e^{-j3\omega_1T_0}}{-8\omega_1^2}\right] \\ & - \eta_{12}\left[\frac{C_2^2(T_1)C_1(T_1)e^{j(2\omega_2T_0+\omega_1T_0)}}{\omega_1^2-(2\omega_2+\omega_1)^2}+\frac{\overline{C_2}^2(T_1)\overline{C_1}(T_1)e^{-j(2\omega_2T_0+\omega_1T_0)}}{\omega_1^2-(2\omega_2+\omega_1)^2}\right] \\ & - \eta_{12}\left[\frac{C_2^2(T_1)\overline{C_1}(T_1)e^{j(2\omega_2T_0-\omega_1T_0)}}{\omega_1^2-(2\omega_2-\omega_1)^2}+\frac{\overline{C_2}^2(T_1)C_1(T_1)e^{-j(2\omega_2T_0-\omega_1T_0)}}{\omega_1^2-(2\omega_2-\omega_1)^2}\right] \\ & - \eta_{13}\left[\frac{C_1^2(T_1)C_2(T_1)e^{j(2\omega_1T_0+\omega_2T_0)}}{\omega_1^2-(2\omega_1+\omega_2)^2}+\frac{\overline{C_1}^2(T_1)\overline{C_2}(T_1)e^{-j(2\omega_1T_0+\omega_2T_0)}}{\omega_1^2-(2\omega_1+\omega_2)^2}\right] \\ & - \eta_{13}\left[\frac{C_1^2(T_1)\overline{C_2}(T_1)e^{j(2\omega_1T_0-\omega_2T_0)}}{\omega_1^2-(2\omega_1-\omega_2)^2}+\frac{\overline{C_1}^2(T_1)C_2(T_1)e^{-j(2\omega_1T_0-\omega_2T_0)}}{\omega_1^2-(2\omega_1-\omega_2)^2}\right] \\ & - \eta_{14}\left[\frac{C_2^3(T_1)e^{j3\omega_2T_0}}{\omega_1^2-9\omega_2^2}+\frac{\overline{C_2}^3(T_1)e^{-j3\omega_2T_0}}{\omega_1^2-9\omega_2^2}\right] \end{aligned} \tag{1.104}$$

$$q_{21} = \left[\frac{-3\eta_{21}C_1^2(T_1)\overline{C_1}(T_1) - 2\eta_{22}C_2(T_1)\overline{C_2}(T_1)C_1(T_1)}{\omega_2^2 - \omega_1^2}\right]e^{j\omega_1 T_0}$$
$$+ \left[\frac{-3\eta_{21}C_1(T_1)\overline{C_1}^2(T_1) - 2\eta_{22}C_2(T_1)\overline{C_2}(T_1)\overline{C_1}(T_1)}{\omega_2^2 - \omega_1^2}\right]e^{-j\omega_1 T_0}$$
$$- \eta_{21}\left[\frac{C_1^3(T_1)e^{j3\omega_1 T_0}}{\omega_2^2 - 9\omega_1^2} + \frac{\overline{C_1}^3(T_1)e^{-j3\omega_1 T_0}}{\omega_2^2 - 9\omega_1^2}\right]$$
$$- \eta_{22}\left[\frac{C_2^2(T_1)C_1(T_1)e^{j(2\omega_2 T_0 + \omega_1 T_0)}}{\omega_2^2 - (2\omega_2 + \omega_1)^2} + \frac{\overline{C_2}^2(T_1)\overline{C_1}(T_1)e^{-j(2\omega_2 T_0 + \omega_1 T_0)}}{\omega_2^2 - (2\omega_2 + \omega_1)^2}\right]$$
$$- \eta_{22}\left[\frac{C_2^2(T_1)\overline{C_1}(T_1)e^{j(2\omega_2 T_0 - \omega_1 T_0)}}{\omega_2^2 - (2\omega_2 - \omega_1)^2} + \frac{\overline{C_2}^2(T_1)C_1(T_1)e^{-j(2\omega_2 T_0 - \omega_1 T_0)}}{\omega_2^2 - (2\omega_2 - \omega_1)^2}\right]$$
$$- \eta_{23}\left[\frac{C_1^2(T_1)C_2(T_1)e^{j(2\omega_1 T_0 + \omega_2 T_0)}}{\omega_2^2 - (2\omega_1 + \omega_2)^2} + \frac{\overline{C_1}^2(T_1)\overline{C_2}(T_1)e^{-j(2\omega_1 T_0 + \omega_2 T_0)}}{\omega_2^2 - (2\omega_1 + \omega_2)^2}\right]$$
$$- \eta_{23}\left[\frac{C_1^2(T_1)\overline{C_2}(T_1)e^{j(2\omega_1 T_0 - \omega_2 T_0)}}{\omega_2^2 - (2\omega_1 - \omega_2)^2} + \frac{\overline{C_1}^2(T_1)C_2(T_1)e^{-j(2\omega_1 T_0 - \omega_2 T_0)}}{\omega_2^2 - (2\omega_1 - \omega_2)^2}\right]$$
$$- \eta_{24}\left[\frac{C_2^3(T_1)e^{j3\omega_2 T_0}}{-8\omega_2^2} + \frac{\overline{C_2}^3(T_1)e^{-j3\omega_2 T_0}}{-8\omega_2^2}\right] \tag{1.105}$$

Substituting Equations (1.97), (1.104), and (1.105) into (1.93) results in the approximate solution of coupled Duffing oscillator (1.92).

1.4 MODAL ANALYSES

Consider an n-DOF linear system:

$$M\ddot{x} + C\dot{x} + Kx = F \tag{1.106}$$

where x is the vector of generalized displacement, F is the vector of external force acting on the system, M is the matrix of inertia, C is the matrix of viscous damping, and K is the matrix of stiffness. By ignoring the damping, free-oscillation equation (1.106) reduces to:

$$M\ddot{x} + Kx = 0 \tag{1.107}$$

The solution of Equation (1.107) is assumed to be:

$$x = \varphi\cos(\omega t - \phi) \tag{1.108}$$

where ω is natural frequency of linear systems (1.106) and (1.107), ϕ is the corresponding phase, and φ is the corresponding mode shape. Substituting Equation (1.108) into (1.107) gives rise to:

$$(K - M\omega^2)\varphi \cos(\omega t - \phi) = 0 \tag{1.109}$$

Because the cosine term in Equation (1.109) is not zero for most times, the solution of Equation (1.109) is:

$$(K - M\omega^2)\varphi = 0 \tag{1.110}$$

From Equation (1.110), the characteristic equation of the linear system is:

$$|K - M\omega^2| = a_0 + a_1\omega^2 + a_2\omega^4 + \ldots + a_n\omega^{2n} = 0 \tag{1.111}$$

Natural frequencies of n-DOF linear system can be derived from Equation (1.111). Then the corresponding mode shapes, φ, can be obtained from Equation (1.110).

Equations (1.110) and (1.111) for estimating natural frequencies and mode shapes are only suitable for undamped linear system with n-DOF. For a damped system (1.106), the natural frequencies and damping ratios can be derived from:

$$|M\lambda^2 + C\lambda + K| = a_0 + a_1\lambda + a_2\lambda^2 + \ldots + a_n\lambda^n = 0 \tag{1.112}$$

where

$$\lambda = -\zeta\omega \pm j\omega\sqrt{1 - \zeta^2} \tag{1.113}$$

2 Vibration Control

Many scientists have worked to provide solutions to the challenging problem posed by the flexible structure. The work can roughly be broken into two categories: feedback control and prefilter technology. The feedback control strategies use measurement and estimation of vibrational state to suppress oscillations in a closed loop, such as linear control, nonlinear control, and intelligent control. However, accurately sensing vibrations is difficulty toward the application of the feedback controller. Meanwhile, the conflict between the computer-based feedback controller and the actions of human operator is also an obstacle.

The prefilter technology modifies the input to create prescribed motions that cause minimal vibrations. Input shaping is a kind of open-loop controller. It can effectively suppress oscillations for many types of flexible dynamic systems including bridge cranes, tower cranes, boom cranes, container cranes, coordinate measurement machines, spacecrafts, robotic arms, robotic work cells, demining robots, micro-milling machines, nano-positioning stages, and linear step motors. The input shaping process is demonstrated as follows. The original command produces an oscillatory response. To eliminate the oscillatory response, the original command is convolved with a series of impulses, called the input shaper, to create a shaped command. The shaped command can move the flexible dynamic system without inducing vibrations. The convolution is performed by simply multiplying the original command by the amplitude of the first impulse, and adding it to the original command multiplied by the amplitude of the other impulses and shifted in time by the corresponding delay period.

Smoothing driving commands can suppress vibrations of the flexible structure caused by commanded motions. The driving command filters through a smoother to create a smoothed command, which moves flexible structures toward the desired position with minimum vibrations. The smoother inherent in the limited response of flexible structures causes a limited response to specified frequencies. The smoother is designed by estimating natural frequencies and damping ratios of flexible structures.

2.1 LINEAR CONTROL

2.1.1 State Feedback

Consider a linear system:

$$\dot{x} = \mathbf{A}x + \mathbf{B}u \tag{2.1}$$

where x is state vector, u is control signal, $\mathbf{A}$ is state matrix, and $\mathbf{B}$ is the input matrix. The controller is chosen as:

DOI: 10.1201/9781003247210-2

$$u = -\mathbf{K}x \tag{2.2}$$

where **K** is the state feedback gain matrix. Because the control signal depends on system state, the controller (2.2) is called state feedback. All state variables are measured or estimated in a closed loop for producing control signal.

Substituting Equation (2.2) into (2.1) gives rise to:

$$\dot{x} = (\mathbf{A} - \mathbf{BK})x \tag{2.3}$$

The solution of Equation (2.3) can be expressed as:

$$x(t) = e^{(\mathbf{A}-\mathbf{BK})}x(0) \tag{2.4}$$

where $x(0)$ is the initial state. Equation (2.4) is the time response of Equation (2.3) resulting from the initial conditions. The stability and transient-response characteristics depend on the eigenvalues of the matrix (**A-BK**). When the matrixes, **A** and **B**, are known, the gain matrix, **K**, can be designed for forcing eigenvalues of the matrix (**A-BK**) to be desired values. The design method is named after the pole placement technique. If the desired eigenvalues (closed-loop poles) are μ_1, μ_2, ..., μ_n, the desired characteristic polynomial is given by:

$$(s - \mu_1)(s - \mu_2) \cdots (s - \mu_n) = |s\mathbf{I} - (\mathbf{A} - \mathbf{BK})| \tag{2.5}$$

Both sides of Equation (2.5) are polynomials in s. Equating the coefficients of the same powers of s on both sides gives rise to the solution of **K**. The method (2.5) is suitable for the low-order system.

In the case of high-order system, the following method may be effective. The gain matrix, **K**, can be derived by:

$$\mathbf{K} = [0 \;\; 0 \;\; \cdots \;\; 1][\mathbf{B} \;\; \mathbf{AB} \;\; \cdots \;\; \mathbf{A}^{n-1}\mathbf{B}]^{-1}\phi \tag{2.6}$$

where

$$\phi = \mathbf{A}^n + \alpha_1\mathbf{A}^{n-1} + \cdots + \alpha_{n-1}\mathbf{A} + \alpha_n\mathbf{I} \tag{2.7}$$

$$(s - \mu_1)(s - \mu_2) \cdots (s - \mu_n) = s^n + \alpha_1 s^{n-1} + \cdots + \alpha_{n-1}s + \alpha_n \tag{2.8}$$

Example 2.1: Design a linear control for bridge crane.

A planar bridge crane can be modeled as a single pendulum on a cart, where the generalized displacements are the position of the trolley, z, and the angular deflection of the payload, δ. The trolley is modeled as a mass with an applied actuator force, F. The payload, which in this case is a sphere, is suspended from a rigid and weightless cable of length, l. A wind disturbance force, w_f, is applied to the payload.

It is assumed that the motion of the cart is unaffected by motion of the payload due to the large mechanical impedance in the drive system, and the suspension cable length does not change during the motion.

Because the controller will ideally keep the angular deflection of the payload small, the equation of motion can be linearized as:

$$l\ddot{\delta} + g\delta = \frac{w_f}{m} + \ddot{z} \tag{2.9}$$

The horizontal swing deflection of payload can be expressed as ($l\sin\delta$). Therefore, using the PD control method, the bridge crane trolley velocity control signal, c_{fb}, can be calculated by:

$$c_{fb} = -b_p(l \sin \delta) - b_d \frac{d(l \sin \delta)}{dt} \tag{2.10}$$

where b_p and b_d are the proportional and differential gains for the bridge crane. Assuming the swing angle, δ, is kept small by the controller, using the small-angle approximations for sine and cosine yields the model:

$$c_{fb} = -b_p l\delta - b_d l\dot{\delta} \tag{2.11}$$

The controller (2.11) can be rewritten as:

$$c_{fb} = -[b_p l \quad b_d l]\begin{bmatrix} \delta \\ \dot{\delta} \end{bmatrix} = -\mathbf{K}x \tag{2.12}$$

Therefore, controller (2.11) is considered as a state feedback control. We get the trolley acceleration control signal by taking the derivative (2.11):

$$\dot{c}_{fb} = \ddot{z} = -b_p l\dot{\delta} - b_d l\ddot{\delta} \tag{2.13}$$

From Equations (2.9) and (2.13), we obtain:

$$(l + b_d l)\ddot{\delta} + b_p l\dot{\delta} + g\delta = \frac{w_f}{m} \tag{2.14}$$

The proportional gain, b_p, and differential gain, b_d, for a bridge crane were designed using the pole placement method. The feedback controller must have low authority so that it does not annoy the human operator. Thus, a system with reasonably small damping (damping ratio of 0.1) and with a reasonably small settling time (less than 10 s) is required. The desired closed-loop poles locations were chosen to be approximately $-0.3 \pm 3i$. Given that the suspension cable length can vary from 0.5 m to 1.6 m on the small-scale crane used for experiments, the average suspension cable length of 1.05 m was used. Thus, the proportional gain, b_p, and differential gain, b_d, were calculated to be 0.6167 and 0.02782, respectively.

2.1.2 Model Reference Control

The dynamic equation of the plant is shown in Equation (2.1). A model reference controller (MRC) regulates the system state by following the state of a prescribed model. The MRC controller includes the prescribed model and asymptotic tracking control law. The prescribed model of the MRC controller is described as:

$$\dot{x}_m = \mathbf{A}_m x_m + \mathbf{B}_m r \tag{2.15}$$

where r is the input of the prescribed model, x_m is the state of prescribed model, $\mathbf{A}_m$ is the state matrix of prescribed model, and $\mathbf{B}_m$ is the input matrix of prescribed model.

The asymptotic tracking control law forces the system state, x, to follow the model state, x_m. The error between the system state, x, and the model state, x_m, is denoted by e. The controller is given by:

$$u = \frac{e^T P(A_m - A)x + e^T P B_m r + (e_1 P_{1,2} + e_2 P_{2,2})^2}{e^T P B} \tag{2.16}$$

where $P_{1,2}$ and $P_{2,2}$ are the entries in the matrix P, e_1 and e_2 are the entries in the matrix e. Additionally, the matrix P should be designed carefully such that $A_m{}^T P + PA_m = -Q$, where Q is a real, positive-definite, symmetric matrix.

2.2 NONLINEAR CONTROL

2.2.1 Lyapunov Stability

If the solution stats at nearby points, it stays nearby. The control system is stable. As time approaches infinity, the solution tends to the equilibrium point. The control system is asymptotically stable.

Let $x = 0$ be an equilibrium point, and D be a domain containing $x = 0$. Let V be a continuously differentiable function such that

$$V(0) = 0 \text{ and } V(\mathrm{x}) > 0 \text{ and } \dot{V} \leq 0$$

Then equilibrium point $x = 0$ is stable. Moreover, if

$$V(0) = 0 \text{ and } V(\mathrm{x}) > 0 \text{ and } \dot{V} < 0$$

then equilibrium point $x = 0$ is asymptotically stable.

Example 2.2: Stability of a damped pendulum.

Consider a damped pendulum:

$$\begin{cases} \dot{x}_1 = x_2 \\ \dot{x}_2 = -\omega^2 \sin x_1 - bx_2 \end{cases} \tag{2.17}$$

where x_1 and x_2 denote the swing angle and swing angular velocity, ω is the linear frequency of single pendulum, and b is the damping coefficient. The equilibrium of (2.17) is zero angle and zero angular velocity of the swing. The energy function is selected as the Lyapunov function:

$$V = \omega^2(1 - \cos x_1) + 0.5x_2^2 \tag{2.18}$$

The Lyapunov function (2.18) is zero at zero time. Meanwhile, the Lyapunov function (2.18) is positive. The derivative of Equation (2.18) is given by:

$$\dot{V} = \omega^2 x_1 \sin x_1 + x_2\dot{x}_2 = -bx_2^2 < 0 \tag{2.19}$$

Therefore, the equilibrium is asymptotically stable.

2.2.2 Sliding Mode Control

Consider a second-order system:

$$\begin{cases} \dot{x}_1 = x_2 \\ \dot{x}_2 = h(x) + g(x)u \end{cases} \tag{2.20}$$

where x is state vector, u is control signal, and h and g are nonlinear functions and satisfy $g > 0$ for all x. A control law can be designed to constrain the system motions to a manifold:

$$s = ax_1 + x_2 = 0 \tag{2.21}$$

The manifold (2.21) is called sliding manifold. The nonlinear functions, h and g, satisfy an inequality:

$$\left| \frac{ax_2 + h(x)}{g(x)} \right| \leq \sigma(x) \tag{2.22}$$

The controller is chosen as:

$$u = -\beta(x)\mathrm{sgn}(s), \ \beta(x) > \sigma(x) \tag{2.23}$$

where

$$\mathrm{sgn}(s) = \begin{cases} 1, \ s > 0 \\ 0, \ s = 0 \\ -1, \ s < 0 \end{cases} \tag{2.24}$$

The controller (2.23) is called sliding mode control. A Lyapunov function is selected as:

$$V = 0.5s^2 = 0.5(ax_1 + x_2)^2 \tag{2.25}$$

The derivative of Equation (2.25) is given by:

$$\dot{V} = s\dot{s} \leq -g|s|(\beta - \sigma) \tag{2.26}$$

The right side of Equation (2.26) is negative. Therefore, the equilibrium is asymptotically stable. The controlled motions contain a reaching phase and a sliding phase. During the reaching phase, trajectories starting off the manifold (2.21) move toward the manifold and reach the manifold. During the sliding phase, trajectories are confined to the manifold and reach equilibrium.

2.3 INTELLIGENT CONTROL

When the plant is challenging to derive mathematical model and experimental tuning is also difficult, the intelligent control may be good choice. The intelligent control includes artificial neural networks and fuzzy logic. The intelligent controller may be applied in two categories: feedback mode and feedforward mode. In the feedback mode, the intelligent controller acts as a control device. In the feedforward mode, the intelligent controller works as a prediction device.

2.3.1 Artificial Neural Networks

Artificial neural networks reflect the dynamic behavior of human brain. They are a subset of machine learning algorithms. Artificial neural networks consist of node layers including input layer, output layer, and more hidden layers. Each node is also called artificial neuron, which connects to other nodes.

The dynamic equation of an artificial neuron is given by:

$$u_i = \Sigma_{j=1}^{n} (w_{ij}x_j - \theta_i), \ \ y_i = f(u_i) \tag{2.27}$$

where x_j is the input signal, w_{ij} denotes weight, θ_i represents bias, y_i is the output signal, and f is the activation function. The dynamic equation (2.27) simulates the

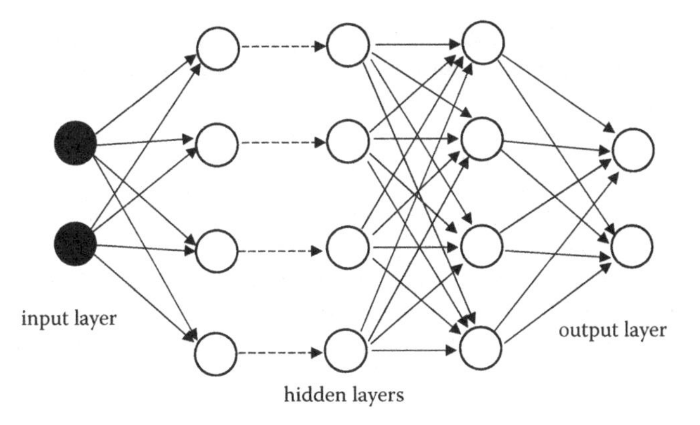

FIGURE 2.1 Artificial neural network.

dynamics of a neuron in the human brain. Some of the activation functions are commonly used:

$$y_i = f(u_i) = \begin{cases} 1, & u_i \geq 0 \\ -1, & u_i < 0, \end{cases} \tag{2.28}$$

$$y_i = f(u_i) = \frac{1}{1 + e^{-u_i}} \tag{2.29}$$

While the human brain contains billion neurons, many artificial neurons arise in an artificial neural network for machine learning. Figure 2.1 shows an artificial neural network including more neurons. The Delta rule is a common method for training neural networks. The objective of machine learning is that the error between the desired output and neuron output approaches zero as time approaches infinity.

The weight error can be written as:

$$\Delta w_{ij}(k + 1) = \eta e_i x_j f' \tag{2.30}$$

where η is the learning rate, e_i is the error between the desired and actual outputs for the ith node, and f' is the derivative of the activation function for the ith node.

More than one hidden layer constructs a multilayer neural network. Nodes in the hidden layer operate on activations from nodes of previous layer and transmit activations forward to nodes of the next layer. Multilayer neural networks are universal approximators but suffer from heavy computational load. Multilayer neural network exhibits a deep-learning algorithm.

2.3.2 Fuzzy Logic

Fuzzy set is different from the classical set. The classical set includes elements that satisfy precise properties of membership. The fuzzy set includes elements that satisfy imprecise properties of membership. Meanwhile, randomness refers to an event that may or may not occur, and fuzziness refers to the boundary of a set that is not precise.

A fuzzy set is denoted by:

$$A = \{(y, \mu_A(y))|y \in U\} \tag{2.31}$$

where A is the fuzzy set, U is the universe of information, and μ_A is the degree of membership of y, and satisfies $\mu_A(y) \in [0, 1]$.

A fuzzy inference system contains fuzzification, reasoning, and defuzzification. Defuzzification maps the quantitative and precise information onto the term-set of a linguistic variable. The centroid technique can be used for defuzzification:

$$\eta = \frac{\int_y y\mu dy}{\int_{y'} \mu dy} \tag{2.32}$$

The fuzzy reasoning is a nonlinear mapping, which derives outputs based on a set of fuzzy rules.

As A and B are both fuzzy sets, fuzzy 'OR' operation can be described as:

$$\mu_{A\cup B} = \mu_A \vee \mu_B = \max\{\mu_A, \mu_B\} \tag{2.33}$$

Fuzzy 'AND' operation can be descried as:

$$\mu_{A\cap B} = \mu_A \wedge \mu_B = \min\{\mu_A, \mu_B\} \tag{2.34}$$

Fuzzy 'NOT' operation can be described as:

$$\mu_{\bar{A}} = 1 - \mu_A \tag{2.35}$$

2.4 INPUT SHAPER

2.4.1 ZV Shaper

The response of an underdamped second-order system from a sequence of n-impulses is given by:

$$f(t) = \sum_k \frac{A_k \cdot \omega \cdot e^{-\zeta\omega(t-\tau_k)}}{\sqrt{1-\zeta^2}} \sin\left[\omega\sqrt{1-\zeta^2}(t-\tau_k)\right] \tag{2.36}$$

where A_k is the kth impulse amplitude, τ_k is the kth impulse time, ω is the frequency of the underdamped second-order system, and ζ is the corresponding damping ratio. The vibration amplitude of Equation (2.36) is:

$$A(t) = \frac{\omega}{\sqrt{1-\zeta^2}} e^{-\zeta\omega t} \sqrt{[S(\omega, \zeta)]^2 + [C(\omega, \zeta)]^2} \tag{2.37}$$

where

$$S(\omega, \zeta) = \sum_k A_k \cdot e^{\zeta\omega\tau_k} \sin(\omega\sqrt{1-\zeta^2}\tau_k) \tag{2.38}$$

$$C(\omega, \zeta) = \sum_k A_k \cdot e^{\zeta\omega\tau_k} \cos(\omega\sqrt{1-\zeta^2}\tau_k) \tag{2.39}$$

The unite-gain constraint should be applied:

$$\sum_k A_k = 1 \tag{2.40}$$

Letting Equations (2.38) and (2.39) be zero, resulting from Equation (2.40) gives rise to a zero-vibration (ZV) shaper:

$$\mathrm{ZV} = \begin{bmatrix} A_k \\ \tau_k \end{bmatrix} = \begin{bmatrix} \frac{1}{[1+K]} & \frac{K}{[1+K]} \\ 0 & 0.5T_1 \end{bmatrix} \tag{2.41}$$

where T_1 is the damped vibration period, and

$$K = e^{(-\pi\zeta/\sqrt{1-\zeta^2})} \tag{2.42}$$

2.4.2 ZVD Shaper

In order to increase the frequency insensitivity, the derivative of (2.38) and (2.39) with respect to frequency should be zero:

$$\sum_k \tau_k A_k \cdot e^{\zeta\omega\tau_k} \sin(\omega\sqrt{1-\zeta^2}\tau_k) = 0 \tag{2.43}$$

$$\sum_k \tau_k A_k \cdot e^{\zeta\omega\tau_k} \cos(\omega\sqrt{1-\zeta^2}\tau_k) = 0 \tag{2.44}$$

Limiting Equations (2.38) and (2.39) to zero and from constraints (2.40), (2.43), and (2.44) yield a zero vibration and derivative (ZVD) shaper:

$$\mathrm{ZVD} = \begin{bmatrix} A_k \\ \tau_k \end{bmatrix} = \begin{bmatrix} \frac{1}{[1+K]^2} & \frac{2K}{[1+K]^2} & \frac{K^2}{[1+K]^2} \\ 0 & 0.5T_1 & T_1 \end{bmatrix} \tag{2.45}$$

2.4.3 EI Shaper

It is challenging to limit vibrations to zero in a real application. Thus, vibrations at the design frequency should be limited to less than a tolerable level:

$$e^{-\zeta\omega\tau_n}\cdot\sqrt{S^2(\omega,\zeta)+C^2(\omega,\zeta)} \le V_{tol} \tag{2.46}$$

where V_{tol} is the tolerable level of vibrations and τ_n is the rise time of the shaper. To increase the frequency insensitivity, the derivative of vibration amplitude with respect to frequency should be limited to zero:

$$S(\omega,\zeta)\cdot\frac{\partial S(\omega,\zeta)}{\partial\omega}+C(\omega,\zeta)\cdot\frac{\partial C(\omega,\zeta)}{\partial\omega}-\zeta\tau_n\cdot[S^2(\omega,\zeta)+C^2(\omega,\zeta)]=0 \tag{2.47}$$

Additional constraints (2.46) and (2.47) for increasing frequency insensitivity yields an extra-insensitivity shaper:

$$\mathrm{EI}=\begin{bmatrix}A_k\\ \tau_k\end{bmatrix}=\begin{bmatrix}A_1 & (1-A_1-A_3) & A_3\\ 0 & t_2 & T_1\end{bmatrix} \tag{2.48}$$

where

$$A_1=0.2497+0.2496V_{tol}+0.8001\zeta+1.233V_{tol}\zeta+0.4960\zeta^2+3.173V_{tol}\zeta^2 \tag{2.49}$$

$$A_3=0.2515+0.2147V_{tol}-0.8325\zeta+1.4158V_{tol}\zeta+0.8518\zeta^2+4.901V_{tol}\zeta^2 \tag{2.50}$$

$$t_2=[0.5+0.4616V_{tol}\zeta+4.262V_{tol}\zeta^2+1.756V_{tol}\zeta^3+8.578V_{tol}^2\zeta-108.6V_{tol}^2\zeta^2 + 337V_{tol}^2\zeta^3]\cdot T_1 \tag{2.51}$$

In the case of zero damping, the EI shaper can be written as:

$$\mathrm{EI}=\begin{bmatrix}A_k\\ \tau_k\end{bmatrix}=\begin{bmatrix}\frac{(1+V_{tol})}{4} & \frac{(1-V_{tol})}{2} & \frac{(1+V_{tol})}{4}\\ 0 & 0.5T_1 & T_1\end{bmatrix} \tag{2.52}$$

2.4.4 MEI Shaper

Vibrations at two modified frequencies should be limited to zero. Then frequency insensitivity would increase. Two modified frequencies are $p\cdot\omega$ and $q\cdot\omega$. The zero-vibration constraints at the modified frequencies are:

$$\sum_k A_k e^{\zeta p\cdot\omega\tau_k}\sin(p\cdot\omega\sqrt{1-\zeta^2}\tau_k)=0,\quad p\le 1 \tag{2.53}$$

$$\sum_k A_k e^{\zeta p\cdot\omega\tau_k}\cos(p\cdot\omega\sqrt{1-\zeta^2}\tau_k)=0,\quad p\le 1 \tag{2.54}$$

$$\sum_k A_k e^{\zeta q \cdot \omega \tau_k} \sin(q \cdot \omega \sqrt{1-\zeta^2}\tau_k) = 0, \quad q \geq 1 \tag{2.55}$$

$$\sum_k A_k e^{\zeta q \cdot \omega \tau_k} \cos(q \cdot \omega \sqrt{1-\zeta^2}\tau_k) = 0, \quad q \geq 1 \tag{2.56}$$

Resulting from constraints (2.53)–(2.56) and unity-gain constraint (2.40) give rise to a modified extra-insensitivity (MEI) shaper:

$$\text{MEI} = \begin{bmatrix} A_k \\ \tau_k \end{bmatrix} = \begin{bmatrix} \frac{1}{[1+K]^2} & \frac{K}{[1+K]^2} & \frac{K}{[1+K]^2} & \frac{K^2}{[1+K]^2} \\ 0 & \frac{T_m}{2q} & \frac{T_m}{2p} & \left(\frac{T_m}{2q} + \frac{T_m}{2p}\right) \end{bmatrix} \tag{2.57}$$

2.5 COMMAND SMOOTHER

2.5.1 One-Piece Smoother

The command smoother is a piecewise function, s_1. If the flexible structure can be modeled as a second-order harmonic oscillator, then the response resulting from the piecewise function, s_1, is:

$$f(t) = \int_{\tau=0}^{+\infty} s_1(\tau) \frac{\omega}{\sqrt{1-\zeta^2}} e^{-\zeta\omega(t-\tau)} \sin(\omega(t-\tau)\sqrt{1-\zeta^2}) \mathrm{d}\tau \tag{2.58}$$

where ω and ζ are the natural frequency and damping ratio of the system. The vibrational amplitude of response (2.58) is:

$$A(t) = \frac{\omega}{\sqrt{1-\zeta^2}} e^{-\zeta\omega t} \sqrt{[S(\omega,\zeta)]^2 + [C(\omega,\zeta)]^2} \tag{2.59}$$

where

$$S(\omega,\zeta) = \int_{\tau=0}^{+\infty} u(\tau) e^{\zeta\omega\tau} \sin(\omega\tau\sqrt{1-\zeta^2}) \mathrm{d}\tau \tag{2.60}$$

$$C(\omega,\zeta) = \int_{\tau=0}^{+\infty} u(\tau) e^{\zeta\omega\tau} \cos(\omega\tau\sqrt{1-\zeta^2}) \mathrm{d}\tau \tag{2.61}$$

If Equations (2.60) and (2.61) are limited to zero, the piecewise function, s_1, would cause zero residual vibrations. Then resulting from Equations (2.60) and (2.61), the smoother is described by:

$$s_1(\tau) = \begin{cases} u_1 e^{-\zeta\omega\tau}, & 0 < T_1 \\ 0 & \text{others} \end{cases} \tag{2.62}$$

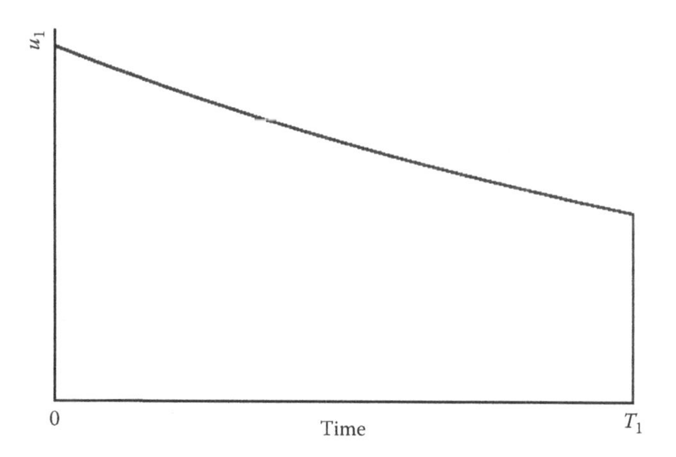

FIGURE 2.2 Curve of one-piece smoother.

where T_1 is a damped oscillation period and the coefficient u_1 is the value of the piecewise function, s_1, at time zero. Equation (2.5) is a continuous piecewise function with one piece, and so is called one-piece smoother. The curve of one-piece smoother is shown in Figure 2.2, which is a function of natural frequency and damping ratio.

Another constraint should be applied to ensure that the smoothed command reaches the same set-point as the original command. The shaping process is to have unity gain in order to satisfy this requirement. Then, the integral of the piecewise function, s_1, is limited to one:

$$\int_{\tau=0}^{+\infty} u(\tau)\mathrm{d}\tau = 1 \tag{2.63}$$

Resulting from Equations (2.60), (2.61), and (2.63), the coefficient, u_1, can be obtained:

$$u_1 = \frac{\zeta\omega}{(1 - M_1)} \tag{2.64}$$

where

$$M_1 = e^{-2\pi\zeta/\sqrt{1-\zeta^2}} \tag{2.65}$$

Resulting from Equations (2.62) and (2.64), the transfer function of the one-piece smoother is given by:

$$s_1(s) = \frac{\zeta\omega(1 - M_1 e^{-2\pi s/(\omega\sqrt{1-\zeta^2})})}{(1 - M_1)(s + \zeta\omega)} \tag{2.66}$$

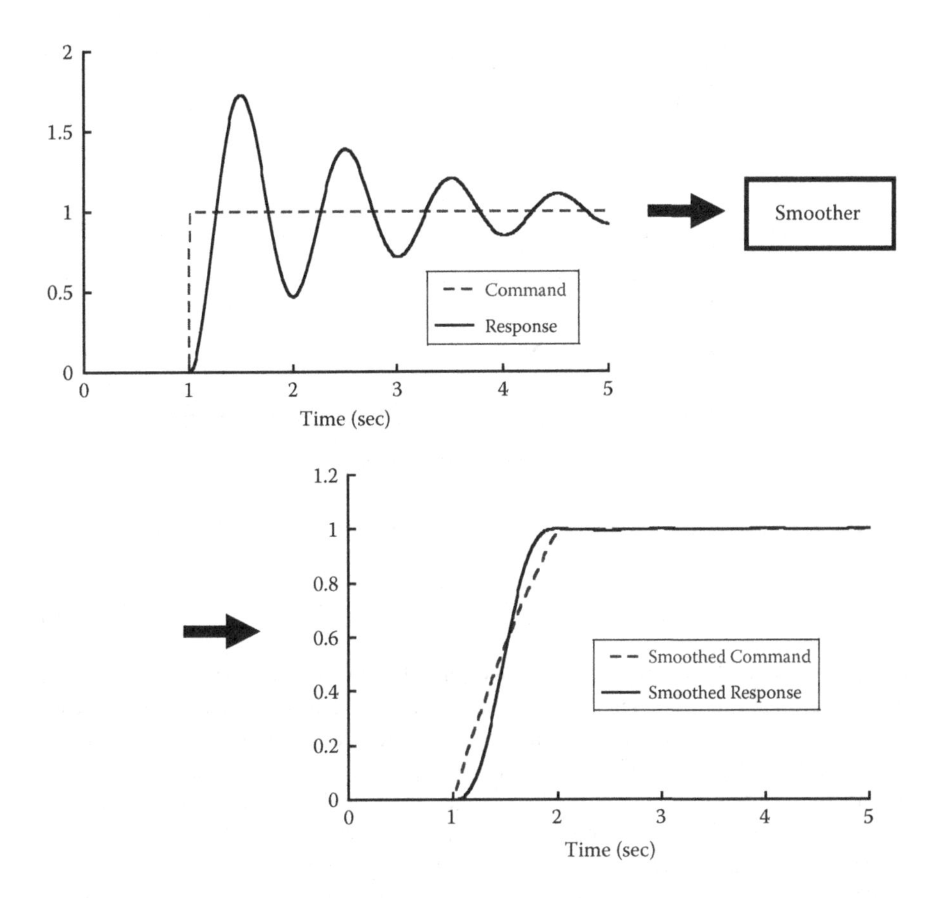

FIGURE 2.3 Command smoothing process.

By using the one-piece smoother (2.66), the vibration amplitude (2.59) will be limited to zero for any arbitrary command. This smoothing process is demonstrated in Figure 2.3. The original command produces an oscillatory response represented by the solid line labeled 'Response'. To eliminate the oscillatory response, the original command filters through the smoother to create the smoothed command shown at the bottom of Figure 2.3. The smoothed command can move the flexible dynamic systems without inducing vibrations.

The frequency and damping ratio of flexible structures may not be known accurately in many cases. It then becomes important to evaluate how this uncertainty can translate into the percentage residual amplitude (PRA). The vibration amplitude at time zero is:

$$A_0 = \frac{\omega}{\sqrt{1-\zeta^2}} \tag{2.67}$$

Dividing Equation (2.2) by Equation (2.10) yields the PRA:

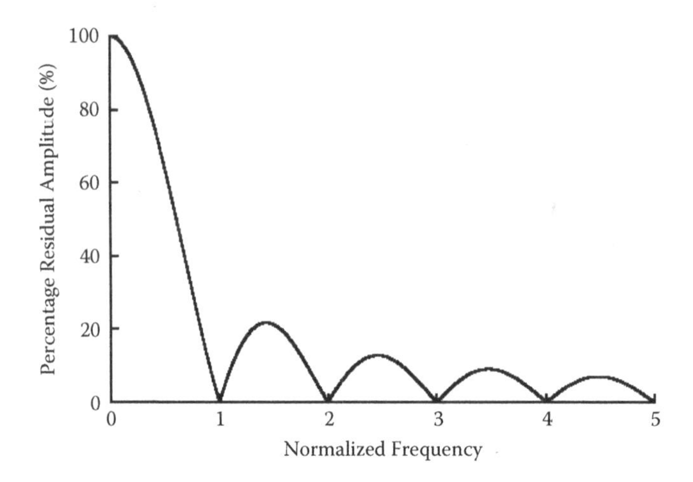

FIGURE 2.4 Frequency sensitivity for one-piece smoother.

$$\text{PRV} = e^{-\zeta\omega t}\sqrt{S(\omega,\zeta)^2 + C(\omega,\zeta)]^2} \tag{2.68}$$

Figure 2.4 shows the frequency sensitivity curve for the one-piece smoother designed for a zero damping ratio. The width of each curve, which lies below a specified vibrational level, is defined as insensitivity. It provides a quantitative measure of robustness. Less than 5% of PRA is the acceptable level of vibrations. The 5% insensitivity of the one-piece smoother ranges from 0.953 to 1.052. The normalized frequency (ω/ω_m) in this section is defined as the ratio of the real frequency to the modeled frequency, where ω_m is the modeled frequency. The PRA is zero at the modeling frequency ($\omega/\omega_m = 1$), while the sensitivity curve reaches a maximum of 100% at the zero normalized frequency. In addition, the PRA for the one-piece smoother at the integer normalized frequency is also zero. Increasing normalized frequency will decrease the peak of PRAs for the one-piece smoother. Additionally, the modeling error in the damping has few impacts on the residual vibration amplitudes.

2.5.2 Two-Pieces Smoother

A new constraint must be added to increase the robustness of the smoother under variations of the frequency and damping ratio. The derivative of Equations (2.60) and (2.61) with respect to ω and ζ should also be set equal to zero. Then changes in the frequency and damping ratio will result in small changes in the residual vibration amplitude. The following equations should be applied in order to satisfy this requirement:

$$\int_{\tau=0}^{+\infty} \tau s_2(\tau)e^{\zeta\omega\tau}\sin(\omega\tau\sqrt{1-\zeta^2})\mathrm{d}\tau = 0 \tag{2.69}$$

$$\int_{\tau=0}^{+\infty} \tau s_2(\tau)e^{\zeta\omega\tau}\cos(\omega\tau\sqrt{1-\zeta^2})\mathrm{d}\tau = 0 \tag{2.70}$$

where s_2 is the two-pieces smoother.

Limiting Equations (2.60) and (2.61) to zero and resulting from constraints (2.69) and (2.70), the two-pieces smoother is given by:

$$s_2(\tau) = \begin{cases} \tau u_2 e^{-\zeta\omega\tau}, & 0 \le \tau \le T_1 \\ (2T_1 - \tau) u_2 e^{-\zeta\omega\tau}, & T_{1\,2} < \tau \le 2T_1 \\ 0 & \text{others} \end{cases} \tag{2.71}$$

where u_2 is the coefficient of the two-pieces smoother. Equation (2.71) is a continuous piecewise function with two pieces, and is called the two-pieces smoother. Resulting from the unity-gain constraint (2.63) for the smoother (2.71), the coefficient, u_2, has the form:

$$u_2 = \frac{\zeta^2\omega^2}{(1 - M_1)^2} \tag{2.72}$$

The curve of the two-pieces smoother, which is a piecewise continuous profile as a function of natural frequency and damping ratio, is shown in Figure 2.5. Resulting from (2.71), the transfer function of the two-pieces smoother is described by:

$$s_2(s) = \frac{\zeta^2\omega^2}{(1 - M_1)^2} \cdot \frac{(1 - M_1 e^{-T_1 s})^2}{(s + \zeta\omega)^2} \tag{2.73}$$

When the modeled frequency and damping ratio are correct, the vibration amplitude (2.59) will be limited to zero for any arbitrary command by using the two-pieces smoother (2.71). The smoother is a combination of multinotch and low-pass filters. It is apparent that the smoother cannot excite flexible dynamic systems which have

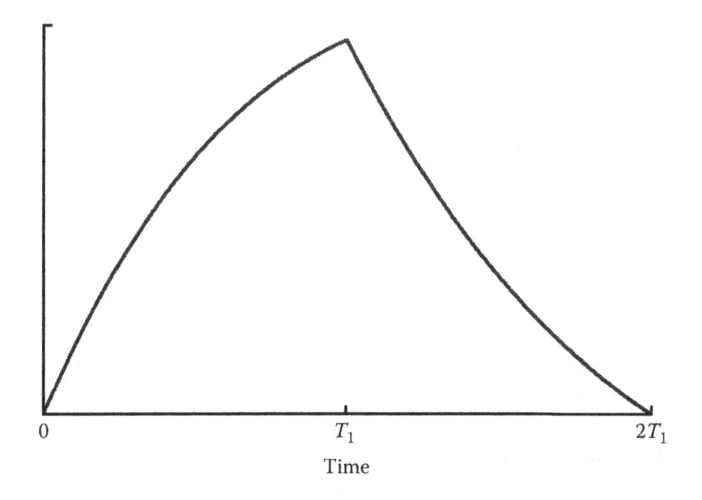

FIGURE 2.5 Curve of the two-pieces smoother.

modes corresponding to the notch frequencies. It is also clear that the smoother cannot excite high frequencies. The magnitude of the smoother is limited to one at the zero frequency. This is because the smoothing process is to have unity gain.

The smoothing process increases the rise time by the duration of the two-pieces smoother. The duration of the two-pieces smoother, R_{s2}, is twice as long as a modeled damped vibration period:

$$R_{s2} = 2T_1 \tag{2.74}$$

The higher-order derivatives of Equations (2.60) and (2.61) with respect to ω and ζ will add more robustness of the smoother. This process has the general form. Limiting the nth derivatives of Equations (2.60) and (2.61) with respect to ω and ζ to zero yields:

$$\int_{\tau=0}^{+\infty} \tau^n s_{nD}(\tau) e^{\zeta\omega\tau} \sin(\omega\tau\sqrt{1-\zeta^2})d\tau = 0 \tag{2.75}$$

$$\int_{\tau=0}^{+\infty} \tau^n s_{nD}(\tau) e^{\zeta\omega\tau} \cos(\omega\tau\sqrt{1-\zeta^2})d\tau = 0 \tag{2.76}$$

where s_{nD} is the smoother with higher-order derivative. Limiting Equations (2.60) and (2.61) to zero and resulting from the unity-gain constrain and constraints (2.75) and (2.76), the transfer function of the smoother with the nth derivative (s_{nD}) with respect to ω and ζ is given by:

$$s_{nD}(s) = \frac{\zeta^{(n+1)}\omega^{(n+1)}}{(1-M_1)^{(n+1)}} \cdot \frac{(1-M_1 e^{-T_1 s})^{(n+1)}}{(s+\zeta\omega)^{(n+1)}} \tag{2.77}$$

The duration of Equation (2.77), R_{snD}, is (n+1) times as long as a modeled damped vibration period:

$$R_{snD} = (n+1)T_1 \tag{2.78}$$

Figure 2.6 shows the frequency sensitivity curve for the two-pieces smoother designed for a zero damping ratio. The PRA is limited to zero when the modeled frequency is correct. The zero-slope at the design frequency is resulting from the derivative constraints (2.75) and (2.76). The 5% insensitivity of the smoother is from 0.81 to infinity. The smoother has more insensitivity at higher frequencies. As the normalized frequency increases, peaks of the PRA for the smoother will decrease. Therefore, the two-pieces smoother can reduce high-frequency vibrations. This performance benefits vibration suppression for multimode systems. While the two-pieces smoother is designed to suppress first-mode vibrations, it will also reduce high-mode vibrations. Meanwhile, the design of the smoother does not need to estimate high-mode frequencies.

Figure 2.6 also shows frequency sensitivity curves for the smoother with the second derivative, and the smoother with the third derivative, designed for a zero

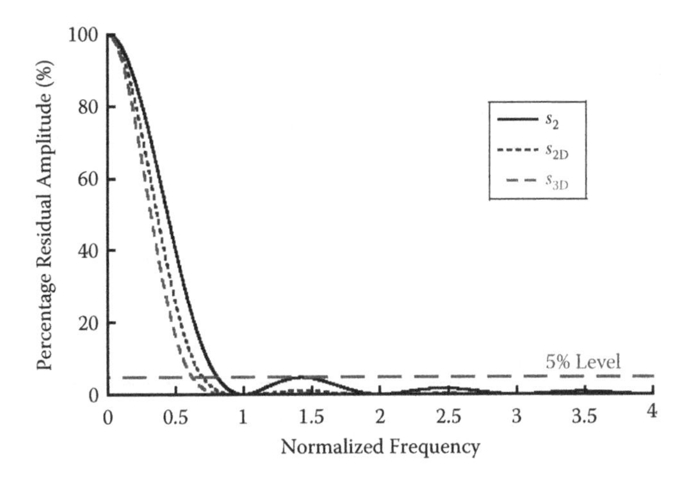

FIGURE 2.6 Frequency sensitivity for high-order derivative smoother.

damping ratio. The 5% frequency insensitivity of the smoother with the second derivative varies from 0.7 to infinity, and that with the third derivative is from 0.63 to infinity. Therefore, taking additional derivatives increases the frequency insensitivity.

Figure 2.7 shows the damping sensitivity curve for the two-pieces smoother designed for a modeled damping ratio of 0.1. The robustness to changes in the damping ratio follows similar trends to that in the frequency. There is one difference that the damping ratio is not normalized. That is because small changes in the damping ratio result in large changes in the normalized damping ratio. When the modeled damping ratio is correct, the PRA is also zero. The derivative constraints also cause a zero-slope performance at the modeling damping ratio. A dramatic reduction in the residual vibration amplitude exists for all values of damping shown in Figure 2.8. Moreover, the insensitivity of the two-pieces smoother tolerates

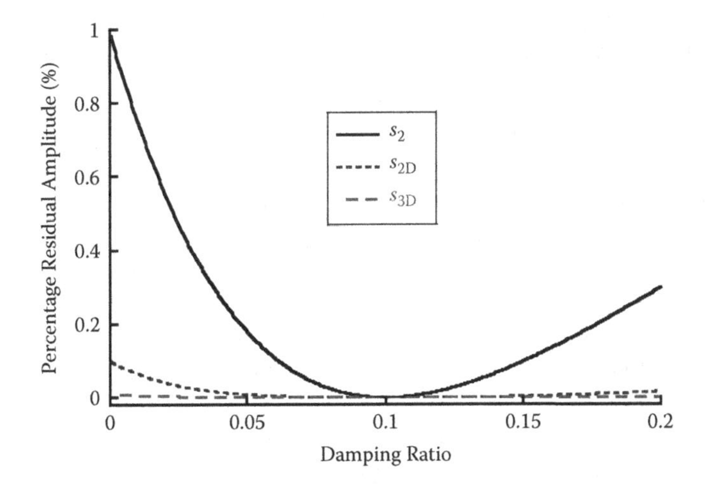

FIGURE 2.7 Damping sensitivity for high-order derivative smoother.

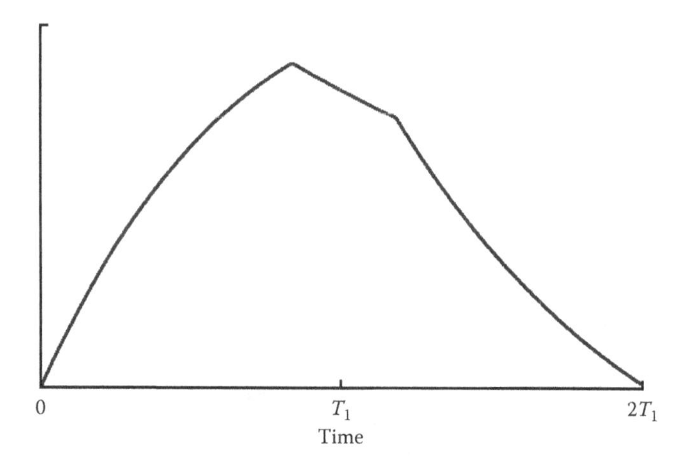

FIGURE 2.8 Graph illustration of three-pieces smoother.

extremely large variations in the damping. Variations in the damping ratio do not have a large effect on residual vibrations as changes in the frequency.

Figure 2.7 also shows damping sensitivity curves for the smoother with the second derivative, and the smoother with the third derivative, designed for a modeled damping ratio of 0.1. Taking additional derivatives will also increase the damping insensitivity. The price for each additional derivative is an increase in the smoothing duration by a modeled damped vibration period.

2.5.3 Three-Pieces Smoother

If Equations (2.60) and (2.61) are limited to zero, the smoother would cause zero residual vibrations. Nevertheless, some degree of uncertainty, which can result from poorly known or time-varying parameters and nonlinearities of the system, occurs in an actual system. Consequently, the smoother could not limit vibrations to zero on a practical system. Therefore, vibrations at the design point could be suppressed to a tolerable level:

$$\left[e^{-\zeta\omega R_{s3}}\sqrt{[S(\omega,\zeta)]^2+[C(\omega,\zeta)]^2}\right]_{\substack{\omega=\omega_m\\ \zeta=\zeta_m}}=V_{tol} \tag{2.79}$$

where ω_m is the modeled frequency, ζ_m is the modeled damping ratio, R_{s3} is the duration of the three-pieces smoother, s_3, and V_{tol} is the tolerable level of vibrations.

Additionally, the smoother should have well robust to the system parameter. Forcing vibrations at the modified frequencies, $p{\cdot}\omega_m$ and $r{\cdot}\omega_m$, to zero, could maximize the robustness to modeling errors in the frequency. The p and r are the amending coefficients. The zero-vibration constraints at the modified frequencies, p and r, are given by:

$$\left[\int_{\tau=0}^{+\infty} s_3(\tau)e^{\zeta\omega\tau}\sin(\omega\sqrt{1-\zeta^2}\tau)\mathrm{d}\tau\right]_{\substack{\omega=p\omega_m\\ \zeta=\zeta_m}}=0 \tag{2.80}$$

$$\left[\int_{\tau=0}^{+\infty} s_3(\tau)e^{\zeta\omega\tau}\cos(\omega\sqrt{1-\zeta^2}\tau)\mathrm{d}\tau\right]_{\substack{\omega=p\omega_m\\ \zeta=\zeta_m}}=0 \tag{2.81}$$

$$\left[\int_{\tau=0}^{+\infty} s_3(\tau)e^{\zeta\omega\tau}\sin(\omega\sqrt{1-\zeta^2}\tau)\mathrm{d}\tau\right]_{\substack{\omega=r\omega_m\\ \zeta=\zeta_m}}=0 \tag{2.82}$$

$$\left[\int_{\tau=0}^{+\infty} s_3(\tau)e^{\zeta\omega\tau}\cos(\omega\sqrt{1-\zeta^2}\tau)\mathrm{d}\tau\right]_{\substack{\omega=r\omega_m\\ \zeta=\zeta_m}}=0 \tag{2.83}$$

where s_3 is the three-pieces smoother.

A zero-slope constraint should be added for increasing the robustness. Then the derivative of Equation (2.79) with respect to frequency should be limited to zero:

$$\left[\frac{d[(e^{-\zeta\omega R_{s3}}S(\omega,\zeta))^2+(e^{-\zeta\omega R_{s3}}C(\omega,\zeta))^2]}{d\omega}\right]_{\substack{\omega=\omega_m\\ \zeta=\zeta_m}}=0 \tag{2.84}$$

In order to ensure the smoother reach the same set-point as the original command, another unit-gain constraint must be added. The integral of the smoother is limited to one to satisfy the requirement. Assuming that p is larger than r, and resulting from constraints (2.80)–(2.83) and unit-gain constraint, the time-optimal solution of the three-pieces smoother is described by:

$$s_3(\tau)=\begin{cases}\mu_3(e^{-r\zeta_m\omega_m\tau}-e^{-p\zeta_m\omega_m\tau}),\ 0\le\tau\le(T_1/p)\\ \mu_3e^{-r\zeta_m\omega_m\tau}(1-\delta_3),\ (T_1/p)<\tau<(T_1/r)\\ \mu_3(\sigma_3e^{-p\zeta_m\omega_m\tau}-\delta_3e^{-r\zeta_m\omega_m\tau}),\ (T_1/r)\le\tau\le(T_1/p+T_1/r)\\ 0,\ \text{others}\end{cases} \tag{2.85}$$

where

$$\delta_3=e^{2\pi(r/p-1)\zeta_m/\sqrt{1-\zeta_m^2}} \tag{2.86}$$

$$\sigma_3=e^{2\pi(p/r-1)\zeta_m/\sqrt{1-\zeta_m^2}} \tag{2.87}$$

$$\mu_3=\frac{pr\zeta_m\omega_m}{(p-r)(1-e^{-2\pi\zeta_m/\sqrt{1-\zeta_m^2}})^2} \tag{2.88}$$

The three-pieces smoother illustrated in Figure 2.8 was derived from Equation (2.85). The curve shows a piecewise continuous profile with three pieces. Convolving the original command with the three-pieces smoother creates a smoothed command. The smoothed command moves the flexible structure inducing tolerable vibrations. Instead of forcing vibrations to zero at the design frequency, the three-pieces

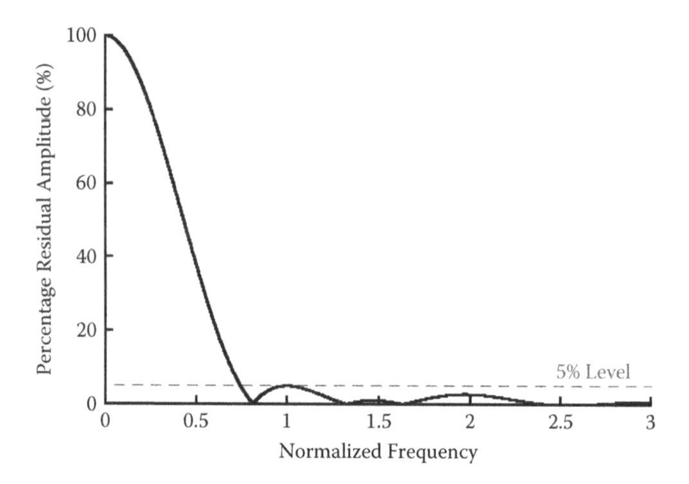

FIGURE 2.9 Frequency sensitivity for three-pieces smoother.

smoother limits vibrations to the tolerable level. Meanwhile, the robustness to modeling errors in the frequency has been increased by forcing vibrations to zero at two modified frequencies.

The zero-derivative constraint ensures that vibrations could be suppressed below the tolerable level between two modified frequencies. The duration of the three-pieces smoother (2.85) is given by:

$$T_{s3} = (1/p + 1/r)\cdot T_1 \tag{2.89}$$

When the tolerable level, V_{tol}, was set to 5%, numerical solutions for undamped systems are $r = 0.7545$ and $p = 1.2277$. When the tolerable level was set to zero, amending coefficients for both undamped and damped systems are $p = r = 1$. Therefore, the zero tolerable level will create a smoother with specified performance, including zero vibration and derivative at the design frequency and damping ratio.

Figure 2.9 shows the frequency sensitivity curve for the three-pieces smoother designed for a zero damping ratio. The 5% insensitivity of the three-pieces smoother ranges from 0.747 to infinity. The three-pieces smoother has more insensitivity at the high frequency, and is less insensitive at the low frequency. As the normalized frequency increases, the magnitude of peaks of the PRA decreases. This low-pass filtering performance will benefit vibration reduction for multimode systems. When the three-pieces smoother is designed to reduce vibrations of the first mode, it will also suppress that of higher modes.

2.5.4 Four-Pieces Smoother

The high mode also has an effect on the system dynamics. Thus, the higher-mode vibrations should be suppressed. In order to create a low-pass filtering effect for the smoother, oscillations at the normalized frequency of two should be suppressed to zero:

$$\int_{\tau=0}^{+\infty} s_4(\tau)e^{2\omega_m\zeta_m\tau}\sin(2\omega_m\sqrt{1-\zeta_m^2}\tau)d\tau = 0 \tag{2.90}$$

$$\int_{\tau=0}^{+\infty} s_4(\tau)e^{2\omega_m\zeta_m\tau}\cos(2\omega_m\sqrt{1-\zeta_m^2}\tau)d\tau = 0 \tag{2.91}$$

where ζ_m is the design damping ratio, ω_m is the design frequency, and s_4 is the four-pieces smoother.

When the design frequency and damping ratio are correct, the constraints (2.90) and (2.91) would suppress vibrations at the normalized frequency of two, $2\omega_m$. Nevertheless, real systems often operate with some degree of uncertainty at the high mode. Therefore, another constraint should be added to increase the robustness to changes in the high-mode frequency. The zero derivative of the percent vibration amplitude with respect to frequency at the normalized frequency of two should also be added:

$$\int_{\tau=0}^{+\infty} \tau s_4(\tau)e^{2\omega_m\zeta_m\tau}\sin(2\omega_m\sqrt{1-\zeta_m^2}\tau)d\tau = 0 \tag{2.92}$$

$$\int_{\tau=0}^{+\infty} \tau s_4(\tau)e^{2\omega_m\zeta_m\tau}\cos(2\omega_m\sqrt{1-\zeta_m^2}\tau)d\tau = 0 \tag{2.93}$$

The smoothed command should reach the same set-point as the original command; thus, the integral of the smoother should also be limited to one. The constraints (2.90)–(2.93) and unit-gain constraint yield a time-optimal solution of the four-pieces smoother:

$$s_4(\tau) = \begin{cases} M_4\tau e^{-2\zeta_m\omega_m\tau}, & 0 \le \tau \le 0.5T \\ M_4[T_1 - K_4^{-1}T_1 - \tau + 2K_4^{-1}\tau]e^{-2\zeta_m\omega_m\tau}, & 0.5T_1 < \tau \le T \\ M_4\cdot[3K_4^{-1}T - K_4^{-2}T - 2K_4^{-1}\tau + K_4^{-2}\tau]e^{-2\zeta_m\omega_m\tau}, & T < \tau \le 1.5T \\ M_4\cdot[2K_4^{-2}T - K_4^{-2}\tau]e^{-2\zeta_m\omega_m\tau}, & 1.5T < \tau \le 2T \end{cases} \tag{2.94}$$

where σ_4 is the coefficient, and the coefficients K_4 and M_4 are:

$$K_4 = e^{(-\pi\zeta_m/\sqrt{1-\zeta_m^2})} \tag{2.95}$$

$$M_4 = 4\zeta_m^2\omega_m^2/(1 + K_4 - K_4^2 - K_4^3)^2 \tag{2.96}$$

Figure 2.10 shows that the four-pieces smoother (2.37) is a piecewise continuous function with four pieces. Continuous transitions between boundary conditions exist in each of pieces. The transient vibrations are also important in many applications. Therefore, the maximum amplitude of transient vibrations should be limited below a tolerable level:

$$\max\left(e^{\omega_m\zeta_m\tau_n}\sqrt{\left[\int_{\tau=0}^{w} s_4 e^{\omega_m\zeta_m\tau}\sin(\omega_m\sqrt{1-\zeta_m^2}\tau)d\tau\right]^2 + \left[\int_{\tau=0}^{w} s_4 e^{\omega_m\zeta_m\tau}\cos(\omega_m\sqrt{1-\zeta_m^2}\tau)d\tau\right]^2}\right) \le V_{tol} \tag{2.97}$$

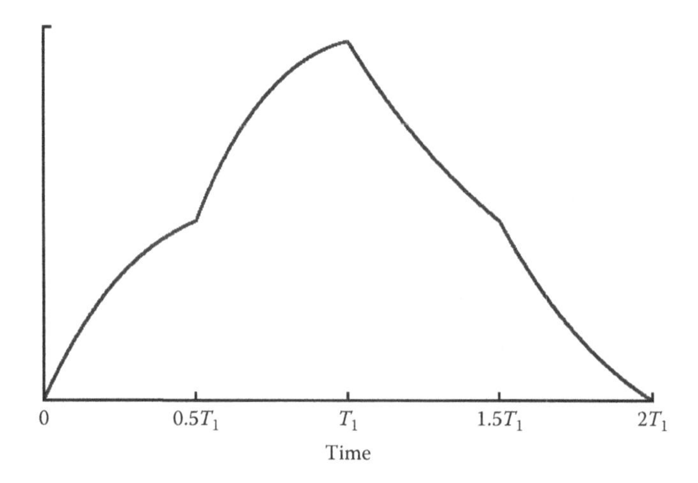

FIGURE 2.10 Curve of four-pieces smoother.

where w is any arbitrary time during the transient stage, τ_n is the duration of the four-pieces smoother, and V_{tol} is the tolerable level of the transient vibrations. For the case of the four-pieces smoother (2.94), the tolerable level of the transient vibrations is 16%, which is a very low level.

The four-pieces smoother is a function of design frequency and damping ratio. However, the frequency might not be known accurately in a real system, then it is important to estimate how the modeling error in the frequency translates into the percent residual amplitude. Figure 2.11 shows the frequency insensitive curve for the four-pieces smoother designed for an undamped system. There are troughs at the design frequency and integer normalized frequencies. The 5% insensitivity of the four-pieces smoother is from 0.81 to infinity. Meanwhile, damping ratio has little impact on vibrations with the four-pieces smoother. This conclusion is fortunate

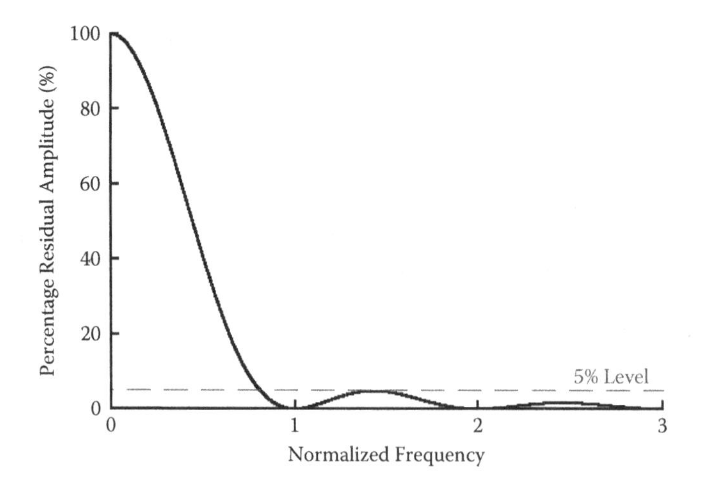

FIGURE 2.11 Frequency sensitivity curve with four-pieces smoother.

because it is generally challenging to estimate the damping ratio accurately under real conditions.

2.5.5 Smoother for Duffing Oscillators

The first-order approximate solution of a damped Duffing oscillator resulting from an impulse function, s_{SD}, is:

$$\begin{aligned} q_k(t) &= \int_{\tau=0}^{+\infty} s_{SD}(\tau)\cdot\psi_1\cdot e^{-\zeta\overline{\omega}(t-\tau)} \sin[\overline{\omega}\sqrt{1-\zeta^2}(t-\tau)+\varphi_1]d\tau \\ &+ \int_{\tau=0}^{+\infty} v(\tau)\cdot\psi_2\cdot e^{-\zeta\overline{\omega}(t-\tau)} \sin[\overline{\omega}\sqrt{1-\zeta^2}(t-\tau)+\varphi_2]d\tau \\ &+ \int_{\tau=0}^{+\infty} v(\tau)\cdot\psi_3\cdot e^{-3\zeta\overline{\omega}(t-\tau)} \sin[3\overline{\omega}\sqrt{1-\zeta^2}(t-\tau)+\varphi_3]d\tau \end{aligned} \tag{2.98}$$

where ψ_1, ψ_2, and ψ_3 are vibration-contribution functions; φ_1, φ_2, and φ_3 are phase functions; and $\bar{\omega}$ is the nonlinear frequency. The parameter, v, is given by:

$$v(\tau) = \int h(t)\cdot s_{SD}(\tau - t)dt; \quad h(t) = \int s_{SD}(y)\cdot s_{SD}(t-y)dy \tag{2.99}$$

The nonlinear frequency, $\bar{\omega}$, is given by:

$$\bar{\omega} = \omega\cdot\sqrt{1 + 0.75e\cdot[q^2 + (\zeta q + \dot{q}/\omega)^2]} \tag{2.100}$$

where ω, e, and q are the linear frequency, nonlinear stiffness parameter, and amplitude of the Duffing oscillator, respectively.

The nonlinear frequency is dependent on the nonlinear stiffness parameter, e, the linear frequency, ω, the time-dependent vibration amplitude, q, and its derivative, $\dot{q}$. As the linear frequency, ω, and nonlinear stiffness parameter, e, increase, the nonlinear frequency increases. Increasing vibration amplitudes increases the nonlinear frequency in the case of $e > 0$ (hardening spring type). Meanwhile, increasing vibration amplitudes decreases the nonlinear frequency in the case of $e < 0$ (softening spring type).

The amplitude of approximate solution (2.98) can be written as:

$$\begin{aligned} A(t) &= \psi_1\cdot e^{-\zeta\overline{\omega}t}\sqrt{[S_1(\zeta,\overline{\omega})]^2 + [C_1(\zeta,\overline{\omega})]^2} \\ &+ \psi_2\cdot e^{-\zeta\overline{\omega}t}\sqrt{[S_2(\zeta,\overline{\omega})]^2 + [C_2(\zeta,\overline{\omega})]^2} \\ &+ \psi_3\cdot e^{-2\zeta\overline{\omega}t}\sqrt{[S_3(\zeta,\overline{\omega})]^2 + [C_3(\zeta,\overline{\omega})]^2} \end{aligned} \tag{2.101}$$

where

$$S_1(\zeta,\overline{\omega}) = \int_{\tau=0}^{+\infty} s_{SD}(\tau)\cdot e^{\zeta\overline{\omega}\tau} \sin[\overline{\omega}\sqrt{1-\zeta^2}\cdot\tau]d\tau \tag{2.102}$$

$$C_1(\zeta,\overline{\omega}) = \int_{\tau=0}^{+\infty} s_{SD}(\tau)\cdot e^{\zeta\overline{\omega}\tau} \cos[\overline{\omega}\sqrt{1-\zeta^2}\cdot\tau]d\tau \tag{2.103}$$

$$S_2(\zeta,\bar{\omega}) = \int_{\tau=0}^{+\infty} v(\tau)\cdot e^{\zeta\bar{\omega}\tau}\sin[\bar{\omega}\sqrt{1-\zeta^2}\cdot\tau]d\tau \tag{2.104}$$

$$C_2(\zeta,\bar{\omega}) = \int_{\tau=0}^{+\infty} v(\tau)\cdot e^{\zeta\bar{\omega}\tau}\cos[\bar{\omega}\sqrt{1-\zeta^2}\cdot\tau]d\tau \tag{2.105}$$

$$S_3(\zeta,\bar{\omega}) = \int_{\tau=0}^{+\infty} v(\tau)\cdot e^{3\zeta\bar{\omega}\tau}\sin[3\bar{\omega}\sqrt{1-\zeta^2}\cdot\tau]d\tau \tag{2.106}$$

$$C_3(\zeta,\bar{\omega}) = \int_{\tau=0}^{+\infty} v(\tau)\cdot e^{3\zeta\bar{\omega}\tau}\cos[3\bar{\omega}\sqrt{1-\zeta^2}\cdot\tau]d\tau \tag{2.107}$$

Limiting Equations (2.102)–(2.107) to zero yields zero vibrations of the approximate response. The modified command should also reach the same position as the original driving command. Thus, a unity gain constraint should be added:

$$\int s_{SD}(\tau) = 1 \tag{2.108}$$

Resulting from Equations (2.102)–(2.107) and constraint (2.108), an impulse function for the single-mode Duffing oscillator (SD) is given by:

$$s_{SD}(\tau) = \begin{bmatrix} \frac{1}{(1+K_5+K_5^3+K_5^4)} & \frac{K_5+K_5^3}{(1+K_5+K_5^3+K_5^4)} & \frac{K_5^4}{(1+K_5+K_5^3+K_5^4)} \\ 0 & 0.5T_5 & T_5 \end{bmatrix} \tag{2.109}$$

where

$$K_5 = e^{(-\pi\zeta/\sqrt{1-\zeta^2})} \tag{2.110}$$

$$T_5 = \frac{2\pi}{\bar{\omega}\cdot\sqrt{1-\zeta^2}} \tag{2.111}$$

The three-impulses SD function is a function of nonlinear frequency, $\bar{\omega}$, and damping ratio, ζ. When the nonlinear frequency and damping ratio are found accurately, the approximate response would be suppressed to zero. Both the nonlinear frequency, $\bar{\omega}$, and the corresponding period, T_5, depend on the vibration amplitude. The duration of the three-impulses SD function is a damped vibration period, T_5.

Vibrations of multimode Duffing oscillators can also be suppressed by the three-impulses SD function. The three-impulses SD function for each mode of the uncoupled Duffing oscillators is designed independently, and then combining those together produces a combination function. Vibrations of multimode Duffing oscillators can be reduced by the combination function. Nevertheless, it is very challenging to estimate high-mode frequencies under practical engineering conditions. However, the modeling error in the high-mode frequency might be large so that designing a suitable controller is impossible.

The low-pass filter benefits vibration suppression of high-mode vibrations. Then vibrations at the even frequency, $2\bar{\omega}$, should also be zero:

$$\int_{\tau=0}^{+\infty} s_{MD}(\tau)\cdot e^{2\zeta\bar{\omega}\tau} \sin\left[2\bar{\omega}\sqrt{1-\zeta^2}\cdot\tau\right]d\tau = 0 \tag{2.112}$$

$$\int_{\tau=0}^{+\infty} s_{MD}(\tau)\cdot e^{2\zeta\bar{\omega}\tau} \cos\left[2\bar{\omega}\sqrt{1-\zeta^2}\cdot\tau\right]d\tau = 0 \tag{2.113}$$

A smoother for multimode Duffing oscillators (MD) can be derived by limiting Equations (2.102)–(2.107) to zero and solving constraints (2.112) and (2.113):

$$s_{MD}(\tau) = \begin{cases} M_5\cdot\tau e^{-2\zeta\bar{\omega}\tau}, & 0 \le \tau \le 0.5T_5 \\ (-0.5K_5^{-1} + 1 - 0.5K_5)M_5\cdot T_5 e^{-2\zeta\bar{\omega}\tau} & 0.5T_5 \le \tau \le T_5 \\ + (K_5^{-1} - 1 + K_5)M_5\cdot\tau e^{-2\zeta\bar{\omega}\tau}, & \\ (1.5K_5^{-1} - 1 + 1.5K_5)M_5\cdot Te^{-2\zeta\bar{\omega}\tau} & T_5 \le \tau \le 1.5T_5 \\ + (-KK_5^{-1} + 1 - K_5)M_5\cdot\tau e^{-2\zeta\bar{\omega}\tau}, & \\ 2M_5\cdot T_5 e^{-2\zeta\bar{\omega}\tau} - 2M_5\cdot\tau e^{-2\zeta\bar{\omega}\tau}, & 1.5T_5 \le \tau \le 2T_5 \end{cases} \tag{2.114}$$

where

$$M_5 = \frac{4\zeta^2\bar{\omega}^2}{(1 + K_5 + K_5^3 + K_5^4)(1 - 2K_5^2 + K_5^4)} \tag{2.115}$$

The profile of the MD smoother (2.115) is shown in Figure 2.12. The MD smoother is a piecewise continuous function involving nonlinear frequency, $\bar{\omega}$, and damping ratio, ζ.

The MD smoother is a combination of multinotch and low-pass filter, while the SD function is a notch filter. The SD function suppresses vibrations of the first-mode Duffing oscillator, but cannot control that of high-mode Duffing oscillators.

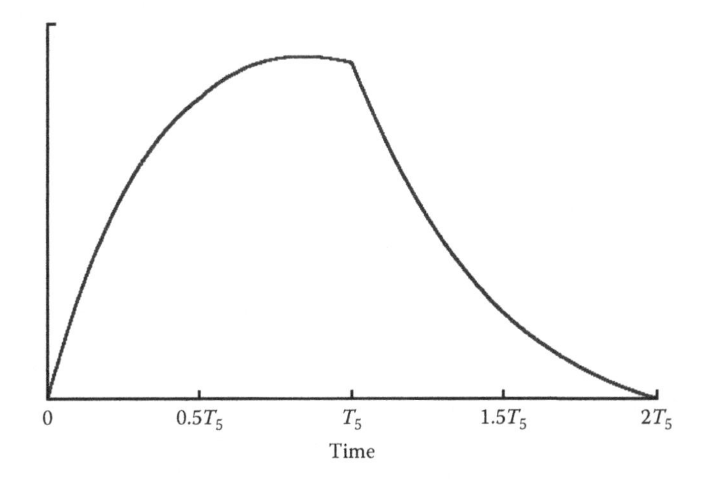

FIGURE 2.12 Profile of the MD smoother.

The high-mode vibrations might also have some effects on the flexible dynamic system. The MD smoother not only suppresses vibrations of the first-mode Duffing oscillator but also attenuates vibrations of high-mode Duffing oscillators. Consequently, the MD smoother does not need to estimate high-mode frequencies. It is generally easier to estimate the first-mode frequency. Thus, this feature exhibits the advantage of control implementation.

The duration of the MD smoother is twice as long as the damped vibration period, T_5. The long duration will slow down the operating speed of flexible structures. Nevertheless, the MD smoother reduces vibrations of flexible structures. Then the flexible structure can move at a higher speed without generating additional vibrations. Additionally, small variations in the system parameter might cause large changes of the high-mode frequency. Robustness in the high-mode frequency must trade off against an increase in rise time.

Both the three-impulse SD function and MD smoother require the use of the nonlinear frequency and damping ratio. However, the nonlinear frequency and damping ratio may not be known accurately. This section uses the dynamics of the single-link flexible manipulator to conduct the numerical verification of the effectiveness and robustness of the three-impulse SD function and MD smoother.

3 Bridge Cranes

Industrial bridge cranes offer material-handling services throughout the world. However, flexible structures degrade their effectiveness and safety. Payload oscillations caused by both intentional motions commanded by the human operator and by external disturbances such as wind are major limitations. Therefore, it is critical to reduce effectively payload oscillations by using the presented smoothers in Chapter 2.

3.1 SINGLE-PENDULUM CRANES

Numerous scientists have focused on providing solutions to the payload oscillation problem posed by the industrial crane [1]. The work can roughly be broken into two categories: feedback controller and input shaping. The feedback control strategies use measurements of the payload swing to reduce vibrations. Masoud presented a feedback controller to suppress oscillations of the rotary crane. A damping effect in the crane system was created by careful selection of the gain and time delay of the controller. The control performance was verified on rotary cranes [2], ship-mounted telescopic cranes [3], and quay-side container cranes [4]. Solihin designed a fuzzy-tuned proportional-integral-derivative controller to suppress vibrations of the gantry crane. The controller is robustness to changing parameters [5].

Input shaping techniques can effectively control the payload swing of many types of industrial cranes including bridge cranes [6,7], tower cranes [8,9], boom cranes [10,11], and container cranes [12]. The first input shaper is called zero vibration (ZV) shaper, which limits vibrations to zero at the design frequency and damping ratio. Since then, robust input shapers have been produced, such as zero vibration and derivative (ZVD) shaper, extra-insensitive (EI) shaper, and specified-insensitive (SI) shaper. The fundamental compromise is that the increased robustness provided by those advanced shapers comes at the cost of additional rise time.

3.1.1 Modeling

Figure 3.1 shows a schematic representation of a bridge crane transporting a point-mass payload. A trolley slides along the bridge in the N_y direction. A payload of mass, m, is attached to the trolley center T^* by a suspension cable S of length, l. The inputs to the model are the acceleration, $\ddot{y}$, of the trolley in the N_x direction. The output is a swing angle α of the suspension cable to orient the pendulum. The suspension cable is assumed to be massless and inelastic. The friction is assumed to be zero because of the small contact area among trolley, payload and cable. The motions of the trolley are unaffected by load motion because of small swing and large impedance in the drives.

DOI: 10.1201/9781003247210-3

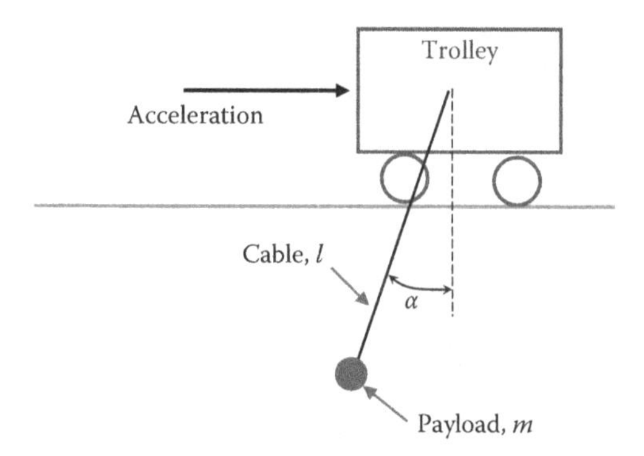

FIGURE 3.1 A planar single-pendulum bridge crane.

The Newtonian reference frame $N_xN_yN_z$ can be converted to the moving Cartesian coordinates $S_xS_yS_z$ in the suspension cable by rotating the swing angle, α. The transformation matrix is:

$$\begin{Bmatrix} \mathbf{N}_y \\ \mathbf{N}_z \end{Bmatrix} = \begin{bmatrix} \cos(\alpha) & -\sin(\alpha) \\ \sin(\alpha) & \cos(\alpha) \end{bmatrix} \begin{Bmatrix} \mathbf{S}_y \\ \mathbf{S}_z \end{Bmatrix} \tag{3.1}$$

The velocity of the trolley center T^* in the Newtonian reference frame is described as:

$$^N\mathbf{v}^{T^*} = \dot{y}\mathbf{N}_y \tag{3.2}$$

The angular velocity of the suspension cable S in the Newtonian reference frame is given by:

$$^N\boldsymbol{\omega}^S = \dot{\alpha}\mathbf{S}_x \tag{3.3}$$

The velocity of the point-mass payload in the Newtonian reference frame is expressed by:

$$^N\mathbf{v}^P = {}^N\mathbf{v}^{T^*} + \dot{\alpha}l\mathbf{S}_y \tag{3.4}$$

The generalized speeds u_i defined in Kane's equation are chosen as:

$$u_1 = \dot{\alpha} \tag{3.5}$$

The corresponding generalized active force can be expressed by:

$$F_1 = -mg\mathbf{N}_z \cdot {}^N\mathbf{v}_1^P = -mglsin(\alpha) \tag{3.6}$$

where $\cdot$ represents the dot product, $^N\mathbf{v}_1^P$ denotes the first partial velocity of the payload. The partial velocity of the payload derives from the derivative of Equation (3.4) with respect to generalized speeds u_i, and the results are given by:

$$ {}^{N}\mathbf{v}_1^P = l\mathbf{S}_y \tag{3.7} $$

The corresponding generalized inertia force can be written as:

$$ F_1^* = -m\,{}^{N}\mathbf{a}^P \cdot {}^{N}\mathbf{v}_1^P \tag{3.8} $$

where ${}^{N}\mathbf{a}^P$ represents the acceleration of the payload. The derivative of Equation (3.4) with respect to time gives rise to the acceleration of the payload.

Forcing the sum of the generalized active force and inertia force to be zero gives rise to dynamic equations of the motion for the model:

$$ l\ddot{\alpha} + g\sin(\alpha) + \ddot{y}\cos(\alpha) = 0 \tag{3.9} $$

The natural frequency of single-pendulum oscillation is:

$$ \omega = \sqrt{\frac{g}{l}} \tag{3.10} $$

where g is the gravitational constant, and l is the cable length.

3.1.2 Oscillation Reduction by Two-Pieces Smoother

Experiments were performed on a planar bridge crane shown in Figure 3.2 to verify some of key results. A Panasonic AC servomotor with an encoder drove the trolley. A personal computer was used for program development and user interface. A DSP-based motion control card (Googol GT-400-SV-PCI) connected the personal computer to a servo amplifier. The original command is sent to the smoother algorithm, and then generates the smoothed command for the drive. A tennis ball was served as the payload. A weightless cable suspended the ball to the trolley.

FIGURE 3.2 A planar single-pendulum bridge crane.

A two-pieces smoother is applied to suppress the payload swing. The two-pieces smoother is described by:

$$u(\tau) = \begin{Bmatrix} \tau u_0 e^{\zeta_k \omega_k \tau}, & 0 \leq \tau \leq T_d \\ (2T_d - \tau) u_0 e^{-\zeta_k \omega_k \tau}, & T_d < \tau \leq 2T_d \\ 0 & \text{others} \end{Bmatrix} \tag{3.11}$$

where

$$u_0 = \frac{\zeta_k^2 \omega_k^2}{(1 - e^{-2\pi\zeta_k/\sqrt{1-\zeta_k^2}})^2} \tag{3.12}$$

A series of experiments were performed by using the two-pieces smoother to move the single-pendulum crane when the modeled suspension length was set to 100 cm (corresponding to natural frequency of 0.50 Hz). Figure 3.3 shows experimental and simulated results of residual vibrations as a function of the suspension length. The suspension cable length varied from 40 cm (corresponding to natural frequency of 0.79 Hz) to 160 cm (corresponding to natural frequency of 0.39 Hz). The experimental results follow the general shape as the simulated curve. The two-pieces smoother suppressed the payload swing near the design frequency of 0.50 Hz. The residual amplitude with the two-pieces smoother increased sharply as the suspension length increased from 100 cm. This is because of more insensitivity at higher frequencies and less insensitivity at low frequencies for the smoother.

The effect when the two-pieces smoother was used to move the trolley 48 cm is shown in Figure 3.4. A trapezoidal velocity profile was used as the original command. The smoothed velocity command was an s-curve profile, which had smooth transitions between boundary conditions.

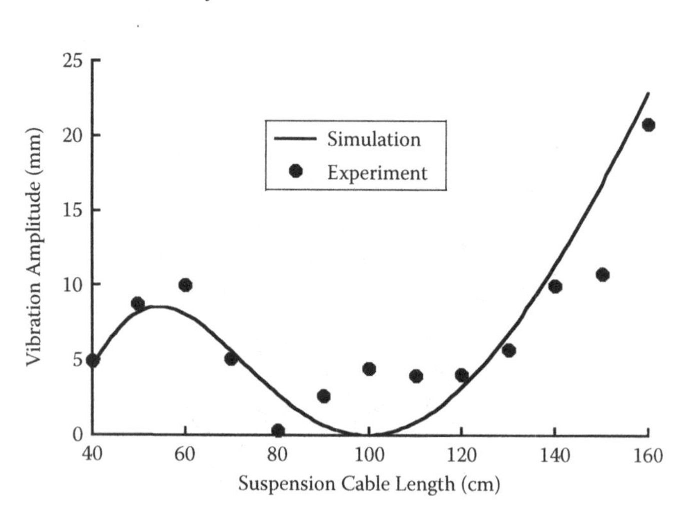

FIGURE 3.3 Residual vibrations as a function of suspension length.

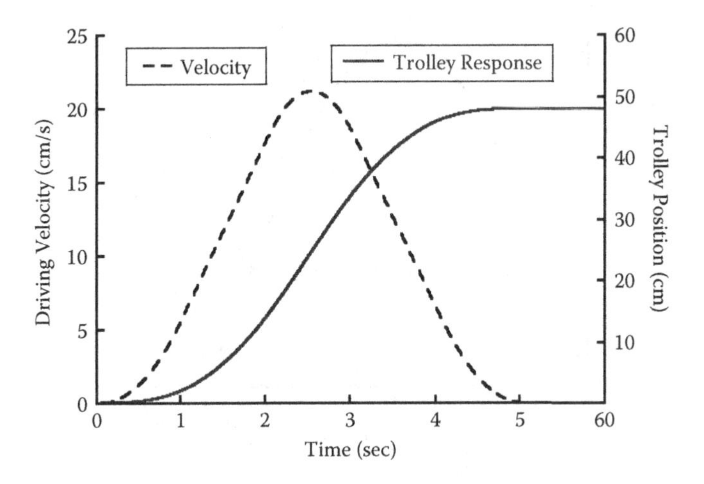

FIGURE 3.4 Trolley response to smoothed commands.

3.1.3 Oscillation Control by Three-Pieces Smoother

Experiments were performed on a planar bridge crane to verify some of key results of the three-pieces smoother. The three-pieces smoother is given by:

$$u(\tau) = \begin{cases} \mu(e^{-r\zeta_m\omega_m\tau} - e^{-p\zeta_m\omega_m\tau}), \ 0 \le \tau \le (T_m/p) \\ \mu e^{-r\zeta_m\omega_m\tau}(1-\delta), \ (T_m/p) < \tau < (T_m/r) \\ \mu(\sigma e^{-p\zeta_m\omega_m\tau} - \delta e^{-r\zeta_m\omega_m\tau}), \ (T_m/r) \le \tau \le (T_m/p + T_m/r) \\ 0, \text{ others} \end{cases} \tag{3.13}$$

where p and r are coefficients, and

$$\delta = e^{2\pi(r/p-1)\zeta_m/\sqrt{1-\zeta_m^2}} \tag{3.14}$$

$$\sigma = e^{2\pi(p/r-1)\zeta_m/\sqrt{1-\zeta_m^2}} \tag{3.15}$$

$$\mu = \frac{pr\zeta_m\omega_m}{(p-r)(1-e^{-2\pi\zeta_m/\sqrt{1-\zeta_m^2}})^2} \tag{3.16}$$

$$T_m = 2\pi/(\omega_m\sqrt{1-\zeta_m^2}) \tag{3.17}$$

The residual amplitude to a set trapezoidal velocity profile for driving distance between 5 cm and 55 cm is shown in Figure 3.5. Without the controller, large oscillations were caused on the planar bridge crane when the trolley was driven by a trapezoidal velocity command. Peaks and troughs in the residual amplitude occur with varying driving distances. When the trolley starts, the acceleration will cause

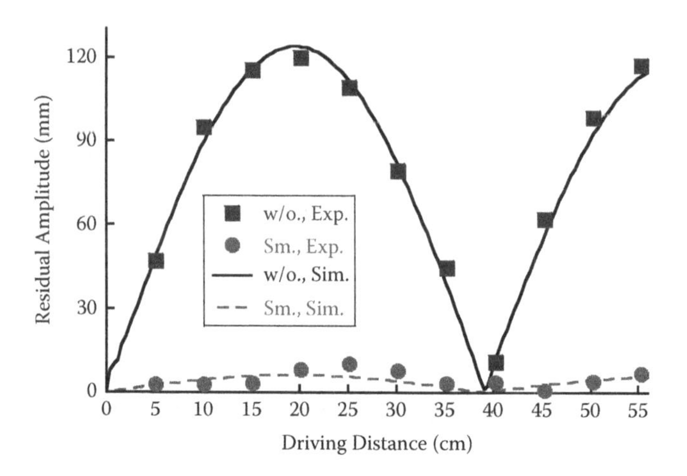

FIGURE 3.5 Residual vibrations induced by driving motions.

large payload swing. Moreover, when the trolley stops, it will also induce additional swing of the payload. Those two oscillations are sometimes out of phase, thereby causing troughs. Sometimes oscillations are in phase, thereby inducing peaks. Figure 3.5 also shows the residual amplitude to the smoothed command. Most of residual vibrations were reduced by the three-pieces smoother. These experiments verified that the three-pieces smoother can also effectively control the payload swing of the single-pendulum crane.

A series of experiments were performed to move planar cranes by using the three-pieces smoother. Figure 3.6 shows residual amplitudes as a function of the suspension cable length. The suspension cable length varies when the modeled suspension length is set to 1 m. The three-pieces smoother suppressed most of the oscillations between 10 cm and 170 cm. The residual amplitudes will increase faster when the suspension length increases from 170 cm. This is because the three-pieces smoother provides

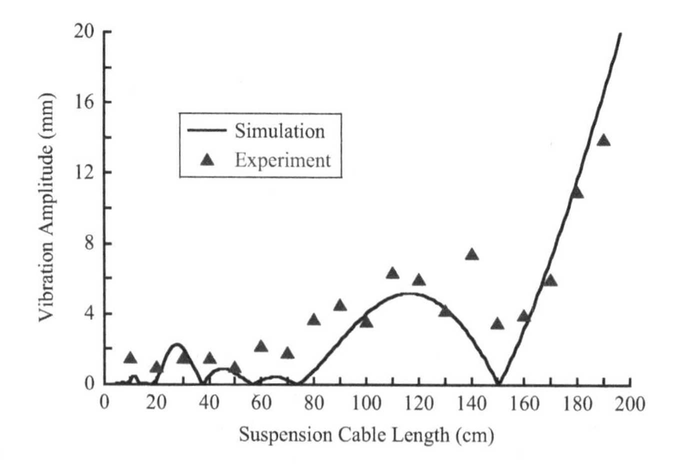

FIGURE 3.6 Residual vibrations from three-pieces smoother.

more insensitivity at higher frequencies and less insensitivity at low frequencies. The experimental findings reported here confirm that the three-pieces smoother can control payload swing of industrial cranes. Furthermore, research findings proved that the three-pieces smoother is more insensitive at higher frequencies.

3.2 DOUBLE-PENDULUM CRANES

Skilled operators reduce much of oscillations of single-pendulum crane manually by causing deceleration oscillations, which cancel oscillations induced during the acceleration. However, the double-pendulum dynamics exhibit in certain types of payloads and hoisting mechanisms. Skilled operators are very challenging to eliminate oscillations manually in this case. Therefore, swing-reduction controllers should be equipped in double-pendulum cranes to achieve safe and efficient operations.

Many scientists have worked to provide solutions to the challenging problem posed by the double-pendulum dynamics. The work can roughly be broken into two categories: feedback control and input shaping. Feedback control strategies use measurement and estimation of system state to suppress vibrations. However, the difficulty of accurately sensing the swing, and the conflict between the computer-based controller and actions of human operator are obstacles toward the application of the feedback controller. Input-shaping technique can effectively reduce payload oscillations of many types of double-pendulum cranes including bridge cranes, tower cranes, boom cranes, and multihoist cranes. While the input shaping technique reduces oscillations induced by intentional motions commanded by the human operator. It cannot reject external disturbances.

A schematic representation of a planar double-pendulum bridge crane is shown in Figure 3.7. A force is applied to the trolley for moving the crane. A hook of mass, m_h, is attached to the trolley by using the suspension cable of length, l_h. The rigging cable of length, l_p, hangs below the hook and supports a payload. The payload is modeled as a point mass, m_p. It is assumed that both the suspension cable length and rigging cable length do not change during the operation.

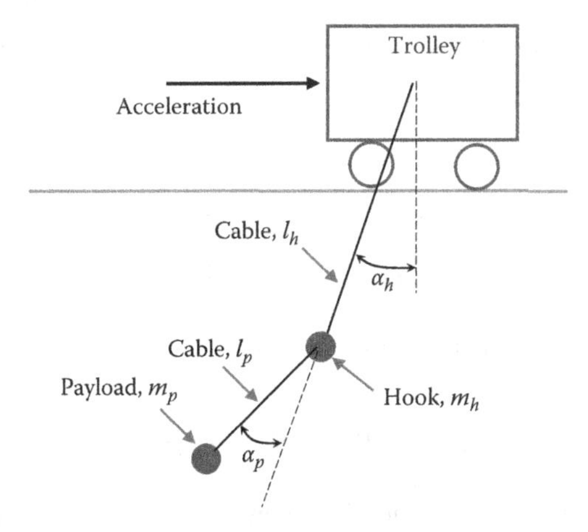

FIGURE 3.7 A planar double-pendulum bridge crane.

The input to the model is the acceleration, $\ddot{y}$, of the trolley in the N_y direction. The outputs are swing angle α_h of the suspension cable to orient the first pendulum, and the swing angle α_p of the rigging cables relative to the suspension cable to orient the second pendulum. The suspension cable and rigging cables are assumed to be massless and inelastic. The hook is assumed to be point mass due to small size. The friction is assumed to be zero because of small contact area among hook, load, and cables. The motions of the trolley are unaffected by payload motion because of small swing and large impedance in the drives.

The Newtonian reference frame $N_xN_yN_z$ can be converted to the moving Cartesian coordinates $S_xS_yS_z$ in the suspension cable by rotating the swing angle, α_h. Then the moving Cartesian coordinates $S_xS_yS_z$ can also be converted to the moving Cartesian coordinates $R_xR_yR_z$ in the rigging cables by rotating the swing angles, α_p. The transformation matrices between the two bodies S (suspension cable) and R (rigging cables) are:

$$\begin{Bmatrix} \mathbf{N}_y \\ \mathbf{N}_z \end{Bmatrix} = \begin{bmatrix} \cos(\alpha_h) & -\sin(\alpha_h) \\ \sin(\alpha_h) & \cos(\alpha_h) \end{bmatrix} \begin{Bmatrix} \mathbf{S}_y \\ \mathbf{S}_z \end{Bmatrix} \tag{3.18}$$

$$\begin{Bmatrix} \mathbf{S}_y \\ \mathbf{S}_z \end{Bmatrix} = \begin{bmatrix} \cos(\alpha_p) & -\sin(\alpha_p) \\ \sin(\alpha_p) & \cos(\alpha_p) \end{bmatrix} \begin{Bmatrix} \mathbf{R}_y \\ \mathbf{R}_z \end{Bmatrix} \tag{3.19}$$

The velocity of the trolley center T^* in the Newtonian reference frame is described as:

$$^N\mathbf{v}^{T^*} = \dot{y}\mathbf{N}_y \tag{3.20}$$

The angular velocity of the suspension cable S in the Newtonian reference frame is given by:

$$^N\boldsymbol{\omega}^S = \dot{\alpha}_h \mathbf{S}_x \tag{3.21}$$

The velocity of the point-mass hook in the Newtonian reference frame is expressed by:

$$^N\mathbf{v}^H = {^N\mathbf{v}^{T^*}} + \dot{\alpha}_h l_h \mathbf{S}_y \tag{3.22}$$

The angular velocity of the load L in the Newtonian reference frame is:

$$^N\boldsymbol{\omega}^L = {^N\boldsymbol{\omega}^S} + \dot{\alpha}_p \mathbf{R}_x \tag{3.23}$$

The velocity of the mass center L^* of the load in the Newtonian reference frame can be written as:

$$^N\mathbf{v}^P = {^N\mathbf{v}^H} - {^N\boldsymbol{\omega}^L} \times l_p \mathbf{R}_z = \dot{y}\mathbf{N}_y + \dot{\alpha}_h l_h \mathbf{S}_y + (\dot{\alpha}_h + \dot{\alpha}_p) l_p \mathbf{R}_y \tag{3.24}$$

where $\times$ denotes the cross product. The generalized speeds u_i defined in Kane's equation are chosen as:

$$u_1 = \dot{\alpha}_h, \quad u_2 = \dot{\alpha}_p \tag{3.25}$$

The corresponding generalized active forces can be expressed by:

$$F_1 = -m_h g l_h \sin(\alpha_h) - m_p g [l_h \sin(\alpha_h) + l_p \sin(\alpha_h + \alpha_p)] \tag{3.26}$$

$$F_2 = -m_p g l_p \sin(\alpha_h + \alpha_p) \tag{3.27}$$

The corresponding generalized inertia forces can be written as:

$$F_1^* = -m_h {}^N\mathbf{a}^H \cdot l_s \mathbf{S}_y - m_p {}^N\mathbf{a}^P \cdot {}^N\mathbf{v}_1^P \tag{3.28}$$

$$F_2^* = -m_h {}^N\mathbf{a}^H \cdot l_s C_{\alpha_x} \mathbf{S}_x - m_p {}^N\mathbf{a}^P \cdot {}^N\mathbf{v}_2^P \tag{3.29}$$

where ${}^N\mathbf{a}^H$ represents the acceleration of the hook and ${}^N\mathbf{a}^P$ denotes the acceleration of the mass center of the load. The derivative of Equation (3.24) with respect to time gives rise to the acceleration ${}^N\mathbf{a}^P$ of the mass center of the load. ${}^N\mathbf{v}_i^P$ is the partial velocity, which derives from the derivative of Equations (3.22) and (3.24) with respect to generalized speeds u_i.

Forcing the sum of the generalized active forces and inertia forces to be zero gives rise to dynamic equations of the motion for the model:

$$\begin{aligned} m_h {}^N\mathbf{a}^H \cdot l_s \mathbf{S}_y + m_p {}^N\mathbf{a}^P \cdot {}^N\mathbf{v}_1^P + m_h g l_h \sin(\alpha_h) \\ + m_p g [l_h \sin(\alpha_h) + l_p \sin(\alpha_h + \alpha_p)] = 0 \end{aligned} \tag{3.30}$$

$$m_h {}^N\mathbf{a}^H \cdot l_s C_{\alpha_x} \mathbf{S}_x + m_p {}^N\mathbf{a}^P \cdot {}^N\mathbf{v}_2^P + m_p g l_p \sin(\alpha_h + \alpha_p) = 0 \tag{3.31}$$

Assuming small oscillations near the equilibria gives rise to a linearized model of double-pendulum cranes. Then linearized equations of motions are given by:

$$\ddot{\alpha}_h = -\frac{g}{l_h}\alpha_h + \frac{gR}{l_h}\alpha_p - \frac{a(t)}{l_h} \tag{3.32}$$

$$\ddot{\alpha}_p = \frac{g}{l_h}\alpha_h - \left(\frac{g}{l_p} + \frac{gR}{l_p} + \frac{gR}{l_h}\right)\alpha_p + \frac{a(t)}{l_h} \tag{3.33}$$

where a is the acceleration of the trolley, R is the ratio of the payload mass to the hook mass, and α_h and α_p are swing angle of the suspension cable and rigging cable, respectively. Then, natural frequencies of the planar double-pendulum bridge crane are given by:

$$\omega_{1,2} = \sqrt{\frac{g}{2}}\sqrt{(1+R)\left(\frac{1}{l_h} + \frac{1}{l_p}\right) \mp \beta} \tag{3.34}$$

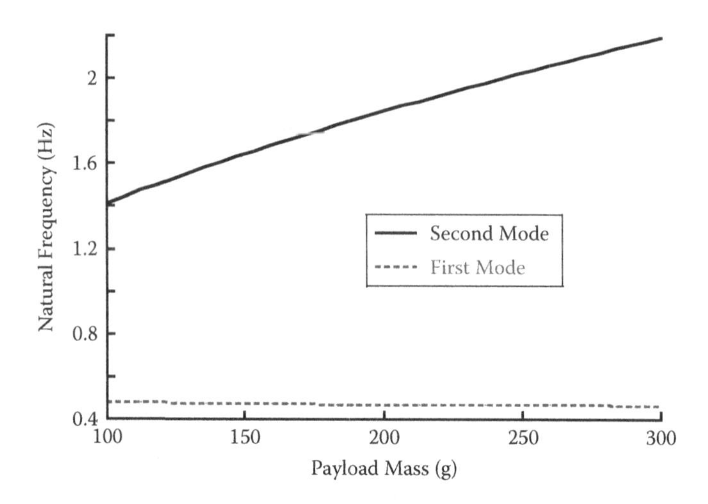

FIGURE 3.8 Variation of first- and second-mode frequencies.

where

$$\beta = \sqrt{(1+R)^2\left(\frac{1}{l_h}+\frac{1}{l_p}\right)^2 - 4\left(\frac{1+R}{l_h l_p}\right)} \tag{3.35}$$

The natural frequencies (3.34) are dependent on the suspension cable length, rigging cable length, and mass ratio. The first- and second-mode frequencies as a function of payload mass ranged from 50 g to 300 g are shown in Figure 3.8 when the suspension length, rigging length, and hook mass were fixed at 60 cm, 60 cm, and 59 g, respectively. The frequency of the first mode ranged from 0.47 Hz to 0.50 Hz, while that of the second mode varied from 1.13 Hz to 2.20 Hz. Such information can be used to design the smoother.

The smoother benefits vibration reduction for double-pendulum cranes. This is because the smoother designed by using the first-mode frequency will also eliminate oscillations at the second-mode frequency. Thus, the second-mode frequency does not need to be estimated. Of course, the first-mode frequency is necessary for the smoother. The damping is assumed to be zero for the double-pendulum crane. When the modeled payload mass is 59 g, a two-pieces smoother for the double-pendulum crane is given by:

$$\text{smoother}(s) = \frac{(1 - 2e^{-2.03s} + e^{-4.06s})}{4.1221s^2} \tag{3.36}$$

A series of experiments were conducted to drive a double-pendulum crane with various mass of the payload. Note that the modeled payload mass was set to 59 g, and the corresponding duration of the two-pieces smoother was 4.06 s. In experiment, the suspension length, rigging length, and hook mass were fixed at 60 cm,

60 cm, and 59 g, respectively. Figure 3.8 shows the residual vibration amplitude of the hook, while that of the payload is shown in Figure 3.10.

The experimental results follow the general shape as the simulated curve. The experimental data were worse than the simulated curve near the modeling payload mass of 59 g (corresponding to the natural frequencies of 0.49 Hz and 1.19 Hz). This is because the complex double-pendulum crane was simplified to a linearized model.

In the case of the payload mass of 300 g, the experimental results shown in Figure 3.10 were better than the simulated curve because the model was undamped, while the actual system had some small amount of damping. The small damping decreases the sensitivity to modeling error. The two-pieces smoother provided a remarkable suppression in payload swing for all values of hook and payload mass as shown in Figures 3.9 and 3.10.

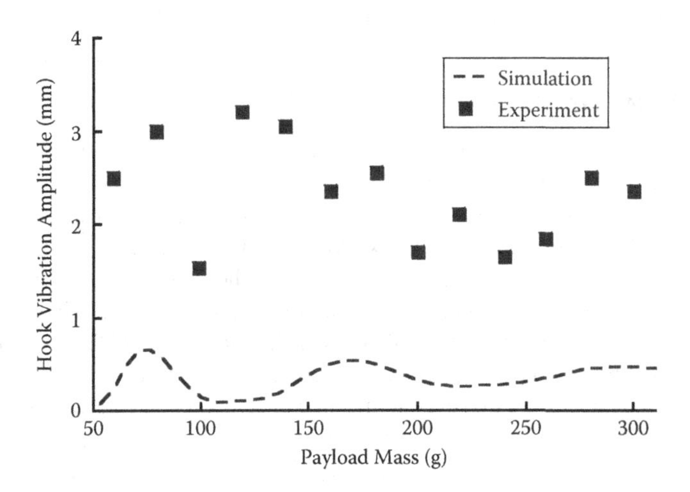

FIGURE 3.9 Hook residual vibrations.

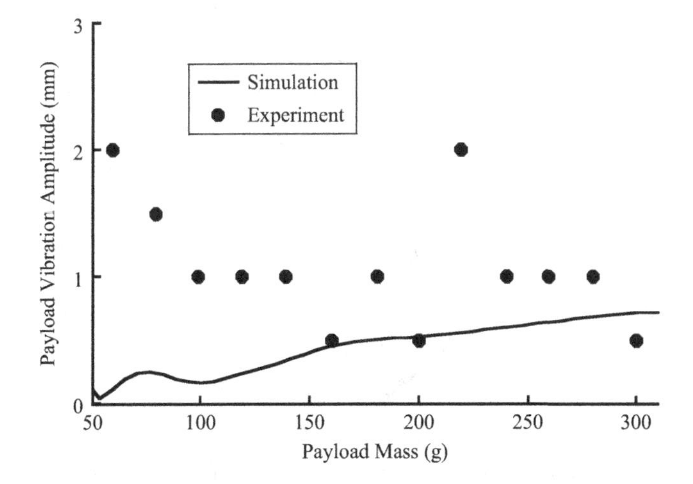

FIGURE 3.10 Payload residual vibrations.

3.3 DISTRIBUTED-MASS PAYLOAD CRANES

Large objects are suspended from the hook by four rigging cables for transporting bulky loads. This configuration of hoisting mechanism can be seen at shipyards and warehouses. In this case, the load can be modeled as distributed-mass payloads. The distributed-mass payload dynamics are more complicated than typically found in point-mass payload dynamics. Difficulty of moving distributed-mass payloads is due to the payload swing toward the driving direction and the payload twisting about rigging cables. Therefore, the challenging factor for controlling payload oscillations is that the payload twisting must be suppressed. However, few attentions have been focused on the control of the payload twisting for industrial cranes. Therefore, dynamics and control of an actual configuration of bridge crane transporting distributed-mass payloads are essential.

3.3.1 Planar Dynamics

A schematic representation of an actual bridge crane transporting a distributed-mass beam is shown in Figure 3.11. A hook of mass, m_h, is attached to the trolley by a massless suspension cable of length, l_h. Two massless rigging cables of length, l_v and l_w, hang below the hook and support a uniformly distributed mass beam of mass, m_p, and length, l_p. The input to the model is the trolley acceleration, a. Outputs are the swing angle of the rigging cable relative to the suspension cable, β, and swing angle of the suspension cable, θ. The motion of the trolley is assumed to be unaffected by payload motions due to the large mechanical impedance in the drive system. The suspension and rigging cable lengths are also assumed to be unchanged during the motion. The damping ratio is assumed to be approximately zero. Using the Kane's method, nonlinear equations of the motion are derived:

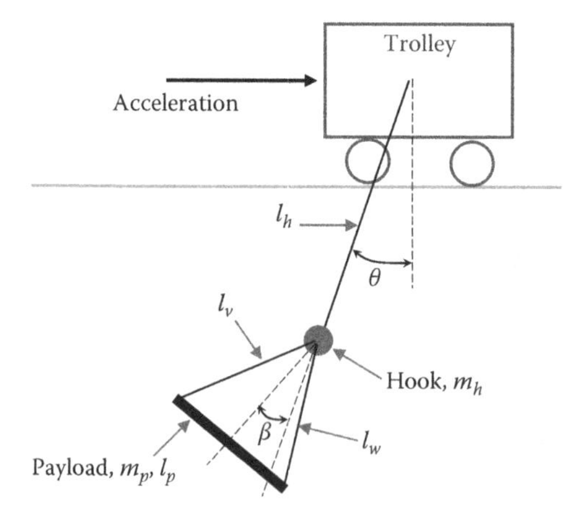

FIGURE 3.11 Model of double-pendulum cranes with distributed-mass beams.

$$R[l_l^2 + l_y(l_y + l_h\cos(\beta))]\ddot{\beta} + [Rl_l^2 + l_h^2 + R(l_y^2 + l_h^2 + 2l_yl_h\cos(\beta))]\ddot{\theta}$$
$$- Rl_yl_h\sin(\beta)\dot{\beta}^2 - 2Rl_yl_h\sin(\beta)\dot{\theta}\dot{\beta} + (gl_h + Rgl_h)\sin(\theta) + Rgl_y\sin(\theta + \beta)$$
$$= -[l_h\cos(\theta) + Rl_h\cos(\theta) + Rl_y\cos(\theta + \beta)]a(t) \tag{3.37}$$

$$[l_l^2 + l_y^2]\ddot{\beta} + [l_l^2 + l_y(l_y + l_h\cos(\beta))]\ddot{\theta} + l_yl_h\sin(\beta)\dot{\theta}^2 + gl_y\sin(\theta + \beta)$$
$$= -[l_y\cos(\theta + \beta)]a(t) \tag{3.38}$$

where

$$l_l = l_p/(2\sqrt{3}) \tag{3.39}$$

$$l_y = \frac{1}{2}\sqrt{2l_v^2 + 2l_w^2 - l_p^2} \tag{3.40}$$

and R is the ratio of the payload mass to the hook mass. It is assumed that the payload swing is small around the equilibrium point, linearized frequencies of this system (3.37) and (3.38) are:

$$\omega_{1,2}^2 = \frac{g(R+1)}{2l_h}(u \mp v) \tag{3.41}$$

where

$$u = \frac{l_y^2 + l_hl_y + l_l^2}{l_y^2 + (R+1)l_l^2} \tag{3.42}$$

$$v = \sqrt{u^2 - \frac{4l_hl_y}{(R+1)(l_y^2 + (R+1)l_l^2)}} \tag{3.43}$$

The natural frequencies are dependent on the mass ratio, suspension length, rigging length, and payload length from Equation (3.41). As the suspension length increases, the first-mode frequency decreases. The payload mass has a small influence on the first-mode frequency. Meanwhile, the suspension length has little effect on the second-mode frequency. Small changes in the mass ratio result in large variations in the second-mode frequency. The second-mode frequency varies more largely than the first-mode frequency.

Both the rigging length and payload length have a relatively small effect on the first-mode frequency because they are commonly shorter than the suspension length in crane applications. However, both the rigging length and payload length have a great influence on the second-mode frequency. The theoretical analyses indicate that the first-mode frequency changes slightly and the second-mode frequency varies sharply. Therefore, the smoother should have more robustness to changes in the second-mode frequency.

Experiments were performed on a double-pendulum bridge crane moving a distributed-mass beam. A slender beam was attached to hook using two rigging cables, and was served as the distributed-mass payload. A CMOS camera was mounted to the trolley to record horizontal displacements of the hook and two red markers on the beam. Averaging the displacement of two red markers yields the position of the beam centroid.

Experimental responses of the hook and payload for a driving distance of 50 cm are shown in Figures 3.12 and 3.13. The hook mass, payload mass, suspension

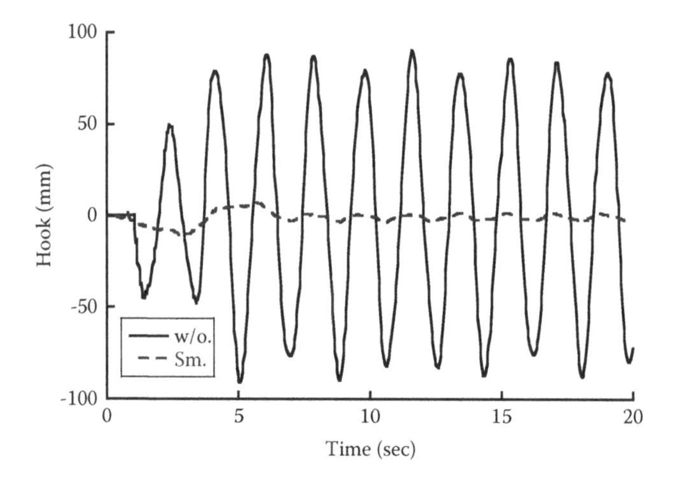

FIGURE 3.12 Experimental responses of hook for 50-cm driving distance.

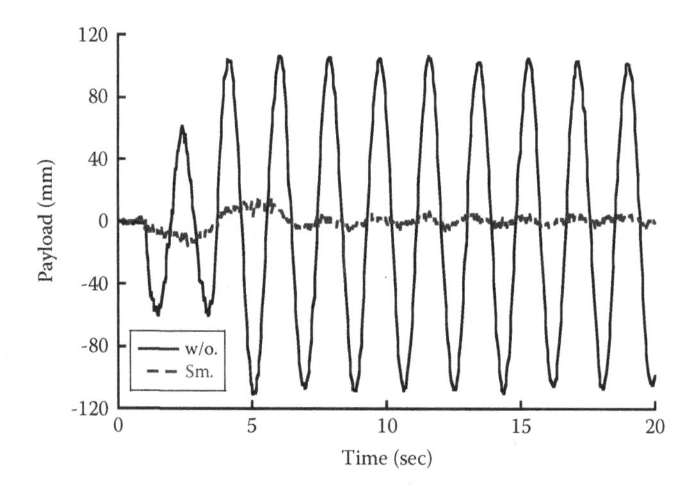

FIGURE 3.13 Experimental responses of payload for 50-cm driving distance.

length, first rigging length, second rigging length, and payload length were held constant at 60 g, 100 g, 70 cm, 30 cm, 30 cm, and 50 cm, respectively. Without the controller, the transient deflection was smaller than the residual amplitude because oscillations caused by the acceleration and deceleration were in phase. The experimental transient deflection and residual amplitude of the payload were 61 mm and 109 mm, respectively. With the two-pieces smoother, the transient deflection and residual amplitude of the payload were 10 mm and 5 mm, respectively. Therefore, the two-pieces smoother dramatically suppresses oscillations of both the hook and payload.

Two sets of experiments were performed to verify the effectiveness of the smoother on suppressing oscillations for variations of system parameters and operation conditions. The first set of experiments examined at the effect of variation in the suspension cable length. The transient and residual amplitudes resulting from these tests are shown in Figures 3.14 and 3.15. The hook mass, payload mass, first rigging length, second rigging length, and payload length were set to be 60 g, 100 g, 30 cm, 30 cm, and 50 cm, respectively. Without the controller, the transient deflection increased with increasing suspension length. A peak in the residual amplitude occurs near the suspension cable length of 65 cm. The experimental results follow the same general shape as the simulated curve. The two-pieces smoother was designed for the modeled suspension length of 90 cm. The two-pieces smoother reduced the transient deflection and residual amplitude by an average of 79.5% and 96.1%, respectively. Therefore, the smoother was robust to changes in the suspension length.

The second set of experiments investigated the effect of variation in the driving distance. The transient deflection and residual amplitude activated for varying driving distances are shown in Figures 3.16 and 3.17. The hook mass, payload mass, suspension length, first rigging length, second rigging length, and payload length were set to be 60 g, 100 g, 70 cm, 30 cm, 30 cm, and 50 cm, respectively. Without the controller, the experimental transient deflection would

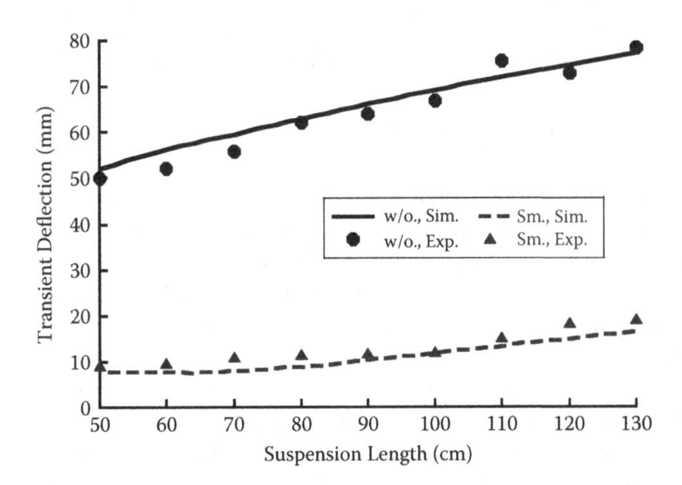

FIGURE 3.14 Transient vibrations for varying suspension length.

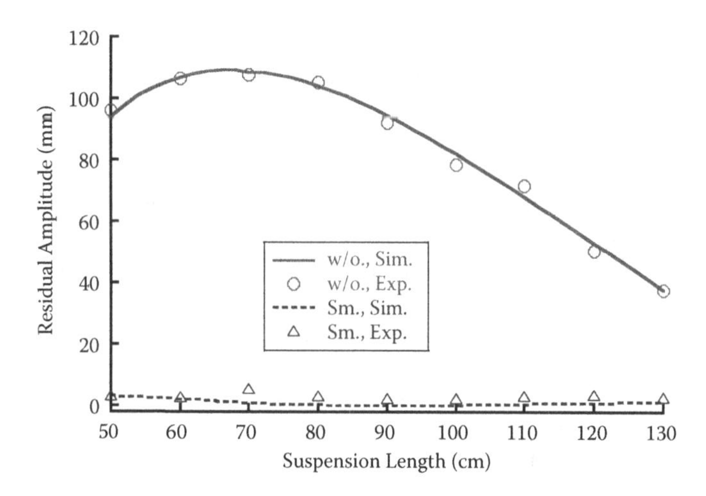

FIGURE 3.15 Residual vibrations for varying suspension length.

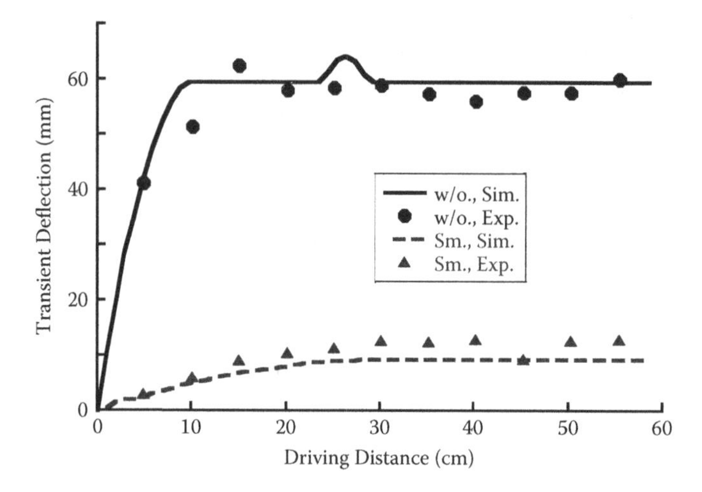

FIGURE 3.16 Transient vibrations induced by driving distance.

increase with increasing the driving distance before 10 cm. After this point, the transient deflection would depend on the interference between the swing caused by the acceleration and deceleration. Once the oscillation amplitude induced by the interference was larger than the swing amplitude caused by the acceleration, a bump would occur in the transient deflection. Peaks and troughs contain in the experimental residual amplitude as the distance varies because the oscillations caused by the acceleration and deceleration are sometimes in phase or sometimes out of phase. The two-pieces smoother reduced the transient deflection and residual amplitude by an average of 81.8% and 94.8%, respectively. Those experiments verified that the smoother can effectively control payload oscillations.

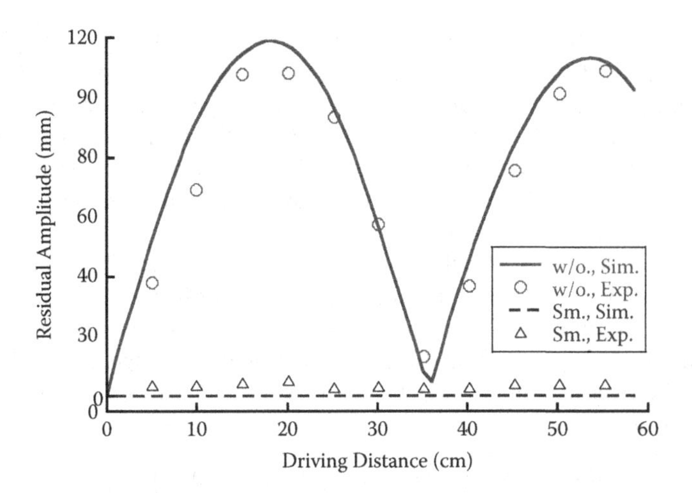

FIGURE 3.17 Residual vibrations induced by driving distance.

3.3.2 Three-Dimensional Dynamics

3.3.2.1 Model

A schematic representation of a bridge crane moving distributed-mass payload is shown in Figure 3.18. The trolley moves along the bridge in the y direction, while the bridge slides along the runway in the x direction. A hook of mass, m_h, is attached to the trolley by a massless suspension cable of length, l_s. Four massless rigging cables of length, l_r, hang below the hook and support a uniformly distributed-mass payload of mass, m_p, length, l_l, width, l_w, and height, l_h. The acceleration of the bridge, a_x, and

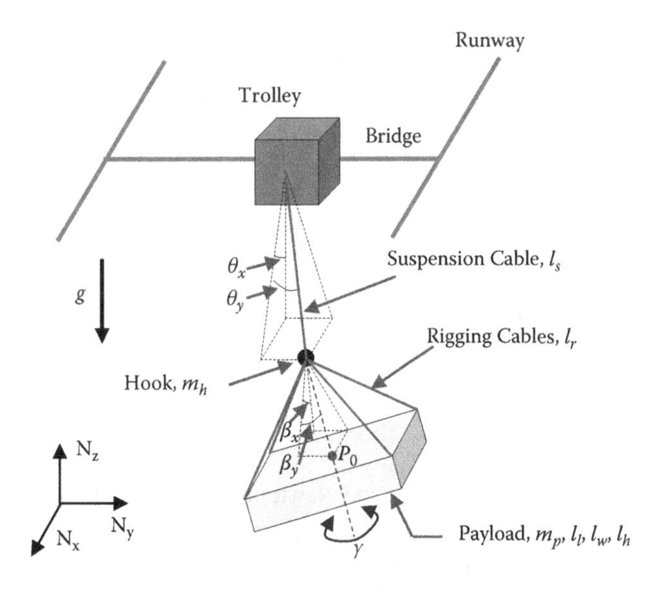

FIGURE 3.18 A bridge crane carrying distributed-mass payload.

the acceleration of the trolley, a_y, are inputs to the model. The swing angles of the suspension cable, θ_x and θ_y, and the payload swing angles relative to the suspension cable, β_x and β_y, and the payload twist angle, γ, are outputs. The model assumes that the trolley is significantly more massive than the hook and payload. The suspension and rigging cables are assumed as massless and rigid. The hook is modeled as a point mass. Equations of the motion for this model can be derived using the Kane method.

The Newtonian reference frame $N_xN_yN_z$ can be converted to the moving Cartesian coordinates $S_xS_yS_z$ in the suspension cable by rotating the swing angles, θ_x and θ_y, respectively. Then the moving Cartesian coordinates $S_xS_yS_z$ can also be converted to the moving Cartesian coordinates $R_xR_yR_z$ in the rigging cables by rotating the swing angles, β_x and β_y, and twist angle, γ, respectively. The transformation matrices among the three bodies S (suspension cable), R (rigging cables), and L (load) are:

$$\begin{Bmatrix} \mathbf{N}_x \\ \mathbf{N}_y \\ \mathbf{N}_z \end{Bmatrix} = \begin{bmatrix} C_{\theta_x} & 0 & S_{\theta_x} \\ 0 & 1 & 0 \\ -S_{\theta_x} & 0 & C_{\theta_x} \end{bmatrix} \begin{bmatrix} 1 & 0 & 0 \\ 0 & C_{\theta_y} & -S_{\theta_y} \\ 0 & S_{\theta_y} & C_{\theta_y} \end{bmatrix} \begin{Bmatrix} \mathbf{S}_x \\ \mathbf{S}_y \\ \mathbf{S}_z \end{Bmatrix} \tag{3.44}$$

$$\begin{Bmatrix} \mathbf{S}_x \\ \mathbf{S}_y \\ \mathbf{S}_z \end{Bmatrix} = \begin{bmatrix} C_{\beta_x} & 0 & S_{\beta_x} \\ 0 & 1 & 0 \\ -S_{\beta_x} & 0 & C_{\beta_x} \end{bmatrix} \begin{bmatrix} 1 & 0 & 0 \\ 0 & C_{\beta_y} & -S_{\beta_y} \\ 0 & S_{\beta_y} & C_{\beta_y} \end{bmatrix} \begin{bmatrix} C_\gamma & -S_\gamma & 0 \\ S_\gamma & C_\gamma & 0 \\ 0 & 0 & 1 \end{bmatrix} \begin{Bmatrix} \mathbf{R}_x \\ \mathbf{R}_y \\ \mathbf{R}_z \end{Bmatrix} \tag{3.45}$$

and the prefixes S_i and C_i throughout the article represent sin(i) and cos(i), respectively. The velocity of the trolley center T^* in the Newtonian reference frame is described as:

$$^N\mathbf{v}^{T^*} = \dot{x}\mathbf{N}_x + \dot{y}\mathbf{N}_y \tag{3.46}$$

The angular velocity of the suspension cable S in the Newtonian reference frame is given by:

$$^N\boldsymbol{\omega}^S = \dot{\theta}_y\mathbf{S}_x - \dot{\theta}_x C_{\theta_y}\mathbf{S}_y + \dot{\theta}_x S_{\theta_y}\mathbf{S}_z \tag{3.47}$$

The velocity of the point-mass hook in the Newtonian reference frame is expressed by:

$$^N\mathbf{v}^H = {}^N\mathbf{v}^{T^*} + \dot{\theta}_x l_s C_{\alpha_y}\mathbf{S}_x - \dot{\theta}_y l_s \mathbf{S}_y \tag{3.48}$$

The angular velocity of the load L in the Newtonian reference frame is:

$$^N\boldsymbol{\omega}^L = {}^N\boldsymbol{\omega}^S + (\dot{\beta}_y C_\gamma + \dot{\beta}_x C_{\beta_y} S_\gamma)\mathbf{R}_x + (\dot{\beta}_x C_{\beta_y} C_\gamma - \dot{\beta}_y S_\gamma)\mathbf{R}_y + (\dot{\gamma} - \dot{\beta}_x S_{\beta_y})\mathbf{R}_z \tag{3.49}$$

The velocity of the mass center L^* of the load in the Newtonian reference frame can be written as:

$$^N\mathbf{v}^{L^*} = {}^N\mathbf{v}^H - {}^N\boldsymbol{\omega}^L \times l_q\mathbf{R}_z \tag{3.50}$$

where $\times$ denotes the cross product and l_q satisfies:

$$l_q = \sqrt{l_r^2 - 0.25l_l^2 - 0.25l_w^2} + 0.5l_h \tag{3.51}$$

The generalized speeds u_i defined in Kane's equation are chosen as:

$$u_1 = \dot{\theta}_x, \quad u_2 = \dot{\theta}_y, \quad u_3 = \dot{\beta}_x, \quad u_4 = \dot{\beta}_y, \quad u_5 = \dot{\gamma} \tag{3.52}$$

The corresponding generalized active forces can be expressed by:

$$F_1 = -m_h g l_s S_{\theta_y} C_{\theta_x} - m_p g \mathbf{N}_z \cdot {}^N\mathbf{v}_1^{L^*} \tag{3.53}$$

$$F_2 = -m_h g l_s C_{\theta_y} S_{\theta_x} - m_p g \mathbf{N}_z \cdot {}^N\mathbf{v}_2^{L^*} \tag{3.54}$$

$$F_3 = -m_p g \mathbf{N}_z \cdot {}^N\mathbf{v}_3^{L^*} \tag{3.55}$$

$$F_4 = -m_p g \mathbf{N}_z \cdot {}^N\mathbf{v}_4^{L^*} \tag{3.56}$$

$$F_5 = 0 \tag{3.57}$$

where $\cdot$ represents the dot product, $^N\mathbf{v}_i^{L^*}$ denotes the ith partial velocity of the mass center of the load. The partial velocity of the mass center of the load $^N\mathbf{v}_i^{L^*}$ derives from the derivative of Equation (3.50) with respect to generalized speeds u_i.

The corresponding generalized inertia forces can be written as:

$$F_1^* = -m_h {}^N\mathbf{a}^H \cdot l_s\mathbf{S}_y - m_p {}^N\mathbf{a}^{L^*} \cdot {}^N\mathbf{v}_1^{L^*} - (\mathbf{I}^L \cdot {}^N\boldsymbol{\alpha}^L + {}^N\boldsymbol{\omega}^L \times \mathbf{I}^L \cdot {}^N\boldsymbol{\omega}^L) \cdot {}^N\boldsymbol{\omega}_1^L \tag{3.58}$$

$$F_2^* = -m_h {}^N\mathbf{a}^H \cdot l_s C_{\theta_y}\mathbf{S}_x - m_p {}^N\mathbf{a}^{L^*} \cdot {}^N\mathbf{v}_2^{L^*} - (\mathbf{I}^L \cdot {}^N\boldsymbol{\alpha}^L + {}^N\boldsymbol{\omega}^L \times \mathbf{I}^L \cdot {}^N\boldsymbol{\omega}^L) \cdot {}^N\boldsymbol{\omega}_2^L \tag{3.59}$$

$$F_3^* = -m_p {}^N\mathbf{a}^{L^*} \cdot {}^N\mathbf{v}_3^{L^*} - (\mathbf{I}^L \cdot {}^N\boldsymbol{\alpha}^L + {}^N\boldsymbol{\omega}^L \times \mathbf{I}^L \cdot {}^N\boldsymbol{\omega}^L) \cdot {}^N\boldsymbol{\omega}_3^L \tag{3.60}$$

$$F_4^* = -m_p {}^N\mathbf{a}^{L^*} \cdot {}^N\mathbf{v}_4^{L^*} - (\mathbf{I}^L \cdot {}^N\boldsymbol{\alpha}^L + {}^N\boldsymbol{\omega}^L \times \mathbf{I}^L \cdot {}^N\boldsymbol{\omega}^L) \cdot {}^N\boldsymbol{\omega}_4^L \tag{3.61}$$

$$F_5^* = -(\mathbf{I}^L \cdot {}^N\boldsymbol{\alpha}^L + {}^N\boldsymbol{\omega}^L \times \mathbf{I}^L \cdot {}^N\boldsymbol{\omega}^L) \cdot {}^N\boldsymbol{\omega}_5^L \tag{3.62}$$

where $^N\mathbf{a}^H$ represents the acceleration of the hook, $^N\mathbf{a}^{L^*}$ denotes the acceleration of the mass center of the load, $^N\boldsymbol{\alpha}^L$ is the angular acceleration of the load, $\mathbf{I}^L$ is the inertia dyadic of the load, and ${}^N\boldsymbol{\omega}_i^L$ represents the ith partial angular velocity of the load. The derivative of Equation (3.50) with respect to time gives rise to the acceleration ${}^N\mathbf{a}^{L^*}$ of the mass center of the load. The angular acceleration ${}^N\boldsymbol{\alpha}^L$ of the load results from the derivative of Equation (3.49) with respect to time. The derivative of Equation (3.49) with respect to generalized speeds u_i yields the partial angular velocity ${}^N\boldsymbol{\omega}_i^L$ of the load.

The inertia dyadic $\mathbf{I}^L$ of the load can be written as:

$$\mathbf{I}^L = I_{xx}\mathbf{L}_x\mathbf{L}_x + I_{yy}\mathbf{L}_y\mathbf{L}_y + I_{zz}\mathbf{L}_z\mathbf{L}_z \tag{3.63}$$

where I_{xx}, I_{yy}, I_{zz} are the mass center principal axis moments of inertia about the $\mathbf{L}_x$, $\mathbf{L}_y$, $\mathbf{L}_z$ axis of the load.

Forcing the sum of the generalized active forces and inertia forces to be zero gives rise to dynamic equations of the motion for the model:

$$\begin{aligned}(\mathbf{I}^L\cdot {}^N\boldsymbol{\alpha}^L + {}^N\boldsymbol{\omega}^L \times \mathbf{I}^L\cdot {}^N\boldsymbol{\omega}^L)\cdot {}^N\boldsymbol{\omega}_1^L + m_h{}^N\mathbf{a}^H\cdot l_s\mathbf{S}_y + m_p{}^N\mathbf{a}^{L^*}\cdot {}^N\mathbf{v}_1^{L^*}\\ + m_pg\mathbf{N}_z\cdot {}^N\mathbf{v}_1^{L^*} + m_hgl_sS_{\theta_y}C_{\theta_x} = 0\end{aligned} \tag{3.64}$$

$$\begin{aligned}(\mathbf{I}^L\cdot {}^N\boldsymbol{\alpha}^L + {}^N\boldsymbol{\omega}^L \times \mathbf{I}^L\cdot {}^N\boldsymbol{\omega}^L)\cdot {}^N\boldsymbol{\omega}_2^L + m_pg\mathbf{N}_z\cdot {}^N\mathbf{v}_2^{L^*} + m_h{}^N\mathbf{a}^H\cdot l_sC_{\alpha_x}\mathbf{S}_x\\ + m_p{}^N\mathbf{a}^{L^*}\cdot {}^N\mathbf{v}_2^{L^*} + m_hgl_sC_{\theta_y}S_{\theta_x} = 0\end{aligned} \tag{3.65}$$

$$(\mathbf{I}^L\cdot {}^N\boldsymbol{\alpha}^L + {}^N\boldsymbol{\omega}^L \times \mathbf{I}^L\cdot {}^N\boldsymbol{\omega}^L)\cdot {}^N\boldsymbol{\omega}_3^L + m_p{}^N\mathbf{a}^{L^*}\cdot {}^N\mathbf{v}_3^{L^*} + m_pg\mathbf{N}_z\cdot {}^N\mathbf{v}_3^{L^*} = 0 \tag{3.66}$$

$$(\mathbf{I}^L\cdot {}^N\boldsymbol{\alpha}^L + {}^N\boldsymbol{\omega}^L \times \mathbf{I}^L\cdot {}^N\boldsymbol{\omega}^L)\cdot {}^N\boldsymbol{\omega}_4^L + m_p{}^N\mathbf{a}^{L^*}\cdot {}^N\mathbf{v}_4^{L^*} + m_pg\mathbf{N}_z\cdot {}^N\mathbf{v}_4^{L^*} = 0 \tag{3.67}$$

$$(\mathbf{I}^L\cdot {}^N\boldsymbol{\alpha}^L + {}^N\boldsymbol{\omega}^L \times \mathbf{I}^L\cdot {}^N\boldsymbol{\omega}^L)\cdot {}^N\boldsymbol{\omega}_5^L = 0 \tag{3.68}$$

Experimental and simulated responses to a set trapezoidal velocity command for driving distance of 55 cm are shown in Figures 3.19 and 3.20. The suspension length, payload length, and payload mass were fixed at 80 cm, 15 cm, and 320 g, respectively. The experimental data follow the same general shape as the simulated curve. The residual amplitude of the payload swing is larger than the transient deflection because the payload swing induced by the deceleration is in phase with that caused by the acceleration. Crane operation will be very challenging when the payload swing and twisting occur. The small-amplitude high-frequency oscillation and the large-amplitude low-frequency oscillation arise in the payload twisting. The small-amplitude high-frequency oscillation of the payload twisting is the only difference between simulated and experimental curves because cables are modeled as massless and rigid, while the actual system has flexible nature of physical structures.

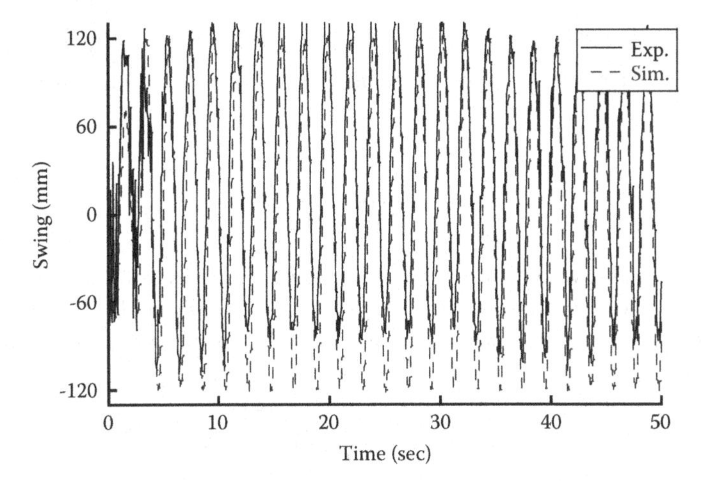

FIGURE 3.19 Experimental responses of swing for payload swing and twisting.

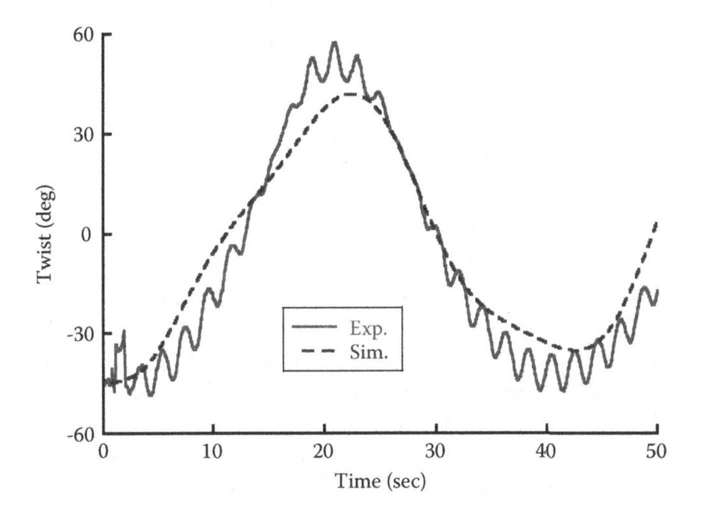

FIGURE 3.20 Experimental responses of twisting for payload swing and twisting.

3.3.2.2 Dynamic Analysis

The payload twisting cannot be excited by accelerations of the bridge and trolley in the case of zero initial angle of the payload twisting. In this specified condition, the model simplifies the dynamic analysis and provides simple estimates of the frequency of the payload swing. Then linearized equations of the motion relating swing angles, θ_x and β_x, to the acceleration of the bridge are given by:

$$(l_p^2 + l_q^2 + l_s l_q)\ddot{\theta}_x + (l_p^2 + l_q^2)\ddot{\beta}_x + g l_q \theta_x + g l_q \beta_x + l_q a_x = 0 \qquad (3.69)$$

$$\begin{aligned}(l_s^2 + Rl_p^2 + Rl_s^2 + Rl_q^2 + 2Rl_s l_q)\ddot{\theta}_x + (Rl_p^2 + Rl_q^2 + Rl_q l_s)\ddot{\beta}_x \\ + g(l_s + Rl_s + Rl_q)\theta_x + gRl_q\beta_x + (Rl_s + Rl_q + l_s)a_x = 0\end{aligned} \tag{3.70}$$

where

$$l_q = \sqrt{l_r^2 - 0.25l_l^2 - 0.25l_w^2} + 0.5l_h \tag{3.71}$$

l_p is the radius of gyration of the payload about the N_y axis through its center of mass P_0, and R is the ratio of the payload mass to the hook mass. Linearized frequencies of the payload swing dynamics modeled in Equations (3.69) and (3.70) are given by:

$$\omega_{1,2} = \sqrt{\frac{g(1 + R)}{2l_s}(u \mp v)} \tag{3.72}$$

where

$$u = \frac{l_q^2 + l_s l_q + l_p^2}{l_q^2 + (R + 1)l_p^2} \tag{3.73}$$

$$v = \sqrt{u^2 - \frac{4l_s l_q}{(R + 1)(l_q^2 + l_p^2 + Rl_p^2)}} \tag{3.74}$$

The first- and second-mode frequencies (3.72) are dependent on the suspension cable length, rigging cable length, mass ratio, and payload size. The second-mode frequency varies more sharply than the first-mode frequency. As a result, the smoother should provide more robustness to variations in the high-mode frequency.

The twisting dynamics of the payload exhibits complicated behavior. The payload swing has a large impact on the frequency of the payload twisting. Increasing the magnitude of the payload swing increases the frequency of the payload twisting. Moreover, the payload size also has some effects on the frequency of the payload twisting. The frequency of the payload twisting is zero when the payload length and width are equal. The frequency of the payload twisting increases with increasing the ratio of the payload length to the payload width.

3.3.2.3 Numerical Verifications

Figures 3.21 and 3.22 show transient and residual amplitudes of the payload swing induced by various suspension length. The payload length and mass were fixed at 15 cm and 320 g, respectively. Without the controller, the transient deflection increases with increasing suspension length. A local maximum in the residual amplitude arises for a suspension length of 50 cm because oscillations caused by the acceleration and deceleration were in phase. The two-pieces smoother was designed when the modeled suspension length was 80 cm. The two-pieces smoother reduced the transient deflection and residual amplitude by an average of 82.2% and 98.9%, respectively.

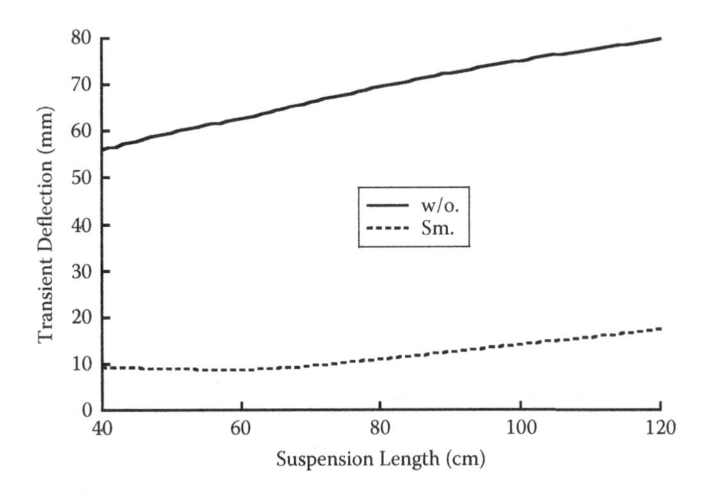

FIGURE 3.21 Transient deflection of swing against suspension cable length.

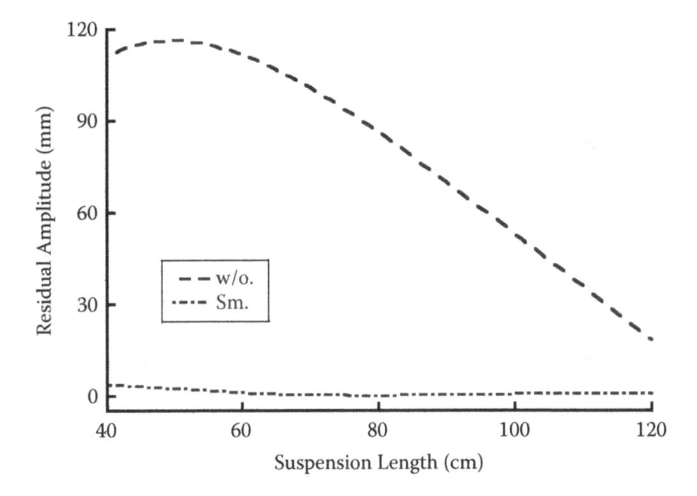

FIGURE 3.22 Residual amplitude of swing against suspension cable length.

Figures 3.23 and 3.24 show transient and residual amplitudes of the payload twisting induced by various suspension length. Without the controller, the transient deflection decreases as the suspension length increases. When the residual amplitude of the payload swing is near its minimum shown in Figure 3.22, a peak in residual amplitude of the payload twisting occurs. This effect can be physically interpreted as the interference between oscillations caused by the acceleration and deceleration. When oscillations caused by the deceleration cancel out oscillations induced by the acceleration, the payload is rotated only by the inertia force caused by transient oscillations of the payload swing. Thus, the spin in one direction causes a sharp increase in the payload twisting. The two-pieces smoother was designed

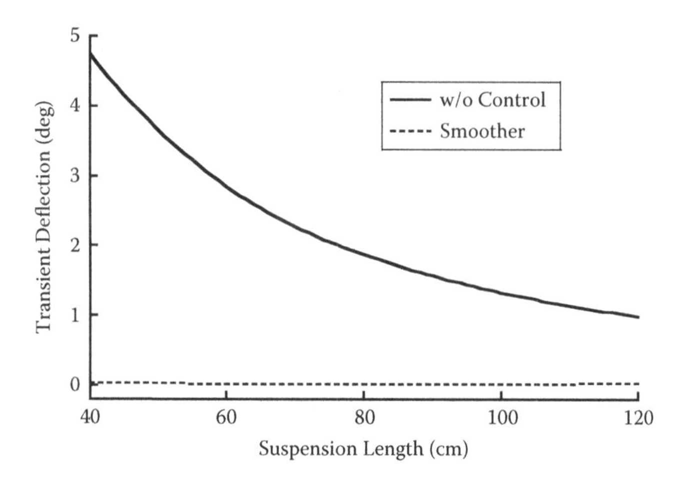

FIGURE 3.23 Transient deflection of twisting against suspension cable length.

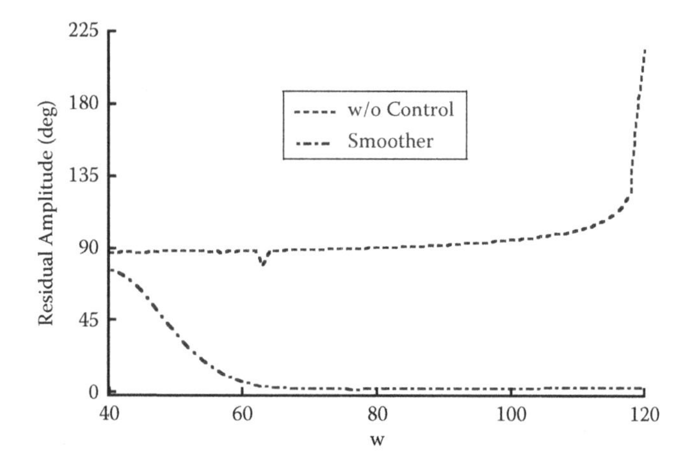

FIGURE 3.24 Residual amplitude of twisting against suspension cable length.

when the modeled suspension length was 80 cm. As the suspension length increases from 40 cm to 60 cm, the controlled residual amplitude of the payload twisting decreases because the payload twisting depends on the size of the payload swing and the duration of the simulation. The two-pieces smoother reduced the transient deflection and residual amplitude by an average of 98.7% and 89.7%, respectively.

The transient deflection and residual amplitude of the payload swing as a function of payload length are shown in Figures 3.25 and 3.26. The suspension length was set to 80 cm. Without the controller, the payload length does not have a large effect on the transient deflection and residual amplitude. The two-pieces smoother was designed when the modeled payload length was set to 15 cm. The two-pieces smoother reduced the transient deflection and residual amplitude by an average of 84.1% and 99.9%, respectively.

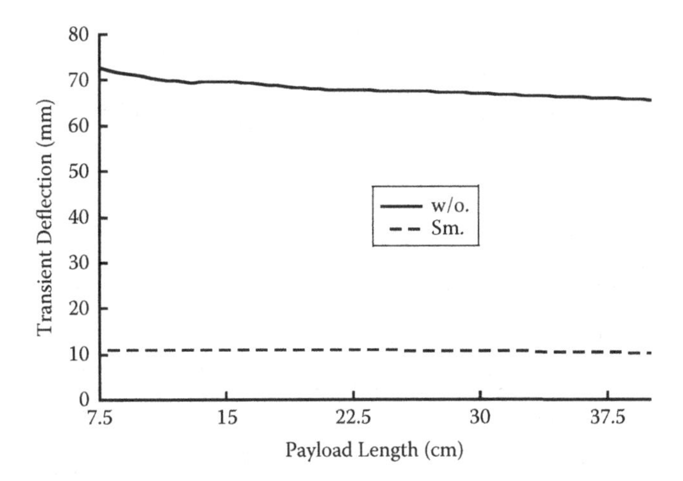

FIGURE 3.25 Transient deflection of swing against payload length.

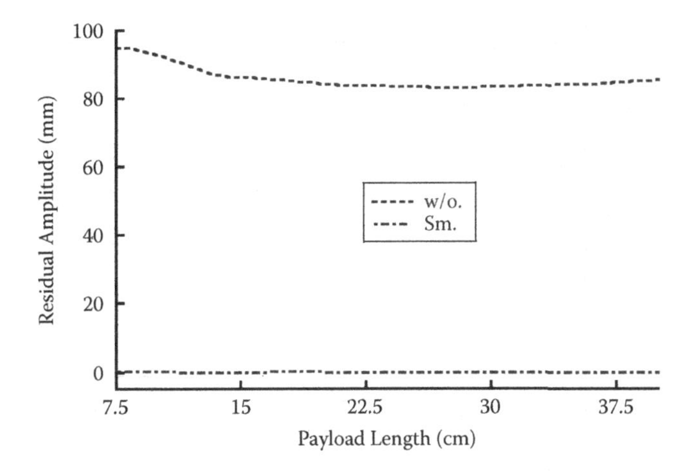

FIGURE 3.26 Residual amplitude of swing against payload length.

The transient deflection and residual amplitude of the payload twisting for various payload lengths are shown in Figures 3.27 and 3.28. Without the controller, when the payload length is set to 7.5 cm, both the transient deflection and the residual amplitude are limited to zero because the payload length and width are equal. After this point, the transient deflection increases as the payload length increases before 17 cm, and then the payload length has small effect on the transient deflection. Meanwhile, the payload size does not have large effects on the residual amplitude of the payload twisting. The two-pieces smoother was designed when the modeled payload length was set to 15 cm. The two-pieces smoother reduced the transient deflection and residual amplitude by an average of 98.2% and 97.2%, respectively. Simulations demonstrated that the smoother can robustly reduce oscillations for a wide range of system parameters.

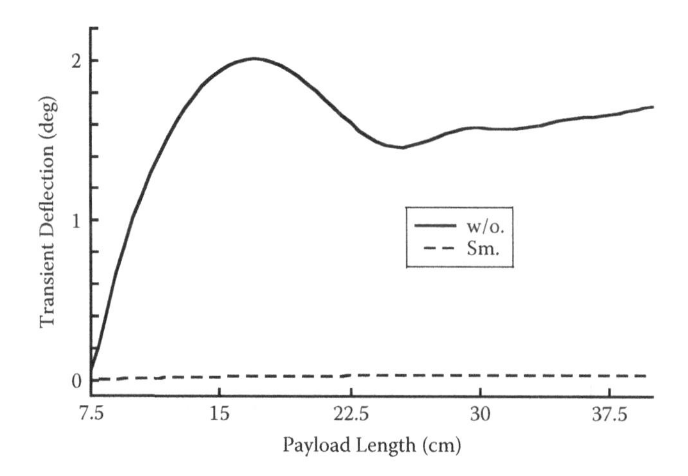

FIGURE 3.27 Transient deflection of twisting against payload length.

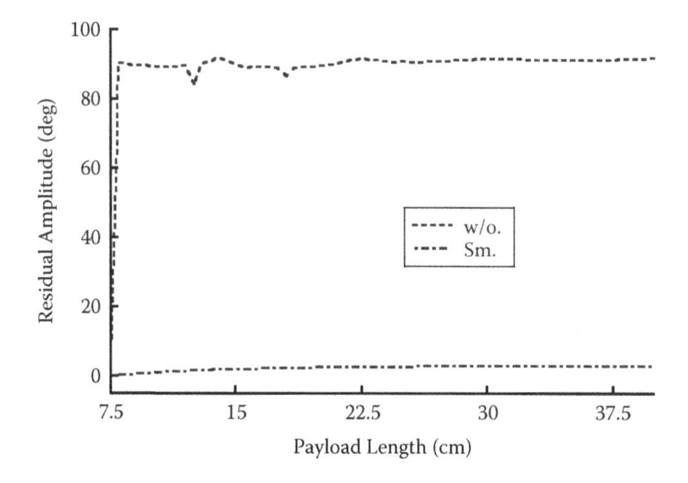

FIGURE 3.28 Residual amplitude of twisting against payload length.

3.3.2.4 Experimental Verifications

Experiments were performed on a bridge crane carrying distributed-mass crates shown in Figure 3.29. A Panasonic AC servomotor with encoder drove the trolley. A digital signal processor (DSP)-based motion control card (Googol GT-400-SV-PCI) connects a personal computer to a servo amplifier. The original command (trapezoidal-velocity profile) is sent to a MATLAB® script in the personal computer, which applies the command smoothing algorithm. A crate was attached to hook using four rigging cables. The payload size is 150 mm × 75 mm × 10 mm, and the payload mass is 320 g. The cables are made of Dyneema super braid fishing line. The displacements of two red markers on the crate were measured by a CMOS camera, which was mounted on the trolley. Note that two red markers were used to calculate the payload swing displacement and the payload twist angle.

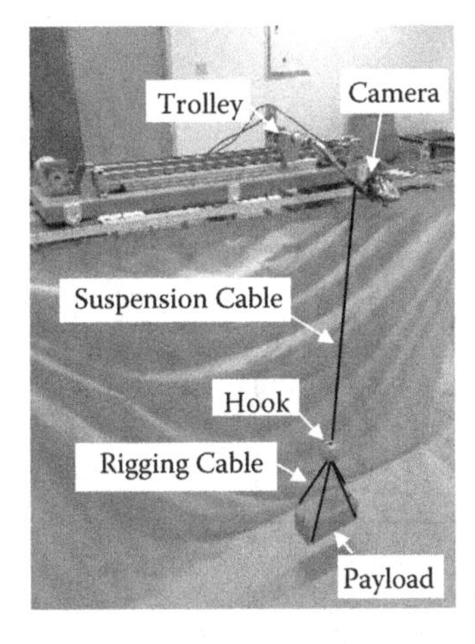

FIGURE 3.29 Bridge crane transporting distributed-mass crates.

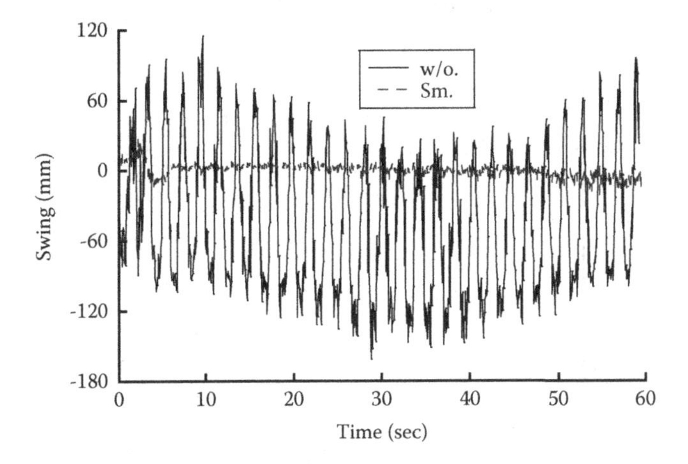

FIGURE 3.30 Experimental responses of swing for 50-cm driving.

Experimental responses of the payload swing and twisting are shown in Figures 3.30 and 3.31. Without the controller, the complex distributed-mass payload dynamics behavior is clearly visible. The two-pieces smoother suppressed oscillations of the payload swing and twisting by an average of 93.4% and 97.8%, respectively. Moreover, the two-pieces smoother reduced both the transient deflection and residual amplitude of the payload swing and twisting. Therefore, the smoother was effective in this case.

One set of experiments was performed to verify the effectiveness of the smoother on suppressing oscillations for variations of system parameters and operation conditions. Variations in the suspension length were selected to exam a wide range of possible dynamics. The residual amplitude of the payload swing and twisting for varying suspension lengths is shown in Figures 3.32 and 3.33, respectively. Without the controller, a peak in the payload swing would occur for the suspension length of 50 cm. The residual amplitude of the payload twisting was

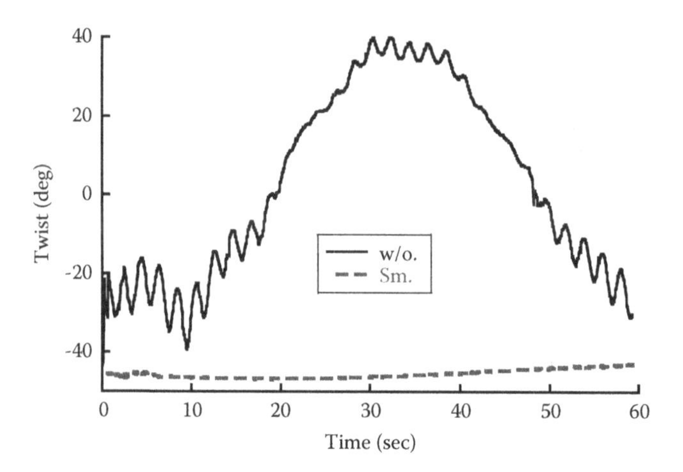

FIGURE 3.31 Experimental responses of twisting for 50-cm driving.

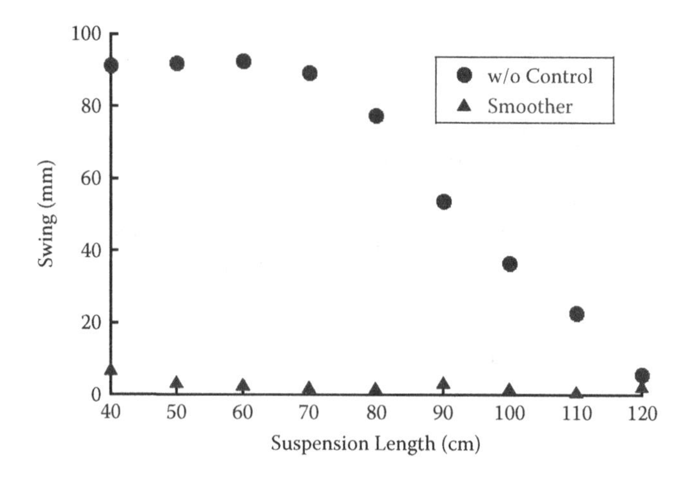

FIGURE 3.32 Experimental residual amplitude of payload swing for varying suspension cable lengths.

approximately 90° for most of suspension length except 120 cm. The experimental data were better than the simulated curve at the suspension length of 120 cm. This is because the model was undamped, while the actual system had some small amount of damping. The damping prevented the inertia force from spinning the payload when oscillations caused by deceleration cancel out oscillations induced by the acceleration. The two-pieces smoother was designed when the modeled suspension length was set to 80 cm. The smoother suppressed experimental residual amplitudes of the payload swing and twisting by an average of 95.2% and 84.9%, respectively. Those experiments proved that the smoother can effectively suppress oscillations of the payload swing and twisting induced by various system parameters as it was predicted by simulations.

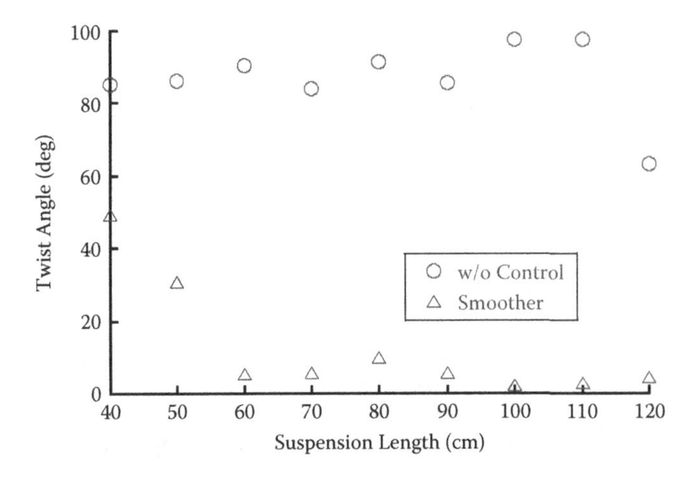

FIGURE 3.33 Experimental residual amplitude of payload twisting for varying suspension cable lengths.

3.4 WIND REJECTION

Many scientists designed feedback controller to damp the payload swing of the crane. However, difficulty of accurately sensing the payload and its velocity is a drawback toward the application of the feedback controller. Additionally, the conflict between the computer-based feedback controller and actions of human operator is also an obstacle. Open-loop control schemes for reducing crane payload oscillations include inverse kinematics, input shaping and command smoothing. However, existing open-loop techniques cannot reject oscillations caused by external wind disturbances. Another method has combined input shaping with feedback control to suppress oscillations caused by both human-operator commands and external disturbances. The payload swing caused by human-operator commands was reduced by input shaping techniques, and that induced by wind disturbances was damped by a low-authority feedback controller.

Accurately sensing payload oscillations is a barrier for the feedback controller, and rejecting the external disturbance is an obstacle toward the application of the previous presented open-loop control scheme. This section will report a new method to control payload oscillations caused by both human-operator commands and wind disturbances without measuring the payload states on-the-fly.

3.4.1 Model

A schematic representation of a bridge crane with distributed-mass beams and wind gusts is shown in Figure 3.34. The trolley moves along the bridge in the x direction. A uniformly distributed-mass beam of mass, m_p, and length, l_p, hangs below the trolley by a massless suspension cable of length, l_h, and two massless rigging cables of length, l_v. A wind force, w_f, acts on the payload in the x direction. The wind travels perpendicular to the initial direction of the edge of the payload length and parallel to the bridge.

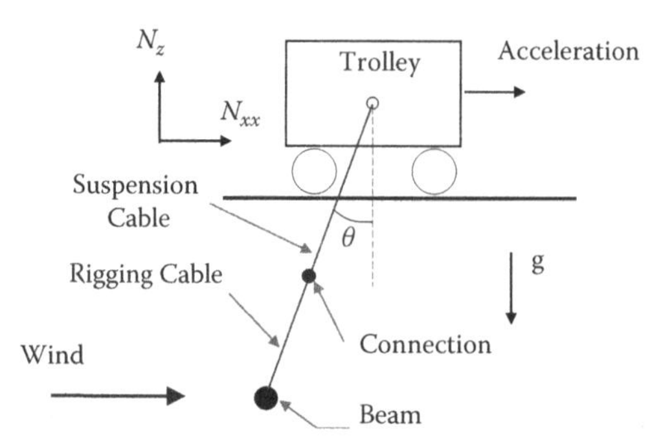

FIGURE 3.34 Model of a bridge crane with distributed-mass beams and wind gusts.

The payload twisting does not occur when the initial direction of the payload length is perpendicular to the wind direction and the bridge. The payload twisting dynamics is very complicated. Thus, the specified direction simplifies the dynamic model. The acceleration of the trolley, a, and the wind force, w_f, are inputs to the model. The swing angle of the suspension cable, θ, is the output. Motions of the payload are assumed to have no effects on the trolley motion due to the large mechanical impedance in the drive system. The damping ratio is approximately zero. The equation of the motion is derived using the Kane's method. The dynamic model relating the swing angle to the trolley acceleration, suspension cable length, and the wind force is:

$$\left(l_h + \sqrt{l_v^2 - 0.25l_p^2}\right) \cdot \ddot{\theta} + 2\dot{l}_h\dot{\theta} + g\sin\theta = \left(a - \frac{w_f}{m_p}\right) \cdot \cos\theta \tag{3.75}$$

Assuming a small angle approximation yields the linearized natural frequency:

$$\omega = \sqrt{g/\left(l_h + \sqrt{l_v^2 - 0.25l_p^2}\right)} \tag{3.76}$$

Thus, the natural frequency is dependent on the suspension cable length, rigging cable length, and payload length. It is clear that the suspension cable length is fundamental, and the payload length also has some effects on the frequency.

Simulated payload oscillations resulting from both a 1.4-m driving motion and a 0.2 N wind disturbance are shown in Figure 3.35. The payload mass, suspension length, rigging length, and payload length were fixed at 0.3 kg, 1.0 m, 0.5 m, and 0.5 m, respectively. The wind gust caused payload oscillations at 0 s, and then the crane accelerated at 4.5 s, which induced another payload oscillation. When the crane decelerated later, additional oscillations were induced.

The system response includes three stages. The time frame before the trolley moves, in which wind gusts cause the payload swing, is defined as the first stage. The time frame when the trolley is in motion is defined as the transient stage. The time frame after the trolley is stopped is defined as the residual stage.

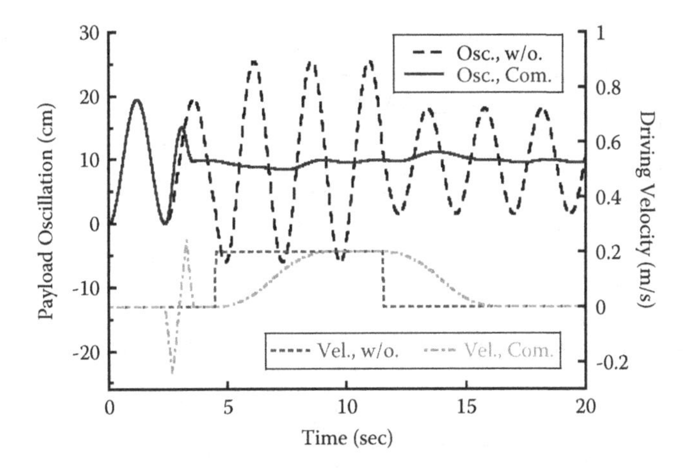

FIGURE 3.35 Oscillations resulting from driving motion and wind disturbance.

The peak-to-peak deflection during the transient stage is referred to as the transient deflection, while the peak-to-peak deflection during the residual stage is defined as the residual amplitude.

In Figure 3.35, oscillations caused by the wind are in phase with that caused by the acceleration. Therefore, oscillations become large during the transient stage. Additionally, oscillations caused by the deceleration and that during the transient stage are out of phase. Thus, oscillations during the residual stage decrease. The transient deflection and residual amplitude caused by both the wind and the original human-operator command were 31.5 cm and 16.4 cm, respectively. The simulated results indicate that the addition of wind disturbances makes the dynamics complicated and operating crane becomes challenging.

3.4.2 Wind Rejection

A combined control technique is designed to control oscillations caused by wind disturbances and human-operator commands. Oscillations induced by wind disturbances are rejected by a wind-rejection command, and the payload swing generated by human-operator commands is eliminated by a smoother in Chapter 2. The combined control scheme is shown in Figure 3.36. The operator produces an original command, c_r, via the control interface. Then the original command (trapezoidal-velocity profile) filters through the smoother to create a smoothed command, c_s. A wind-rejection command, c_w, and the smoothed command, c_s, work together to move the trolley toward the desired position without inducing payload oscillations. The payload equilibrium angle, U_m, was utilized to design the wind-rejection command. Moreover, the equilibrium angle can be easily estimated by the operator with his/her eyes.

Oscillations around the equilibrium angle are assumed to be small for the crane dynamics shown in Equation (3.75). During the crane motion, the suspension length does not change. Then, Equation (3.75) can be simplified as:

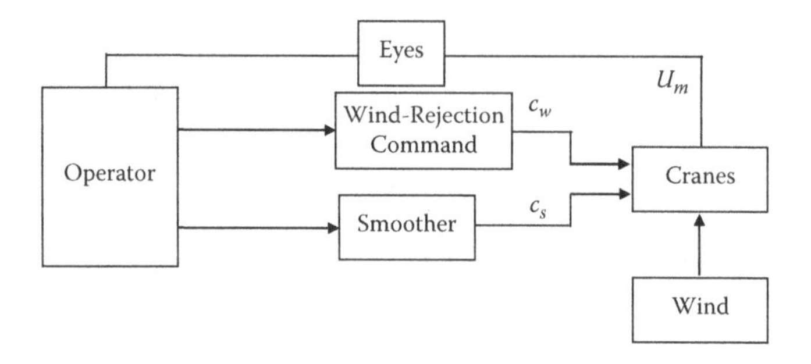

FIGURE 3.36 Combined control scheme.

$$\ddot{\theta} + \omega^2\theta = \omega^2\left(\frac{a}{g} - \frac{w_f}{m_p g}\right) \tag{3.77}$$

Therefore, Equation (3.77) can be considered as a simplified model for a bridge crane with distributed-mass beams and wind gusts. The new equilibrium angle for this simplified model is given by:

$$U = \frac{w_f}{m_p g} \tag{3.78}$$

The design of the wind-rejection command is very challenging for the crane dynamics shown in Equation (3.75). Therefore, a disturbance-rejection command was obtained for a second-order system subject to external step disturbance. Then, the disturbance-rejection command was refined as a wind-rejection command for the bridge crane with distributed-mass beams and wind gusts. The resulting wind-rejection command will be used to test its performance on a crane dynamics shown in Equation (3.75).

The harmonic response of the second-order system from the disturbance-rejection command, c, and the external step disturbance, D, is given by:

$$\begin{aligned} f(t) \;=\; & \int_{\tau=0}^{+\infty} D\frac{\omega}{\sqrt{1-\zeta^2}}e^{-\zeta\omega(t-\tau)}\sin(\omega\sqrt{1-\zeta^2}(t-\tau))d\tau \\ & + \int_{\tau=0}^{+\infty} c(\tau)\frac{\omega}{\sqrt{1-\zeta^2}}e^{-\zeta\omega(t-\tau)}\sin(\omega\sqrt{1-\zeta^2}(t-\tau))d\tau \end{aligned} \tag{3.79}$$

where ω is undamped natural frequency of the second-order system, and ζ is the corresponding damping ratio. The vibration amplitude of response (3.79) is:

$$A(t) = \frac{\omega}{\sqrt{1-\zeta^2}}e^{-\zeta\omega t}\sqrt{[S(\zeta,\omega,D)]^2 + [C(\zeta,\omega,D)]^2} \tag{3.80}$$

where

$$S(\zeta,\omega,D) = \int_{\tau=0}^{+\infty} c(\tau)e^{\zeta\omega\tau}\cos(\omega\sqrt{1-\zeta^2}\tau)d\tau - \zeta D/\omega \tag{3.81}$$

$$C(\zeta, \omega, D) = \int_{\tau=0}^{+\infty} c(\tau)e^{\zeta\omega\tau} \sin(\omega\sqrt{1-\zeta^2}\tau)d\tau + \sqrt{1-\zeta^2}D/\omega \quad (3.82)$$

The disturbance-rejection command, c, could cancel oscillations resulting from the external disturbance when Equations (3.81) and (3.82) are limited to zero. A constraint must be satisfied to ensure that the velocity of the system can be limited to zero after the disturbance-rejection command:

$$\int_{\tau=0}^{\infty} c(\tau)d\tau = 0 \quad (3.83)$$

Limiting equations (3.81) and (3.82) to zero and resulting from constraint (3.83) yield a disturbance-rejection command:

$$c(\tau) = \begin{cases} A_1 e^{-\zeta_m\omega_m\tau}, & \Delta \le \tau < (\Delta + T_m/8) \\ -A_2 e^{-\zeta_m\omega_m\tau}, & (\Delta + T_m/8) \le \tau < (\Delta + T_m/4) \\ -A_1 e^{-\zeta_m\omega_m(\tau - T_m/4)}, & (\Delta + T_m/4) \le \tau < (\Delta + 3T_m/8) \\ A_2 e^{-\zeta_m\omega_m(\tau - T_m/4)}, & (\Delta + 3T_m/8) \le \tau \le (\Delta + T_m/2) \end{cases} \quad (3.84)$$

where

$$A_1 = \frac{D_m\sqrt{1-\zeta_m^2}\{[\zeta_m + (\sqrt{2}-1)\sqrt{1-\zeta_m^2}] - [(\sqrt{2}-1)\zeta_m - \sqrt{1-\zeta_m^2}]e^{\zeta_m\pi/\sqrt{1-\zeta_m^2}}\}}{(2-\sqrt{2})(e^{\zeta_m\pi/\sqrt{1-\zeta_m^2}}+1)} \quad (3.85)$$

$$A_2 = \frac{D_m\sqrt{1-\zeta_m^2}\{[(\sqrt{2}-1)\zeta_m + \sqrt{1-\zeta_m^2}] - [\zeta_m - (\sqrt{2}-1)\sqrt{1-\zeta_m^2}]e^{\zeta_m\pi/\sqrt{1-\zeta_m^2}}\}}{(2-\sqrt{2})(e^{\zeta_m\pi/\sqrt{1-\zeta_m^2}}+1)} \quad (3.86)$$

$$\Delta = nT_m \quad n = 0, 1, 2\cdots \quad (3.87)$$

where D_m is the modeled external disturbance and Δ is the initial time of the disturbance-rejection command. When the modeled damping ratio, ζ_m, modeled natural frequency, ω_m, modeled damped oscillation period, T_m, modeled external disturbance, and the initial time of the disturbance-rejection command are correct, oscillations induced by the external disturbance will be limited to zero using the disturbance-rejection command, c. The duration of the disturbance-rejection command is one half of the damped oscillation period.

Disturbance-rejection command (3.84) and simplified dynamics (3.77) and (3.78) yield a wind-rejection command:

$$c_w(\tau) = \begin{cases} -(\sqrt{2}+1)gU_m/2, & \Delta \le \tau < (\Delta + T_m/8) \\ (\sqrt{2}+1)gU_m/2, & (\Delta + T_m/8) \le \tau < (\Delta + 3T_m/8) \\ -(\sqrt{2}+1)gU_m/2, & (\Delta + 3T_m/8) \le \tau < (\Delta + T_m/2) \end{cases} \quad (3.88)$$

where U_m is the equilibrium angle for the bridge crane with distributed-mass beams and wind gusts.

The wind-rejection command is dependent on the modeled natural frequency, ω_m, equilibrium angle, U_m, and the initial time, Δ. However, a great wind force might cause actuator saturation. The wind force is assumed to be small such that the magnitude of the wind-rejection command is within the bound of the actuator. When operators have the visibility of the payload, both the equilibrium angle, U_m, and the initial time, Δ, can be estimated by eyes. The maximum deflection of the payload can be easily estimated by the human-operator with his/her eyes. Then half of the maximum deflection is corresponding to the equilibrium position of the payload. Meanwhile, the time for zero payload deflection is corresponding to the initial time, Δ. Therefore, both the initial time, Δ, and the equilibrium angle, U_m, do not need to be estimated on-the-fly.

Oscillations resulting from both wind disturbances and human-operator commands could be eliminated under the action of the wind-rejection command, c_w, and the smoothed command, c_s. Figure 3.35 also shows the effectiveness of the combined control scheme. The combined command drives the trolley inducing minimal vibrations. With the combined command, the transient deflection and residual amplitude were suppressed to 15.0 cm and 0.4 cm, respectively. The simulated process clearly verifies the effectiveness of the combined control scheme.

3.4.3 Experimental Results

Two sets of experiments were performed to verify the effectiveness of the presented method on suppressing payload oscillations for variations of system parameters. The first set of experiments investigated the effect of variation in the wind forces. The payload length and mass were 0.4 m and 0.182 kg, respectively. Adjusting the distance between the fan and payload varies the wind force on the payload. As the distance increased, the equilibrium position of the payload and wind force decreased. The distances were set to 0.5, 0.6, 0.7, 0.8, and 0.9 m. The estimates of the new equilibrium position corresponding to the distances were approximately 10.23 cm, 9.80 cm, 9.33 cm, 8.80 cm, and 8.16 cm, respectively. According to the equilibrium position equation, corresponding wind forces were 0.145 N, 0.139 N, 0.132 N, 0.125 N, and 0.116 N, respectively. The driving distance for the desired task was fixed at 0.8 m. The combined controller was designed when the modeled wind force was set to 0.132 N.

Experimental transient deflection and residual amplitude from those tests are shown in Figures 3.37 and 3.38. Without the controller, increasing wind force increased the transient deflection and residual amplitude. Experimental results match the general trend as the simulated curve. With the combined controller, reduction in the transient deflection and residual amplitude was remarkable. The experimental

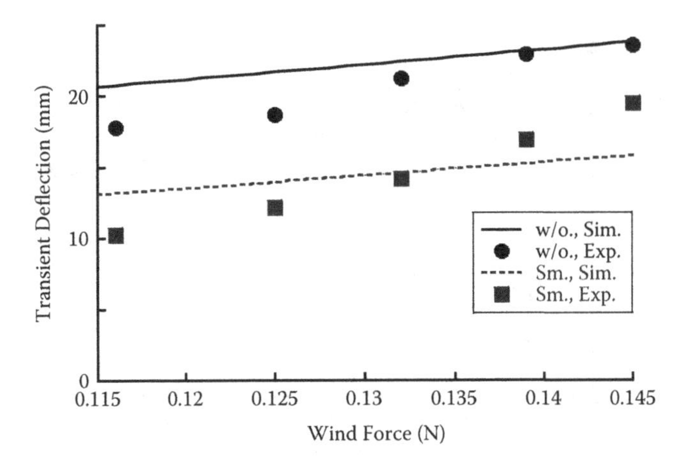

FIGURE 3.37 Experimental transient amplitudes induced by wind force.

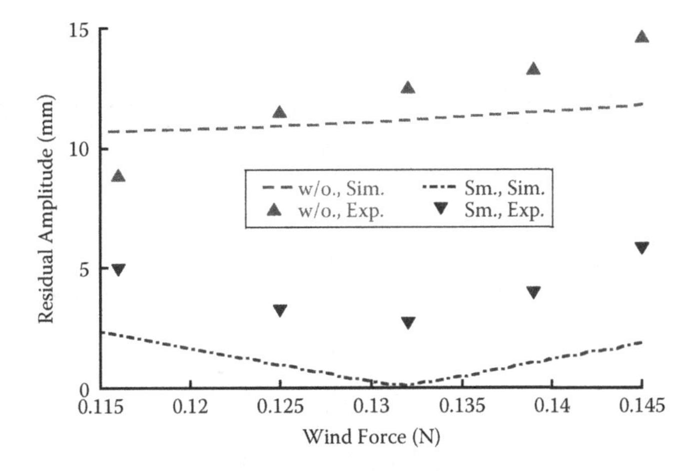

FIGURE 3.38 Experimental residual amplitudes induced by wind force.

residual amplitude was not eliminated to zero at the modeled wind force. This is because the actual system had estimation error of the initial time, while the initial time of the wind-rejection command in the simulation was correct. Increasing modeling errors decreased the performance of vibration reduction. The combined controller reduced the transient deflection by an average of 29.9%, and residual amplitude by an average of 66.3%. Therefore, the combined controller was effective at each of tested wind forces.

Another set of experiments was conducted to reduce payload swing from varying payload lengths as shown in Figures 3.39 and 3.40. Without the controller, the experimental transient deflection and residual amplitude decreased with increasing payload length. The combined control arithmetic was applied when the modeled payload length was set to 0.4 m. Both the transient deflection and residual amplitude

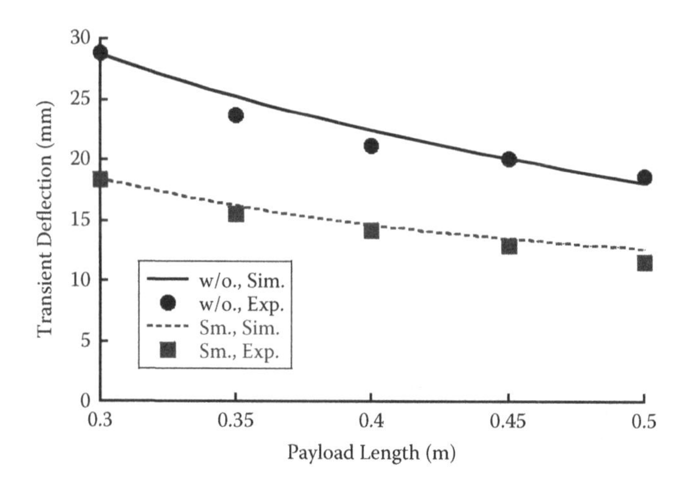

FIGURE 3.39 Experimental transient amplitudes induced by payload length.

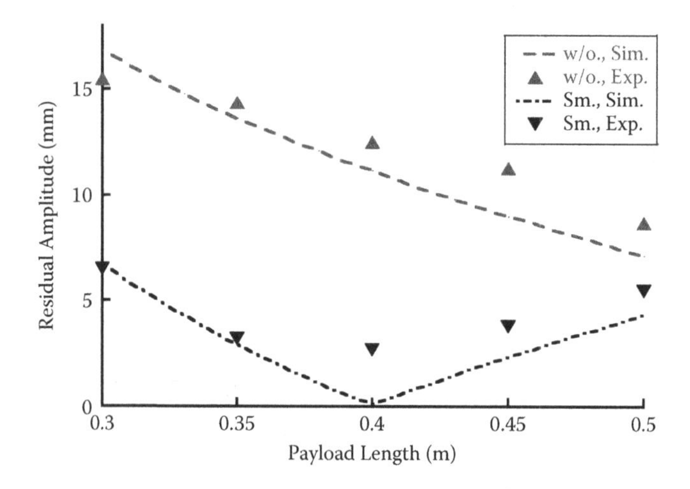

FIGURE 3.40 Experimental residual amplitudes induced by payload length.

had a dramatical reduction. The residual amplitude at the modeled payload length was not attenuated to zero because of estimation errors of the initial time of the wind-rejection command. The combined controller reduced the transient deflection by an average of 35.4%, and the residual amplitude by an average of 65.2%. Those experiments clearly verified that the combined control technique can effectively suppress oscillations induced by both wind disturbances and human-operator commands.

3.5 ECCENTRIC-LOAD DYNAMICS

The eccentric-load dynamics are more complicated than the dynamics of the point-mass load and uniformly distributed-mass load. The eccentric distance between mass center and centroid causes very complex dynamic behavior in the heterogeneous or non-uniform loads. In the case of the point-mass or uniform loads, the eccentric distance between mass center and centroid is zero. Furthermore, natural frequencies of the swing and twisting of the eccentric load are different from those of the point-mass and uniform loads. Although heterogeneous or non-uniform loads are widely transported by industrial cranes, no effects have been focused on eccentric-load dynamics.

3.5.1 Modeling

Figure 3.41 shows a schematic representation of a bridge crane transporting a heterogeneous or non-uniform load. A trolley slides along the bridge in the N_y direction and the bridge moves along the runway in the N_x direction. A hook of mass, m_h, is attached to the trolley center T^* by a suspension cable S of length, l_s. Multiple rigging cables R hang below the hook and connect to a heterogeneous or non-uniform load L. The mass center of the load is denoted by L^*, and the centroid of the load is denoted by $L^{\#}$. The distance between mass center, L^*, and centroid, $L^{\#}$, of the load is l_e. Meanwhile, the distance from the hook to the mass center, L^*, of the load is l_x.

The inputs to the model are the acceleration, $\ddot{y}$, of the trolley in the N_y direction and the acceleration, $\ddot{x}$, of the bridge in the N_x direction. The outputs are two swing angles α_x and α_y of the suspension cable to orient the first pendulum, two swing angles β_x and β_y of the rigging cables relative to the suspension cable to orient the second pendulum, and one twist angle γ of the load about the rigging cables. The suspension cable and rigging cables are assumed to be massless and inelastic. The hook is assumed to be point mass due to small size. The friction is assumed to be zero because of small contact area among hook, load, and cables. The motions of the trolley and bridge are unaffected by load motion because of small swing and large impedance in the drives.

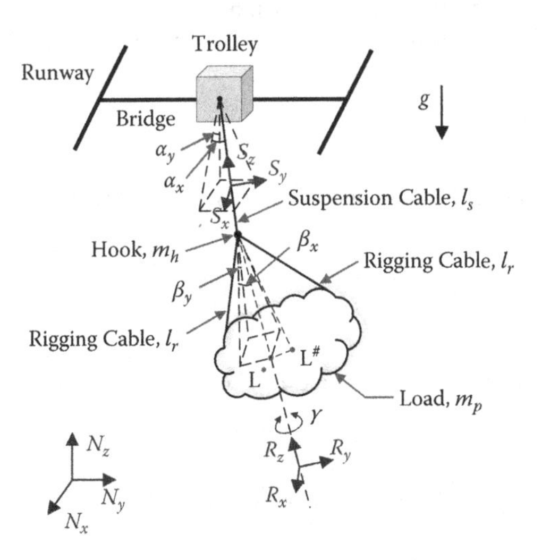

FIGURE 3.41 Model of a bridge crane transporting a heterogeneous or non-uniform load.

The Newtonian reference frame $N_xN_yN_z$ can be converted to the moving Cartesian coordinates $S_xS_yS_z$ in the suspension cable by rotating the swing angles, α_y and α_x, respectively. Then the moving Cartesian coordinates $S_xS_yS_z$ can also be converted to the moving Cartesian coordinates $R_xR_yR_z$ in the rigging cables by rotating the swing angles, β_y and β_x, and twist angle, γ, respectively. The transformation matrices among the three bodies S (suspension cable), R (rigging cables), and L (load) are:

$$\begin{Bmatrix} \mathbf{N}_x \\ \mathbf{N}_y \\ \mathbf{N}_z \end{Bmatrix} = \begin{bmatrix} C_{\alpha_y} & 0 & -S_{\alpha_y} \\ 0 & 1 & 0 \\ S_{\alpha_y} & 0 & C_{\alpha_y} \end{bmatrix} \begin{bmatrix} 1 & 0 & 0 \\ 0 & C_{\alpha_x} & -S_{\alpha_x} \\ 0 & S_{\alpha_x} & C_{\alpha_x} \end{bmatrix} \begin{Bmatrix} \mathbf{S}_x \\ \mathbf{S}_y \\ \mathbf{S}_z \end{Bmatrix} \tag{3.89}$$

$$\begin{Bmatrix} \mathbf{S}_x \\ \mathbf{S}_y \\ \mathbf{S}_z \end{Bmatrix} = \begin{bmatrix} C_{\beta_y} & 0 & -S_{\beta_y} \\ 0 & 1 & 0 \\ S_{\beta_y} & 0 & C_{\beta_y} \end{bmatrix} \begin{bmatrix} 1 & 0 & 0 \\ 0 & C_{\beta_x} & -S_{\beta_x} \\ 0 & S_{\beta_x} & C_{\beta_x} \end{bmatrix} \begin{bmatrix} C_\gamma & -S_\gamma & 0 \\ S_\gamma & C_\gamma & 0 \\ 0 & 0 & 1 \end{bmatrix} \begin{Bmatrix} \mathbf{R}_x \\ \mathbf{R}_y \\ \mathbf{R}_z \end{Bmatrix} \tag{3.90}$$

$$\begin{Bmatrix} \mathbf{R}_x \\ \mathbf{R}_y \\ \mathbf{R}_z \end{Bmatrix} = \begin{bmatrix} 1 & 0 & 0 \\ 0 & C_\delta & -S_\delta \\ 0 & S_\delta & C_\delta \end{bmatrix} \begin{Bmatrix} \mathbf{L}_x \\ \mathbf{L}_y \\ \mathbf{L}_z \end{Bmatrix} \tag{3.91}$$

where

$$\delta = \arcsin(l_e/l_x) \tag{3.92}$$

and the prefixes S_i and C_i throughout the article represent sin(i) and cos(i), respectively. Rotating a constant angle δ will convert the moving Cartesian coordinates $R_xR_yR_z$ to the moving Cartesian coordinates $L_xL_yL_z$ in the load. Note that the angle δ is created by the eccentric distance, l_e.

The velocity of the trolley center T^* in the Newtonian reference frame is described as:

$${}^N\mathbf{v}^{T^*} = \dot{x}\mathbf{N}_x + \dot{y}\mathbf{N}_y \tag{3.93}$$

The angular velocity of the suspension cable S in the Newtonian reference frame is given by:

$${}^N\boldsymbol{\omega}^S = \dot{\alpha}_x\mathbf{S}_x - \dot{\alpha}_y C_{\alpha_x}\mathbf{S}_y + \dot{\alpha}_y S_{\alpha_x}\mathbf{S}_z \tag{3.94}$$

The velocity of the point-mass hook in the Newtonian reference frame is expressed by:

$${}^N\mathbf{v}^H = {}^N\mathbf{v}^{T^*} + \dot{\alpha}_y l_s C_{\alpha_x}\mathbf{S}_x + \dot{\alpha}_x l_s\mathbf{S}_y \tag{3.95}$$

The angular velocity of the load L in the Newtonian reference frame is:

$$^N\boldsymbol{\omega}^L = {}^N\boldsymbol{\omega}^S + (\dot{\beta}_x C_\gamma + \dot{\beta}_y C_{\beta_x} S_\gamma)\mathbf{R}_x + (\dot{\beta}_y C_{\beta_x} C_\gamma - \dot{\beta}_x S_\gamma)\mathbf{R}_y + (\dot{\gamma} - \dot{\beta}_y S_{\beta_x})\mathbf{R}_z \tag{3.96}$$

The velocity of the centroid $L^\#$ of the load in the Newtonian reference frame can be described as:

$$^N\mathbf{v}^{L^\#} = {}^N\mathbf{v}^H - {}^N\boldsymbol{\omega}^L \times l_x C_\delta \mathbf{R}_z \tag{3.97}$$

where × denotes the cross product in this chapter.

The generalized speeds u_i defined in Kane's equation are chosen as:

$$u_1 = \dot{\alpha}_x, \quad u_2 = \dot{\alpha}_y, \quad u_3 = \dot{\beta}_x, \quad u_4 = \dot{\alpha}_y, \quad u_5 = \dot{\gamma} \tag{3.98}$$

The corresponding generalized active forces can be expressed by:

$$F_1 = -m_h g l_s S_{\alpha_x} C_{\alpha_y} - m_p g \mathbf{N}_z \cdot {}^N\mathbf{v}_1^{L^*} \tag{3.99}$$

$$F_2 = -m_h g l_s C_{\alpha_x} S_{\alpha_y} - m_p g \mathbf{N}_z \cdot {}^N\mathbf{v}_2^{L^*} \tag{3.100}$$

$$F_3 = -m_p g \mathbf{N}_z \cdot {}^N\mathbf{v}_3^{L^*} \tag{3.101}$$

$$F_4 = -m_p g \mathbf{N}_z \cdot {}^N\mathbf{v}_4^{L^*} \tag{3.102}$$

$$F_5 = 0 \tag{3.103}$$

where $\cdot$ represents the dot product, $^N\mathbf{v}_i^{L^*}$ denotes the ith partial velocity of the mass center of the load. The partial velocity of the mass center of the load $^N\mathbf{v}_i^{L^*}$ derives from the derivative of Equation (3.97) with respect to generalized speeds u_i, and the results are given by:

$$^N\mathbf{v}_1^{L^*} = l_s \mathbf{S}_y + (l_x S_\gamma C_{\beta_y} + l_x S_{\beta_x} S_{\beta_y} C_\gamma)\mathbf{R}_x + (l_x C_{\beta_y} C_\gamma - l_x S_{\beta_x} S_{\beta_y} C_\gamma)\mathbf{R}_y \tag{3.104}$$

$$\begin{aligned} ^N\mathbf{v}_2^{L^*} = \; & l_s C_{\alpha_x} \mathbf{S}_x + (l_x S_{\alpha_x} S_{\beta_y} S_\gamma + l_x C_\gamma C_{\alpha_x} C_{\beta_x} - l_x C_\gamma S_{\alpha_x} S_{\beta_x} C_{\beta_y})\mathbf{R}_x \\ & + (l_x S_{\alpha_x} S_{\beta_y} C_\gamma - l_x S_\gamma C_{\alpha_x} C_{\beta_x} + l_x S_\gamma S_{\alpha_x} S_{\beta_x} C_{\beta_y})\mathbf{R}_y \end{aligned} \tag{3.105}$$

$$^N\mathbf{v}_3^{L^*} = l_x S_\gamma \mathbf{R}_x + l_x C_\gamma \mathbf{R}_y \tag{3.106}$$

$$^N\mathbf{v}_4^{L^*} = l_x C_{\beta_x} C_\gamma \mathbf{R}_x - l_x C_{\beta_x} S_\gamma \mathbf{R}_y \tag{3.107}$$

$$^N\mathbf{v}_5^{L^*} = 0 \tag{3.108}$$

The corresponding generalized inertia forces can be written as:

$$F_1^* = -m_h {}^N\mathbf{a}^H \cdot l_s\mathbf{S}_y - m_p {}^N\mathbf{a}^{L^*} \cdot {}^N\mathbf{v}_1^{L^*} - (\mathbf{I}^L \cdot {}^N\boldsymbol{\alpha}^L + {}^N\boldsymbol{\omega}^L \times \mathbf{I}^L \cdot {}^N\boldsymbol{\omega}^L) \cdot {}^N\boldsymbol{\omega}_1^L \tag{3.109}$$

$$F_2^* = -m_h {}^N\mathbf{a}^H \cdot l_s C_{\alpha_x}\mathbf{S}_x - m_p {}^N\mathbf{a}^{L^*} \cdot {}^N\mathbf{v}_2^{L^*} - (\mathbf{I}^L \cdot {}^N\boldsymbol{\alpha}^L + {}^N\boldsymbol{\omega}^L \times \mathbf{I}^L \cdot {}^N\boldsymbol{\omega}^L) \cdot {}^N\boldsymbol{\omega}_2^L \tag{3.110}$$

$$F_3^* = -m_p {}^N\mathbf{a}^{L^*} \cdot {}^N\mathbf{v}_3^{L^*} - (\mathbf{I}^L \cdot {}^N\boldsymbol{\alpha}^L + {}^N\boldsymbol{\omega}^L \times \mathbf{I}^L \cdot {}^N\boldsymbol{\omega}^L) \cdot {}^N\boldsymbol{\omega}_3^L \tag{3.111}$$

$$F_4^* = -m_p {}^N\mathbf{a}^{L^*} \cdot {}^N\mathbf{v}_4^{L^*} - (\mathbf{I}^L \cdot {}^N\boldsymbol{\alpha}^L + {}^N\boldsymbol{\omega}^L \times \mathbf{I}^L \cdot {}^N\boldsymbol{\omega}^L) \cdot {}^N\boldsymbol{\omega}_4^L \tag{3.112}$$

$$F_5^* = -(\mathbf{I}^L \cdot {}^N\boldsymbol{\alpha}^L + {}^N\boldsymbol{\omega}^L \times \mathbf{I}^L \cdot {}^N\boldsymbol{\omega}^L) \cdot {}^N\boldsymbol{\omega}_5^L \tag{3.113}$$

where ${}^N\mathbf{a}^H$ represents the acceleration of the hook, ${}^N\mathbf{a}^{L^*}$ denotes the acceleration of the mass center of the load, ${}^N\boldsymbol{\alpha}^L$ is the angular acceleration of the load, $\mathbf{I}^L$ is the inertia dyadic of the load, and ${}^N\boldsymbol{\omega}_i^L$ represents the ith partial angular velocity of the load.

The derivative of Equation (3.97) with respect to time gives rise to the acceleration ${}^N\mathbf{a}^{L^*}$ of the mass center of the load. The angular acceleration ${}^N\boldsymbol{\alpha}^L$ of the load results from the derivative of Equation (3.96) with respect to time. The derivative of Equation (3.96) with respect to generalized speeds u_i yields the partial angular velocity ${}^N\boldsymbol{\omega}_i^L$ of the load:

$$\begin{aligned} {}^N\boldsymbol{\omega}_1^L &= (C_{\beta_y}C_\gamma - S_{\beta_x}S_{\beta_y}S_\gamma)\mathbf{L}_x - (S_\delta S_{\beta_y}C_{\beta_x} + C_\delta S_\gamma C_{\beta_y} + C_\delta S_{\beta_x}S_{\beta_y}C_\gamma)\mathbf{L}_y \\ &\quad - (C_\delta S_{\beta_y}C_{\beta_x} - S_\delta S_\gamma C_{\beta_y} - S_\delta S_{\beta_x}S_{\beta_y}C_\gamma)\mathbf{L}_z \end{aligned} \tag{3.114}$$

$$\begin{aligned} {}^N\boldsymbol{\omega}_2^L &= (S_{\alpha_x}S_{\beta_y}C_\gamma - S_\gamma C_{\alpha_x}C_{\beta_x} + S_\gamma S_{\alpha_x}S_{\beta_x}C_{\beta_y})\mathbf{L}_x \\ &\quad + (S_\delta S_{\beta_x}C_{\alpha_x} + S_\delta S_{\alpha_x}C_{\beta_x}C_\delta - C_\delta S_{\alpha_x}S_{\beta_y}S_\gamma - C_\delta C_\gamma C_{\alpha_x}C_{\beta_x} + C_\delta C_\gamma S_{\alpha_x}S_{\beta_x}C_{\beta_y})\mathbf{L}_y \\ &\quad + (C_\delta S_{\beta_x}C_{\alpha_x} + C_\delta S_{\alpha_x}C_{\beta_x}C_\delta - S_\delta S_{\alpha_x}S_{\beta_y}S_\gamma - S_\delta C_\gamma C_{\alpha_x}C_{\beta_x} + S_\delta C_\gamma S_{\alpha_x}S_{\beta_x}C_{\beta_y})\mathbf{L}_z \end{aligned} \tag{3.115}$$

$${}^N\boldsymbol{\omega}_3^L = C_\gamma\mathbf{L}_x - C_\delta S_\gamma\mathbf{L}_y + S_\delta S_\gamma\mathbf{L}_z \tag{3.116}$$

$${}^N\boldsymbol{\omega}_4^L = -S_\gamma C_{\beta_x}\mathbf{L}_x + (S_\delta S_{\beta_x} - C_\delta C_{\beta_x}C_\gamma)\mathbf{L}_y + (C_\delta S_{\beta_x} + S_\delta C_{\beta_x}C_\gamma)\mathbf{L}_z \tag{3.117}$$

$${}^N\boldsymbol{\omega}_5^L = S_\delta\mathbf{L}_y + S_\delta\mathbf{L}_z \tag{3.118}$$

The inertia dyadic $\mathbf{I}^L$ of the load can be written as:

$$\mathbf{I}^L = I_{xx}\mathbf{L}_x\mathbf{L}_x + I_{yy}\mathbf{L}_y\mathbf{L}_y + I_{zz}\mathbf{L}_z\mathbf{L}_z \tag{3.119}$$

where I_{xx}, I_{yy}, I_{zz} are the mass center principal axis moments of inertia about the $\mathbf{L}_x$, $\mathbf{L}_y$, $\mathbf{L}_z$ axis of the load.

Forcing the sum of the generalized active forces and inertia forces to be zero gives rise to dynamic equations of the motion for the model in Figure 3.41:

$$\begin{aligned}&(\mathbf{I}^L\cdot {}^N\boldsymbol{\alpha}^L + {}^N\boldsymbol{\omega}^L \times \mathbf{I}^L\cdot {}^N\boldsymbol{\omega}^L)\cdot {}^N\boldsymbol{\omega}_1^L + m_h{}^N\mathbf{a}^H\cdot l_s\mathbf{S}_y + m_p{}^N\mathbf{a}^{L^*}\cdot {}^N\mathbf{v}_1^{L^*}\\&\quad + m_p g\mathbf{N}_z\cdot {}^N\mathbf{v}_1^{L^*} + m_h g l_s S_{\alpha_x} C_{\alpha_y} = 0\end{aligned} \tag{3.120}$$

$$\begin{aligned}&(\mathbf{I}^L\cdot {}^N\boldsymbol{\alpha}^L + {}^N\boldsymbol{\omega}^L \times \mathbf{I}^L\cdot {}^N\boldsymbol{\omega}^L)\cdot {}^N\boldsymbol{\omega}_2^L + m_p g\mathbf{N}_z\cdot {}^N\mathbf{v}_2^{L^*} + m_h{}^N\mathbf{a}^H\cdot l_s C_{\alpha_x}\mathbf{S}_x\\&\quad + m_p{}^N\mathbf{a}^{L^*}\cdot {}^N\mathbf{v}_2^{L^*} + m_h g l_s C_{\alpha_x} S_{\alpha_y} = 0\end{aligned} \tag{3.121}$$

$$(\mathbf{I}^L\cdot {}^N\boldsymbol{\alpha}^L + {}^N\boldsymbol{\omega}^L \times \mathbf{I}^L\cdot {}^N\boldsymbol{\omega}^L)\cdot {}^N\boldsymbol{\omega}_3^L + m_p{}^N\mathbf{a}^{L^*}\cdot {}^N\mathbf{v}_3^{L^*} + m_p g\mathbf{N}_z\cdot {}^N\mathbf{v}_3^{L^*} = 0 \tag{3.122}$$

$$(\mathbf{I}^L\cdot {}^N\boldsymbol{\alpha}^L + {}^N\boldsymbol{\omega}^L \times \mathbf{I}^L\cdot {}^N\boldsymbol{\omega}^L)\cdot {}^N\boldsymbol{\omega}_4^L + m_p{}^N\mathbf{a}^{L^*}\cdot {}^N\mathbf{v}_4^{L^*} + m_p g\mathbf{N}_z\cdot {}^N\mathbf{v}_4^{L^*} = 0 \tag{3.123}$$

$$(\mathbf{I}^L\cdot {}^N\boldsymbol{\alpha}^L + {}^N\boldsymbol{\omega}^L \times \mathbf{I}^L\cdot {}^N\boldsymbol{\omega}^L)\cdot {}^N\boldsymbol{\omega}_5^L = 0 \tag{3.124}$$

3.5.2 Dynamics

3.5.2.1 Double-Pendulum Swing

The load twisting cannot occur when the bridge cranes undergo planar motions and the initial twist angle is a specified value. Under those conditions, the three-dimensional dynamic Equations (3.120)–(3.124) can be reduced to be a planar model. When the bridge is motionless and the acceleration of the trolley drives the model, the outputs of the planar double-pendulum model are only two swing angles. Resulting from Equations (3.120)–(3.124), the dynamic equations of the planar double-pendulum model relating the swing angles, α_x and β_x, to the trolley acceleration, $\ddot{y}$, are given by:

$$\begin{aligned}&(I_{xx} + m_h l_s^2 + m_p l_s^2 + m_p l_x^2 + 2m_p l_s l_x C_{\beta_x})\ddot{\alpha}_x + (I_{xx} + m_p l_x^2 + m_p l_x l_s C_{\beta_x})\ddot{\beta}_x\\&\quad - m_p l_s l_x S_{\beta_x}(\dot{\alpha}_x + \dot{\beta}_x)^2 + m_p l_s l_x S_{\beta_x}\dot{\alpha}_x^2 + g(m_h l_s S_{\alpha_x} + m_p l_s S_{\alpha_x} + m_p l_x S_{(\alpha_x+\beta_x)})\\&\quad - (m_p l_x C_{(\alpha_x+\beta_x)} + m_h l_s C_{\alpha_x} + m_p l_s C_{\alpha_x})\ddot{y} = 0\end{aligned} \tag{3.125}$$

$$\begin{aligned}&(I_{xx} + m_p l_x^2 + m_p l_x l_s C_{\beta_x})\ddot{\alpha}_x + (I_{yy} + m_p l_x^2)\ddot{\beta}_x\\&\quad + m_p l_s l_x S_{\beta_x}\dot{\alpha}_x^2 + g m_p l_x S_{(\alpha_x+\beta_x)} - m_p l_x C_{(\alpha_x+\beta_x)}\ddot{y} = 0\end{aligned} \tag{3.126}$$

Two linearized frequencies can be obtained from planar double-pendulum dynamics (3.125) and (3.126) by assuming small oscillations around the zero swing. The equation for estimating two linearized frequencies can be expressed as:

$$\omega_\alpha = \sqrt{\frac{g(\mu_1 - \sqrt{\mu_2 + \mu_3})}{2l_s(I_{xx}m_h + I_{xx}m_p + m_h m_p l_x^2)}} \tag{3.127}$$

$$\omega_\beta = \sqrt{\frac{g(\mu_1 + \sqrt{\mu_2 + \mu_3})}{2l_s(I_{xx}m_h + I_{xx}m_p + m_h m_p l_x^2)}} \tag{3.128}$$

where

$$\mu_1 = I_{xx}(m_h + m_p) + m_p l_x(m_h l_s + m_h l_x + m_p l_s + m_p l_x) \tag{3.129}$$

$$\mu_2 = (m_h + m_p)\begin{bmatrix} I_{xx}^2 m_h + I_{xx}^2 m_p + m_p^3 l_x^4 + m_p^3 l_s^2 l_x^2 + 2m_p^3 l_s l_x^3 \\ + m_h m_p^2 l_x^4 + 2I_{xx} m_p^2 l_x^2 - 2I_{xx} m_p^2 l_s l_x \end{bmatrix} \tag{3.130}$$

$$\mu_3 = (m_h + m_p)[m_h m_p^2 l_s^2 l_x^2 - 2m_h m_p^2 l_s l_x^3 + 2I_{xx} m_h m_p l_x^2 - 2I_{xx} m_h m_p l_s l_x] \tag{3.131}$$

Two frequencies in Equations (3.127) and (3.128) can be applied to analyze the double-pendulum effects for the full nonlinear dynamics (3.120)–(3.124). The low frequency, ω_α, exhibits the first-mode swing, which is more related with the motion of the swing angles, α_x and α_y, of the suspension cable. The high frequency, ω_β, displays the second-mode swing, which is more related with the motion of the swing angles, β_x, β_y, of the rigging cables.

3.5.2.2 Load Twisting

The double-pendulum effects can be neglected when the hook mass is much less than the payload mass. By ignoring the hook mass, the three-dimensional dynamic equations (3.120)–(3.124) can be reduced to be a simplified single-pendulum model with load swing and twisting dynamics. A free-oscillation model can be obtained from the simplified single-pendulum model by assuming small oscillations of the load swing:

$$\begin{aligned} &[I_{xx}C_\gamma + I_{yy}C_\delta(S_\delta\alpha_x - C_\delta C_\gamma) - I_{zz}S_\delta(C_\gamma S_\delta + C_\delta\alpha_x)]S_\gamma\ddot{\alpha}_x \\ &- [I_{zz}C_\delta(C_\gamma S_\delta + C_\delta\alpha_x) + I_{yy}S_\delta(S_\delta\beta_x - C_\delta C_\gamma)]\ddot{\gamma} - gm_p(l_s + l_x)\alpha_y \\ &- [m_p(l_s + l_x)^2 + I_{xx}S_\gamma^2 + I_{yy}(S_\delta\alpha_x - C_\delta C_\gamma)^2 + I_{zz}(C_\gamma S_\delta + C_\delta\alpha_x)^2]\ddot{\alpha}_y = 0 \end{aligned} \tag{3.132}$$

$$\begin{aligned} &[I_{xx} + m_p(l_s + l_x)^2 - S_\gamma^2(I_{xx} - I_{yy} + I_{yy}S_\delta^2 - I_{zz}S_\delta^2)]\ddot{\alpha}_x \\ &- [I_{xx}C_\gamma + I_{yy}C_\delta(S_\delta\alpha_x - C_\delta C_\gamma) - I_{zz}S_\delta(C_\gamma S_\delta + C_\delta\alpha_x)]S_\gamma\ddot{\alpha}_y \\ &- \ddot{\gamma}C_\delta S_\delta S_\gamma(I_{yy} - I_{zz}) + gm_p(l_s + l_x)\alpha_x = 0 \end{aligned} \tag{3.133}$$

$$\begin{aligned} &(I_{yy} - I_{zz})C_\delta S_\delta S_\gamma\ddot{\alpha}_x - [(I_{yy} - I_{zz})S_\delta^2 + I_{zz}]\ddot{\gamma} \\ &- [I_{zz}C_\delta(C_\gamma S_\delta + C_\delta\alpha_x) + I_{yy}S_\delta(S_\delta\alpha_x - C_\delta C_\gamma)]\ddot{\alpha}_y = 0 \end{aligned} \tag{3.134}$$

Then from the free-oscillation model (3.132)–(3.134), two frequencies can be derived:

$$\omega_{\alpha,x} = \sqrt{\frac{gm_p(l_s + l_x)}{m_p(l_s + l_x)^2 + I_{xx}}} \tag{3.135}$$

$$\omega_{\alpha,y} = \sqrt{\frac{gm_p(l_s + l_x)[I_{zz} + (I_{yy} - I_{zz})S_\delta^2]}{I_{yy}I_{zz} + m_p(l_s + l_x)^2[I_{zz} + (I_{yy} - I_{zz})S_\delta^2]}} \tag{3.136}$$

The two frequencies in Equations (3.135) and (3.136) are both corresponding to the first-mode swing, and are linked to the motion of the swing angles, α_x and α_y, of the suspension cable. The frequency, $\omega_{\alpha,x}$, is related with the motion of the first-mode swing angle, α_x, of the suspension cable, while the frequency, $\omega_{\alpha,y}$, is related with the motion of the first-mode swing angle, α_y.

Figure 3.42 shows two first-mode frequencies, $\omega_{\alpha,x}$, $\omega_{\alpha,y}$, defined in Equations (3.135) and (3.136) for various eccentric distances. The suspension cable length, l_s, rigging cable length, l_r, hook mass, m_h, load mass, m_p, load length, l_p, and moment of inertia, I_{xx}, I_{yy}, I_{zz}, were set to 495 mm, 550 mm, 32 g, 169 g, 992 mm, 151.6644 kg cm^2, 0.0639 kg cm^2, and 151.6644 kg cm^2, respectively. Increasing eccentric distance decreases two frequencies, $\omega_{\alpha,y}$, $\omega_{\alpha,y}$, slightly. The first- and second-mode frequencies, ω_α, ω_β, defined in Equations (3.127) and (3.128) for varying eccentric distances are also shown in Figure 3.42. The first-mode frequency, ω_α, ranges between two first-mode frequencies, $\omega_{\alpha,x}$ and $\omega_{\alpha,y}$, and is very close to the first-mode frequency, $\omega_{\alpha,x}$. The second-mode frequency, ω_β, decreases sharply with increasing eccentric distance.

The free-oscillation response of the swing angles, α_x and α_y, near the zero swing can be considered as the sum of response of harmonic oscillators with two frequencies in Equations (3.135) and (3.136). Then from Equation (3.134), the free-oscillation equation of the load twisting can be obtained:

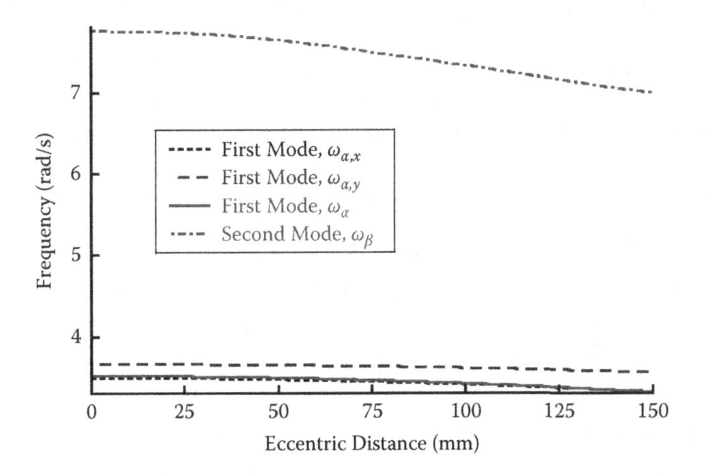

FIGURE 3.42 Swing frequencies for various eccentric distances.

$$\begin{aligned}\ddot{\gamma} = & B_1 \sin[(\omega_{\alpha,y} - \omega_{\alpha,x})t + \varphi_1] + B_2 \sin(\omega_{\alpha,x}t + \varphi_2) + B_3 \sin(\omega_{\alpha,y}t + \varphi_3) \\ & + B_4 \sin(2\omega_{\alpha,x}t + \varphi_4) + B_5 \sin[(\omega_{\alpha,x} + \omega_{\alpha,y})t + \varphi_5] + B_6 \sin(2\omega_{\alpha,y}t + \varphi_6) + B_7\end{aligned} \tag{3.137}$$

where $B_{i=1-7}$ are vibration-contribution coefficients, and $\varphi_{i=1-6}$ are corresponding phases. Resulting from Equation (3.137), the twisting dynamics contain six frequencies, $(\omega_{\alpha,y}\text{-}\omega_{\alpha,x})$, $(\omega_{\alpha,x})$, $(\omega_{\alpha,y})$, $(2\omega_{\alpha,x})$, $(\omega_{\alpha,y} + \omega_{\alpha,x})$, and $(2\omega_{\alpha,y})$. Notice that swing frequencies and twisting frequencies both come from the small-swing assumption. Therefore, other higher frequencies might arise in the swing and twisting dynamics because of nonlinearity.

3.5.3 Control

A discontinuous piecewise (DP) smoother is given by:

$$c(\tau) = \begin{cases} \dfrac{\zeta_m \omega_m e^{-\zeta_m \omega_m \tau}}{(1+K)(1-K^h)}, & 0 \le \tau \le hT \\ 0, & hT < \tau < T \\ \dfrac{\zeta_m \omega_m e^{-\zeta_m \omega_m \tau}}{(1+K)(1-K^h)}, & T \le \tau \le (1+h)T \\ 0, & \text{others} \end{cases} \tag{3.138}$$

where

$$K = e^{\frac{-\pi\zeta_m}{\sqrt{1-\zeta_m^2}}} \tag{3.139}$$

$$T = \frac{\pi}{\omega_m \sqrt{1 - \zeta_m^2}} \tag{3.140}$$

where ζ_m is the design damping ratio and h is the modified factor. Convolving two DP smoothers with different frequencies, $\omega_{\alpha,x}$ and $\omega_{\alpha,y}$, together generates a hybrid smoother with the multinotch and low-pass filtering effect. The multinotch filtering effect suppresses the first-mode swing, while the low-pass filtering effect reduces the second-mode swing. Meanwhile, reducing the swing amplitude decreases the amplitude of the load twisting.

The control architecture is shown in Figure 3.43. The operating commands filter through two DP smoothers, which are designed for each of the two frequencies, $\omega_{\alpha,x}$ and $\omega_{\alpha,y}$, of the first-mode swing. The smoothed commands will be used to move the industrial cranes having heterogeneous or non-uniform loads. The structure parameters of the industrial crane are applied to predict the two frequencies of the first-mode swing from Equations (3.135) and (3.136).

3.5.4 Experiments

Experimental validation is performed on a Cartesian coordinate manipulator suspending a heterogeneous bar, as shown in Figure 3.44. The trolley was driven by two

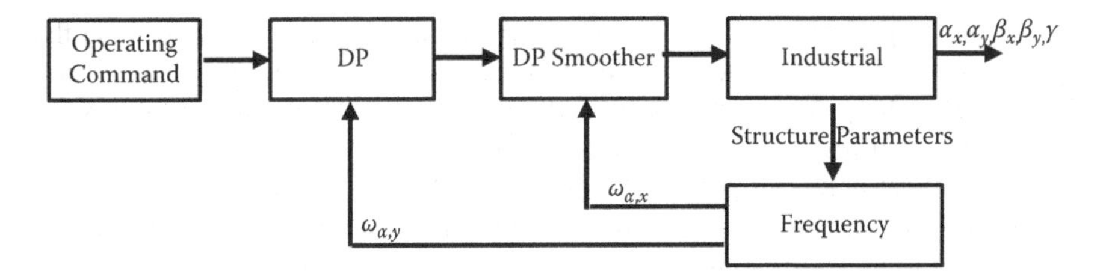

FIGURE 3.43 Control architecture.

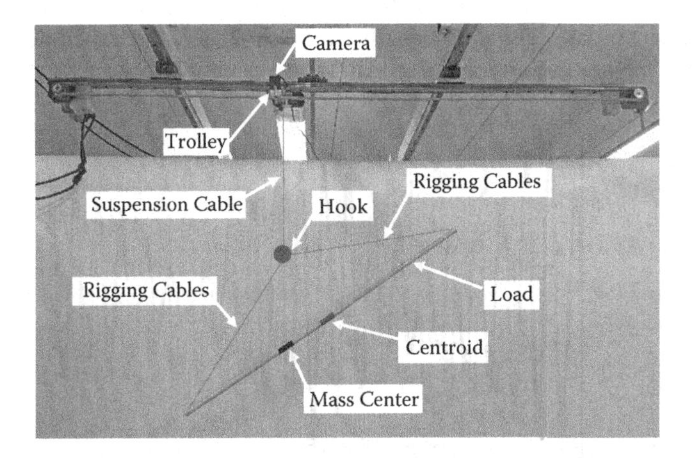

FIGURE 3.44 A Cartesian coordinate manipulator suspending a heterogeneous bar.

servo motors with encoders along Cartesian coordinates. A small-size hook was attached to the trolley by a suspension cable, and supports a heterogeneous bar by two rigging cables. Some part of the bar was made of aluminum alloy, and other part of the bar was made of polymer. Changing the combining ratio varies the eccentric distance between mass center and centroid of the load. The motions of the hook and load do not affect the movement of the trolley due to the high-ratio geared drives.

A computer is applied for operator interface and implementation of the control methods for the drives. A motion control card connects the computer to the motor amplifier. The mass center and centroid of the load were marked. One camera is installed on the trolley for recording the vibrational deflection of the mass center and centroid of the load. Meanwhile, the twist angle and twist angular velocity can also be estimated via the two markers using the inverse tangent function.

The original driving commands are trapezoidal velocity profile, which will be sent to two DP smoothers to create modified commands for moving the trolley. A proportional-integral-derivative controller implanted in the motion control card forces the trolley motions to follow the modified commands. The suspension cable length, rigging cable length, hook mass, load mass, load length, maximum driving velocity, and acceleration in experiments were fixed at 495 mm, 550 mm, 32 g, 169 g, 992 mm, 20 cm/s, and 2 m/s^2, respectively. The dynamic model (3.120)–(3.124) was applied for simulating the dynamic behavior in this section.

The first experiment was performed to validate the theoretical findings: (1) the dynamic model is effective for predicting the experimental response; (2) the method for estimating frequency of the swing and twisting is also effective for forecasting the experimental results. The eccentric distance, l_e, moment of inertia, I_{xx}, I_{yy}, I_{zz}, and initial twist angle were fixed at 150 mm, 151.6644 kg cm^2, 0.0639 kg cm^2, 151.6644 kg cm^2, and –159.2840°, respectively. A trapezoidal velocity command was applied to move the trolley for 25 cm. Figure 3.45 displays the simulated and experimental deflections of mass center of the load in the N_x direction and N_y direction from these tests, while Figure 3.46 shows that of centroid of the load. The simulated curves resulting from the dynamic model (3.120)–(3.124) closely match the experimental data. The difference between simulated and experimental swing comes from the nonlinearity and uncertainty of system dynamics.

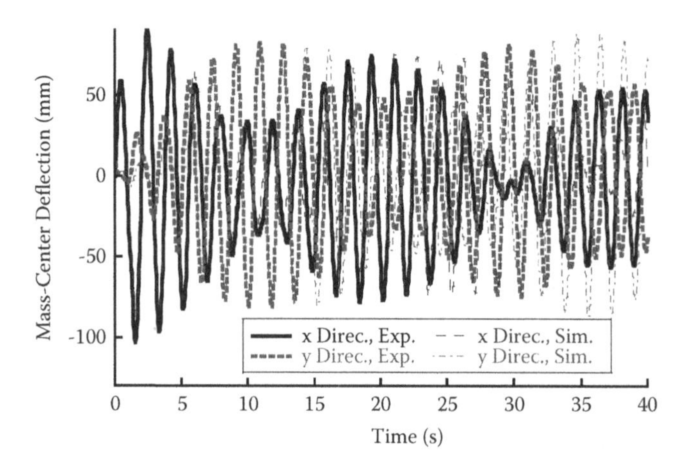

FIGURE 3.45 Experimental verification of mass-center deflection of the load.

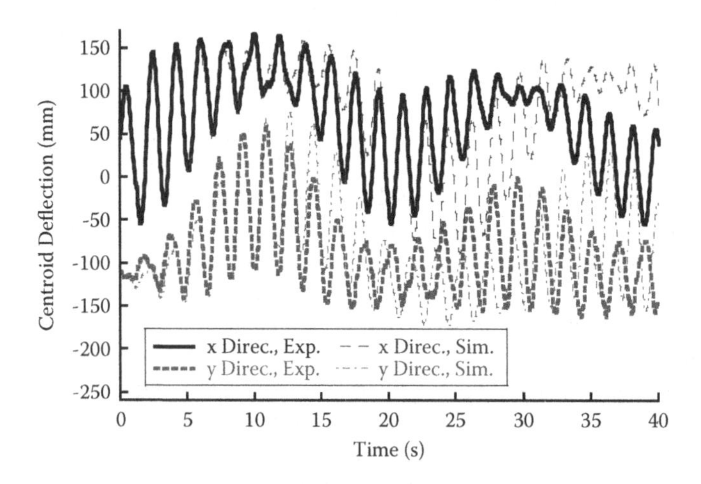

FIGURE 3.46 Experimental verification of centroid deflection of the load.

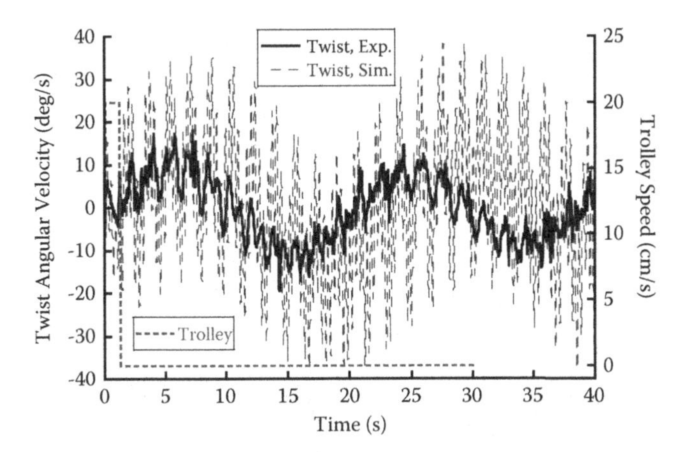

FIGURE 3.47 Experimental verification of twist angular velocity of the load.

Resulting from Equations (3.135) and (3.136), the estimation of the two frequencies, $\omega_{a,x}$, $\omega_{a,y}$, of the first-mode swing is 3.3160 rad/s and 3.5540 rad/s. Two frequencies of 3.2670 rad/s and 3.5506 rad/s arose in the experimental responses of mass center and centroid of the load. Therefore, the experiment obviously validates the effectiveness of the frequency prediction of swing defined in Equations (3.135) and (3.136).

The transient deflection is defined as the maximum deflection when the trolley moves. Meanwhile, the residual amplitude is referred to as the maximum deflection after the trolley stops. The experimental transient deflection and residual amplitude of the mass-center deflection in Figure 3.45 were 60.4 mm and 105.3 mm, respectively.

The simulated and experimental results of the twist angular velocity are shown in Figure 3.47. Resulting from Equation (3.137), the estimation of the twisting frequency includes 0.2380 rad/s, 3.3160 rad/s, 3.5540 rad/s, 6.632 rad/s, 6.8700 rad/s, and 7.1080 rad/s. The first three frequencies of the twisting have been found from the experimental data of the twist angular velocity. The difference between simulated and experimental results of the twist angular velocity is high-frequency vibrations in the simulated curve. The frequency of high-frequency vibrations is 19 rad/s approximately. This is because the damping from air in experiments reduced the high-frequency vibrations, while the damping in simulation is assumed to be zero. The experimental transient deflection and residual amplitude of the twist angular velocity in Figure 3.47 were 8.5°/s, and 19.5°/s, respectively.

The second experiment was performed to investigate the dynamic behavior of the model and the effectiveness of the presented control method for suppressing oscillations. The eccentric distance, l_e, moment of inertia, I_{xx}, I_{yy}, I_{zz}, were fixed at 150 mm, 151.6644 kg cm^2, 0.0639 kg cm^2, and 151.6644 kg cm^2, respectively. The transient deflection and residual amplitude of mass-center deflection for various driving displacements are shown in Figures 3.48 and 3.49, while those of twist

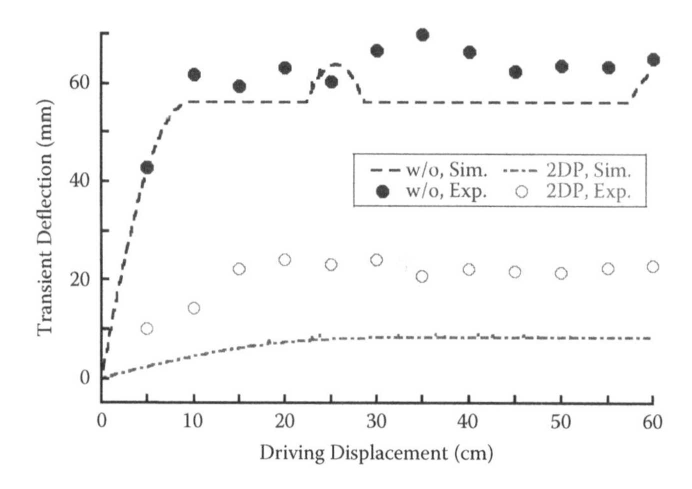

FIGURE 3.48 Experimental transient deflection of swing induced by driving distances.

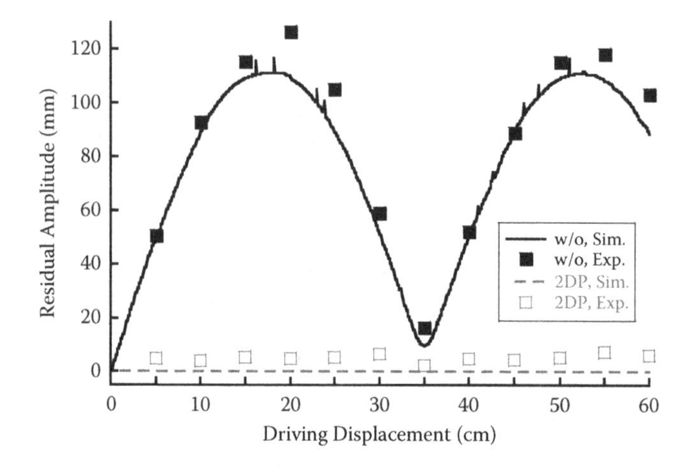

FIGURE 3.49 Experimental residual amplitude of swing induced by driving distances.

angular velocity are displayed in Figures 3.50 and 3.51. In the absence of controller, a trough in the mass-center deflection and twist angular velocity occurred near the driving displacement of 35 cm. This displacement is corresponding to the frequencies of the first-mode swing. More troughs arose in the simulated results of the twist angular velocity shown in Figure 3.51. This phenomenon can also be physically interpreted as the vibrations with the high frequency of 19 rad/s. The experimental data match the simulated curves very well. Thus, the analytical model is effective to predict the dynamic behavior in experiments.

The two DP smoothers were designed using the two first-mode swing frequencies, $\omega_{a,x}$ and $\omega_{a,y}$, of 3.3160 rad/s and 3.5540 rad/s. The modified factor, h, was selected to be 1 for wide range in frequency insensitivity. The experimental results are independent on the driving displacements because most oscillations were

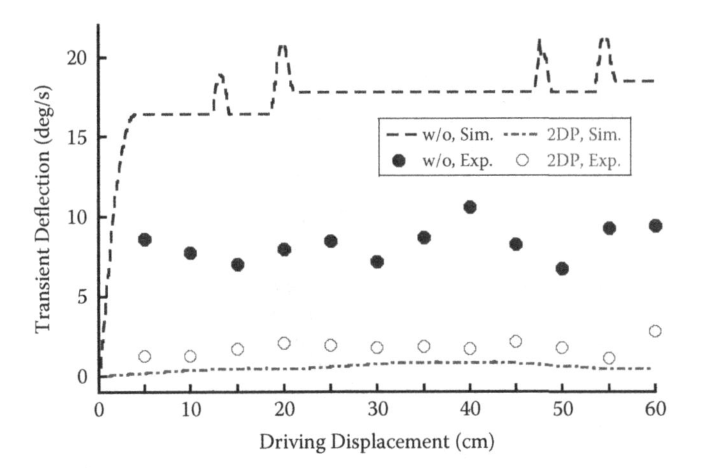

FIGURE 3.50 Experimental transient deflection of twisting induced by driving distances.

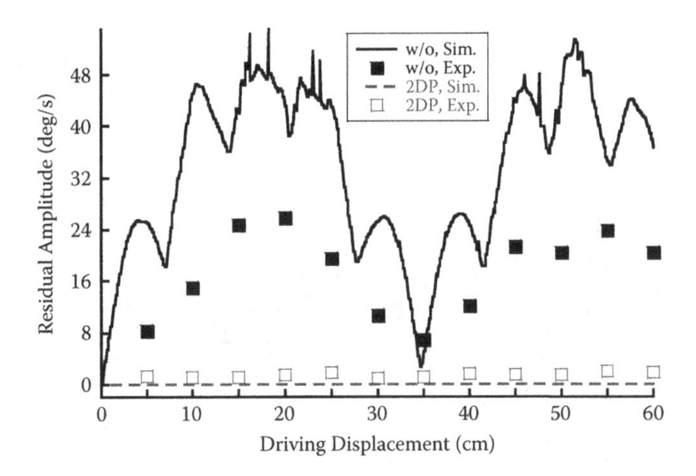

FIGURE 3.51 Experimental residual amplitude of twisting induced by driving distances.

suppressed. The two DP smoothers attenuated the experimental transient and residual amplitudes of mass-center deflection by an average of 66.8% and 93.0%, and reduced those of twist angular velocity of 78.4% and 91.0%. Therefore, the two DP smoothers are effective to control swing and twisting of the load.

The third experiment was performed to present the time response. The trapezoidal velocity command filtered through the two DP smoothers for a sliding displacement of 25 cm. The eccentric distance, l_e, moment of inertia, I_{xx}, I_{yy}, I_{zz}, were fixed at 150 mm, 151.6644 kg cm^2, 0.0639 kg cm^2, and 151.6644 kg cm^2, respectively. The smoothed velocity profile had smooth transitions between boundary conditions. Meanwhile, Figures 3.52 and 3.53 display the experimental response of mass center and twist angular velocity of the load when the eccentric distance was 150 mm. The transient deflection and residual amplitude of mass-center oscillations

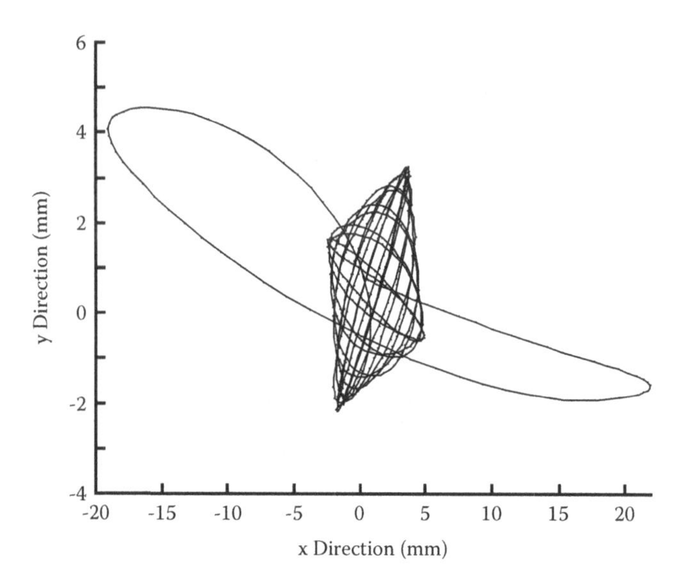

FIGURE 3.52 Experimental response of mass-center deflection of the load viewed from above.

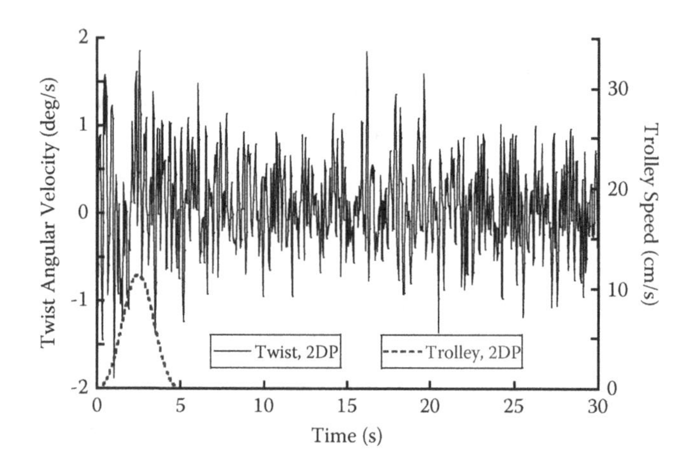

FIGURE 3.53 Experimental response of twist angular velocity of the load.

in Figure 3.52 were 23.1 mm, and 5.3 mm, respectively. In addition, the transient deflection and residual amplitude of twist angular velocity in Figure 3.53 were 1.9°/s and 1.8°/s, respectively.

The design of the DP smoothers is dependent on the eccentric distance. However, the eccentric distance is very challenging to measure or estimate in a real application. Therefore, the two DP smoothers should exhibit good control performance to large modeling errors in the eccentric distance. The last experiment was performed to study the robustness of the control method to changes in the eccentric distance. The trapezoidal velocity command filtered through the two DP smoothers for a sliding displacement of 25 cm.

Figures 3.54 and 3.55 display the experimental residual amplitudes of mass center and twist angular velocity of the load. The eccentric distance ranges from 0 mm to 150 mm. The design eccentric distance for the two DP smoothers was held constant at 75 mm for all cases shown in Figures 3.54 and 3.55. The residual amplitudes of the mass-center deflection and twist angular velocity at the eccentric distance of 75 mm are the lowest because of the smallest modeling error at this point. Increasing the modeling errors in the eccentric distance increases the residual amplitudes. Nevertheless, the residual amplitudes at the eccentric distances of 0 mm and 150 mm were mitigated to be low levels. The two DP smoothers suppressed

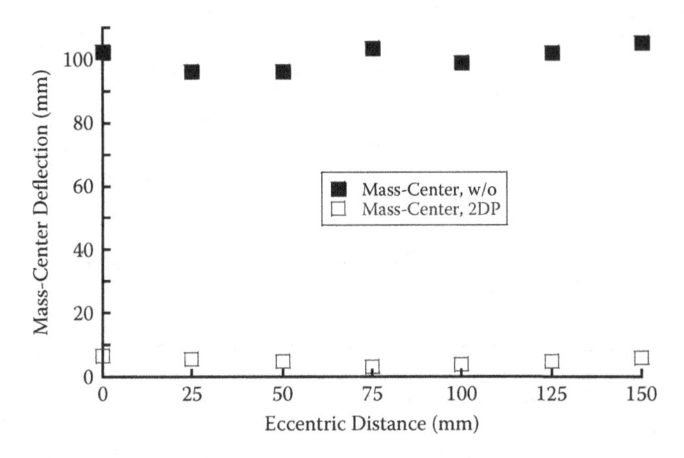

FIGURE 3.54 Experimental residual amplitudes of mass-center deflection for varying eccentric distances.

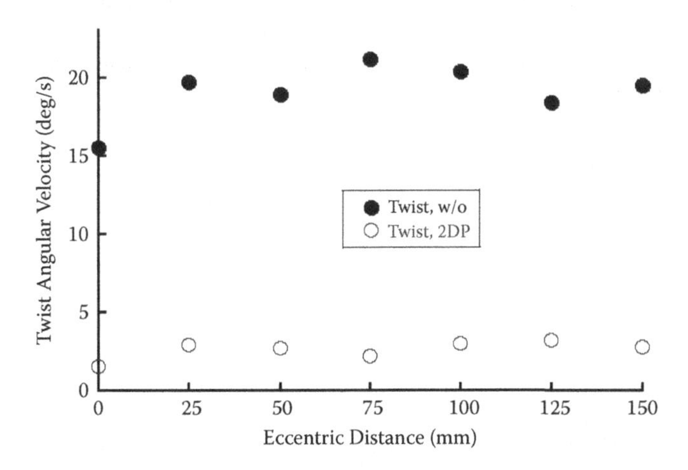

FIGURE 3.55 Experimental residual amplitudes of twist angular velocity for varying eccentric distances.

residual amplitudes of mass-center deflection and twist angular velocity by an average of 95.1% and 86.4%, respectively. Therefore, the experiments shown in Figures 3.54 and 3.55 validate the robustness of the two DP smoothers.

REFERENCES

[1] L. Ramli, Z. Mohamed, A. Abdullahi, et al., Control strategies for crane systems: a comprehensive review, *Mechanical Systems and Signal Processing*, 95 (2017) 1–23.

[2] Z. Masoud, A. Nayfeh, A. Al-Mousa, Delayed position-feedback controller for the reduction of payload pendulations of rotary cranes, *Journal of Vibration and Control*, 9 (1–2) (2003) 257–277.

[3] Z. Masoud, M. Daqaq, N. Nayfeh, Pendulation reduction on small ship-mounted telescopic cranes, *Journal of Vibration and Control*, 10 (8) (2004) 1167–1179.

[4] Z. Masoud, A. Nayfeh, N. Nayfeh, Sway reduction on quay-side container cranes using delayed feedback controller: simulations and experiments, *Journal of Vibration and Control*, 11 (8) (2005) 1103–1122.

[5] M. Solihin, W. Legowo, A. Legowo, Fuzzy-tuned PID anti-swing control of automatic gantry crane, *Journal of Vibration and Control*, 16 (1) (2010) 127–145.

[6] G.P. Starr, Swing-free transport of suspended objects with a path-controlled robot manipulator, *Journal of Dynamic Systems, Measurement, and Control*, 107 (1) (1985) 97–100.

[7] K. Sorensen, W. Singhose, S. Dickerson, A controller enabling precise positioning and sway reduction in bridge and gantry cranes, *Control Engineering Practice*, 15 (7) (2007) 825–837.

[8] J. Lawrence, W. Singhose, R. Weiss, et al., An Internet-driven tower crane for dynamics and controls education, Proceedings of the 7th IFAC Symposium Advances in Control Education, Madrid, Spain, 2006, pp. 511–516.

[9] A. Elbadawy, M. Shehata, Anti-sway control of marine cranes under the disturbance of a parallel manipulator, *Nonlinear Dynamics*, 82 (1–2) (2015) 415–434.

[10] E. Maleki, W. Singhose, Dynamics and control of a small-scale boom crane, *Journal of Computational and Nonlinear Dynamics*, 6 (3) (2011) 031015.

[11] G.G. Parker, K. Groom, J. Hurtado, et al., Command shaping boom crane control system with nonlinear inputs, Proceedings of IEEE Conference of Control Applications, Kohala Coast, USA, 1999, 1774–1778.

[12] D. Blackburn, W. Singhose, J. Kitchen, et al., Command shaping for nonlinear crane dynamics, *Journal of Vibration and Control*, 16 (4) (2010) 477–501.

4 Tower Cranes

Tower cranes are broadly applied in construction throughout the world [1]. Tower cranes transport payloads by translating a trolley in and out along the jib, slewing a jib about the tower, and hoisting a suspension cable. Nonetheless, payload oscillations induced by motions commanded by the human operator are a major drawback for efficient and safe manipulation. Operators usually address oscillatory problems manually by driving slowly and waiting for undesirable oscillations to decay.

Payloads hang below the hook by rigging cables, which are utilized for transporting bulky objects. The dynamics are very complex due to the payload swing and twisting about the rigging cables. Additionally, operation tasks may be more challenging because even skilled operators cannot control the payload twisting manually. Therefore, there is essential to research twisting dynamics and control in tower cranes moving bulky payloads.

Mounting numbers of the work have focused on the single-pendulum and double-pendulum dynamics of tower cranes moving point-mass payloads [2–27]. The point-mass payload dynamics only exhibit payload swing and do not include payload twisting because of ignoring the payload size. The corresponding control methods include linear control [2–4], nonlinear control [5–15], intelligent control [16–19], inverse kinematics [20,21], feedforward control [22], and input shaper [23–27]. However, all abovementioned literature did not concentrate on the payload twisting dynamics. Therefore, their presented schemes cannot control the payload twisting.

The payload twisting dynamics in bridge cranes have been studied in Chapter 3. Of course, the dynamics of the payload twisting of bridge cranes is similar to that of tower cranes during pure radial motions. However, the dynamics of the payload twisting in tower cranes during slewing motions is much more complicated. This chapter presents a model of tower cranes transporting distributed-mass beams. The dynamic behavior of the payload twisting in tower cranes during slewing motions is also studied. A demonstration that the smoother described above is also effective on oscillation reduction for these types of cranes.

4.1 SINGLE-PENDULUM DYNAMICS

Figure 4.1 shows a schematic representation of a tower crane transporting a slender bar. While a jib J rotates by an angle of θ about the N_3 direction, a trolley A moves with a position of r along the jib in the J_1 direction. The inertial coordinates $N_1N_2N_3$ can be converted to the moving Cartesian coordinates $J_1J_2J_3$ by rotating the driving angle, θ. Two massless cables of length, l_r, suspend below the trolley and support a uniformly distributed-mass bar P of mass, m_p, and length, l_p. The swing angles of the cable are defined as β_1 and β_2. When the bar length is small, the load can be modeled as a point-mass model. Therefore, the load twisting is ignored by the small-size assumption of the load. Then the Cartesian coordinates $J_1J_2J_3$ of the jib

DOI: 10.1201/9781003247210-4

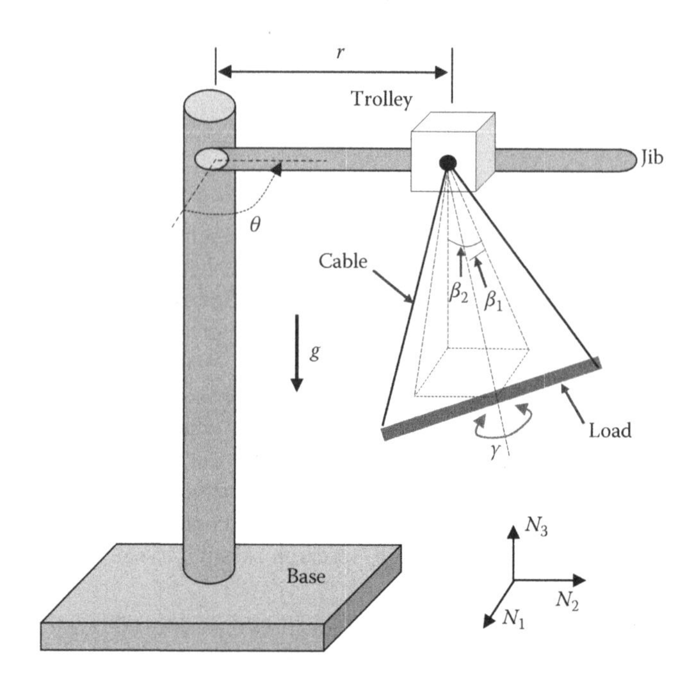

FIGURE 4.1 A model of a tower crane carrying a slender bar.

can also be converted to the moving coordinates $P_1P_2P_3$ of the bar by rotating the swing angles β_2 and β_1, respectively.

It is assumed that motions of the tower crane are unaffected by motions of the bar, the cable length does not change during the motions, and the damping ratio is approximately zero. The angular acceleration, $\ddot{\theta}$, of the jib and the acceleration, $\ddot{r}$, of the trolley are inputs to the model. The swing angles, β_1 and β_2, are the outputs.

The Kane's equation will be applied to derive the dynamic model in Figure 4.1. The generalized coordinates, q_k, are chosen as β_2, β_1. Meanwhile, the generalized speeds, u_k, are selected as $\dot{\beta}_2$ and $\dot{\beta}_1$:

$$u_1 = \dot{\beta}_2,\ u_2 = \dot{\beta}_1 \tag{4.1}$$

The basis transformation matrix between the two bodies J (jib) and P (bar) are:

$$\begin{Bmatrix} \mathbf{N}_1 \\ \mathbf{N}_2 \\ \mathbf{N}_3 \end{Bmatrix} = \begin{bmatrix} \cos(\theta) & \sin(\theta) & 0 \\ -\sin(\theta) & \cos(\theta) & 0 \\ 0 & 0 & 1 \end{bmatrix} \begin{Bmatrix} \mathbf{J}_1 \\ \mathbf{J}_2 \\ \mathbf{J}_3 \end{Bmatrix} \tag{4.2}$$

$$\begin{Bmatrix} \mathbf{J}_1 \\ \mathbf{J}_2 \\ \mathbf{J}_3 \end{Bmatrix} = \begin{bmatrix} \cos(\beta_2) & 0 & -\sin(\beta_2) \\ 0 & 1 & 0 \\ \sin(\beta_2) & 0 & \cos(\beta_2) \end{bmatrix} \begin{bmatrix} 1 & 0 & 0 \\ 0 & \cos(\beta_1) & \sin(\beta_1) \\ 0 & -\sin(\beta_1) & \cos(\beta_1) \end{bmatrix} \begin{Bmatrix} \mathbf{P}_1 \\ \mathbf{P}_2 \\ \mathbf{P}_3 \end{Bmatrix} \tag{4.3}$$

The angular velocity of jib in the inertial coordinates is given by:

$$ {}^{N}\boldsymbol{\omega}^{J} = \dot{\theta}\mathbf{J}_3 \tag{4.4} $$

The angular acceleration of jib in the inertial coordinates is given by:

$$ {}^{N}\boldsymbol{\alpha}^{J} = \ddot{\theta}\mathbf{J}_3 \tag{4.5} $$

The velocity of trolley center of mass A^* in the inertial coordinates is:

$$ {}^{N}\mathbf{v}^{A^*} = \dot{r}\mathbf{J}_1 + r\dot{\theta}\mathbf{J}_2 \tag{4.6} $$

The acceleration of trolley center of mass A^* in the inertial coordinates is described as:

$$ {}^{N}\mathbf{a}^{A^*} = (\ddot{r} - r\dot{\theta}^2)\mathbf{J}_1 + (2\dot{r}\dot{\theta} + r\ddot{\theta})\mathbf{J}_2 \tag{4.7} $$

The velocity of bar center of mass P^* in the inertial coordinates is:

$$ {}^{N}\mathbf{v}^{P^*} = {}^{N}\mathbf{v}^{A^*} - l_y[\cos(\beta_1)u_1 + \sin(\beta_1)\cos(\beta_2)\dot{\theta}]\mathbf{P}_1 - l_y[\sin(\beta_2)\dot{\theta} - u_2]\mathbf{P}_2 \tag{4.8} $$

where

$$ l_y = \sqrt{l_r^2 - 0.25l_p^2} \tag{4.9} $$

The acceleration of the bar center of mass P^* in the inertial coordinates is:

$$ \begin{aligned} {}^{N}\mathbf{a}^{P^*} = {} & {}^{N}\mathbf{a}^{A^*} - l_y[\cos(\beta_1)\dot{u}_1 + \sin(\gamma)\dot{u}_2 + \sin(\beta_1)\cos(\beta_2)\ddot{\theta}]\mathbf{P}_1 \\ & - l_y[\sin(\beta_2)\ddot{\theta} - \dot{u}_2]\mathbf{P}_2 + {}^{N}\boldsymbol{\omega}^{P} \times ({}^{N}\mathbf{v}^{P^*} - {}^{N}\mathbf{v}^{A^*}) \end{aligned} \tag{4.10} $$

The generalized active forces are expressed as:

$$ F_k = -m_p g\mathbf{N}_3 \cdot {}^{N}\mathbf{v}_k^{P^*}, \quad k = 1, 2 \tag{4.11} $$

where g is the gravity acceleration. Meanwhile, the generalized inertia forces are given by:

$$ F_k^* = -m_p \cdot {}^{N}\mathbf{a}^{P^*} \cdot {}^{N}\mathbf{v}_k^{P^*}, \quad k = 1, 2 \tag{4.12} $$

where ${}^{N}\mathbf{v}_k^{P^*}$ is the kth partial velocity of the bar center. They are expressed as:

$$ {}^{N}\mathbf{v}_1^{P^*} = \frac{\partial\, {}^{N}\mathbf{v}^{P^*}}{\partial u_1} \tag{4.13} $$

$$ {}^{N}\mathbf{v}_2^{P^*} = \frac{\partial\, {}^{N}\mathbf{v}^{P^*}}{\partial u_2} \tag{4.14} $$

The sum of generalized active forces and generalized inertia forces is constrained to be zero. Then nonlinear equations of the motion for the model in Figure 4.1 are given by:

$$ \begin{aligned} &l_y\ddot{\beta}_2 \cos\beta_1 + g\sin\beta_2 + l_y\ddot{\theta}\sin\beta_1\cos\beta_2 - \ddot{r}\cos\beta_2 - 2l_y\dot{\beta}_1\dot{\beta}_2\sin\alpha_y \\ &\quad + r\dot{\theta}^2\cos\beta_2 + 2l_y\dot{\theta}\dot{\beta}_1\cos\beta_2\cos\beta_1 - l_y\dot{\theta}^2\sin\beta_2\cos\beta_2\cos\beta_1 = 0 \end{aligned} \tag{4.15} $$

$$ \begin{aligned} &l_y\ddot{\beta}_1 + g\sin\beta_1\cos\beta_2 + \ddot{\theta}(r\cos\beta_1 - l_y\sin\beta_2) + \ddot{r}\sin\beta_2\sin\beta_1 \\ &\quad + l_y\dot{\beta}_2^2\sin\beta_1\cos\beta_1 \\ &- r\dot{\theta}^2\sin\beta_2\sin\beta_1 - 2l_y\dot{\theta}\dot{\beta}_2\cos^2\beta_1\cos\beta_2 - l_y\dot{\theta}^2\sin\beta_1\cos\beta_1\cos^2\beta_2 \\ &\quad + 2\dot{r}\dot{\theta}\cos\beta_1 = 0 \end{aligned} \tag{4.16} $$

The dynamic model (4.15) and (4.16) does not include motions of the payload hoisting. This is because both jib slewing motions about the base and trolley sliding motions along the jib cause large oscillations of the payload twisting. However, hoisting motions of the payload have slight impacts on the payload twisting dynamics.

4.2 TWISTING DYNAMICS

4.2.1 Modeling

In many cases, the load size cannot be ignored. The angle between the jib and the long axis of the bar is defined as the payload twist angle, γ. Then the Cartesian coordinates $J_1J_2J_3$ of the jib can also be converted to the moving coordinates $P_1P_2P_3$ of the bar by rotating the swing angles, β_2 and β_1, and twist angle, γ, respectively.

It is assumed that motions of the tower crane are unaffected by motions of the bar, the cable length does not change during the motions, and the damping ratio is approximately zero. The angular acceleration, $\ddot{\theta}$, of the jib and the acceleration, $\ddot{r}$, of the trolley are inputs to the model. The swing angles, β_1 and β_2, and twist angle, γ, are the outputs.

The Kane's equation will be applied to derive the dynamic model in Figure 4.1. The generalized coordinates, q_k, are chosen as β_2, β_1, and γ. Meanwhile, the generalized speeds, u_k, are selected as $\dot{\beta}_2$, $\dot{\beta}_1$, and $\dot{\gamma}$. The basis transformation matrices between the two bodies J (jib) and P (bar) are:

$$ u_1 = \dot{\beta}_2,\ u_2 = \dot{\beta}_1,\ u_3 = \dot{\gamma} \tag{4.17} $$

$$ u_1 = \dot{\beta}_2,\ u_2 = \dot{\beta}_1,\ u_3 = \dot{\gamma} \tag{4.18} $$

The basis transformation matrices between the two bodies J (jib) and P (bar) are:

$$\begin{Bmatrix} \mathbf{N}_1 \\ \mathbf{N}_2 \\ \mathbf{N}_3 \end{Bmatrix} = \begin{bmatrix} \cos(\theta) & \sin(\theta) & 0 \\ -\sin(\theta) & \cos(\theta) & 0 \\ 0 & 0 & 1 \end{bmatrix} \begin{Bmatrix} \mathbf{J}_1 \\ \mathbf{J}_2 \\ \mathbf{J}_3 \end{Bmatrix} \tag{4.19}$$

$$\begin{Bmatrix} \mathbf{J}_1 \\ \mathbf{J}_2 \\ \mathbf{J}_3 \end{Bmatrix} = \begin{bmatrix} \cos(\beta_2) & 0 & -\sin(\beta_2) \\ 0 & 1 & 0 \\ \sin(\beta_2) & 0 & \cos(\beta_2) \end{bmatrix} \begin{bmatrix} 1 & 0 & 0 \\ 0 & \cos(\beta_1) & \sin(\beta_1) \\ 0 & -\sin(\beta_1) & \cos(\beta_1) \end{bmatrix} \begin{bmatrix} \cos(\gamma) & \sin(\gamma) & 0 \\ -\sin(\gamma) & \cos(\gamma) & 0 \\ 0 & 0 & 1 \end{bmatrix} \begin{Bmatrix} \mathbf{P}_1 \\ \mathbf{P}_2 \\ \mathbf{P}_3 \end{Bmatrix} \tag{4.20}$$

The angular velocity of jib in the inertial coordinates is given by:

$$^{N}\boldsymbol{\omega}^{J} = \dot{\theta}\mathbf{J}_3 \tag{4.21}$$

The angular acceleration of jib in the inertial coordinates is given by:

$$^{N}\boldsymbol{\alpha}^{J} = \ddot{\theta}\mathbf{J}_3 \tag{4.22}$$

The velocity of trolley center of mass A^* in the inertial coordinates is:

$$^{N}\mathbf{v}^{A^*} = \dot{r}\mathbf{J}_1 + r\dot{\theta}\mathbf{J}_2 \tag{4.23}$$

The acceleration of trolley center of mass A^* in the inertial coordinates is described as:

$$^{N}\mathbf{a}^{A^*} = (\ddot{r} - r\dot{\theta}^2)\mathbf{J}_1 + (2\dot{r}\dot{\theta} + r\ddot{\theta})\mathbf{J}_2 \tag{4.24}$$

The angular velocity of the bar in the inertial coordinates is:

$$\begin{aligned} ^{N}\boldsymbol{\omega}^{P} &= {}^{N}\boldsymbol{\omega}^{J} + [\sin(\gamma)\cos(\beta_1)u_1 + \cos(\gamma)u_2]P_1 \\ &+ [\cos(\beta_1)\cos(\gamma)u_1 - \sin(\gamma)u_2]P_2 + [u_3 - \sin(\beta_1)u_1]P_3 \end{aligned} \tag{4.25}$$

The angular acceleration of the bar in the inertial coordinates is:

$$\begin{aligned} ^{N}\boldsymbol{\alpha}^{P} &= {}^{N}\boldsymbol{\alpha}^{J} + [\sin(\gamma)\cos(\beta_1)\dot{u}_1 + \cos(\gamma)\dot{u}_2]\mathbf{P}_1 \\ &+ [\cos(\beta_1)\cos(\gamma)\dot{u}_1 - \sin(\gamma)\dot{u}_2]\mathbf{P}_2 + [\dot{u}_3 - \sin(\beta_1)\dot{u}_1]\mathbf{P}_3 \end{aligned} \tag{4.26}$$

The velocity of bar center of mass P^* in the inertial coordinates is:

$$\begin{aligned} ^{N}\mathbf{v}^{P^*} &= {}^{N}\mathbf{v}^{A^*} - l_y \begin{bmatrix} \cos(\beta_1)\cos(\gamma)u_1 - \sin(\beta_2)\sin(\gamma)\dot{\theta} \\ +\sin(\gamma)u_2 + \sin(\beta_1)\cos(\beta_2)\cos(\gamma)\dot{\theta} \end{bmatrix} \mathbf{P}_1 \\ &- l_y \begin{bmatrix} \cos(\beta_1)\sin(\gamma)u_1 + \sin(\beta_2)\cos(\gamma)\dot{\theta} \\ -\cos(\gamma)u_2 + \sin(\beta_1)\cos(\beta_2)\sin(\gamma)\dot{\theta} \end{bmatrix} \mathbf{P}_2 \end{aligned} \tag{4.27}$$

where

$$l_y = \sqrt{l_r^2 - 0.25 l_p^2} \tag{4.28}$$

The acceleration of the bar center of mass P^* in the inertial coordinates is:

$$\begin{aligned} {}^N\mathbf{a}^{P^*} &= {}^N\mathbf{a}^{A^*} - l_y \begin{bmatrix} \cos(\beta_1)\cos(\gamma)\dot{u}_1 + \sin(\gamma)\dot{u}_2 - \sin(\beta_2)\sin(\gamma)\ddot{\theta} \\ +\sin(\beta_1)\cos(\beta_2)\cos(\gamma)\ddot{\theta} \end{bmatrix} \mathbf{P}_1 \\ &- l_y \begin{bmatrix} \cos(\beta_1)\sin(\gamma)\dot{u}_1 - \cos(\gamma)\dot{u}_2 + \sin(\beta_2)\cos(\gamma)\ddot{\theta} \\ +\sin(\beta_1)\cos(\beta_2)\sin(\gamma)\ddot{\theta} \end{bmatrix} \mathbf{P}_2 \\ &+ {}^N\boldsymbol{\omega}^P \times \left({}^N\mathbf{v}^{P^*} - {}^N\mathbf{v}^{A^*}\right) \end{aligned} \tag{4.29}$$

The generalized active forces are expressed as:

$$\begin{cases} F_1 = -m_p g l_y [\cos(\beta_1)\cos(\gamma) + \cos(\beta_1)\sin(\gamma)]; \\ F_2 = -m_p g l_y [\sin(\gamma) - \cos(\gamma)]; \\ F_3 = 0. \end{cases} \tag{4.30}$$

where g is the gravity acceleration. Meanwhile, the generalized inertia forces are given by:

$$F_k^* = -m_p \cdot {}^N\mathbf{a}^{P^*} \cdot {}^N\mathbf{v}_k^{P^*} - \left(\mathbf{I}^{P/P^*} \cdot {}^N\boldsymbol{\alpha}^P + {}^N\boldsymbol{\omega}^P \times \mathbf{I}^{P/P^*} \cdot {}^N\boldsymbol{\omega}^P\right) \cdot {}^N\boldsymbol{\omega}_k^P \tag{4.31}$$

where $\mathbf{I}^{P/P^*}$ is the inertia dyadic of the bar about the mass center P^*, ${}^N\mathbf{v}_k^{P^*}$ and ${}^N\boldsymbol{\omega}_k^P$ are the kth partial velocity of the bar center and kth partial angular velocities of the bar. They are expressed as:

$${}^N\mathbf{v}_1^{P^*} = -l_y \cos(\beta_1)\cos(\gamma)\mathbf{P}_1 - l_y \cos(\beta_1)\sin(\gamma)\mathbf{P}_2 \tag{4.32}$$

$${}^N\mathbf{v}_2^{P^*} = -l_y \sin(\gamma)\mathbf{P}_1 + l_y\cos(\gamma)\mathbf{P}_2 \tag{4.33}$$

$${}^N\mathbf{v}_3^{P^*} = 0 \tag{4.34}$$

$${}^N\boldsymbol{\omega}_k^P = \sin(\gamma)\cos(\beta_1)\mathbf{P}_1 + \cos(\beta_1)\cos(\gamma)\mathbf{P}_2 - \sin(\beta_1)\mathbf{P}_3 \tag{4.35}$$

$${}^N\boldsymbol{\omega}_k^P = \cos(\gamma)\mathbf{P}_1 - \sin(\gamma)\mathbf{P}_2 \tag{4.36}$$

$${}^N\boldsymbol{\omega}_k^P = \mathbf{P}_3 \tag{4.37}$$

The sum of generalized active forces and generalized inertia forces is constrained to be zero. Then nonlinear equations of the motion for the model in Figure 4.1 are given by:

$$
\begin{aligned}
&(12l_y^2\cos^2(\beta_1) + l_p^2[\sin^2(\beta_1) + \cos^2(\beta_1)\cos^2(\gamma)])\ddot{\beta}_2 \\
&\quad - l_p^2\sin(\gamma)\cos(\beta_1)\cos(\gamma)\ddot{\beta}_1 - l_p^2\sin(\beta_1)\ddot{\gamma} + 12gl_y\sin(\beta_2)\cos(\beta_1) \\
&\quad + \frac{1}{2}l_p^2\sin(\beta_1)\sin(2\gamma)\dot{\beta}_1^{\,2} + [l_p^2\sin^2(\gamma) - 12l_y^2]\sin(2\beta_1)\dot{\beta}_1\dot{\beta}_2 \\
&\quad - l_p^2\cos(\beta_1)\sin(2\gamma)\cos(\beta_2)\dot{\beta}_2\dot{\gamma} - 2l_p^2\cos(\beta_1)\cos^2(\gamma)\dot{\beta}_1\dot{\gamma} - 12l_y\cos(\beta_1)\cos(\beta_2)\ddot{r} \\
&\quad + \begin{bmatrix} 6l_y^2\sin(2\beta_1)\cos(\beta_2) + \frac{1}{2}l_p^2\cos(\beta_1)\sin(\beta_2)\sin(2\gamma) \\ +\frac{1}{2}l_p^2\sin(2\beta_1)\cos(\beta_2)\cos^2(\gamma) - \frac{1}{2}l_p^2\sin(2\beta_1)\cos(\beta_2) \end{bmatrix}\ddot{\theta} \\
&\quad + \begin{bmatrix} 6l_y^2\sin(2\beta_2)\cos^2(\beta_1) + 12l_y\cos(\beta_1)\cos(\beta_2)r \\ \quad + \frac{1}{2}l_p^2\sin(2\beta_2)\cos^2(\gamma) \\ -\frac{1}{2}l_p^2\sin(\beta_1)\sin(2\gamma)\cos(2\beta_2) - \frac{1}{2}l_p^2\sin(2\beta_2)\sin^2(\beta_1)\sin^2(\gamma) \end{bmatrix}\dot{\theta}^2 \\
&- l_p^2\sin(\beta_2)\sin(\beta_1)\sin(2\gamma)\dot{\beta}_1\dot{\theta} + l_p^2\begin{bmatrix} 2\sin(\beta_2)\cos(\beta_1)\cos^2(\gamma) \\ -\frac{1}{2}\sin(2\beta_1)\sin(2\gamma)\cos(\beta_2) \end{bmatrix}\dot{\gamma}\dot{\theta} = 0
\end{aligned}
\tag{4.38}
$$

$$
\begin{aligned}
&\frac{1}{2}l_p^2\sin(2\gamma)\cos(\beta_1)\ddot{\beta}_2 - (12l_y^2 + l_p^2\sin^2(\gamma))\ddot{\beta}_1 \\
&\quad + \left[\frac{1}{2}l_p^2\sin(2\beta_1)\sin^2(\gamma) - 12l_y^2\sin(\beta_1)\cos(\beta_2)\right]\dot{\beta}_2^{\,2} \\
&\quad - 12gl_y\sin(\beta_1)\cos(\beta_2) - 2l_p^2\sin^2(\gamma)\cos(\beta_1)\dot{\beta}_2\dot{\gamma} - l_p^2\sin(2\gamma)\dot{\beta}_1\dot{\gamma} \\
&\quad - 12l_y\sin(\beta_1)\sin(\beta_2)\ddot{r} + \begin{bmatrix} l_p^2\sin(\beta_2)\sin^2(\gamma) + 12l_y^2\sin(\beta_2) \\ +\frac{1}{2}l_p^2\sin(\beta_1)\sin(2\gamma)\cos(\beta_2) - 12l_y\cos(\beta_1)r \end{bmatrix}\ddot{\theta} \\
&\quad + \begin{bmatrix} \frac{1}{4}l_p^2\sin(2\beta_2)\sin(2\gamma)\cos(\beta_1) + 12l_y\sin(\beta_1)\sin(\beta_2)r \\ -\frac{1}{2}l_p^2\sin(2\beta_1)\sin^2(\gamma)\cos^2(\beta_2) + 6l_y^2\sin(2\beta_1)\cos^2(\beta_2) \end{bmatrix}\dot{\theta}^2 - 24l_y\cos(\beta_1)\dot{r}\dot{\theta} \\
&\quad + \begin{bmatrix} 2l_p^2\sin^2(\beta_1)\sin^2(\gamma)\cos(\beta_2) \\ -l_p^2\sin(\beta_1)\sin(\beta_2)\sin(2\gamma) \\ +24l_y^2\cos(\beta_2)\cos^2(\beta_1) \end{bmatrix}\dot{\beta}_2\dot{\theta} - \begin{bmatrix} 2l_p^2\sin(\beta_1)\sin^2(\gamma)\cos(\beta_2) \\ -l_p^2\sin(\beta_2)\sin(2\gamma) \end{bmatrix}\dot{\gamma}\dot{\theta} = 0
\end{aligned}
\tag{4.39}
$$

$$\begin{aligned}&\sin(\beta_1)\ddot{\beta}_2 - \ddot{\gamma} - \cos(\beta_1)\cos(\beta_2)\ddot{\theta} - \frac{1}{2}\sin(2\gamma)\cos^2(\beta_1)\dot{\beta}_2^{\,2} + \frac{1}{2}\sin(2\gamma)\dot{\beta}_1^{\,2}\\&+ 2\sin^2(\gamma)\cos(\beta_1)\dot{\beta}_1\dot{\beta}_2 + [2\sin(\beta_1)\cos(\beta_2)\sin^2(\gamma) - \sin(\beta_2)\sin(2\gamma)]\dot{\beta}_1\dot{\theta}\\&+ \begin{bmatrix}\sin(\beta_2)\sin(\gamma)\\ +\sin(\beta_1)\cos(\beta_2)\cos(\gamma)\end{bmatrix}\begin{bmatrix}\sin(\beta_2)\cos(\gamma)\\ -\sin(\beta_1)\sin(\gamma)\cos(\beta_2)\end{bmatrix}\dot{\theta}^2\\&+ \begin{bmatrix}\sin(\beta_2)\cos(\beta_1)\cos(2\gamma) - \frac{1}{2}\sin(2\gamma)\sin(2\beta_1)\cos(\beta_2)\\ -\sin(\beta_2)\cos(\beta_1)\end{bmatrix}\dot{\beta}_2\dot{\theta} = 0\end{aligned} \tag{4.40}$$

4.2.2 Dynamic Analysis

Resulting from the dynamic model (4.38), (4.39), and (4.40), the equilibrium points are zero swing angles, arbitrary twist angle, and zero angular velocity of the swing and twisting. A simplified free-oscillation model can be derived by assuming small oscillations of the swing around the abovementioned equilibria:

$$(12l_y^2 + l_p^2[\beta_1^{\,2} + \cos^2(\gamma)])\ddot{\beta}_2 - l_p^2\sin(\gamma)\cos(\gamma)\ddot{\beta}_1 + 12gl_y\beta_2 - l_p^2\beta_1\ddot{\gamma} = 0 \tag{4.41}$$

$$\frac{1}{2}l_p^2\sin(2\gamma)\ddot{\beta}_2 - [12l_y^2 + l_p^2\sin^2(\gamma)]\ddot{\beta}_1 - 12gl_y\beta_1 = 0 \tag{4.42}$$

$$\beta_1\ddot{\beta}_2 - \ddot{\gamma} = 0 \tag{4.43}$$

Then two linearized frequencies of the payload swing can be derived from the simplified model (4.41), (4.42), and (4.43):

$$\begin{cases}\omega_1 = \sqrt{g/[l_y + l_p^2/(12l_y)]}\\ \omega_2 = \sqrt{g/l_y}\end{cases} \tag{4.44}$$

where l_y is the distance between the suspension point A^* and the payload center P^*, and the distance l_y is defined in Equation (4.28). Therefore, frequencies of the swing angles, β_1 and β_2, in the radial and tangential directions can be estimated from Equation (4.44).

The dynamics of the payload swing near the equilibria can be approximated as two harmonic oscillators with linearized frequencies (4.44). Then resulting from Equation (4.43), the free-oscillation dynamics of the payload twisting near the equilibrium points can be described as:

$$\begin{aligned}\ddot{\gamma} = &A_1\sin[(\omega_2 - \omega_1)t + \varphi_1] + A_2\sin[(\omega_2 + \omega_1)t + \varphi_2]\\&+ A_3\sin(2\omega_1 t + \varphi_3) + A_4\sin(2\omega_2 t + \varphi_4)\end{aligned} \tag{4.45}$$

where A_1, A_2, A_3, A_4 are vibration-contribution functions, and φ_1, φ_2, φ_3, φ_4 are corresponding phase functions. The vibration-contribution functions and phase functions depend on the system parameters and vibration states. The linearized frequencies involved in the twisting model (4.45) include $(\omega_2 - \omega_1)$, $(\omega_2 + \omega_1)$, $(2\omega_1)$, and $(2\omega_2)$.

Figure 4.2 shows the two swing frequencies, ω_1 and ω_2, and Figure 4.3 shows four twisting frequencies, $(\omega_2 - \omega_1)$, $(\omega_2 + \omega_1)$, $(2\omega_1)$, and $(2\omega_2)$, for various distances, l_y, when the bar length was fixed at 78.5 cm. The first swing frequency, ω_1, increased with increasing the distance, l_y, before 22.7 cm. After this distance, the first swing frequency, ω_1, decreased slowly. Meanwhile, increasing the distance, l_y, decreased the second swing frequency, ω_2. Furthermore, as the distance, l_y, increased, the linearized frequencies $(\omega_2 - \omega_1)$ and $(\omega_2 + \omega_1)$ of the twisting decreased. After the distance, l_y, of 30 cm, the first twisting frequency $(\omega_2 - \omega_1)$ is much lower than the last three twisting frequencies, $(\omega_2 + \omega_1)$, $(2\omega_1)$, and $(2\omega_2)$, and the last three twisting

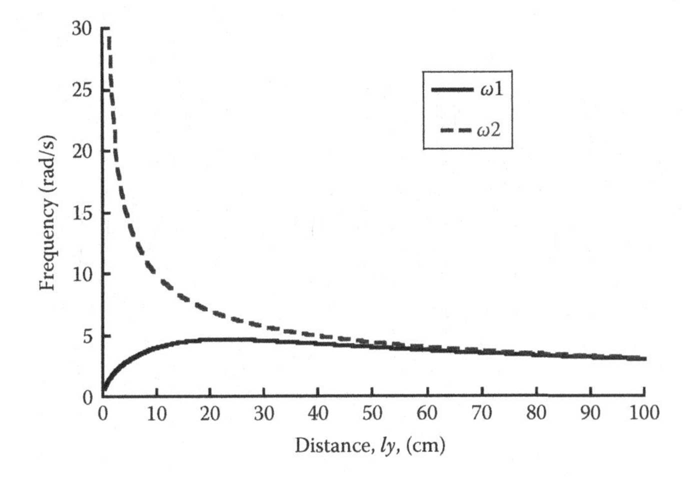

FIGURE 4.2 Linearized swing frequencies.

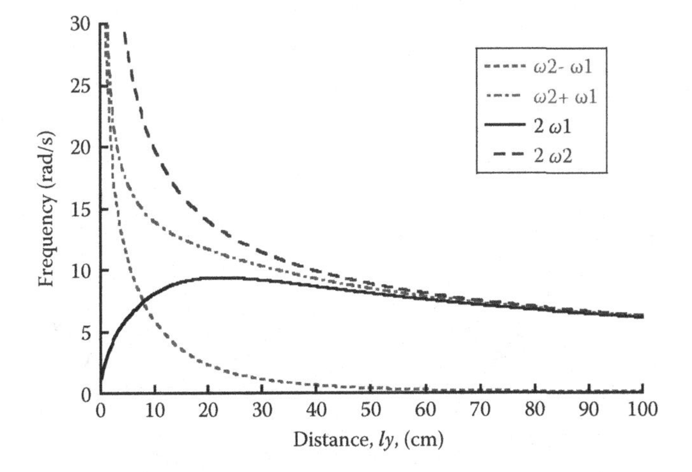

FIGURE 4.3 Linearized swing and twisting frequencies.

frequencies are approximately twice times as large as the swing frequencies, ω_1 and ω_2. This is because the swing frequencies, ω_1 and ω_2, are slightly different after the distance, l_y, of 30 cm. Near the distance, l_y, of 13 cm, the swing frequency, ω_1, approached the twisting frequency $(\omega_2 - \omega_1)$ and the swing frequency, ω_2, reached the twisting frequency, $(2\omega_1)$. Therefore, the internal resonance between the swing and twisting might arise at this point. In the case of large distance, l_y, the swing frequencies, ω_1 and ω_2, will approach the same value such that the first twisting frequency $(\omega_2 - \omega_1)$ will approach zero.

Figures 4.4 and 4.5 show a simulated free-oscillation response from the dynamic model (4.38), (4.39), and (4.40) when cable length, bar length, initial swing angle, β_1, initial swing angle, β_2, and initial twist angle, γ, were set to 100 cm, 100 cm, 1°, 0°, and 45°, respectively. The oscillational frequencies of

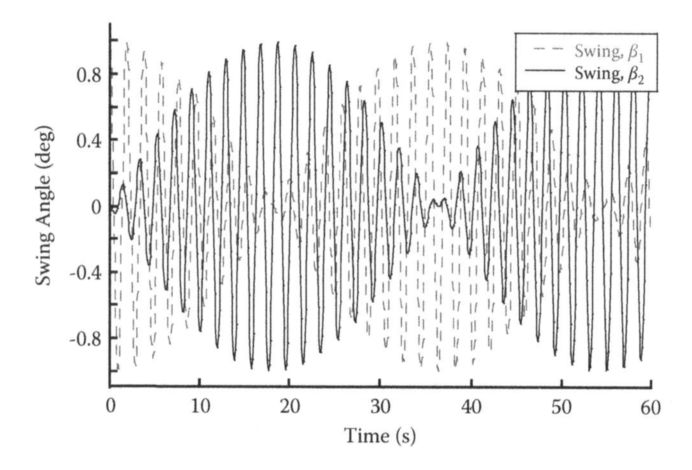

FIGURE 4.4 Simulated results to a non-zero initial condition.

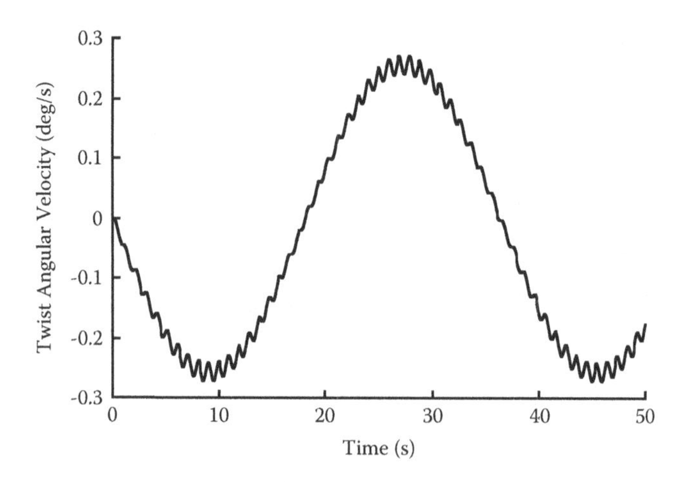

FIGURE 4.5 Simulated results to a non-zero initial condition.

the swing angles, β_1 and β_2, in Figure 4.4 are predicted accurately by the two linearized frequencies, ω_1 and ω_2, of the swing to be 3.1929 rad/s and 3.3657 rad/s. The two frequencies of the swing are slightly different in this case. Therefore, this results in beat phenomenon between two oscillations of the swing angles, β_1 and β_2. Meanwhile, the oscillation frequencies of the twist angular velocity in Figure 4.5 can be estimated correctly by the four linearized frequencies $(\omega_2 - \omega_1)$, $(\omega_2 + \omega_1)$, $(2\omega_1)$, and $(2\omega_2)$ of the twisting to be 0.1727 rad/s, 6.5586 rad/s, 6.3859 rad/s, and 6.7313 rad/s, respectively. The first twisting frequency, $(\omega_2 - \omega_1)$, is equal to the beat frequency in the swing.

Note that the four twisting frequencies $(\omega_2 - \omega_1)$, $(\omega_2 + \omega_1)$, $(2\omega_1)$, and $(2\omega_2)$ are linearized frequencies by assuming small oscillations, which are only dependent on the system parameters. Furthermore, the nonlinear frequencies of the payload twisting are more complicated, and also depend on vibrational amplitude and initial states because of strongly nonlinear dynamical behavior. As the swing angles increase, the nonlinear frequencies of the twist angular velocity increase. Meanwhile, the initial twist angles also have a large influence on the nonlinear frequencies of the twisting [28–30].

4.2.3 Oscillation Control

A discontinuous piecewise (DP) smoother with two pieces will be applied to control swing and twisting of the load. The DP smoother is given by:

$$c(\tau) = \begin{cases} \dfrac{\zeta_m \omega_m e^{-\zeta_m \omega_m \tau}}{\left[1 - e^{-0.5hT\zeta_m\omega_m} + e^{-0.5T\zeta_m\omega_m} - e^{-0.5(1+h)T\zeta_m\omega_m}\right]}, & 0 \le \tau \le 0.5hT \\ 0, \ 0.5hT < \tau < 0.5T & \\ \dfrac{\zeta_m \omega_m e^{-\zeta_m \omega_m \tau}}{\left[1 - e^{-0.5hT\zeta_m\omega_m} + e^{-0.5T\zeta_m\omega_m} - e^{-0.5(1+h)T\zeta_m\omega_m}\right]}, & 0.5T \le \tau \le 0.5(1+h)T \\ 0, \quad \text{others} & \end{cases} \tag{4.46}$$

where

$$T = \frac{2\pi}{\omega_m \sqrt{1 - \zeta_m^2}} \tag{4.47}$$

ω_m is the design frequency, ζ_m is the design damping ratio, h is the modified factor for the duration of the DP smoother, and satisfies $0 \le h \le 1$. Additionally, the duration of the DP smoother (4.46) is $0.5(1 + h)T$.

The theoretical findings in [28–30] indicate that the amplitude of the payload twisting can be reduced by decreasing the swing amplitude. In addition, the swing dynamics near the equilibria can be approximated as two harmonic oscillators with linearized frequencies, ω_2 and ω_1. Therefore, the original commands are convolved with a DP smoother with the first linearized frequency, ω_1, of the swing, and then convolved with another DP smoother with the second linearized frequency, ω_2, to produce accelerations of the jib and trolley. The smoothed command will drive the tower cranes toward the desired position with minimal swing and twisting.

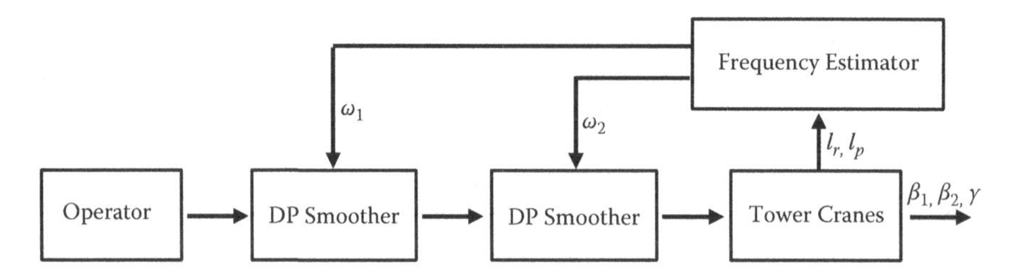

FIGURE 4.6 Controlling tower cranes via two DP smoothers.

Figure 4.6 show the control architecture of tower cranes. The cable length and payload length were applied to estimate the linearized swing frequencies, ω_1 and ω_2, for two DP smoothers. Due to the notch filtering effect of two DP smoothers at the two swing frequencies, the payload swing could be suppressed to the minimum or completely suppressed. Furthermore, the oscillation suppression of the swing would also result in the magnitude attenuation of the twisting.

When the cable length and bar length were fixed at 64.5 cm and 78.5 cm, the linearized frequencies of the swing are 4.0032 rad/s and 4.3780 rad/s. For the modified factor, h of 0, 5% frequency insensitivity for the two DP smoothers varies from 3.561 rad/s to 4.821 rad/s. For the modified factor, h of 0.5, 5% frequency insensitivity for the two DP smoothers is from 3.516 rad/s to 4.914 rad/s. For the modified factor, h of 1, 5% frequency insensitivity for the two DP smoothers ranges from 3.366 rad/s to infinity. Increasing modified factor h increases the rise time and frequency insensitivity. Additionally, the combination of two DP smoothers produces a wider insensitive range in the frequency. The wide insensitivity can reject the oscillations induced by the modeling error in the design frequency. The error comes from the difference between the linearized frequency and nonlinear frequency of the swing.

4.2.4 Computational Dynamics

The dynamic behavior of the payload twisting and the effectiveness of the two DP smoothers will be verified in the simulations. The nonlinear dynamic model (4.38), (4.39), and (4.40) is applied to simulate the various working conditions and system parameters in this section. The motions of the tower crane include the slewing motions of the jib and sliding motions of the trolley. Slewing motions of the jib is defined as the rotation of the jib about the tower base. Sliding motions of the trolley occur when the trolley moves along the jib.

The baseline command is the trapezoidal velocity profile. The nominal slewing velocity and acceleration of the jib are 20°/s and 67°/s^2, while the nominal sliding velocity and acceleration of the trolley are 20 cm/s and 2 m/s^2. The initial twist angle in the simulations was fixed at 45°. The swing deflection is defined as the displacement deflection of the mass center of the payload relative to the trolley. The residual amplitude is referred to as the maximum deflection after the jib or trolley stops.

4.2.4.1 Slewing

The simulated residual amplitudes of the swing deflection and twist angular velocity induced by various slewing displacements are shown in Figures 4.7 and 4.8.

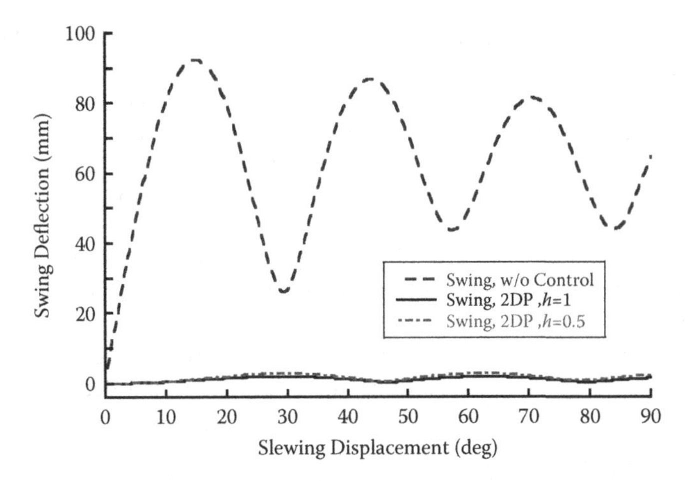

FIGURE 4.7 Residual amplitudes of swing deflection induced by jib slewing displacements.

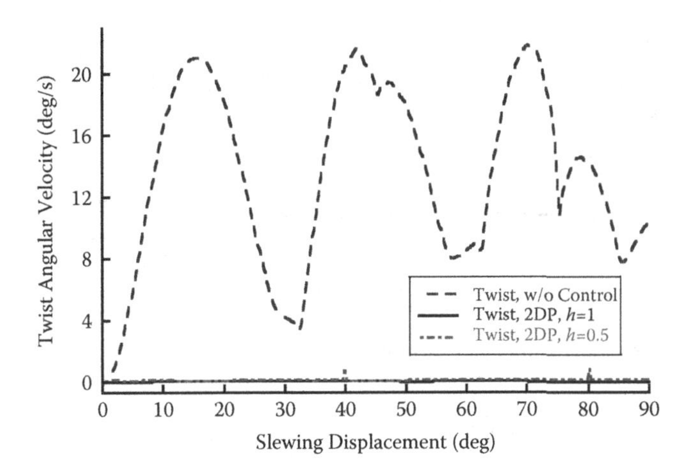

FIGURE 4.8 Residual amplitudes of twist angular velocity induced by jib slewing displacements.

The cable length, bar length, and trolley position were set to be 64.5 cm, 78.5 cm, and 64 cm, respectively. In the absence of control, peaks and troughs arose in the swing because of the interference between oscillations caused by the acceleration and deceleration of the slewing motions. In addition, the distance between peaks and troughs is approximately corresponding to the linearized swing frequency. The uncontrolled swing deflection reached local maximum or minimum, thereby causing peaks and troughs in the twist angular velocity. This phenomenon can be physically interpreted as the coupling effect between the swing and twisting. The uncontrolled residual amplitudes of the twisting contain narrow gaps at the slewing displacement of 45.4 cm and 75.4 cm because of nonlinearity.

When two DP smoothers are implemented, residual amplitudes of the swing deflection and twist angular velocity are independent on the slewing displacement. In the case of modified factor h of 1, the residual amplitudes of the swing and twisting are <2 mm and $<0.04°$/s for all the displacement tested. In the case of modified factor h of 0.5, that of the swing and twisting are <3 mm and $<0.77°$/s. Therefore, the residual oscillations of the swing and twisting are both greatly suppressed for all the slewing motions.

Figures 4.9 and 4.10 show the simulated residual amplitudes of the swing deflection and twist angular velocity for various cable lengths. The bar length, jib slewing displacement, and trolley position were held constant at 78.5 cm, 80°, and 64 cm, respectively. Peaks and troughs arose in the swing because of the interference between oscillations induced by the acceleration and deceleration of the

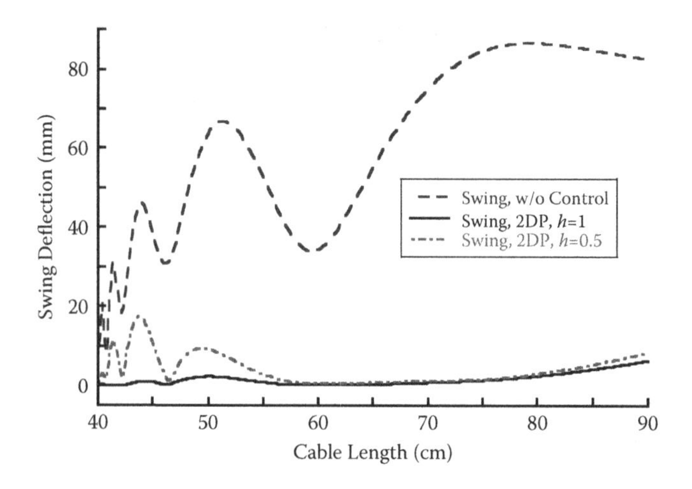

FIGURE 4.9 Residual amplitudes of swing deflection induced by cable lengths.

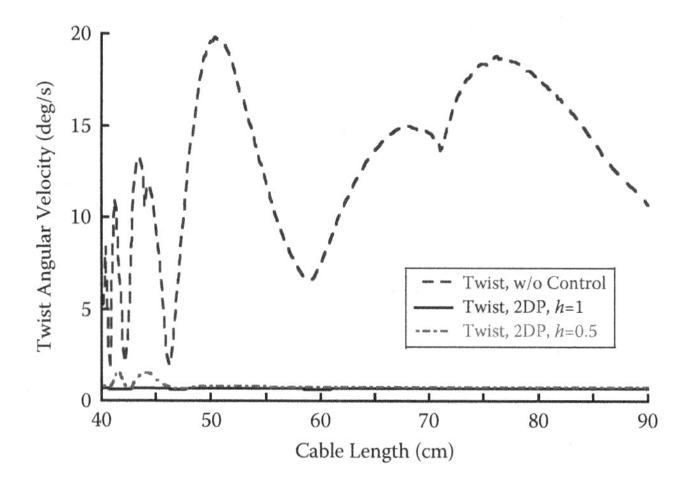

FIGURE 4.10 Residual amplitudes of twist angular velocity induced by cable lengths.

slewing motions. Peaks and troughs in the twisting occurred at the same cable length as the swing dynamics because of the coupling impacts. A narrow gap in the twisting arose at the length of 71 cm because of nonlinearity. As the cable length increased, peaks and troughs were spaced farther apart in both swing and twisting. This is because the swing frequency decreased with increasing the cable length.

The cable length ranged from 40 cm to 90 cm, while the design length for the two DP smoothers was fixed at 64.5 cm. The two DP smoothers suppressed the residual amplitude of the swing to be near zero at the design length of 64.5 cm. As the modeling error in the cable length increased, the residual amplitude of the swing increased. The two DP smoothers with the modified factor h of 1 reduced residual amplitudes of the swing and twisting for all the cable lengths by an average of 97.3% and 95.0%, respectively. The two DP smoothers with the modified factor h of 0.5 attenuated residual amplitudes of the swing and twisting by an average of 93.5% and 93.8%, respectively. Two DP smoothers with the higher modified factor can suppress more swing and twisting because of providing more robustness in the frequency. However, two DP smoothers with modified factor h of 1 and 0.5 both mitigated the residual oscillations of the swing and twisting to a low level. Therefore, two DP smoothers are robust to the modeling error in the cable length.

4.2.4.2 Sliding

The simulated residual amplitudes of the swing deflection for varying trolley sliding displacements are shown in Figure 4.11, and those of twist angular velocity are shown in Figure 4.12. The cable length and bar length were fixed at 64.5 cm and 78.5 cm, respectively. The uncontrolled dynamics is similar to the complicated behavior observed in jib slewing motions. The trolley sliding motions have little impacts on the system dynamics when the two DP smoothers are utilized. In the case of the modified factor h of 1, the residual amplitudes of the swing and twisting are <0.04 mm and <0.04°/s for all the trolley sliding displacement tested. The residual amplitudes of the

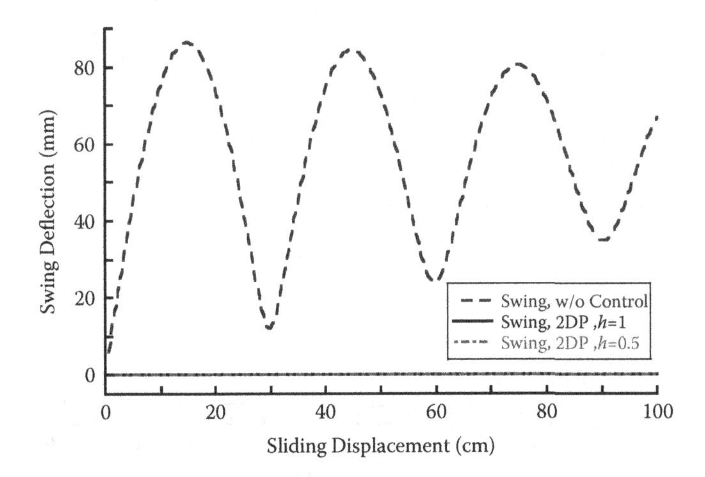

FIGURE 4.11 Residual amplitudes of swing deflection induced by trolley sliding displacements.

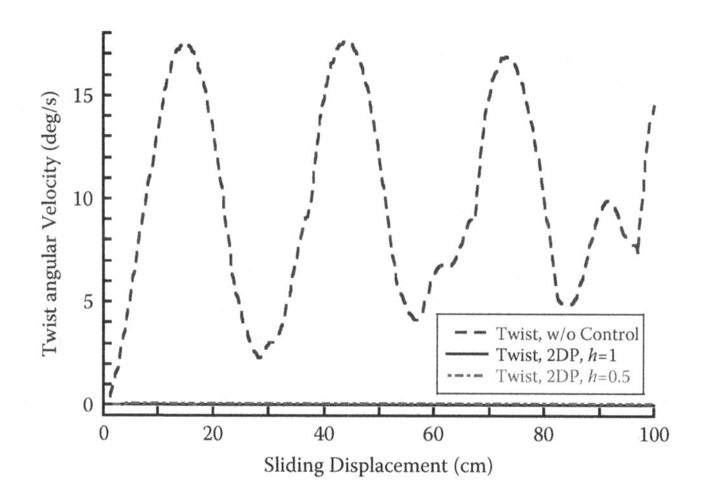

FIGURE 4.12 Residual amplitudes of twist angular velocity induced by trolley sliding displacements.

swing and twisting are <0.11 mm and <0.11°/s in the case of the modified factor h of 0.5. Therefore, two DP smoothers are effective to control the swing and twisting during trolley sliding motions.

The simulated residual amplitudes of the swing deflection and twist angular velocity for various bar lengths are shown in Figures 4.13 and 4.14 when the cable length and trolley sliding displacement were held constant at 64.5 cm and 45 cm, respectively. Without the controller, a trough in the swing occurred at the bar length of 115 cm because of the interference between oscillations caused by the acceleration and deceleration of sliding motions. The interaction between the swing and twisting also caused a trough in the twisting. When the design length is fixed at

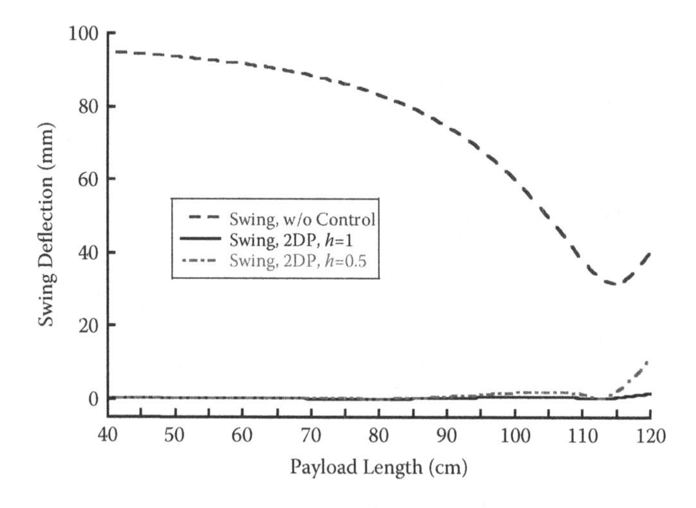

FIGURE 4.13 Residual amplitudes of swing deflection induced by bar lengths.

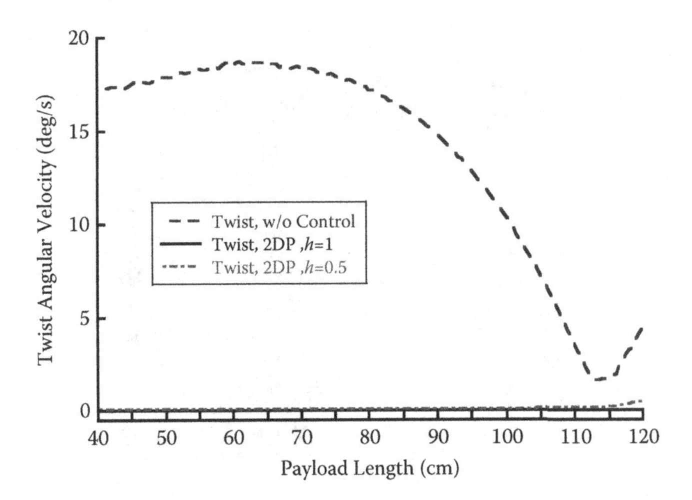

FIGURE 4.14 Residual amplitudes of twist angular velocity induced by bar lengths.

78.5 cm, the bar length varied from 40 cm to 120 cm. The two DP smoothers with the modified factor h of 1 reduced residual oscillations of the swing for all payload length by an average of 99.6%, and that of the twisting by an average of 99.8%. Two DP smoothers with the modified factor h of 0.5 attenuated residual oscillations of the swing and twisting by an average of 98.6% and 99.2%, respectively. The two DP smoothers with the higher modified factor suppress more swing and twisting. This is because increasing modified factor increases the robustness in the frequency. However, two DP smoothers with the modified factor h of both 1 and 0.5 suppressed the swing and twisting to a low level. Therefore, the control method exhibits good robustness to the modeling error in the payload length.

4.2.5 Experimental Verification

Experiments were conducted using a small-scale tower crane carrying a slender payload shown in Figure 4.15. The base of the tower crane is approximately 50 × 50 cm. A host computer is used to implement the control method and operator interface. An amplifier connects the computer to the servo motors. Motors with encoders rotate the jib arm through a maximum angle of ±155° about the tower, and drive the trolley with a maximum displacement of 80 cm along the jib.

Two cables with the same length were made of Dyneema rope, and were connected the payload to the trolley. An aluminum bar served as the payload. The mass, length, and diameter of the bar were 265 g, 78.5 cm, and 12 mm, respectively. The trolley position along the jib was held constant at 64 cm. The jib mounted a camera for recording the deflection of two markers on the bar. The middle point between the two markers is the bar center. Therefore, the swing deflection of the bar can be estimated by the average of the displacement of two markers. Moreover, the two markers can also be applied to estimate the twist angle and twist angular velocity by using the inverse tangent function.

FIGURE 4.15 A small-scale tower crane carrying a slender bar.

Experiments were performed to verify key theoretical results: (1) the dynamic equations (4.38), (4.39), and (4.40) are appropriate for modeling of tower cranes suspending slender bars, and (2) frequency-estimation method can predict accurately the twisting frequency. Figure 4.16 shows simulated and experimental responses of the bar swing to a non-zero initial condition, while that of the twist angular velocity are shown in Figure 4.17. The cable lengths were fixed at 64.5 cm. The initial deflection of the payload center in the radial and tangential directions are 16.7 mm and 38.6 mm, respectively. Meanwhile, the initial twist angle and initial twist angular velocity were 10° and 0°/s, respectively. The low frequency in the

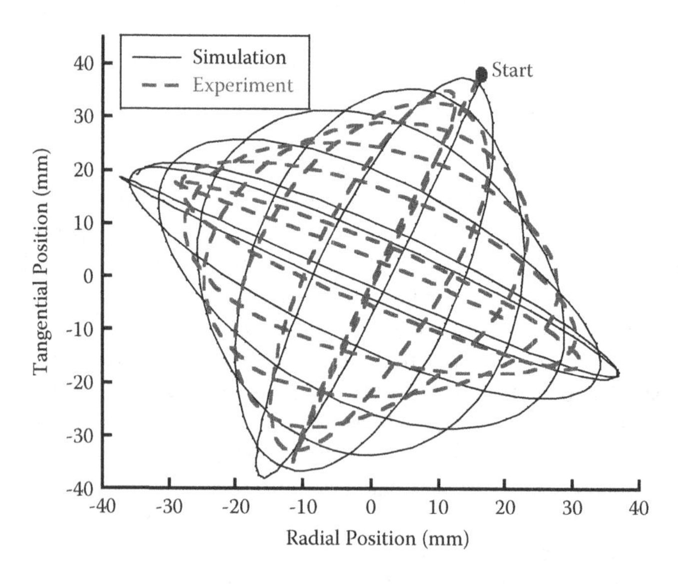

FIGURE 4.16 Simulated and experimental responses of swing to an initial condition viewed from above.

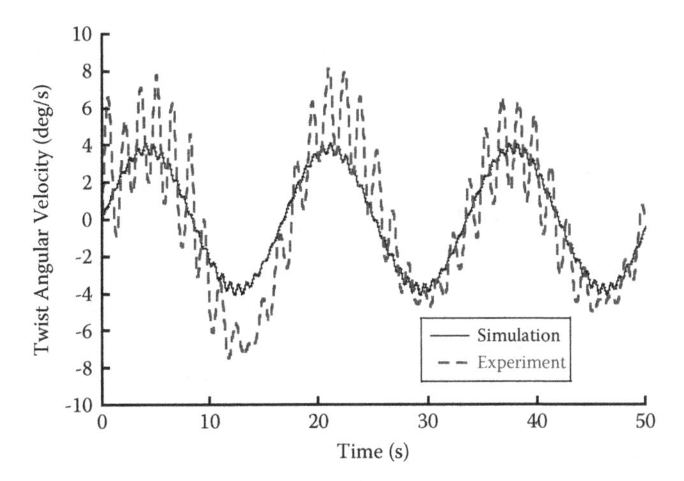

FIGURE 4.17 Simulated and experimental responses of twisting to an initial condition.

experimental response of the twisting is 0.38 rad/s, while the high frequency is 4.2 rad/s approximately. Meanwhile, four twisting frequencies $(\omega_2 - \omega_1)$, $(\omega_2 + \omega_1)$, $(2\omega_1)$, and $(2\omega_2)$ are estimated to be 0.38 rad/s, 8.4 rad/s, 8.0 rad/s, and 8.8 rad/s, respectively. The low frequency of the twisting has been predicted very well, while the estimation in the high frequencies of the twisting has a large error. The error between estimates and experimental results is primary due to the assumption of uniformly distributed-mass bar in the dynamic model, while bar in experiments is non-uniformly distributed-mass load. In summary, experimental data follow similar trends to the simulated curve in both swing and twisting. Therefore, the dynamic model is effective to describe the dynamics of tower cranes carrying slender bars, and the estimation method is also effective to predict the frequency of the payload twisting.

Another set of experiments was performed to validate the effectiveness of the control method at suppressing the swing and twisting. Figures 4.18 and 4.19 indicate the experimental responses of the swing and twisting to a baseline trapezoidal velocity command and smoothed commands by two types of smoother. The cable length was fixed at 64.5 cm. The baseline trapezoidal velocity command slews the jib for 80°. Without the controller, the corresponding residual amplitude of the swing deflection was 82.9 mm, and that of the twist angular velocity was 27.0°/s.

The two DP smoother should apply the two swing frequencies, ω_1 and ω_2. In the case of undamped system (industrial crane), the rise time of the TP smoother is $4\pi/\omega_1$, and that of two DP smoothers is $(1 + h)\pi/\omega_1 + (1 + h)\pi/\omega_2$. Thus, the rise time of the TP smoother is longer than the two DP smoothers because the modified factor, h, varies between zero and one. The rise time of TP smoother, two DP smoother with the modified factor h of 1, two DP smoother with the modified factor h of 0.5 are 3.14 s, 3.00 s, and 2.25 s, respectively.

With the modified factor h of 1, in two DP smoothers, the residual amplitudes of the swing deflection and twist angular velocity were 6.2 mm and 7.8°/s. With the modified factor h of 0.5, the residual amplitudes of the swing deflection and twist

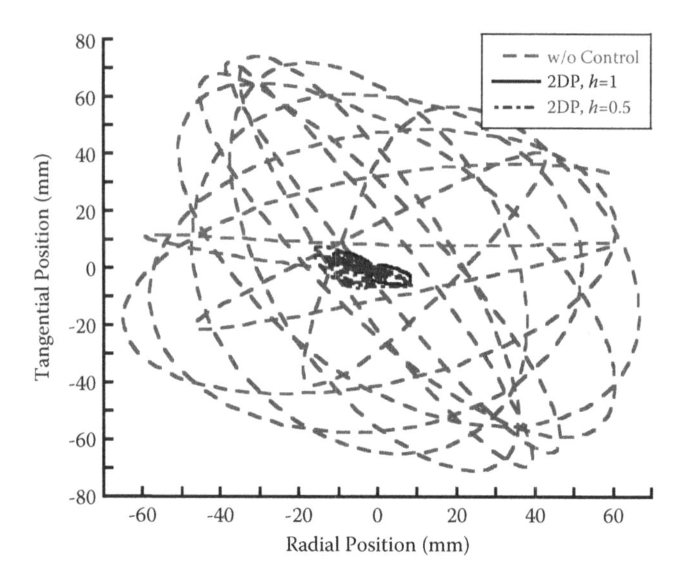

FIGURE 4.18 Experimental responses of swing for a slewing distance of 80° viewed from above.

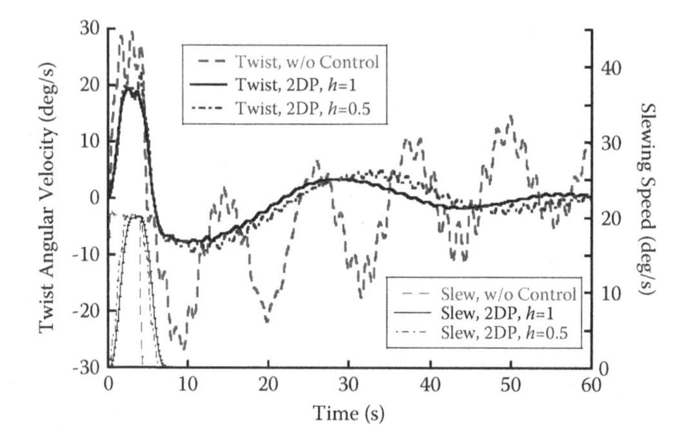

FIGURE 4.19 Experimental responses of twisting for a slewing distance of 80°.

angular velocity were 13.5 mm and 9.5°/s. Two DP smoothers with the higher modified factor suppressed more oscillations because the higher modified factor remains wider frequency insensitivity and longer rise time.

Last experiments were performed to demonstrate the robustness of two DP smoothers to changes in the system parameters. The DP smoother depends on the estimation of the swing frequency. Cable length has a large influence on the swing frequency. However, cable length may not be known accurately under a real application. Therefore, it is essential to study the control performs in the presence of modeling errors in the cable length. Figure 4.20 shows the residual amplitudes of

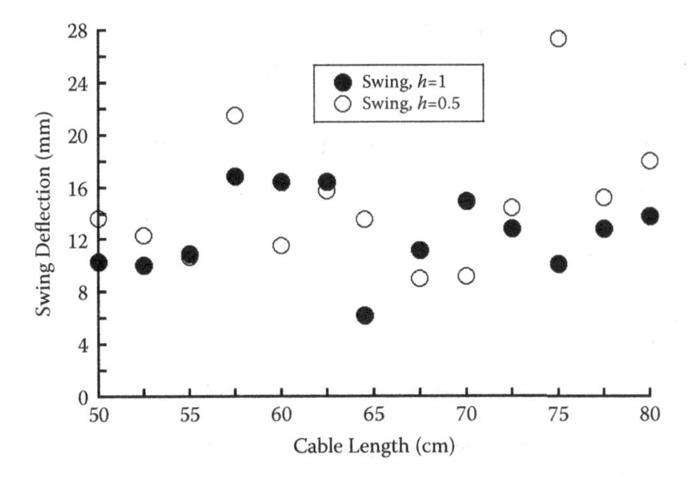

FIGURE 4.20 Experimental residual amplitudes of swing under different cable lengths for a fixed design length of 64.5 cm.

the swing deflection under different cable lengths, and Figure 4.21 shows those of twist angular velocity. The cable length varied from 50 cm to 80 cm, while the design length was fixed at 64.5 cm. The jib was slewed 80°. Note that the uncontrolled residual amplitudes of the swing deflection and twist angular velocity were 82.9 mm, and 27.0°/s, respectively. In the case of the modified factor h of 1, the two DP smoothers attenuated the experimental amplitudes of the swing and twisting by an average of 84.9% and 68.7% for all the cable lengths. In the case of the modified factor h of 0.5, the experimental residual amplitudes of the swing and twisting were suppressed by an average of 82.2% and 68.3%. Therefore, two DP smoothers with the modified factor of 1 and 0.5 both reduced the swing and

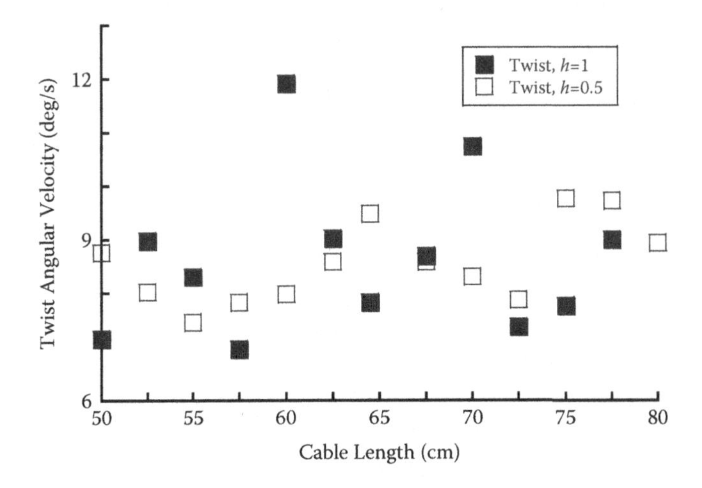

FIGURE 4.21 Experimental residual amplitudes of twisting under different cable lengths for a fixed design length of 64.5 cm.

twisting to a low level in experiments. Furthermore, the two DP smoothers exhibit good robustness to the modeling error in the cable length.

4.3 DOUBLE-PENDULUM AND TWISTING DYNAMICS

Many scientists have studied on the dynamics and control of tower cranes with point-mass payloads. The payload twisting cannot be captured in the point-mass payload model because of neglecting the payload size. The control methods for tower cranes with point-mass payloads include closed-loop controllers and open-loop controllers. Closed-loop controllers measure or estimate the payload swing to reduce oscillations in a feedback loop, including gain-scheduled control, fuzzy control, H_∞ control, neural network control, sliding mode control, LQR control, path-following control, adaptive control, and predictive control. However, accurately measuring the payload swing on-the-fly is difficult. Open-loop controllers modify the prescribed commands for swing suppression, including inverse kinematics, smooth commands, and input shaping. Nevertheless, all the above-mentioned works have focused on tower cranes with point-mass payload dynamics.

4.3.1 Modeling

Figure 4.22 shows a schematic representation of a tower crane transporting a distributed-mass beam. While a trolley moves radially with a position of r along the jib, a jib arm rotates by an angle of θ about the tower. A hook of mass, m_h, is attached to the trolley by a massless suspension cable of length, l_s. Two massless rigging cables of length, l_r, hangs below the hook and supports a uniformly distributed-mass beam of mass, m_p, and length, l_p. The payload twist angle, γ, is defined as the angle between the radial jib direction and the long axis of the beam.

The model assumes that the motion of the tower crane is unaffected by the motion of the hook and payload because of the high-ratio geared drives and small

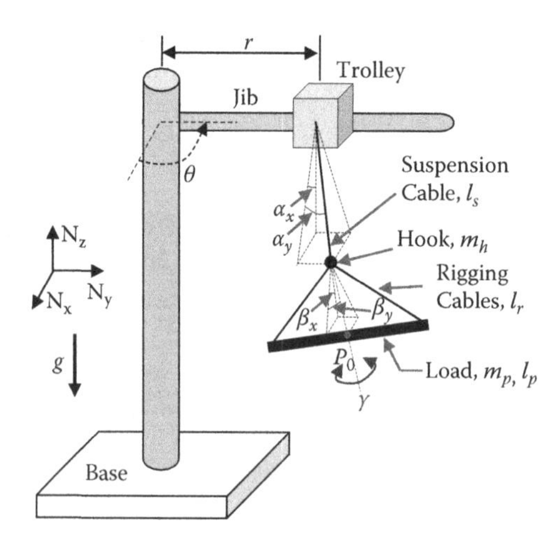

FIGURE 4.22 Model of a tower crane carrying a distributed-mass beam.

swing. It is also assumed that the damping ratio is approximately zero, the hook is modeled as a point mass, and the length of the suspension cable does not change during the motions. The angular acceleration of the jib and acceleration of the trolley are inputs to the model. The swing angles of the suspension cable α_x and α_y, the swing angles relative to the suspension cable β_x and β_y, and the payload twist angle γ are the outputs.

The nonlinear equations of motion for the model in Figure 4.22 were derived by using Kane's method. The generalized coordinates, q_k, are chosen as α_x, α_y, β_x, β_y, and γ. Meanwhile, the generalized speeds, u_k, are selected as:

$$u_1 = \dot{\alpha}_x,\ u_2 = \dot{\alpha}_y,\ u_3 = \dot{\beta}_x,\ u_4 = \dot{\beta}_y,\ u_5 = \dot{\gamma} \tag{4.48}$$

The basis transformation matrices between the two bodies J (jib), S (suspension cable) and P (bar) are:

$$\begin{Bmatrix} \mathbf{N}_x \\ \mathbf{N}_y \\ \mathbf{N}_z \end{Bmatrix} = \begin{bmatrix} \cos(\theta) & \sin(\theta) & 0 \\ -\sin(\theta) & \cos(\theta) & 0 \\ 0 & 0 & 1 \end{bmatrix} \begin{Bmatrix} \mathbf{J}_x \\ \mathbf{J}_y \\ \mathbf{J}_z \end{Bmatrix} \tag{4.49}$$

$$\begin{Bmatrix} \mathbf{J}_x \\ \mathbf{J}_y \\ \mathbf{J}_z \end{Bmatrix} = \begin{bmatrix} \cos(\alpha_x) & 0 & -\sin(\alpha_x) \\ 0 & 1 & 0 \\ \sin(\alpha_x) & 0 & \cos(\alpha_x) \end{bmatrix} \begin{bmatrix} 1 & 0 & 0 \\ 0 & \cos(\alpha_y) & -\sin(\alpha_y) \\ 0 & \sin(\alpha_y) & \cos(\alpha_y) \end{bmatrix} \begin{Bmatrix} \mathbf{S}_x \\ \mathbf{S}_y \\ \mathbf{S}_z \end{Bmatrix} \tag{4.50}$$

$$\begin{Bmatrix} \mathbf{S}_x \\ \mathbf{S}_y \\ \mathbf{S}_z \end{Bmatrix} = \begin{bmatrix} \cos(\beta_x) & 0 & -\sin(\beta_x) \\ 0 & 1 & 0 \\ \sin(\beta_x) & 0 & \cos(\beta_x) \end{bmatrix} \begin{bmatrix} 1 & 0 & 0 \\ 0 & \cos(\beta_y) & -\sin(\beta_y) \\ 0 & \sin(\beta_y) & \cos(\beta_y) \end{bmatrix} \begin{bmatrix} \cos(\gamma) & -\sin(\gamma) & 0 \\ \sin(\gamma) & \cos(\gamma) & 0 \\ 0 & 0 & 1 \end{bmatrix} \begin{Bmatrix} \mathbf{P}_x \\ \mathbf{P}_y \\ \mathbf{P}_z \end{Bmatrix} \tag{4.51}$$

The angular velocity of jib in the inertial coordinates is given by:

$${}^{N}\boldsymbol{\omega}^{J} = \dot{\theta}\mathbf{J}_3 \tag{4.52}$$

The angular acceleration of jib in the inertial coordinates is given by:

$${}^{N}\boldsymbol{\alpha}^{J} = \ddot{\theta}\mathbf{J}_3 \tag{4.53}$$

The velocity of trolley center of mass A^* in the inertial coordinates is:

$${}^{N}\mathbf{v}^{A^*} = \dot{r}\mathbf{J}_1 + r\dot{\theta}\mathbf{J}_2 \tag{4.54}$$

The acceleration of trolley center of mass A^* in the inertial coordinates is described as:

$${}^{N}\mathbf{a}^{A^*} = (\ddot{r} - r\dot{\theta}^2)\mathbf{J}_1 + (2\dot{r}\dot{\theta} + r\ddot{\theta})\mathbf{J}_2 \tag{4.55}$$

The angular velocity of the suspension cable S in the Newtonian reference frame is given by:

$$ {}^{N}\boldsymbol{\omega}^{S} = {}^{N}\boldsymbol{\omega}^{J} + \dot{\alpha}_x \mathbf{S}_x - \dot{\alpha}_y \cos(\alpha_x) \mathbf{S}_y + \dot{\alpha}_y \sin(\alpha_x) \mathbf{S}_z \tag{4.56} $$

The velocity of the point-mass hook in the Newtonian reference frame is expressed by:

$$ {}^{N}\mathbf{v}^{H} = {}^{N}\mathbf{v}^{A^*} + \dot{\alpha}_y l_s \cos(\alpha_x) \mathbf{S}_x + \dot{\alpha}_x l_s \mathbf{S}_y \tag{4.57} $$

The angular velocity of the load L in the Newtonian reference frame is:

$$ \begin{aligned} {}^{N}\boldsymbol{\omega}^{P} = {} & {}^{N}\boldsymbol{\omega}^{S} + (\cos(\gamma)\dot{\beta}_x + \cos(\beta_x)\sin(\gamma)\dot{\beta}_y)\mathbf{P}_x \\ & + (\cos(\beta_x)\cos(\gamma)\dot{\beta}_y - \sin(\gamma)\dot{\beta}_x)\mathbf{R}_y + (\dot{\gamma} - \sin(\beta_x)\dot{\beta}_y)\mathbf{R}_z \end{aligned} \tag{4.58} $$

The velocity of the mass center of the load in the Newtonian reference frame can be described as:

$$ {}^{N}\mathbf{v}^{P^*} = {}^{N}\mathbf{v}^{H} - {}^{N}\boldsymbol{\omega}^{P} \times l_x \mathbf{R}_z \tag{4.59} $$

where $\times$ denotes the cross product in this chapter,
where

$$ l_y = \sqrt{l_r^2 - 0.25 l_p^2} \tag{4.60} $$

The generalized active forces are expressed as:

$$ F_1 = -m_h g l_s S_{\alpha_x} C_{\alpha_y} - m_p g \mathbf{N}_z \cdot {}^{N}\mathbf{v}_1^{P^*} \tag{4.61} $$

$$ F_2 = -m_h g l_s C_{\alpha_x} S_{\alpha_y} - m_p g \mathbf{N}_z \cdot {}^{N}\mathbf{v}_2^{P^*} \tag{4.62} $$

$$ F_3 = -m_p g \mathbf{N}_z \cdot {}^{N}\mathbf{v}_3^{P^*} \tag{4.63} $$

$$ F_4 = -m_p g \mathbf{N}_z \cdot {}^{N}\mathbf{v}_4^{P^*} \tag{4.64} $$

$$ F_5 = 0 \tag{4.65} $$

where g is the gravity acceleration. Meanwhile, the generalized inertia forces are given by:

$$ F_k^* = -m_h \cdot {}^{N}\mathbf{a}^{H} . {}^{N}\mathbf{v}_k^{H} - m_p \cdot {}^{N}\mathbf{a}^{P^*} . {}^{N}\mathbf{v}_k^{P^*} - \left(\mathbf{I}^{P/P^*} . {}^{N}\boldsymbol{\alpha}^{P} + {}^{N}\boldsymbol{\omega}^{P} \times \mathbf{I}^{P/P^*} . {}^{N}\boldsymbol{\omega}^{P}\right) . {}^{N}\boldsymbol{\omega}_k^{P} \tag{4.66} $$

where $\mathbf{I}^{P/P^*}$ is the inertia dyadic of the payload about the mass center P^*, ${}^{N}\mathbf{a}^{H}$ is the acceleration of the hook, ${}^{N}\mathbf{v}_k^{H}$ is the kth partial velocity of the hook, ${}^{N}\mathbf{a}^{P^*}$ is the acceleration of the mass center of the payload, ${}^{N}\mathbf{v}_k^{P^*}$ is the kth partial velocities of the mass center of the payload, ${}^{N}\boldsymbol{\omega}_k^{P}$ is the kth partial angular velocity of the payload, and ${}^{N}\boldsymbol{\alpha}^{P}$ is the angular acceleration of the payload.

The sum of generalized active forces and generalized inertia forces is constrained to be zero. Then nonlinear equations of the double-pendulum motions for the model are given by:

$$M \cdot \begin{pmatrix} \ddot{\alpha}_x \\ \ddot{\alpha}_y \\ \ddot{\beta}_x \\ \ddot{\beta}_y \\ \ddot{\gamma} \end{pmatrix} + f\begin{pmatrix} \alpha_x,\ \alpha_y,\ \beta_x,\ \beta_y,\ \gamma,\ \theta,\ r, \\ \dot{\alpha}_x,\ \dot{\alpha}_y,\ \dot{\beta}_x,\ \dot{\beta}_y,\ \dot{\gamma},\ \dot{\theta},\ \dot{r},\ \ddot{\theta},\ \ddot{r} \end{pmatrix} = 0 \tag{4.67}$$

where the mass matrix is M, and the column matrix of gravity terms, centrifugal and Coriolis terms, and control input terms is f.

A trapezoidal-velocity profile was used to drive both the model (4.67) and the small-scale tower crane shown in Figure 4.23 through a slew of 80° in order to verify the nonlinear equations of motion. The jib accelerates at time zero. Then, the jib decelerates 4 seconds later. Both the acceleration and deceleration induce both the payload swing and twisting. The maximum slewing speed between the acceleration and deceleration is 20°/s. Figure 4.24 shows the experimental and simulated responses of the payload swing. The position deflection of the mass center of the payload relative to the trolley is defined as the payload swing displacement. In addition, Figure 4.25 shows the results of the angular velocity of the payload twist.

FIGURE 4.23 A testbed of a tower crane transporting a distributed-mass beam.

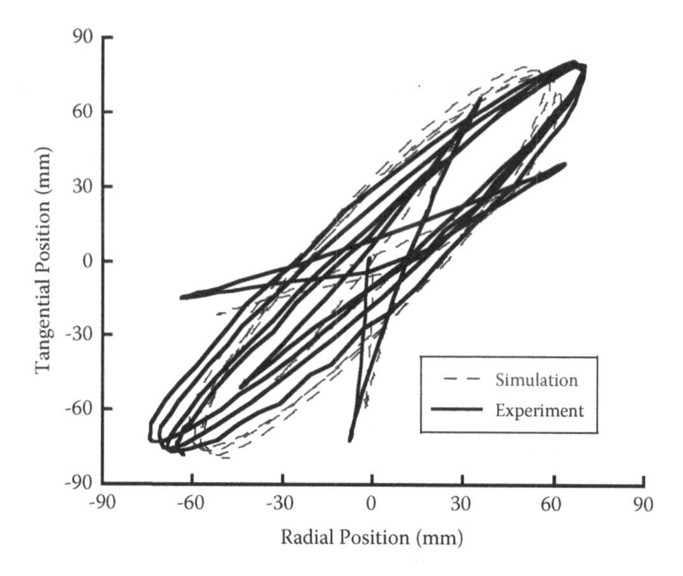

FIGURE 4.24 Experimental verification of payload swing.

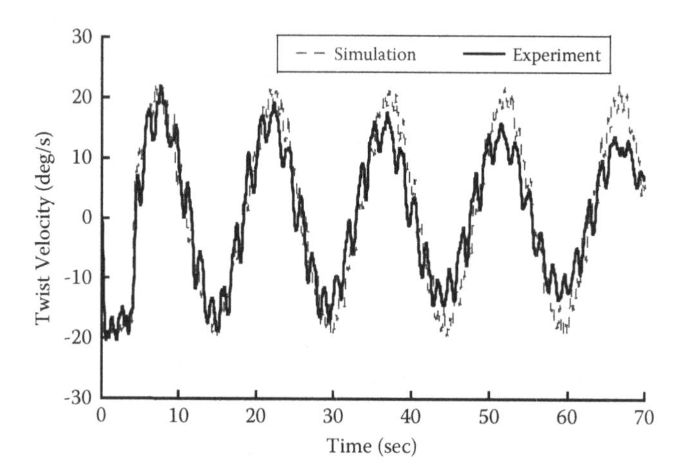

FIGURE 4.25 Experimental verification of payload twisting.

The experimental data show fairly good agreement with the nonlinear simulations. The difference between the simulated and experimental results is caused by the friction and air drag. The decay of the experimental response revealed such difference, which is most easily seen in the twist result in Figure 4.25.

4.3.2 Dynamics

The dynamics of tower cranes during pure radial motions are similar to that of bridge cranes. The payload twisting cannot be excited by pure radial motions in tower cranes when the initial twist angle is limited to zero. Thus, the simplified equations during the pure radial motions are:

$$\begin{bmatrix} [cR^2 + l_s^2 + c(l_y^2 + l_s^2 + 2l_y l_s \cos(\beta_y))] & c[R^2 + l_y(l_y + l_s \cos(\beta_y))] \\ [R^2 + l_y(l_y + l_s \cos(\beta_y))] & [R^2 + l_y^2] \end{bmatrix} \begin{pmatrix} \ddot{\alpha}_y \\ \ddot{\beta}_y \end{pmatrix}$$
$$+ \begin{bmatrix} -cl_y l_s \sin(\beta_y) \\ 0 \end{bmatrix} \cdot \dot{\beta}_y^2 + \begin{bmatrix} 0 \\ l_y l_s \sin(\beta_y) \end{bmatrix} \dot{\alpha}_y^2 + \begin{bmatrix} -2cl_y l_s \sin(\beta_y) \\ 0 \end{bmatrix} \dot{\alpha}_y \dot{\beta}_y$$
$$+ \begin{bmatrix} l_s(1 + c)\sin(\alpha_y) + cl_y \sin(\alpha_y + \beta_y) \\ l_y \sin(\alpha_y + \beta_y) \end{bmatrix} g$$
$$+ \begin{bmatrix} [l_s \cos(\alpha_y) + cl_s \cos(\alpha_y) + cl_y \cos(\alpha_y + \beta_y)] \\ [l_y \cos(\alpha_y + \beta_y)] \end{bmatrix} \ddot{r} = 0 \tag{4.68}$$

where the ratio of the payload mass to the hook mass is c. The coefficients R and l_y are given by:

$$R = l_p/(2\sqrt{3}) \tag{4.69}$$

$$l_y = \sqrt{l_r^2 - 0.25 l_p^2} \tag{4.70}$$

A linearized model can be derived from the abovementioned simplified model to approximate the pure radial motions by assuming small oscillations around the equilibrium position. Then the linearized natural frequencies of the payload swing during pure radial motions are given by:

$$\omega_{1,2}^2 = \frac{g(c + 1)}{2l_s}(u \mp v) \tag{4.71}$$

where

$$u = \frac{l_y^2 + l_s l_y + R^2}{l_y^2 + (c + 1)R^2} \tag{4.72}$$

$$v = \sqrt{u^2 - \frac{4 l_s l_y}{(c + 1)(l_y^2 + (c + 1)R^2)}} \tag{4.73}$$

The natural frequencies of the payload swing during slewing motions can also be estimated by Equation (4.71). This is because the payload swing exhibits weakly nonlinear dynamic behavior, especially when tower cranes move slowly. The swing frequencies depend on the mass ratio, suspension cable length, rigging cable length, and payload length. The mass ratio, cable length, and payload size have larger effects on the second-mode swing frequency than the first-mode swing frequency. The payload size and mass ratio only have minor impacts on the first-mode swing frequency. Meanwhile, increasing cable length decreases the first-mode swing frequency slightly, and decreases the second-mode swing frequency sharply.

The hook mass is set to be zero, and the swing angles are assumed to be small around the equilibrium point. Therefore, a simplified model for the twisting dynamics can be derived:

$$\ddot{\gamma} + [\alpha_y \sin(2\gamma) + \alpha_x - \alpha_x \cos(2\gamma)]\dot{\alpha}_x\dot{\theta} + [\alpha_x \sin(2\gamma) - 2\alpha_y \sin^2\gamma]\dot{\alpha}_y\dot{\theta}$$
$$+ [\alpha_x \sin\gamma + \alpha_y \cos\gamma][\alpha_y \sin\gamma - \alpha_x \cos\gamma]\cdot(\dot{\theta})^2 + \ddot{\theta} - \alpha_y\cdot\ddot{\alpha}_x = 0 \tag{4.74}$$

The twist acceleration is dependent on slewing motions and payload swing. The slewing motion is external excitation, while the payload swing is parametric excitation. The twisting response is similar to a harmonic motion in the residual stage (no external excitation). The payload twists about the swing direction back and forth. This is because the payload swing results in the twist acceleration. The magnitudes of the twist acceleration are dependent on the amplitude of the payload swing. The sign of the twist acceleration depends on the payload position. Zero swing of the payload causes zero acceleration of the payload twisting. The inertia effect will rotate the payload in one direction under conditions of constant twist angular velocity.

The twist frequency for various amplitudes of the payload swing is shown in Figure 4.26. The swing amplitude has a great impact on the twist frequency. The twist frequency increases as swing amplitude increases. This effect can be physically interpreted as the interference between the sign and magnitude of the twist acceleration. As the amplitude of the payload swing increase, the payload rotates faster before changing the sign of the twist acceleration. Furthermore, the twist frequency is limited to zero in the case of zero swing because zero payload swing

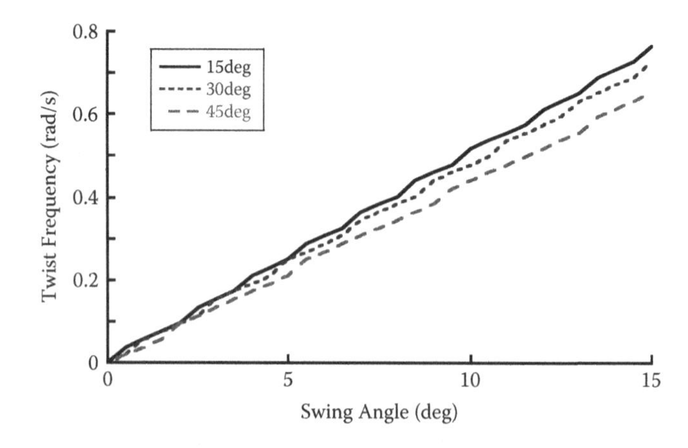

FIGURE 4.26 Twist frequency versus swing angle.

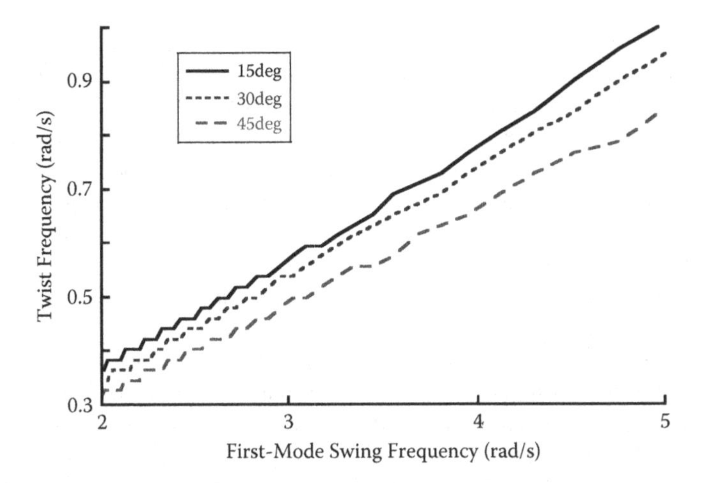

FIGURE 4.27 Twist frequency versus first-mode swing frequency.

results in zero twist acceleration. Thus, the payload will spin in one direction on condition of a constant angular velocity of the payload twist. Therefore, decreasing twist frequency results from reduction of the amplitude of the swing.

The initial twist angle also has a great effect on the payload twisting dynamics. The payload twisting dynamics for initial twist angles of 15°, 30°, and 45° are also shown in Figure 4.26. The twist frequency decreases with increasing the initial twist angle. This effect can also be interpreted as the interference between the sign and magnitude of the twist acceleration. When the initial twist angle increases, the payload rotates further before changing the sign of the twist acceleration. Because the complicated dynamical behavior of the payload twisting is sensitive to the initial twist angle, the payload twisting displays strongly nonlinear dynamical behavior.

The twist frequency for various first-mode frequency of the payload swing is shown in Figure 4.27. The first-mode swing frequency is larger than the twist frequency. The swing frequency also has a great impact on the twist frequency. The twist frequency also increases as swing frequency increases. Therefore, decreasing twist frequency results from reduction of frequency of the swing.

4.3.3 Experimental Validation

The testbed shown in Figure 4.23 was used to verify the dynamical behavior of the model and effectiveness of the smoother. The position of the trolley along the jib was set to be 75 cm. A tennis ball and a slender beam served as the hook and payload, respectively. Both the suspension cable and rigging cables were made of Dyneema fishing line. The hook mass, payload mass, payload length, suspension length, and rigging length were fixed at 32 g, 155 g, 29.7 cm, 54 cm, and 16 cm, respectively.

To record the displacements of two markers on the payload, the jib mounted to a camera, which captured video at 30 frames/s. The displacements of two markers were used to measure the payload swing displacement and twist angle. Averaging the displacements of the two markers derived the payload swing displacement. The twist angular velocity can be estimated by using the inverse tangent function.

The experimental control architecture is shown in Figure 4.28. A baseline trapezoidal-velocity profile is produced via the control interface. The command is then modified by the two-pieces smoother to produce a smoothed command for slewing the jib. The design frequency for the smoother was estimated by the suspension cable length, l_s, rigging cable length, l_r, payload length, l_p, and mass ratio, c. The design damping ratio is fixed at zero in the experiments.

The experimental response of the payload swing caused by the baseline trapezoid and smoothed commands are shown in Figure 4.29. The jib was slewed 80° using a maximum velocity of 20°/s. The residual amplitude is defined as the maximum deflection after the jib stops. The residual amplitude of the uncontrolled swing was 106 mm, while residual amplitude with the smoother was only 9.9 mm.

Figure 4.30 shows the slewing velocity profile and the experimental angular velocity of the payload twist. The smooth transitions between boundary conditions for the smoothed velocity profile reduce swing and twisting of the payload. Without the controller, the residual amplitude of the twist angular velocity was 21.9°/s. The residual amplitude with the smoother was 6.8°/s. Meanwhile, the experimental

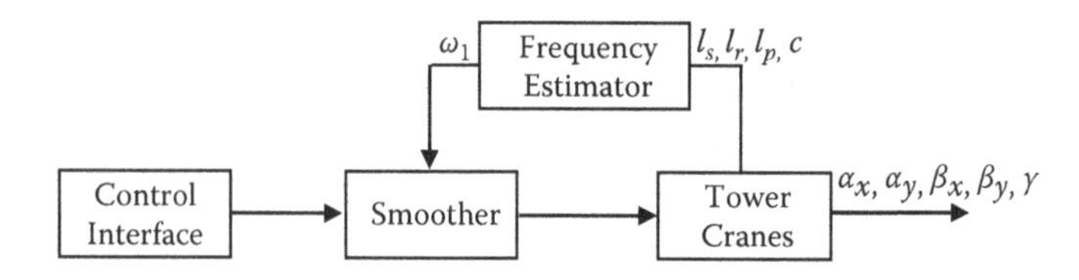

FIGURE 4.28 Control architecture for tower cranes.

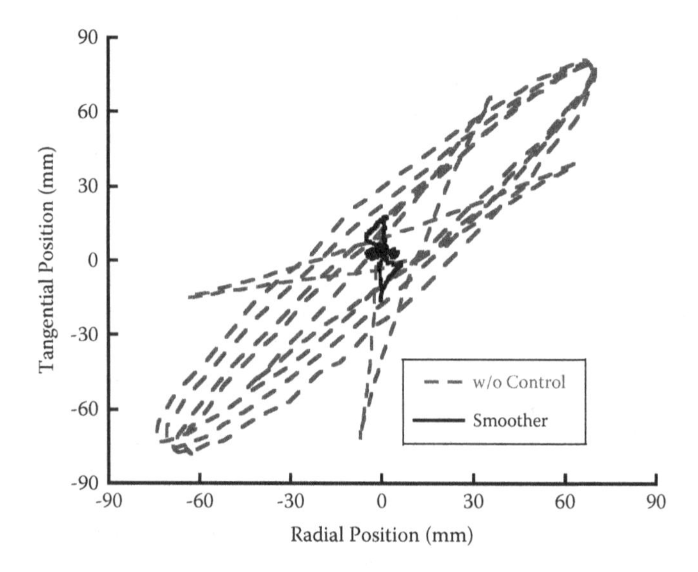

FIGURE 4.29 Experimental responses of payload swing.

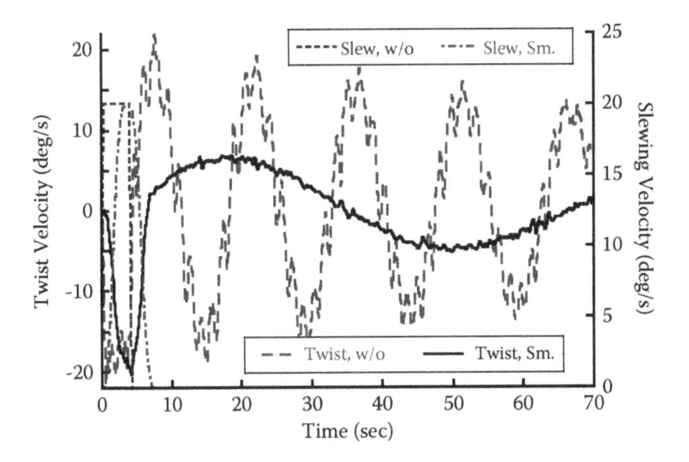

FIGURE 4.30 Experimental responses of payload twisting.

frequency of large-amplitude low-frequency twist oscillations without the controller was 0.43 rad/s, while that with the smoother was only 0.1 rad/s. Thus, the smoother benefits reduction of the twist frequency. Such a reduction would largely improve the safety of a real crane transporting a heavy and bulky payload.

Figure 4.31 shows the experimental and simulated results of the residual amplitudes of the payload swing. Peaks and troughs arose in the amplitude of the payload swing when using the unsmoothed trapezoidal commands. The design frequency and damping ratio of the smoother were set to be 4.032 rad/s and 0, respectively. The smoother eliminated most of the payload swing such that the controlled results were nearly independent of the slewing motions. The experimental results with the smoother were somewhat worse than the simulated results. This is because small modeling errors and uncertainty exist in the overall system.

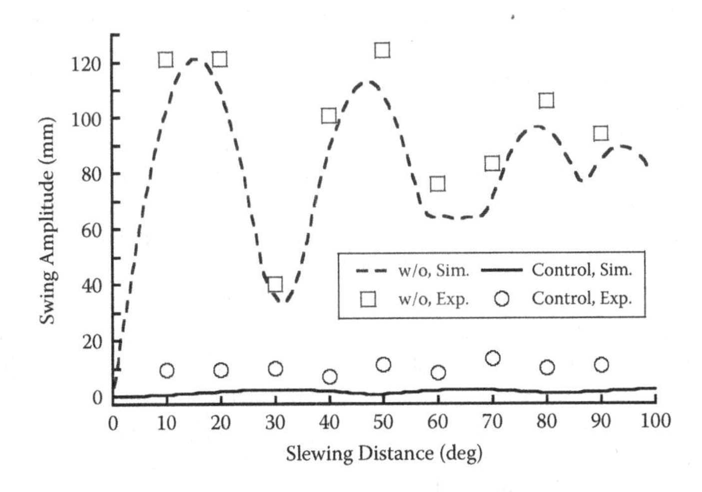

FIGURE 4.31 Experimental responses of payload swing as a function of slewing distances.

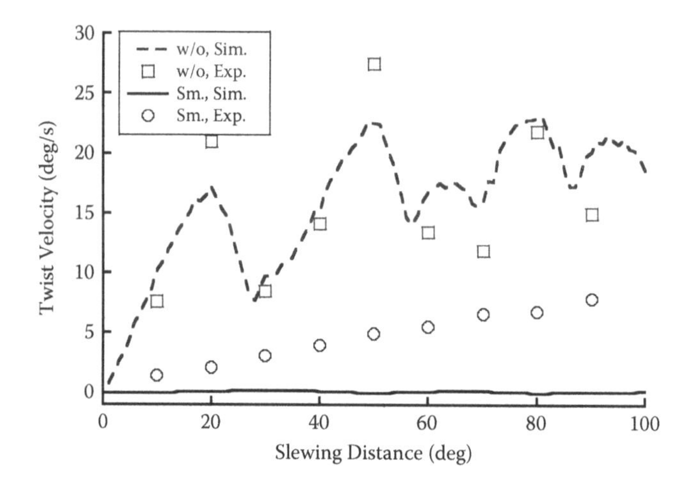

FIGURE 4.32 Experimental responses of payload twisting velocity as a function of slewing distances.

The results of the amplitude of the payload twisting are shown in Figure 4.32. As the rotating distance changed, peaks and troughs arose in the amplitude of the payload twisting. The magnitude of the twist acceleration reached a large value when the swing amplitude reached a maximum. Therefore, peaks in the twist amplitude occurred. As the swing amplitudes increase, the amplitude of the payload twisting increases. With the smoother, the experimental twist velocity increases as the slewing distance increases, because the model is undamped while the real system has some small amount of damping. The small damping corrupts the oscillation reduction for long slewing distances. However, the experimental data follow the general shape of the simulated results. The smoother attenuated the experimental swing amplitude and twist amplitude by an average of 89.8% and 71.7%, respectively.

The results of the frequency of the payload twisting are shown in Figure 4.33. By using the fast Fourier transform, the twist frequency was derived. As the rotating distance changed, peaks and troughs also arose in the frequency of the payload twisting. The magnitude of the twist acceleration reached a large value when the swing amplitude reached a maximum. Therefore, peaks in the twist frequency occurred. As the swing amplitudes increase, the frequency of the payload twisting increase. The smoother attenuated the experimental twist frequency by an average of 71.0%. The experimental data verified the theoretical dynamical behavior and the effectiveness of the two-pieces smoother.

Figures 4.34 and 4.35 show the experimental payload swing and twisting resulting from different modeling errors in the frequency. The ratio of the modeled frequency to the real frequency is defined as the normalized frequency. The design frequency of 4.032 rad/s corresponds to the normalized frequency of one. In the case of normalized frequency of one, the residual amplitudes of the swing and twisting were 9.9 mm and 6.8°/s, respectively. The residual amplitudes of the swing and twisting for the frequency of 0.6 were 10.1 mm and 7.6°/s, and those for the normalized frequency of 1.4 were 36.8 mm and 9.0°/s.

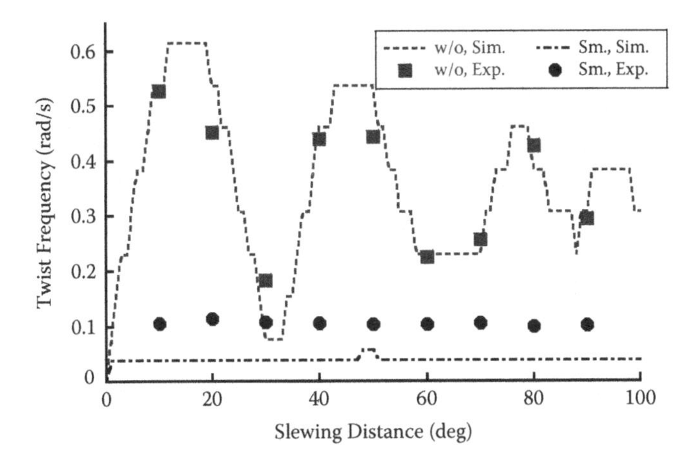

FIGURE 4.33 Experimental responses of payload twisting frequency as a function of slewing distances.

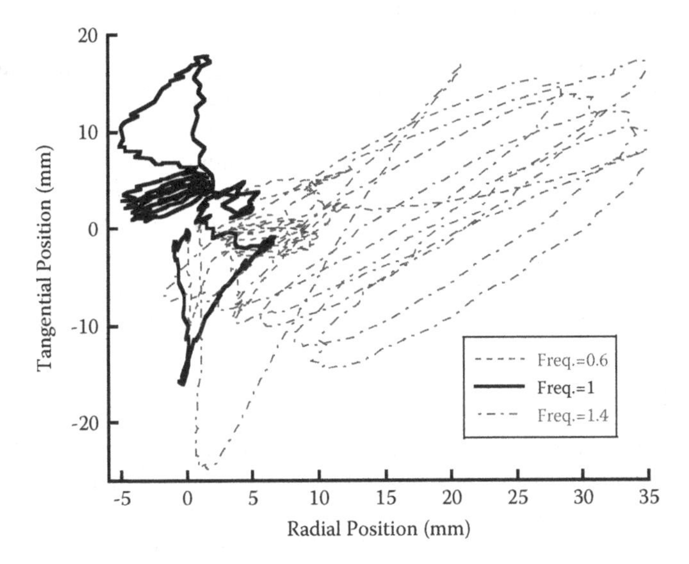

FIGURE 4.34 Experimental responses of payload swing for various modeling errors in the frequency.

The experimental results clearly indicate that the payload swing increases with increasing modeling error. The payload swing exhibits insensitivity to the modeling error at low normalized frequencies due to the low-pass filtering effect. The experimental results show that the controlled twisting response is insensitive to modeling errors in the frequency. The smoother suppresses both the payload swing and twisting to a very low level, and provides good insensitivity to modeling errors in the frequency.

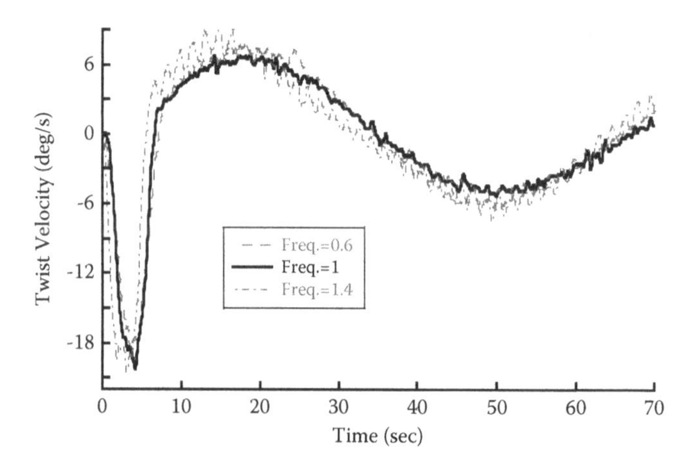

FIGURE 4.35 Experimental responses of payload twisting for various modeling errors in the frequency.

4.4 JIB-PENDULUM DYNAMICS

Mounting numbers of the work have directed at the jib-pendulum dynamics for tower cranes [31–35], beam-pendulum dynamics for bridge cranes [36–40], and pendulum-beam dynamics in other types of applications [40–47]. While jib/beam vibrations have been applied to derive the pendulum dynamics, both swing and twisting of the load have not been involved in the jib/beam dynamics. Therefore, the coupled behaviors of jib-pendulum dynamics are weak, and the previous model should be improved. In addition, the frequency analyses of coupled jib-pendulum systems did not report in the prior literature. Furthermore, the corresponding control methods are challenging to reduce oscillations of jib-pendulum dynamics.

4.4.1 Modeling

A schematic representation of a tower crane with flexible jib carrying a distributed-mass load is shown in Figure 4.36. A tower base slews a lightweight jib, J, of length, L, to a prescribed angular displacement, θ, about the N_3 direction. In the ideal condition, the movement of the jib tracks the specified angular displacement, θ. However, the flexible nature of the lightweight jib often causes undesirable vibrations during the manipulation. The vibrational deflection of a point on the jib at a distance, x, from the base in the J_2 direction is denoted by $w(x,t)$. The vibrational deflection of the jib in the J_1 and J_3 direction is neglected in this chapter due to geometrically constrained structure in the tower crane.

A trolley of mass, m_A, slides with a position of r along the jib in the J_1 direction. Two massless cables with the same length, l_r, suspend below the trolley and carry a uniformly distributed-mass load, P. The load is a uniformly distributed-mass bar of mass, m_p, and length, l_p. The cross-section of the thin load is assumed to be small. The swing angles of the cables are denoted by β_1 and β_2, while the twist angle of the load is referred to as γ.

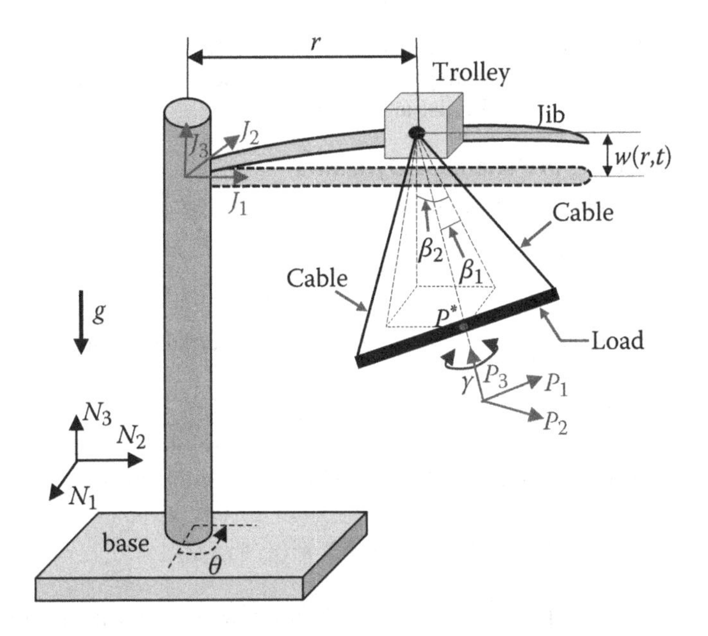

FIGURE 4.36 Model of a tower crane with flexible jib transporting a thin bar.

Rotating the slewing angle, θ, converts the inertial coordinates $N_1N_2N_3$ to the moving Cartesian coordinates $J_1J_2J_3$. In addition, the moving coordinates $J_1J_2J_3$ of the jib can also be converted to the moving Cartesian coordinates $P_1P_2P_3$ of the load by rotating swing angles, β_2 and β_1, and twist angle, γ, respectively. The basis transformation matrices between the bodies J (jib) and P (load) are:

$$\begin{Bmatrix} \mathbf{N}_1 \\ \mathbf{N}_2 \\ \mathbf{N}_3 \end{Bmatrix} = \begin{bmatrix} C_\theta & S_\theta & 0 \\ -C_\theta & C_\theta & 0 \\ 0 & 0 & 1 \end{bmatrix} \begin{Bmatrix} \mathbf{J}_1 \\ \mathbf{J}_2 \\ \mathbf{J}_3 \end{Bmatrix} \tag{4.75}$$

$$\begin{Bmatrix} \mathbf{J}_1 \\ \mathbf{J}_2 \\ \mathbf{J}_3 \end{Bmatrix} = \begin{bmatrix} C_{\beta_2} & 0 & -S_{\beta_2} \\ 0 & 1 & 0 \\ S_{\beta_2} & 0 & C_{\beta_2} \end{bmatrix} \begin{bmatrix} 1 & 0 & 0 \\ 0 & C_{\beta_1} & S_{\beta_1} \\ 0 & -S_{\beta_1} & C_{\beta_1} \end{bmatrix} \begin{bmatrix} C_\gamma & S_\gamma & 0 \\ -S_\gamma & C_\gamma & 0 \\ 0 & 0 & 1 \end{bmatrix} \begin{Bmatrix} \mathbf{P}_1 \\ \mathbf{P}_2 \\ \mathbf{P}_3 \end{Bmatrix} \tag{4.76}$$

where the following abbreviations are used:

$$\begin{cases} S_\theta \triangleq \sin(\theta),\ C_\theta \triangleq \cos(\theta),\ S_{\beta_1} \triangleq \sin(\beta_1),\ C_{\beta_1} \triangleq \cos(\beta_1), \\ S_{\beta_2} \triangleq \sin(\beta_2),\ C_{\beta_2} \triangleq \cos(\beta_2),\ S_\gamma \triangleq \sin(\gamma),\ C_\gamma \triangleq \cos(\gamma). \end{cases} \tag{4.77}$$

It is assumed that the movement of the tower base is unaffected by the motion of the jib and suspended load due to the large inertia of the base and small oscillations of jib and load. The cable length is assumed to be unchanged during the operation. The inputs to the model are the angular acceleration, $\ddot{\theta}$, of the slewing jib and the acceleration, $\ddot{r}$, of the sliding trolley. The outputs to the model are jib deflection, w, swing angles, β_2 and β_1, and twist angle, γ.

4.4.1.1 Modeling of Pendulum Dynamics

The angular velocity of the jib in the inertial coordinates is described as:

$$^{N}\boldsymbol{\omega}^{J} = \dot{\theta}\mathbf{N}_3 \tag{4.78}$$

Meanwhile, the angular acceleration of the jib in the inertial coordinates is given by:

$$^{N}\boldsymbol{\alpha}^{J} = \ddot{\theta}\mathbf{N}_3 \tag{4.79}$$

The velocity of trolley centroid, A^{*}, in the inertial coordinates is expressed as:

$$^{N}\mathbf{v}^{A^{*}} = (\dot{r} - w\dot{\theta})\mathbf{J}_1 + (\dot{w} + r\dot{\theta})\mathbf{J}_2 \tag{4.80}$$

Meanwhile, the acceleration of trolley centroid in the inertial coordinates is given by:

$$^{N}\mathbf{a}^{A^{*}} = (\ddot{r} - r\dot{\theta}^{2} - w\ddot{\theta} - 2\dot{w}\dot{\theta})\mathbf{J}_1 + (2\dot{r}\dot{\theta} + r\ddot{\theta} + \ddot{w} - w\dot{\theta}^{2})\mathbf{J}_2 \tag{4.81}$$

The angular velocity of the load in the inertial coordinates is:

$$^{N}\boldsymbol{\omega}^{P} = \dot{\theta}\mathbf{N}_3 + \left(S_\gamma C_{\beta_1}\dot{\beta}_2 + C_\gamma\dot{\beta}_1\right)P_1 + \left(C_{\beta_1}C_\gamma\dot{\beta}_2 - S_\gamma\dot{\beta}_1\right)P_2 + \left(\dot{\gamma} - S_{\beta_1}\dot{\beta}_2\right)P_3 \tag{4.82}$$

Meanwhile, the angular acceleration of the load in the inertial coordinates is:

$$^{N}\boldsymbol{\alpha}^{P} = \ddot{\theta}\mathbf{N}_3 + \left(S_\gamma C_{\beta_1}\ddot{\beta}_2 + C_\gamma\ddot{\beta}_1\right)\mathbf{P}_1 + \left(C_{\beta_1}C_\gamma\ddot{\beta}_2 - S_\gamma\ddot{\beta}_1\right)\mathbf{P}_2 + \left(\ddot{\gamma} - S_{\beta_1}\ddot{\beta}_2\right)\mathbf{P}_3 \tag{4.83}$$

The velocity of load centroid, P^{*}, in the inertial coordinates is:

$$\begin{aligned} ^{N}\mathbf{v}^{P^{*}} = {} & ^{N}\mathbf{v}^{A^{*}} - l_y\left(C_{\beta_1}C_\gamma\dot{\beta}_2 - \dot{\theta}S_{\beta_2}S_\gamma + S_\gamma\dot{\beta}_1 + \dot{\theta}S_{\beta_1}C_{\beta_2}C_\gamma\right)\mathbf{P}_1 \\ & - l_y\left(C_{\beta_1}S_\gamma\dot{\beta}_2 + \dot{\theta}S_{\beta_2}C_\gamma - C_\gamma\dot{\beta}_1 + \dot{\theta}S_{\beta_1}C_{\beta_2}S_\gamma\right)\mathbf{P}_2 \end{aligned} \tag{4.84}$$

where

$$l_y = \sqrt{l_r^2 - 0.25l_p^2} \tag{4.85}$$

Meanwhile, the acceleration of load centroid in the inertial coordinates is:

$$
\begin{aligned}
{}^{N}\mathbf{a}^{P^*} = & {}^{N}\mathbf{a}^{A^*} - l_y\left(C_{\beta_1}C_\gamma\ddot{\beta}_2 - \ddot{\theta}S_{\beta_2}S_\gamma + S_\gamma\ddot{\beta}_1 + \ddot{\theta}S_{\beta_1}C_{\beta_2}C_\gamma\right)\mathbf{P}_1 \\
& - l_y\left(C_{\beta_1}S_\gamma\ddot{\beta}_2 + \ddot{\theta}S_{\beta_2}C_\gamma - C_\gamma\ddot{\beta}_1 + \ddot{\theta}S_{\beta_1}C_{\beta_2}S_\gamma\right)\mathbf{P}_2 \\
& + {}^{N}\boldsymbol{\omega}^{P} \times \left({}^{N}\mathbf{v}^{P^*} - {}^{N}\mathbf{v}^{A^*}\right)
\end{aligned} \tag{4.86}
$$

The pendulum dynamics will be derived using Kane's equation. The generalized speed is chosen as angular velocity, $\dot{\beta}_2$, $\dot{\beta}_1$, $\dot{\gamma}$, of the swing and twisting, respectively. The generalized active forces are expressed as:

$$
\begin{cases}
F_1 = -m_p g l_y C_{\beta_1}(S_\gamma + C_\gamma), \\
F_2 = -m_p g l_y (S_\gamma - C_\gamma), \\
F_3 = 0,
\end{cases} \tag{4.87}
$$

where g is the gravity acceleration. Meanwhile, the generalized inertia forces are given by:

$$
F_k^* = -m_p \cdot {}^{N}\mathbf{a}^{P^*} \cdot {}^{N}\mathbf{v}_k^{P^*} - \left(\mathbf{I}^{P/P^*} \cdot {}^{N}\boldsymbol{\alpha}^{P} + {}^{N}\boldsymbol{\omega}^{P} \times \mathbf{I}^{P/P^*} \cdot {}^{N}\boldsymbol{\omega}^{P}\right) \cdot {}^{N}\boldsymbol{\omega}_k^{P} \tag{4.88}
$$

where $\cdot$ is dot product and $\times$ is the cross product, $\mathbf{I}^{P/P*}$ is the inertia dyadic of the load about the centroid, ${}^{N}\mathbf{v}_k{}^{P^*}$ and ${}^{N}\boldsymbol{\omega}_k{}^{P}$ are the kth partial velocity of the load center and kth partial angular velocity of the load. Additionally, the partial velocities of the load center are described as:

$$
\begin{cases}
{}^{N}\mathbf{v}_1^{P^*} = -l_y \cos\beta_1 \cos\gamma \mathbf{P}_1 - l_y \cos\beta_1 \sin\gamma \mathbf{P}_2, \\
{}^{N}\mathbf{v}_2^{P^*} = -l_y \sin\gamma \mathbf{P}_1 + l_y \cos\gamma \mathbf{P}_2, \\
{}^{N}\mathbf{v}_3^{P^*} = 0.
\end{cases} \tag{4.89}
$$

The partial angular velocities of the load are expressed as:

$$
\begin{cases}
{}^{N}\boldsymbol{\omega}_1^{P} = S_\gamma C_{\beta_1}\mathbf{P}_1 + C_{\beta_1}C_\gamma \mathbf{P}_2 - S_{\beta_1}\mathbf{P}_3, \\
{}^{N}\boldsymbol{\omega}_2^{P} = C_\gamma \mathbf{P}_1 - S_\gamma \mathbf{P}_2, \\
{}^{N}\boldsymbol{\omega}_3^{P} = \mathbf{P}_3.
\end{cases} \tag{4.90}
$$

Constraining the sum of generalized active forces (4.87) and generalized inertia forces (4.88) to be zero gives rise to the nonlinear equations of the pendulum dynamics:

$$
\begin{aligned}
& m_p g l_y C_{\beta_1}(S_\gamma + C_\gamma) + m_p \cdot {}^{N}\mathbf{a}^{P^*} \cdot {}^{N}\mathbf{v}_1^{P^*} \\
& + \left(\mathbf{I}^{P/P^*} \cdot {}^{N}\boldsymbol{\alpha}^{P} + {}^{N}\boldsymbol{\omega}^{P} \times \mathbf{I}^{P/P^*} \cdot {}^{N}\boldsymbol{\omega}^{P}\right) \cdot {}^{N}\boldsymbol{\omega}_1^{P} = 0
\end{aligned} \tag{4.91}
$$

$$
m_p g l_y (S_\gamma - C_\gamma) + m_p \cdot {}^{N}\mathbf{a}^{P^*} \cdot {}^{N}\mathbf{v}_2^{P^*} + \left(\mathbf{I}^{P/P^*} \cdot {}^{N}\boldsymbol{\alpha}^{P} + {}^{N}\boldsymbol{\omega}^{P} \times \mathbf{I}^{P/P^*} \cdot {}^{N}\boldsymbol{\omega}^{P}\right) \cdot {}^{N}\boldsymbol{\omega}_2^{P} = 0 \tag{4.92}
$$

$$
\left(\mathbf{I}^{P/P^*} \cdot {}^{N}\boldsymbol{\alpha}^{P} + {}^{N}\boldsymbol{\omega}^{P} \times \mathbf{I}^{P/P^*} \cdot {}^{N}\boldsymbol{\omega}^{P}\right) \cdot {}^{N}\boldsymbol{\omega}_3^{P} = 0 \tag{4.93}
$$

4.4.1.2 Modeling of Jib Dynamics

The force equation of the motion of a small element at the distance, x, on the jib measured from the base is given by:

$$\rho_J dx \frac{\partial^2 w(x,t)}{\partial t^2} + dV(x,t) - f_e(x,t)dx - \rho_J x \frac{d^2\theta}{dt^2} dx = 0; \quad 0 \le x \le L \tag{4.94}$$

where ρ_J denotes the linear mass density of the jib, $V(x,t)$ denotes the shear force on the cross-section of the jib, dx is the small change of the distance x, dV represents the small change of the shear force V, and $f_e(x,t)$ denotes the pendulum-induced force acting on the per unit length of the jib. The distributed force, $f_e(x,t)$, is expressed as:

$$f_e(x,t) = \begin{cases} F_e(t), & at \quad x = r \\ 0, & \text{others} \end{cases} \tag{4.95}$$

where $F_e(t)$ is the pendulum-induced force acting on the jib in the J_2 direction. The distributed force, $f_e(x,t)$, is equal to the point force, $F_e(t)$, at the distance, r, and zero otherwise. Besides, the point force, $F_e(t)$, caused by the pendulum motions is defined as:

$$F_e(t) = \left(m_A \,{}^N\mathbf{a}^{A^*} + 2m_A \,{}^N\boldsymbol{\omega}^J \times {}^N\mathbf{v}^{A^*} + T\right)\cdot \mathbf{J}_2 \tag{4.96}$$

where T is the cable tension force. The tension force of the cable satisfies:

$$T = \left[\left(m_p \,{}^N\mathbf{a}^{P^*} + 2m_p \,{}^N\boldsymbol{\omega}^P \times {}^N\mathbf{v}^{P^*} - m_p g N_3\right)\cdot P_3\right] P_3 \tag{4.97}$$

The moment equation of the motion of the small element is described by:

$$dM(x,t) - [V(x,t) + dV(x,t)]dx + [\rho x\ddot{\theta} dx + f_e(x,t)dx]\frac{1}{2}dx = 0 \tag{4.98}$$

where $M(x,t)$ denotes the bending moment on the cross-section of the jib, dM is the small change of the bending moment, $M(x,t)$. The relationship between the bending moment and jib deflection is:

$$M(x,t) = EI_J \frac{\partial^2 w(x,t)/\partial x^2}{[1 + (\partial w(x,t)/\partial x)^2]^{1.5}} \tag{4.99}$$

where E is the Young's modulus, and I_J is the moment of inertia of the cross-section of the jib. Expanding (4.99) and then ignoring the high-order power terms give rise to:

$$M(x,t) = EI_J \frac{\partial^2 w(x,t)}{\partial x^2}[1 - 1.5(\partial w(x,t)/\partial x)^2] \tag{4.100}$$

Substituting Equations (4.98) and (4.100) into (4.94) and then neglecting the high-order power terms of dx give rise to:

$$\begin{aligned} &EI_J \frac{\partial^4 w(x,t)}{\partial x^4} - 1.5EI_J \frac{\partial^2\left(\frac{\partial^2 w(x,t)}{\partial x^2}\left(\frac{\partial w(x,t)}{\partial x}\right)^2\right)}{\partial x^2} + \rho_J \frac{\partial^2 w(x,t)}{\partial t^2} \\ &- f_e(x,t) - \rho_J x \frac{d^2\theta}{dt^2} = 0 \end{aligned} \tag{4.101}$$

The boundary conditions of the flexible jib at the tower base and free end are expressed by:

$$w(x,t)|_{x=0} = 0 \tag{4.102}$$

$$\frac{\partial w(x,t)}{\partial x}|_{x=0} = 0 \tag{4.103}$$

$$\frac{\partial}{\partial x}\left[EI_J \frac{\partial^2 w(x,t)}{\partial x^2}\right]|_{r=l_B} = 0 \tag{4.104}$$

$$EI_J \frac{\partial^2 w(x,t)}{\partial x^2}|_{x=L} = 0 \tag{4.105}$$

By the mode superposition method, the jib deflection is assumed as:

$$w(x,t) = \sum_{k=1}^{+\infty} [\varphi_k(x) q_k(t)] \tag{4.106}$$

where q_k is the time-dependent function of the kth vibrational model of the flexible jib, and φ_k is the corresponding mode shape. Substituting Equation (4.106) into (4.102), (4.103), (4.104), and (4.105) yields the natural frequency and mode shape:

$$\omega_k^2 = \beta_k^4 \frac{EI_J}{\rho_J} \tag{4.107}$$

$$\varphi_k(x) = [\sin(\beta_k x) - \sinh(\beta_k x)] - \lambda_k[\cos(\beta_k x) - \cosh(\beta_k x)] \tag{4.108}$$

$$\lambda_k = \frac{\sin(\beta_k L) + \sinh(\beta_k L)}{\cos(\beta_k L) + \cosh(\beta_k L)} \tag{4.109}$$

$$1 + \cos(\beta_k l_B)\cosh(\beta_k l_B) = 0 \tag{4.110}$$

where ω_k is the natural frequency of the kth vibrational mode of the jib. Substituting Equation (4.106) into (4.101) yields:

$$\sum_{k=1}^{+\infty}\left[\varphi_k\frac{d^2q_k}{dt^2}\right]+\frac{EI_J}{\rho_J}\sum_{k=1}^{+\infty}\left[\frac{\partial^4\varphi_k}{\partial x^4}q_k\right]-\frac{1.5EI_J}{\rho_J}\sum_{k=1}^{+\infty}\left[\frac{\partial^2\left(\frac{\partial^2\varphi_k}{\partial x^2}\left(\frac{\partial\varphi_k}{\partial x}\right)^2\right)}{\partial x^2}q_k^3\right]$$
$$-\frac{f_e}{\rho_J}-x\frac{d^2\theta}{dt^2}=0 \tag{4.111}$$

Multiplying Equation (4.111) by the mode shape, φ_k, integrating over the jib length, and then ignoring the modal coupling between different vibrational modes of the jib result in:

$$\frac{d^2q_k}{dt^2}+\omega_k^2q_k-\frac{1.5\int_0^L\frac{\partial^2\left(\frac{\partial^2\varphi_k}{\partial x^2}\left(\frac{\partial\varphi_k}{\partial x}\right)^2\right)}{\partial x^2}dx}{\int_0^L\frac{\partial^4\varphi_k}{\partial x^4}dx}\omega_k^2q_k^3-\frac{\varphi_k(r)F_e}{\rho_J\int_0^L\varphi_k\varphi_k dr}-\frac{\int_0^L x\varphi_k dx}{\int_0^L\varphi_k\varphi_k dx}\frac{d^2\theta}{dt^2}=0 \tag{4.112}$$

Including the proportional damping into Equation (4.112) yields the dynamic equation of the kth vibrational mode of the jib:

$$\frac{d^2q_k}{dt^2}+2\zeta_k\omega_k\frac{dq_k}{dt}+\omega_k^2q_k-\frac{1.5\int_0^L\frac{\partial^2\left(\frac{\partial^2\varphi_k}{\partial x^2}\left(\frac{\partial\varphi_k}{\partial x}\right)^2\right)}{\partial x^2}dx}{\int_0^L\frac{\partial^4\varphi_k}{\partial x^4}dx}\omega_k^2q_k^3$$
$$-\frac{\varphi_k(r)F_e}{\rho_J\int_0^L\varphi_k\varphi_k dr}-\frac{\int_0^L x\varphi_k dx}{\int_0^L\varphi_k\varphi_k dx}\frac{d^2\theta}{dt^2}=0 \tag{4.113}$$

The coupled jib-pendulum dynamics in a tower crane with flexible jib carrying a distributed-mass load are described by swing and twisting (4.91), (4.92), and (4.93), pendulum-induced force (4.96), mode superposition (4.106), mode shape (4.108), and time-dependent function (4.113).

4.4.2 Dynamics

Letting the velocity and acceleration of the jib deflection and that of load swing and twisting be zero in the full dynamic model (4.91), (4.92), (4.93), (4.96), (4.106), (4.108), and (4.113) results in the equilibrium points. The equilibria of tower cranes having flexible jib and carrying distributed-mass loads are zero jib deflection, zero swing angles, and arbitrary twist angle. By assuming small oscillations near the equilibria, the dynamic model (4.91), (4.92), (4.93), (4.96), (4.106), (4.108), and (4.113) yields a free-oscillation linearized model:

$$\ddot{q}_k + \omega_k^2 q_k - \kappa\left(m_p g\beta_1 + m_A \sum_{k=1}^{+\infty} [\varphi_k(r)\ddot{q}_k]\right) = 0 \quad (4.114)$$

$$l_p^2 S_\gamma C_\gamma \ddot{\beta}_1 - \left(12l_y^2 + l_p^2 C_\gamma^2\right)\ddot{\beta}_2 - 12gl_y\beta_2 = 0 \quad (4.115)$$

$$\left(12l_y^2 + l_p^2 S_\gamma^2\right)\ddot{\beta}_1 - l_p^2 S_\gamma C_\gamma \ddot{\beta}_2 + 12gl_y\beta_1 + 12l_y \sum_{k=1}^{+\infty} [\varphi_k(r)\ddot{q}_k] = 0 \quad (4.116)$$

$$\ddot{\gamma} - \beta_1\ddot{\beta}_2 = 0 \quad (4.117)$$

where

$$\kappa = \frac{\varphi_k(r)}{\rho_J \int_0^L \varphi_k\varphi_k dr} \quad (4.118)$$

Equation (4.114) describes the free-oscillation linearized equation of jib deflection, Equations (4.115) and (4.116) report that of load swing, and Equation (4.117) defines that of the load twisting. Dynamics of jib deflection contain an infinite number of vibrational modes. By ignoring the high modes in the jib dynamics, the free-oscillation linearized model (4.114), (4.115), and (4.116) results in a simplified model:

$$[1 - m_A\kappa\varphi_1(r)]\ddot{q}_1 + \omega_1^2 q_1 - m_p g\kappa\beta_1 = 0 \quad (4.119)$$

$$l_p^2 S_\gamma C_\gamma \ddot{\beta}_1 - \left(12l_y^2 + l_p^2 C_\gamma^2\right)\ddot{\beta}_2 - 12gl_y\beta_2 = 0 \quad (4.120)$$

$$\left(12l_y^2 + l_p^2 S_\gamma^2\right)\ddot{\beta}_1 - l_p^2 S_\gamma C_\gamma \ddot{\beta}_2 + 12gl_y\beta_1 + 12l_y\varphi_1(r)\ddot{q}_1(t) = 0 \quad (4.121)$$

The simplified model (4.119), (4.120), and (4.121) gives rise to an equation for estimating three coupled frequencies, Ω, in tower cranes having flexible jib and carrying distributed-mass loads:

$$l_y^2\left(12l_y^2 + l_p^2\right)[m_A\kappa\varphi_1(r) - 1]\Omega^6 + \begin{pmatrix} gl_y\left(24l_y^2 + l_p^2\right)[m_A\kappa\varphi_1(r) - 1] \\ -l_y^2\left(12l_y^2 + l_p^2\right)\omega_1^2 \\ +m_p gl_y\kappa\varphi_1(r)(12l_y^2 + l_p^2 C_\gamma^2) \end{pmatrix}\Omega^4$$
$$+ gl_y\left[12gl_y - 12gl_y(m_A + m_p)\kappa\varphi_1(r) + \omega_1^2(24l_y^2 + l_p^2)\right]\Omega^2 - 12g^2l_y^2\omega_1^2 = 0 \quad (4.122)$$

The three frequencies resulting from (4.122) are dependent on the jib material, the moment of inertia, I_J, of the jib, trolley position, r, trolley mass, m_A, load mass, m_p, cable

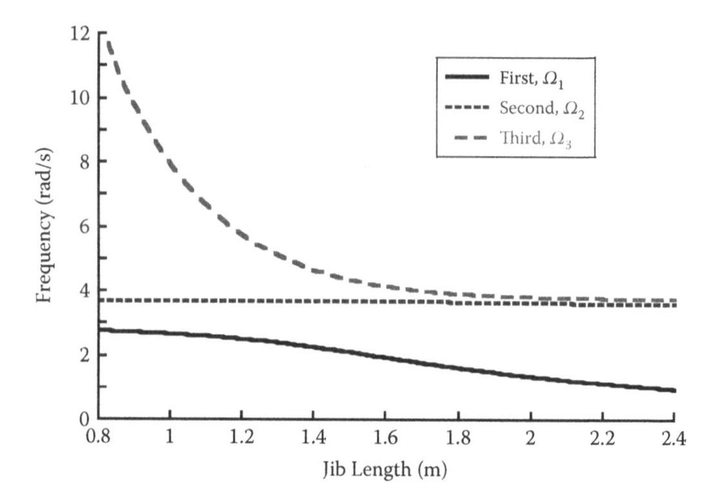

FIGURE 4.37 Frequencies of jib-pendulum dynamics versus jib length.

length, l_r, load length, l_p, and twist angle, γ. The system parameters used in the numerical simulation are the Young's modulus of the jib material, moment of inertia of the cross-section of the uniform jib, linear mass density of the jib, trolley mass, trolley displacement, cable length, load length, load mass, and initial twist angle are 206 GPa, 4.4367 mm^4, 0.3476 kg/m, 10 g, 80 cm, 80 cm, 78.5 cm, 265 g, and 60°, respectively.

Figure 4.37 displays three frequencies of the jib-pendulum dynamics for various jib length. As the cable length increases, the first frequency, Ω_1, and the third frequency, Ω_3, decrease, and the second frequency, Ω_2, keeps constant. The third frequency is more related with the jib motions, while the other low frequencies are more related with the pendulum motions. The third frequency approaches the second frequency after the cable length of 2 m, thereby arising internal resonances.

4.4.3 Oscillation Suppression

When the modified factor satisfies $0 \le h < 0.5$, the hybrid piecewise (HP) smoother is:

$$c(\tau) = \begin{cases} \mu^2(\tau)e^{-\zeta\omega\tau}, & 0 \le \tau \le 0.5hT \\ \mu^2(-\tau + hT)e^{-\zeta\omega\tau}, & 0.5hT < \tau \le hT \\ 0, & hT < \tau \le 0.5T \\ \mu^2(2\tau - T)e^{-\zeta\omega\tau}, & 0.5T < \tau \le (0.5T + 0.5hT) \\ \mu^2(-2\tau + T + 2hT)e^{-\zeta\omega\tau}, & (0.5T + 0.5hT) < \tau \le (0.5T + hT) \\ 0, & (0.5T + hT) < \tau \le T \\ \mu^2(\tau - T)e^{-\zeta\omega\tau}, & T < \tau \le (T + 0.5hT) \\ \mu^2(-\tau + T + hT)e^{-\zeta\omega\tau}, & (T + 0.5hT) < \tau \le (T + hT) \\ 0, & \text{others} \end{cases} \tag{4.123}$$

When the modified factor satisfies $0.5 \le h \le 1$, the HP smoother is:

$$c(\tau) = \begin{cases} \mu^2(\tau)e^{-\zeta_m\omega_m\tau}, & 0 \le \tau \le 0.5hT \\ \mu^2(-\tau + hT)e^{-\zeta_m\omega_m\tau}, & 0.5hT < \tau \le 0.5T \\ \mu^2(\tau + hT - T)e^{-\zeta_m\omega_m\tau}, & 0.5T < \tau \le hT \\ \mu^2(2\tau - T)e^{-\zeta_m\omega_m\tau}, & hT < \tau \le (0.5T + 0.5hT) \\ \mu^2(-2\tau + T + 2hT)e^{-\zeta_m\omega_m\tau}, & (0.5T + 0.5hT) < \tau \le T \\ \mu^2(-\tau + 2hT)e^{-\zeta_m\omega_m\tau}, & T < \tau \le (0.5T + hT) \\ \mu^2(\tau - T)e^{-\zeta_m\omega_m\tau}, & (0.5T + hT) < \tau \le (T + 0.5hT) \\ \mu^2(-\tau + T + hT)e^{-\zeta_m\omega_m\tau}, & (T + 0.5hT) < \tau \le (T + hT) \\ 0, & \text{others} \end{cases} \quad (4.124)$$

The rise time of the HP smoother (4.123) and (4.124) is $(1 + h)T$. While the HP smoother (4.123) is a discontinuous piecewise function with three discontinuous pieces, the HP smoother (4.124) is a continuous piecewise with eight pieces. The transfer function of the HP smoother (4.123) and (4.124) can be described as:

$$\mathrm{HP}(s) = \frac{\mu^2 \begin{bmatrix} 1 - 2e^{-h\pi\zeta_m/\sqrt{1-\zeta_m^2}}e^{-(0.5hT)s} + 2e^{-\pi\zeta_m/\sqrt{1-\zeta_m^2}}e^{-(0.5T)s} \\ + e^{-2h\pi\zeta_m/\sqrt{1-\zeta_m^2}}e^{-(hT)s} - 4e^{-(1+h)\pi\zeta_m/\sqrt{1-\zeta_m^2}}e^{-(0.5T+0.5hT)s} \\ + e^{-2\pi\zeta_m/\sqrt{1-\zeta_m^2}}e^{-(T)s} + 2e^{-(1+2h)\pi\zeta_m/\sqrt{1-\zeta_m^2}}e^{-(0.5T+hT)s} \\ - 2e^{-(2+h)\pi\zeta_m/\sqrt{1-\zeta_m^2}}e^{-(T+0.5hT)s} + e^{-(2+2h)\pi\zeta_m/\sqrt{1-\zeta_m^2}}e^{-(1+h)Ts} \end{bmatrix}}{(s + \zeta_m\omega_m)^2} \quad 4.125$$

The low-pass filtering effect inherent in the HP smoother will attenuate the high-mode oscillations. For the case of high normalized frequency ($\omega > \omega_m$), the percentage vibrational amplitude should be limited to be less than a tolerable level, V_{tol}, of the vibration:

$$\left(\sqrt{\left(\int_{\tau=0}^{+\infty}\left[ce^{\zeta\omega\tau}\sin\left(\omega\sqrt{1-\zeta^2}\tau\right)d\tau\right]\right)^2 + \left(\int_{\tau=0}^{+\infty}\left[ce^{\zeta\omega\tau}\cos\left(\omega\sqrt{1-\zeta^2}\tau\right)d\tau\right]\right)^2}\right)_{\substack{\omega>\omega_m \\ \zeta=\zeta_m}} \le V_{tol} \quad (4.126)$$

For a tolerable vibration level, the modified factor, h, of the HP smoother can be calculated from Equation (4.126). When the tolerable level, V_{tol}, was set to 5%, 10%, and 15%, the modified factor, h, of the HP smoother were 0.98963, 0.85815, 0.77380, respectively. Figure 4.38 shows the frequency-sensitive curves of the HP smoother. The HP smoother exhibits the notch and low-pass filtering effects. In the case of modified factor h of 0.98963, a peak value of 5% arises at the normalized frequency of 1.44. In the case of modified factor h of 0.85815, frequency-sensitive curve occurs at another peak with the value of 10% at the normalized frequency of 1.54. In the case of modified factor h of 0.77380, another peak value of 15% arises at the normalized frequency of 1.61.

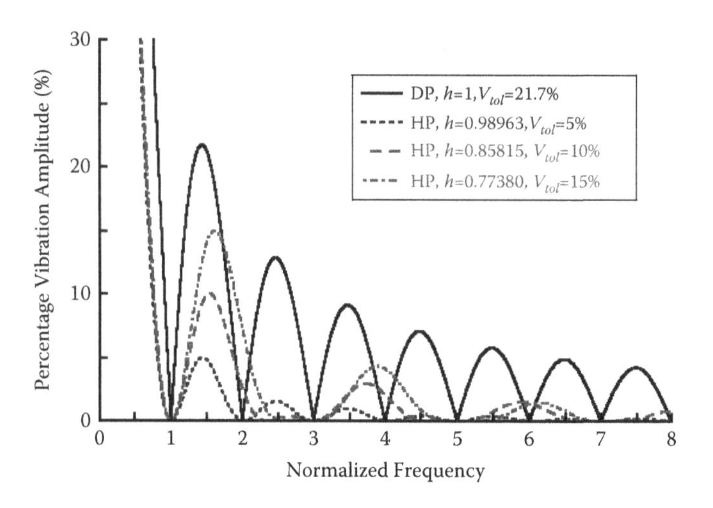

FIGURE 4.38 Frequency-sensitive curves of DP and HP smoothers.

The notch filtering effect inherent in the HP smoother can suppress oscillations of the fundamental mode, while the low-pass filtering effect embedded in the HP smoother can reduce oscillations of high modes. Oscillation reduction of high modes depends on the modified factor, h. Increasing the modified factor increases oscillation attenuation of high modes, but increases the rise time.

4.4.4 Experimental Validation

A small-scale tower crane testbed shown in Figure 4.39 was applied to validate the dynamic behavior of the nonlinear model described in Equations (4.91), (4.92),

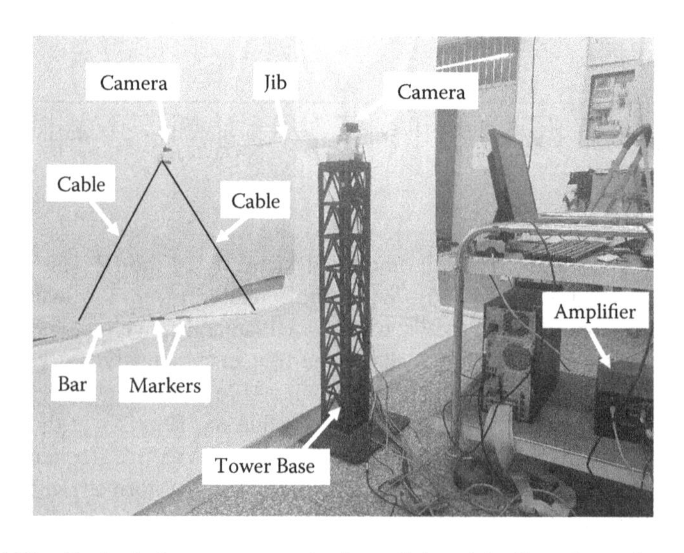

FIGURE 4.39 Testbed of a tower crane having a lightweight jib and carrying a thin bar.

(4.93), (4.96), (4.106), (4.108), and (4.113) and effectiveness and robustness of control method given in Equations (4.123) and (4.124). A rectangular link was connected on the tower base to serve as the jib. The length, height, and breadth of the jib were 86.5 cm, 4 cm, and 1.1 mm, respectively. A motor with encoders mounted on the base slews the jib about the tower. Two Dyneema cables with the same length of 80 cm were connected to the jib at the displacement of 80 cm, and supported a thin bar. The bar served as the load, and the mass, length, and diameter of the load were 265 g, 78.5 cm, and 12 mm, respectively.

A motion control card attached a host computer to the amplifier of the motor. The host computer was used as the operator interface and implementation of the control algorithm. A proportional-integral-derivative controller inherent in the motion control card forces the slewing velocity profile of the jib to follow the operator's commands in the host computer.

One camera fixed at the base for recording the jib deflection, while the other camera mounted on the jib for measuring two markers on the bar. Averaging the deflections of two markers gives rise to the swing deflection of the mass center of the bar. Meanwhile, the inverse tangent function was applied to estimate the twist angular velocity of the bar.

Figures 4.40–4.42 show free-oscillation response when the initial deflection of the load in the radial direction, that in the tangential direction, initial twist angle, and initial jib deflection were set to −47.7 mm, 29.8 mm, 45°, and 0.8 mm, respectively. Three theoretic frequencies of 2.75 rad/s, 3.71 rad/s, and 10.58 rad/s can be calculated from Equation (4.122). The first frequency of 2.75 rad/s and the third frequency of 10.58 rad/s can be found from the response of the jib deflection shown in Figure 4.42. The load deflection in the radial direction in Figure 4.40 is related with the second frequency, while that in the tangential direction is related with the first frequency. The three frequencies and beat frequencies arose in the response of twist angular velocity in Figure 4.41. The experimental response in

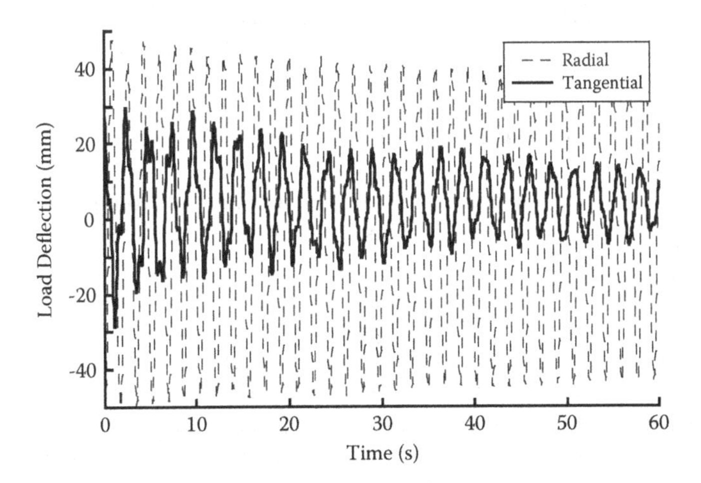

FIGURE 4.40 Free-oscillation experimental response of load deflection.

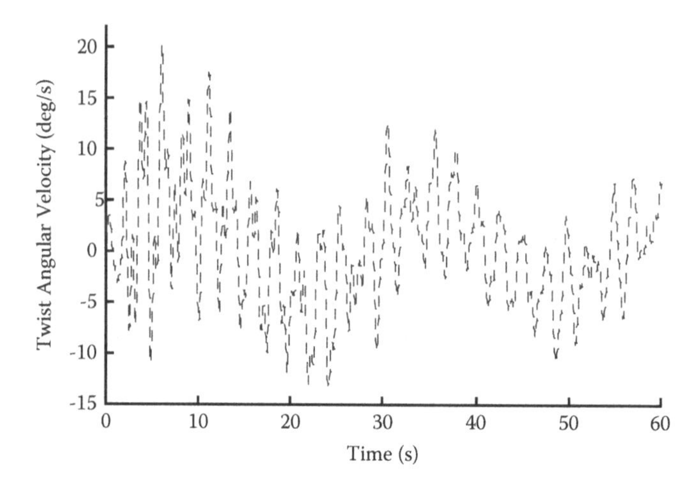

FIGURE 4.41 Free-oscillation experimental response of twist angular velocity.

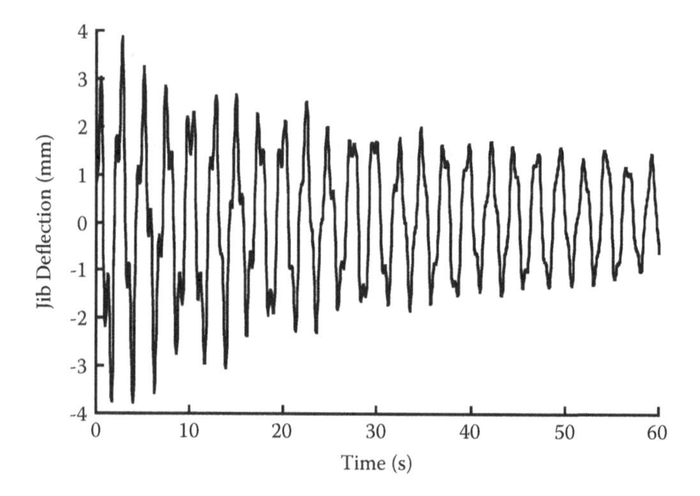

FIGURE 4.42 Free-oscillation experimental response of jib deflection.

Figures 4.40–4.42 demonstrated that the frequency estimation given in Equation (4.122) is effective. Thus, experiments verified the effectiveness of the dynamic behavior of the model.

Figures 4.43–4.47 show experimental response to a trapezoidal velocity profile for slewing the jib of 15°. The maximum slewing speed is 10°/s. The transient deflection is defined as the maximum deflection as the motor slews, while the residual amplitude is referred as to the maximum deflection after the slewing speed reaches zero. The residual amplitudes of the load deflection in the radial direction, load deflection in tangential direction, twist angular velocity, and jib deflection are 50.0 mm, 17.1 mm, 22.7°/s, and 9.1 mm, respectively. The residual amplitude is the interference between oscillations caused by the slewing acceleration and

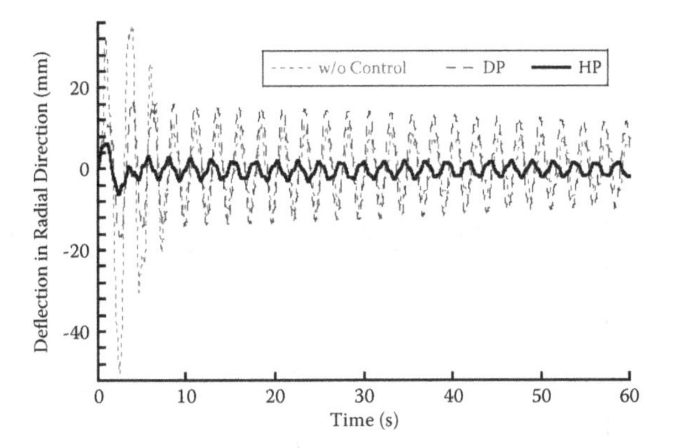

FIGURE 4.43 Experimental response of load deflection in radial direction.

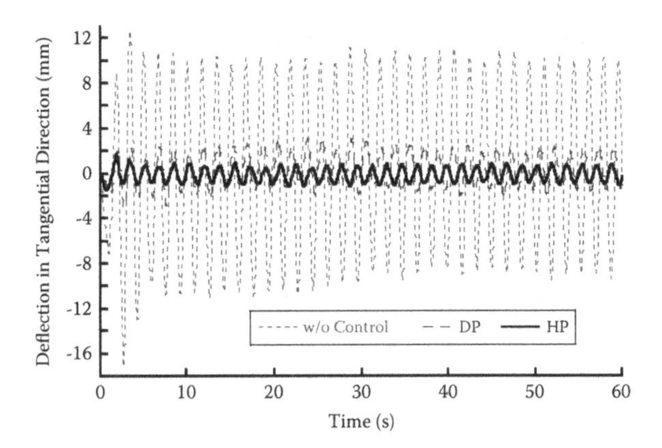

FIGURE 4.44 Experimental response of load deflection in tangential direction.

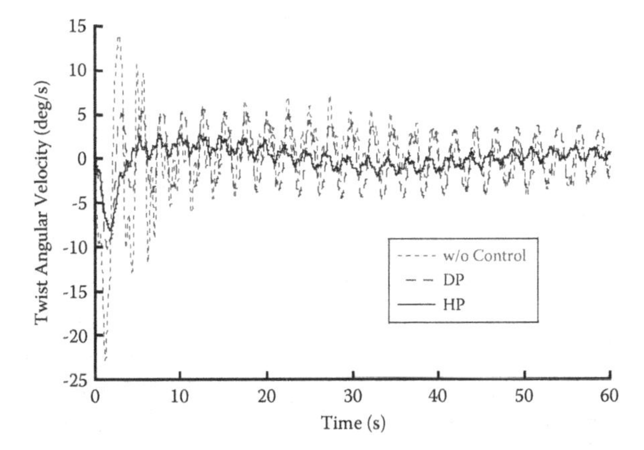

FIGURE 4.45 Experimental response of twist angular velocity.

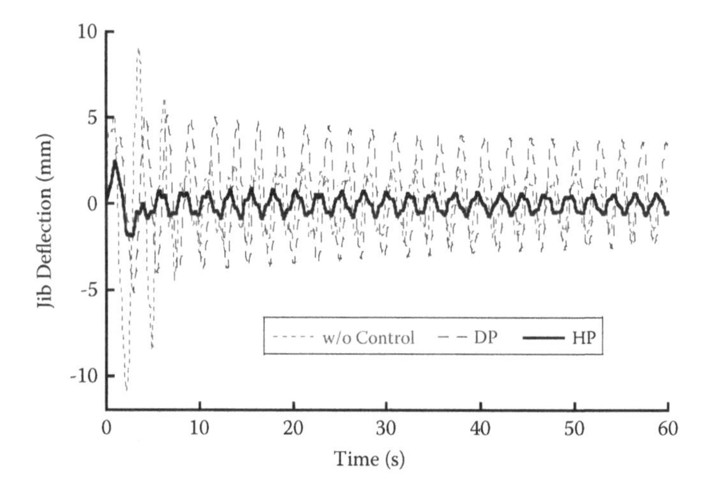

FIGURE 4.46 Experimental response of jib deflection.

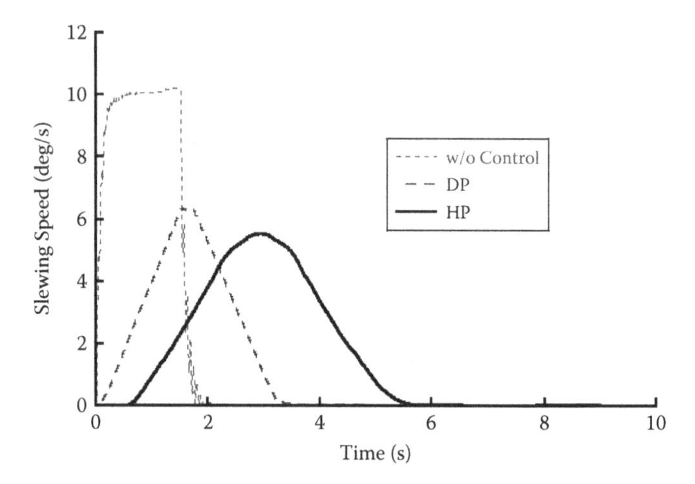

FIGURE 4.47 Experimental response of slewing speed.

deceleration. Three frequencies, 2.75 rad/s, 3.71 rad/s, and 10.58 rad/s, which were calculated from Equation (4.122), can also be found from the residual oscillations in Figures 4.43–4.47.

When the HP smoother is implemented, the experimental response under the same conditions is also shown in Figures 4.43–4.47. Given that the tolerable level, V_{tol}, was selected as 10% for low-pass filtering effect, the modified factor h of 0.85815 was used. Therefore, the rise time of the HP smoother is 4.25 second in this case. The trapezoidal velocity profile filters through the HP smoother to generate a smooth-transition profile. The HP-smoothed residual amplitudes of the load deflection in the radial direction, load deflection in tangential direction, twist angular velocity, and jib deflection are 3.1 mm, 1.2 mm, 2.8°/s, and 0.9 mm, respectively.

The experimental response with the discontinuous piecewise (DP) smoother is also shown in Figures 4.43–4.47 for the purpose of comparison. The corresponding rise time of the DP smoother is 2.28 seconds. The DP-smoothed residual amplitudes of the load deflection in the radial direction, load deflection in tangential direction, twist angular velocity, and jib deflection are 16.7 mm, 3.1 mm, 5.6°/s, and 5.1 mm, respectively. Therefore, the HP smoother suppressed more oscillations than the DP smoother.

Figures 4.48 and 4.49 show experimental residual amplitudes of load deflection, twist angular velocity, and jib deflection induced by slewing displacement. The slewing displacement ranged from 0° to 60°. The HP smoother attenuated residual

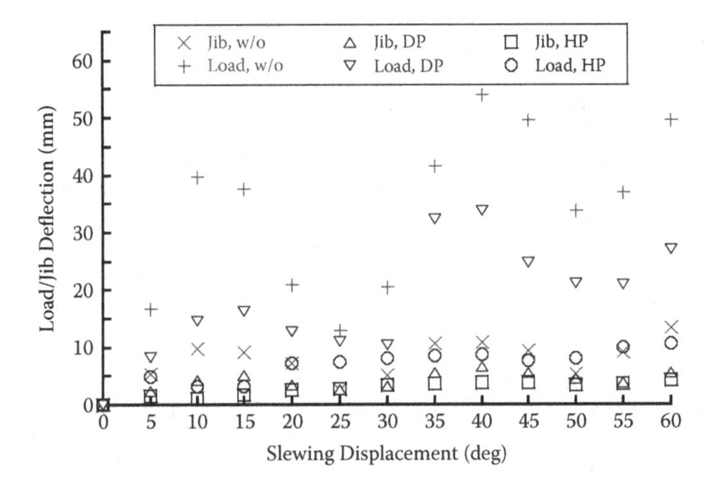

FIGURE 4.48 Experimental vibrational amplitudes of load and jib deflection for various slewing displacements.

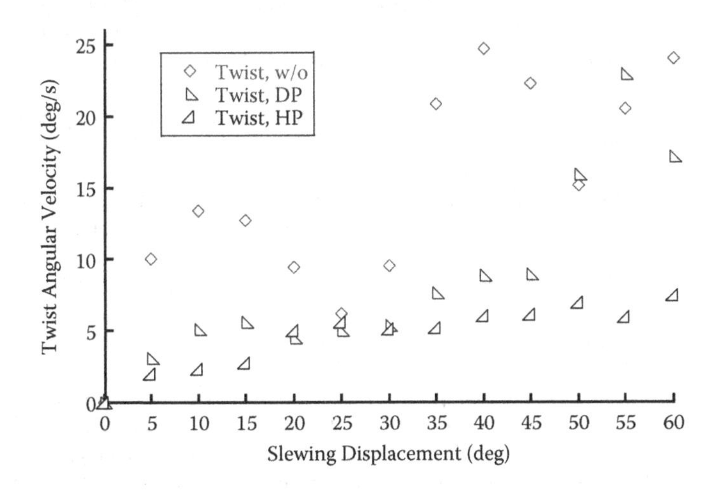

FIGURE 4.49 Experimental vibrational amplitudes of twist angular velocity for various slewing displacements.

oscillations of load deflection, twist angular velocity, and jib deflection by an average of 78.3%, 70.1%, and 66.8%, while the DP smoother reduced those by an average of 48.8%, 50.6%, and 52.8%. Therefore, the HP smoother is more effective to suppress oscillations of jib deflection, load swing, and load twisting than the DP smoother.

REFERENCES

[1] H. Kenan, O. Azeloğlu, Design of scaled down model of a tower crane mast by using similitude theory, *Engineering Structures*, 220 (2020) 110985.

[2] H. Omar, A. Nayfeh, Gain scheduling feedback control of tower cranes with friction compensation, *Journal of Vibration and Control*, 10 (2) (2004) 269–289.

[3] H. Coral-Enriquez, S. Pulido-Guerrero, J. Cortes-Romero, Robust disturbance rejection based control with extended-state resonant observer for sway reduction in uncertain tower-cranes, *International Journal of Automation and Computing*, 16 (6) (2019) 812–827.

[4] F. Rauscher, O. Sawodny, Modeling and control of tower cranes with elastic structure, *IEEE Transactions on Control Systems Technology*, doi:10.1109/TCST.2019.2961639

[5] M. Zhang, Y. Zhang, H. Ouyang, et al., Adaptive integral sliding mode control with payload sway reduction for 4-DOF tower crane systems, *Nonlinear Dynamics*, 99 (2020) 2727–2741.

[6] H. Ouyang, Z. Tian, L. Yu, et al., Adaptive tracking controller design for double-pendulum tower cranes, *Mechanism and Machine Theory*, 153 (2020) 103980.

[7] L. Aboserre, A. El-Badawy, Robust integral sliding mode control of tower cranes, *Journal of Vibration and Control*, doi:10.1177/1077546320938183

[8] Y. Qian, Y. Fang, Switching logic-based nonlinear feedback control of offshore ship-mounted tower cranes: a disturbance observer-based approach, *IEEE Transactions on Automation Science and Engineering*, 16 (3) (2019) 1125–1136.

[9] H. Chen, N. Sun, Nonlinear control of underactuated systems subject to both actuated and unactuated state constraints with experimental verification, *IEEE Transactions on Industrial Electronics*, 67 (9) (2020) 7702–7714.

[10] H. Chen, Y. Fang, N. Sun, An adaptive tracking control method with swing suppression for 4-DOF tower crane systems, *Mechanical Systems and Signal Processing*, 123 (2019) 426–442.

[11] R. Ngabesong, M. Yilmaz, Parametric and linear parameter varying modeling and optimization of uncertain crane systems, *International Journal of Dynamics and Control*, 7 (2) (2019) 430–438.

[12] V. Pyrhonen, M. Vilkko, Composite nonlinear feedback control of a JIB trolley of a tower crane behaviors, European Control Conference, Naples, Italy, pp. 1124–1129, 2019.

[13] T.A. Le, V.H. Dang, D.H. Ko, T.N. An, S.G. Lee, Nonlinear controls of a rotating tower crane in conjunction with trolley motion, *Proceedings of the Institution of Mechanical Engineers, Part I: Journal of Systems and Control Engineering*, 227(5) (2013) 451–460.

[14] A.T. Le, S.G. Lee, 3D cooperative control of tower cranes using robust adaptive techniques, *Journal of the Franklin Institute*, 354 (18) (2017) 8333–8357.

[15] P.V. Trieu, H.M. Cuong, H.Q. Dong, N.H. Tuan, L.A. Tuan, Adaptive fractional-order fast terminal sliding mode with fault-tolerant control for underactuated mechanical systems: application to tower cranes, *Automation in Construction*, 123 (2021) 103533.

[16] S. Duong, E. Uezato, H. Kinjo, et al., A hybrid evolutionary algorithm for recurrent neural network control of a three-dimensional tower crane, *Automation in Construction*, 23 (2012) 55–63.
[17] J. Matuško, Š. Ileš, F. Kolonić, et al., Control of 3D tower crane based on tensor product model transformation with neural friction compensation, *Asian Journal of Control*, 17 (2) (2015) 443–458.
[18] T. Wu, M. Karkoub, W. Yu, et al., Anti-sway tracking control of tower cranes with delayed uncertainty using a robust adaptive fuzzy control, *Fuzzy Sets and Systems*, 290 (1) (2016) 118–137.
[19] R. Roman, R. Precup, E. Petriu, Hybrid data-driven fuzzy active disturbance rejection control for tower crane systems, *European Journal of Control*, doi:10.1016/j.ejcon.2020.08.001
[20] P. Shen, R. Caverly, Noncolocated passivity-based control of a 2 DOF tower crane with a flexible hoist cable, American Control Conference, Denver, CO, USA, pp. 5046–5051, 2020.
[21] H. Ouyang, Z. Tian, L. Yu, et al., Motion planning approach for payload swing reduction in tower cranes with double-pendulum effect, *Journal of the Franklin Institute*, 357 (13) (2020) 8299–8320.
[22] S. Bonnabel, X. Claeys, The industrial control of tower cranes: an operator-in-the-loop approach, *IEEE Control Systems*, 40 (5) (2020) 27–39.
[23] S. Fasih, Z. Mohamed, A. Husain, et al., Payload swing control of a tower crane using a neural network-based input shaper, *Measurement and Control*, 53 (7–8) (2020) 1171–1182.
[24] D. Blackburn, W. Singhose, J. Kitchen, et al., Command shaping for nonlinear crane dynamics, *Journal of Vibration and Control*, 16 (4) (2010) 477–501.
[25] J. Lawrence, W. Singhose, Command shaping slewing motions for tower cranes, *Journal of Vibration and Acoustics*, 132 (2010) 011002.
[26] J. Vaughan, D. Kim, W. Singhose, Control of tower cranes with double-pendulum payload dynamics, *IEEE Transactions on Control Systems Technology*, 18 (6) (2010) 1345–1358.
[27] A. Elbadawy, M. Shehata, Anti-sway control of marine cranes under the disturbance of a parallel manipulator, *Nonlinear Dynamics*, 82 (1–2) (2015) 415–434.
[28] J. Huang, X. Xie, Z. Liang, Control of bridge cranes with distributed-mass payload dynamics, *IEEE/ASME Transactions on Mechatronics*, 20 (1) (2015) 481–486.
[29] J. Peng, J. Huang, W. Singhose, Payload twisting dynamics and oscillation suppression of tower cranes during slewing motions, *Nonlinear Dynamics*, 98 (2) (2019) 1041–1048.
[30] J. Ye, J. Huang, Analytical analysis and oscillation control of payload twisting dynamics in a tower crane carrying a slender payload, *Mechanical Systems and Signal Processing*, 158, (2021) 107763.
[31] K. Takagi, H. Nishimura, Control of a jib-type crane mounted on a flexible structure, *IEEE Transactions on Control Systems Technology*, 11 (1) (2003) 32–42.
[32] W. Yang, Z. Zhang, R. Shen, Modeling of system dynamics of a slewing flexible beam with moving payload pendulum, *Mechanics Research Communications*, 34 (3) (2007) 260–266.
[33] F. Rauscher, O. Sawodny, An elastic jib model for the slewing control of tower cranes, *IFAC-PapersOnLine*, 50 (1) (2017) 9796–9801.
[34] S. Kimmerle, M. Gerdts, R. Herzog, Optimal control of an elastic crane-trolley-load system-a case study for optimal control of coupled ODE-PDE systems, *Mathematical and Computer Modelling of Dynamical Systems*, 24 (2) (2018) 182–206.
[35] F. Rauscher, Florentin, O. Sawodny, Modeling and control of tower cranes with elastic structure, *IEEE Transactions on Control Systems Technology*, 29 (1) (2020) 64–79.

[36] R. Struble, J. Heinbockel, Resonant oscillations of a beam-pendulum system, *Journal of Applied Mechanics*, (1963) 181–188.
[37] D. Oguamanam, J. Hansen, G. Heppler, Dynamic response of an overhead crane system, *Journal of Sound and Vibration*, 213 (5) (1998) 889–906.
[38] W. He, Vertical dynamics of a single-span beam subjected to moving mass-suspended payload system with variable speeds, *Journal of Sound and Vibration*, 418 (2018) 36–54.
[39] H. Liu, W. Cheng, Y. Li, Dynamic responses of an overhead crane's beam subjected to a moving trolley with a pendulum payload, *Shock and Vibration*, 2019 (2019) 1291652.
[40] J. Xu, C. Chen, Y. Sun, et al., Nonlinear dynamic analysis of coupled vibration of beam and pendulum, IEEE/ASME International Conference on Advanced Intelligent Mechatronics, Hong Kong, China, July, 2019, pp. 702–707.
[41] J. Pan, W. Qin, W. Deng, et al., Harvesting weak vibration energy by integrating piezoelectric inverted beam and pendulum, *Energy*, 227 (2021) 120374.
[42] X. Fu, Dynamic trajectory and parametric resonance in coupled pendulum-beam system, *American Journal of Physics*, 88 (8) (2020) 625–639.
[43] H. Li, W. Qin, Nonlinear dynamics of a pendulum-beam coupling piezoelectric energy harvesting system, *The European Physical Journal Plus*, 134 (12) (2019) 595.
[44] T.J.B. Li, T. Liu, H. Qiu, Measure synchronization and clustering in a coupled-pendulum system suspended from a common beam, *Chaos: An Interdisciplinary Journal of Nonlinear Science*, 29 (9) (2019) 093131.
[45] J. Pan, W. Qin, W. Deng, H. Zhou, Harvesting base vibration energy by a piezoelectric inverted beam with pendulum, *Chinese Physics B*, 28 (1) (2019) 017701.
[46] M. El-Raheb, Effect of cable flexibility on transient response of a beam–pendulum system, *Journal of Sound and Vibration*, 307 (3–5) (2007) 834–848.
[47] I. Cicek, A. Ertas, Experimental investigation of beam-tip mass and pendulum system under random excitation, *Mechanical Systems and Signal Processing*, 16 (6) (2002) 1059–1072.

5 Aerial Cranes

5.1 HELICOPTERS SLUNG LOADS

Helicopters suspending heavy, bulky loads are now widely used in construction, transportation, and rescue missions. The helicopters working as aerial cranes provide essential material-handling services throughout the world, such as delivering cargoes to ships at sea. Unfortunately, the slung load fastened by a cable to the aircraft constitutes a flexible system that degrades the effectiveness and safety. Both the pilot's commanded motions and external disturbances induce unwanted oscillations. Therefore, the study of dynamics and control of aerial cranes is essential for safe and efficient transportation.

Prior literature has discussed the solutions to the challenging problems posed by the suspended-load swing of aerial cranes. Some techniques use active mechanisms directly attached to the load [1–3]. However, the active techniques add complexity to the entire system. In order to reduce the swing, many examples show that feedback control schemes using measurement and estimation of the suspended-load oscillations are effective [4–13]. Unfortunately, accurately sensing the suspended load is difficult, and the feedback controller can conflict with the actions of the pilot.

Command shaping techniques have been used to reduce the oscillations by filtering the pilot commands to create a low-swing motion [14–16]. In practice, the input shaper includes a series of impulses, which can reduce the residual oscillations, but they are less effective at reducing the transient oscillations. Significant efforts have concentrated on modeling the aerial crane systems [17–22] and the construction of experimental test rigs [23]. All of the above-mentioned works focus on the single-pendulum and point-mass load dynamics at low speed.

Mounting numbers of examples show that aerial cranes transport bulky loads, which should be modeled as a distributed mass. In the recent literature, little attention has been directed at the distributed-mass load dynamics and control of aerial cranes. The distributed-mass loads exhibit complicated dynamical behavior that is a challenging feature of the slung loads. Furthermore, the coupling effect between the oscillations of the distributed-mass loads and aerial vehicles cannot be neglected. Therefore, it is essential to study the distributed-mass load dynamics and control in aerial cranes.

5.1.1 Planar Dynamics

5.1.1.1 Modeling

Figure 5.1 shows a schematic representation of an aerial vehicle transporting a distributed-mass beam undergoing planar motions. The aerial vehicle flies along

DOI: 10.1201/9781003247210-5

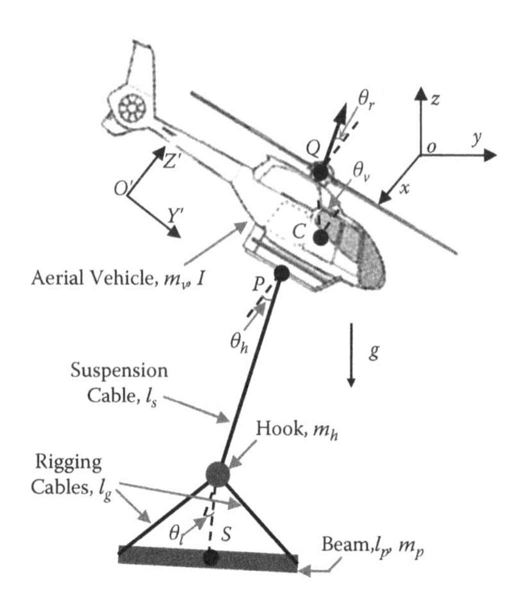

FIGURE 5.1 Model of an aerial vehicle transporting a distributed-mass beam.

the y direction in near-hover operation. The mass of the aerial vehicle is m_v, and the moment of inertia about the x-axis is I. Motion of the aerial vehicle is divided into the displacement along the y direction, and the attitude angle, θ_v, about the x-axis in the Inertial Newtonian frame. Let $O'X'Y'Z'$ be the moving Cartesian coordinates fixed to the vehicle so that the x-axis is parallel to the X'-axis. The inertial coordinates $oxyz$ can be converted to the moving coordinates $O'X'Y'Z'$ by rotating the attitude, θ_v, about the x-axis. The distance between the vehicle center of gravity, C, and the main rotor hub, Q, is a. A thrust angle, θ_r, is defined as the angle between the thrust force produced by the main rotor and the Z' direction.

The distance between the vehicle center of gravity, C, and the load suspension point, P, is b. A massless suspension cable of length, l_s, which is assumed to be inelastic, hangs below the aerial vehicle and supports a hook of mass, m_h. The swing angle of the suspension cable, θ_h, is defined as the angle between the suspension cable and the Z' direction. A uniformly distributed-mass beam of mass, m_p, and length, l_p, is attached to the hook by two rigging cables of length, l_g. The centroid of the beam is located at the point, S. The load swing angle relative to the suspension cable is θ_l.

The input to the near-hover model is the thrust angle, θ_r. The outputs are the vehicle displacement, y, the vehicle attitude, θ_v, and the swing angles, θ_h, and θ_l. The hook is modeled as a point mass. The model also assumes that the aerodynamic effects on the suspended load are negligible [16]. The initial direction of the beam length coincides with the oyz plane such that the motion of the aerial crane in Figure 5.1 cannot cause beam twisting about the rigging cables. This simplifies the system to a planar dynamic model.

The kinetic energy of the model in Figure 5.1 is:

$$T = \frac{1}{2}m_v\dot{y}^2 + \frac{1}{2}I_x\dot{\theta}_v^2 + \frac{1}{2}m_h\begin{bmatrix}\dot{y} + l_s(\dot{\theta}_h - \dot{\theta}_v)\cos(\theta_h - \theta_v) \\ -b\cdot\dot{\theta}_v\cdot\cos\theta_v\end{bmatrix}^2$$
$$+\frac{1}{2}m_h[l_s\cdot(\dot{\theta}_h - \dot{\theta}_v)\cdot\sin(\theta_h - \theta_v) - b\cdot\dot{\theta}_v\cdot\sin\theta_v]^2$$
$$+\frac{1}{2}m_p\begin{bmatrix}\dot{y} + l_s\cdot(\dot{\theta}_h - \dot{\theta}_v)\cdot\cos(\theta_h - \theta_v) - b\cdot\dot{\theta}_v\cdot\cos\theta_v \\ + l_y\cdot(\dot{\theta}_h + \dot{\theta}_l - \dot{\theta}_v)\cdot\cos(\theta_h + \theta_l - \theta_v)\end{bmatrix}^2$$
$$+\frac{1}{2}m_h\begin{bmatrix}l_s\cdot(\dot{\theta}_h - \dot{\theta}_v)\cdot\sin(\theta_h - \theta_v) - b\cdot\dot{\theta}_v\cdot\sin\theta_v \\ + l_y\cdot(\dot{\theta}_h + \dot{\theta}_l - \dot{\theta}_v)\cdot\sin(\theta_h + \theta_l - \theta_v)\end{bmatrix}^2 \quad (5.1)$$

where

$$l_y = \sqrt{l_g^2 - 0.25l_p^2} \quad (5.2)$$

While the vehicle centroid is defined as the zero potential energy surface, the potential energy is:

$$V = -m_p g[b\cos\theta_v + l_s\cos(\theta_h - \theta_v) + l_y\cos(\theta_h + \theta_l - \theta_v)]$$
$$- m_h g[b\cos\theta_v + l_s\cos(\theta_h - \theta_v)] \quad (5.3)$$

where g is the gravitational constant.

By using the generalized Lagrange method, the nonlinear equations of motion for the model shown in Figure 5.1 are derived as:

$$\begin{cases} h_1\cdot\ddot{\theta}_v + h_2\cdot\ddot{y} + h_3\cdot\ddot{\theta}_h - 0.5h_4\cdot\ddot{\theta}_l - h_5 = 0 \\ h_6\cdot\ddot{\theta}_v + h_1\cdot\ddot{y} + h_7\cdot\ddot{\theta}_h + h_8\cdot\ddot{\theta}_l - h_9 = 0 \\ h_7\cdot\ddot{\theta}_v + h_3\cdot\ddot{y} + h_{10}\cdot\ddot{\theta}_h + h_{11}\cdot\ddot{\theta}_l - h_{12} = 0 \\ h_8\cdot\ddot{\theta}_v - 0.5h_4\cdot\ddot{y} + h_{11}\cdot\ddot{\theta}_h + h_{14}\cdot\ddot{\theta}_l - h_{13} = 0 \end{cases} \quad (5.4)$$

where h_1, h_2, h_3, h_4, h_5, h_6, h_7, h_8, h_9, h_{10}, h_{11}, h_{12}, h_{13}, and h_{14} are coefficients. Note that the coefficients h_5 and h_9 depend on the thrust angle θ_r. The dynamical model (5.4) includes four system states, y, θ_v, θ_h, θ_l, and one input, θ_r. Therefore, it is an underactuated system that requires a sophisticated control system. The control objectives consist of two parts, attitude control of the aerial vehicle and swing suppression of the distributed-mass load. To achieve the objectives, a combination of feedback and command-shaping control is presented in this chapter. A feedback controller regulates the vehicle attitude by following the states of a prescribed model and reducing the tracking errors, while a command-shaping controller attenuates the distributed-mass load swing by smoothing the pilot commands.

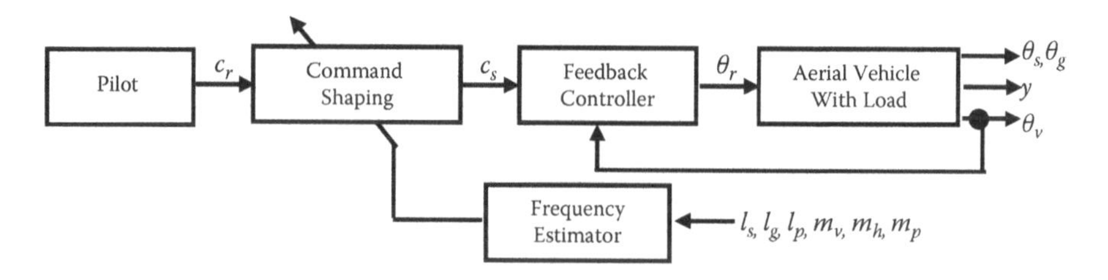

FIGURE 5.2 Combined feedforward and feedback control architecture.

5.1.1.2 Control

The configuration of the combined feedback and command-shaping control is shown in Figure 5.2. The pilot generates a pilot command, c_r, via the control stick. A command-shaping controller modifies the pilot command to produce a smoothed command, c_s. The measurement of the vehicle attitude, θ_v, is used to adjust the thrust angle, θ_r, in a feedback control loop. Then, the attitude, θ_v, is forced to track the smoothed command, c_s, by the feedback controller. The system frequency is estimated by system parameters and is used to design the command-shaping controller. The smoothed command drives the aerial vehicle toward the desired position without inducing oscillations of the distributed-mass load.

5.1.1.2.1 Model-Following Control for Attitude Regulation

The aerial crane dynamics shown in Equation (5.4) is too complex to analytically design an MFC controller. Therefore, an MFC controller for a simplified model without the sling load was created. Then, the result was refined as needed for the aerial crane. The controller will be used on aerial crane dynamics shown in Equation (5.4) to test its performance.

Ignoring the swing angles, θ_h and θ_l, in the dynamics model (5.4) yields a simplified model without the sling load:

$$\begin{cases} I\cdot\ddot{\theta}_v\cdot \cos(\theta_r + \theta_v) = a\cdot g\cdot m_v\cdot \sin(\theta_r) \\ \ddot{y} = g\cdot \tan(\theta_r + \theta_v) \end{cases} \tag{5.5}$$

Both the vehicle attitude, θ_v, and the thrust angle, θ_r, are assumed to be small so that Equation (5.5) yields a linearized model. From the linearized model, an explicit MFC controller can be derived [16]:

$$\begin{cases} \ddot{\theta}_{vm} + 2\zeta_p\omega_p\dot{\theta}_{vm} + \omega_p^2\theta_{vm} = \omega_p^2\cdot c_s \\ \theta_r = I\cdot\ddot{\theta}_{vm}/a/g/m_v + k_d\cdot(\dot{\theta}_{vm} - \dot{\theta}_v) + k_p\cdot(\theta_{vm} - \theta_v) \end{cases} \tag{5.6}$$

where θ_{vm} is model output, ζ_p is the damping ratio of the prescribed model, ω_p is the frequency of the prescribed model, and k_p and k_d are positive control gains. The first equation in Equation (5.6) is the prescribed model, while the second one describes the asymptotic tracking control law.

The damping ratio, ζ_p, and the frequency, ω_p, of the prescribed model were designed by using the pole placement method. The prescribed model with reasonable damping ratio (= 0.707) and reasonable settling time (≤2 seconds) is required. The desired closed-loop poles were calculated to be approximately −2 ± 2i. Thus, the frequency, ω_p, was 2.83 rad/s.

Suitable tracking errors can be achieved by designing control gains k_p and k_d. Referencing the eigenvalues discussed in the literature [16], and considering the coupling effect between the vehicle and the externally slung load decrease the real part of the eigenvalues and increase the system damping. Thus, the prescribed eigenvalues of the tracking controller are selected to be −18 ± 8.7i. Given that the vehicle mass, m_v, moment of inertia, I, and distance, a, in the simulations are 6000 kg, 17450 kg m^2, and 3.5 m, respectively. Then, the control gains k_p and k_d were calculated by the pole placement technique to be 33.916 and 3.053, respectively.

5.1.1.2.2 Smoothing Pilot Commands for Swing Reduction

After modifying the vehicle's attitude with the MFC controller, it is useful to design a command-shaping controller from a simplified model. Ignoring the vehicle attitude, θ_v, in the dynamics model (5.4) yields a simplified kinetic equation of the double-pendulum load swing:

$$\begin{cases} [e\cdot l_s\cdot \cos\theta_h + c\cdot l_s\cdot \cos\theta_h + c\cdot l_y\cdot \cos(\theta_h+\theta_l)]\cdot\ddot{\theta}_h + c\cdot l_y\cdot\cos(\theta_h+\theta_l)\cdot\ddot{\theta}_l \\ \quad + (1+e+c)\cdot\ddot{y} \\ - e\cdot l_s\cdot \sin\theta_h\cdot\dot{\theta}_h^2 - c\cdot[l_y\cdot\sin(\theta_h+\theta_l)\cdot[\dot{\theta}_h+\dot{\theta}_l]^2 + l_s\cdot\sin\theta_h\cdot\dot{\theta}_h^2] = 0 \\ (c\cdot R^2 + e\cdot l_s^2 + c\cdot l_s^2 + c\cdot l_y^2 + 2c\cdot l_s\cdot l_y\cdot\cos\theta_l)\cdot\ddot{\theta}_h + c\cdot(R^2 + l_y^2 + l_s\cdot l_y\cdot\cos\theta_l)\cdot\ddot{\theta}_l \\ - c\cdot l_s\cdot l_y\cdot\sin\theta_l\cdot(\dot{\theta}_l^2 + 2\dot{\theta}_h\dot{\theta}_l) + [e\cdot l_s\cdot\cos\theta_h + c\cdot l_s\cdot\cos\theta_h + c\cdot l_y\cdot\cos(\theta_h+\theta_l)]\cdot\ddot{y} \\ + g\cdot c\cdot l_y\cdot\sin(\theta_h+\theta_l) + g\cdot(e+c)\cdot l_s\cdot\sin\theta_h = 0 \\ (R^2 + l_y^2 + l_s\cdot l_y\cdot\cos\theta_l)\cdot\ddot{\theta}_h + (R^2 + l_y^2)\cdot\ddot{\theta}_l + l_y\cdot\cos(\theta_h+\theta_l)\cdot\ddot{y} \\ + l_s\cdot l_y\cdot\sin\theta_l\cdot\dot{\theta}_h^2 + g\cdot l_y\cdot\sin(\theta_h+\theta_l) = 0 \end{cases} \tag{5.7}$$

where

$$R = l_p/(2\sqrt{3}) \tag{5.8}$$

e is the ratio of the hook mass to the vehicle mass and c is the ratio of the load mass to the vehicle mass. Assuming small oscillations around the equilibrium angles, the linearized frequencies of the load swing are:

$$\omega_{1,2} = \sqrt{\frac{g\cdot(w \mp v)}{2l_s}} \tag{5.9}$$

where

$$w = \frac{1}{e \cdot l_y^2 + (c + e) \cdot R^2}((e^2 + e + c + e \cdot c) \cdot l_y^2 + (e + c) \cdot l_y \cdot l_s + (e^2 + c^2 + 2e \cdot c + c + e) \cdot R^2) \quad (5.10)$$

$$v = \sqrt{w^2 - \frac{4(e^2 + c^2 + 2e \cdot c + e + c) \cdot l_y \cdot l_s}{e \cdot l_y^2 + (e + c) \cdot R^2}} \quad (5.11)$$

The technique for smoothing pilot commands is used to limit the suspended-load swing. A smoothing function inherent in the limited motion of the aerial vehicle results in a limited response to oscillations of the suspended load. Thus, the command-shaping controller is a smoothing function, which is a piecewise continuous function.

The model (5.7) can be approximated near the equilibrium angles as two uncoupled second-order systems with the frequencies shown in Equation (5.9). A four-pieces smoother was applied to reduce swing:

$$u(\tau) = \begin{cases} M(1 + \sigma)\tau e^{-2\zeta_m \omega_m \tau}, \quad 0 \le \tau \le 0.5T_m \\ M[(1 + \sigma + \sigma K - K)T_m + (2K - 2K\sigma - \sigma - 1)\tau]e^{-2\zeta_m \omega_m \tau}, \\ 0.5T_m < \tau \le T_m \\ M[K(3 - K\sigma - 3\sigma - K)T_m + K(K + \sigma K + 2\sigma - 2)\tau]e^{-2\zeta_m \omega_m \tau}, \\ T_m < \tau \le 1.5T_m \\ M[2K^2(1 + \sigma)T_m - K^2(1 + \sigma)\tau]e^{-2\zeta_m \omega_m \tau}, \quad 1.5T_m < \tau \le 2T_m \end{cases} \quad (5.12)$$

where ω_m, ζ_m, and T_m are the modeled frequency, damping ratio, and damped vibrational period of first-mode swing. The coefficients K and M in Equation (5.12) are given by:

$$K = e^{2\pi\zeta_m/\sqrt{1-\zeta_m^2}} \quad (5.13)$$

$$M = \zeta_m^2 \omega_m^2/(1 - K^{-1})^2 \quad (5.14)$$

A local extreme value in the percent vibration amplitude exists around the modeled frequency. The frequency at the local extremum is defined as $v \cdot \omega_m$, which satisfies $p \le v \le q$. Limiting the percent vibration amplitude at the local extremum to a tolerable level, V_{tol}, yields:

$$\left[\sqrt{\left[\int_{\tau=0}^{+\infty} u(\tau)e^{\zeta\omega\tau}\sin(\omega\sqrt{1-\zeta^2}\tau)d\tau\right]^2 + \left[\int_{\tau=0}^{+\infty} u(\tau)e^{\zeta\omega\tau}\cos(\omega\sqrt{1-\zeta^2}\tau)d\tau\right]^2}\right]_{\substack{\omega = v \cdot \omega_m \\ \zeta = \zeta_m}} \le V_{tol} \quad (5.15)$$

Another constraint can be applied to ensure that the derivative of the percent vibration amplitude at the local extremum with respect to frequency is limited to zero. To satisfy this requirement, the zero derivative constraint is given by:

$$\left[\frac{d\left(\left[\int_{\tau=0}^{+\infty} u(\tau)e^{\zeta\omega\tau}\sin(\omega\sqrt{1-\zeta^2}\tau)d\tau\right]^2+\left[\int_{\tau=0}^{+\infty} u(\tau)e^{\zeta\omega\tau}\cos(\omega\sqrt{1-\zeta^2}\tau)d\tau\right]^2\right)}{d\omega}\right]_{\substack{\omega=\nu\cdot\omega_m\\ \zeta=\zeta_m}} = 0 \tag{5.16}$$

Solving the constraints (5.15) and (5.16) yields the solution of the coefficient, σ, in Equation (5.12). The closed-form solution cannot be obtained because of nonlinearity. Instead, substituting Equation (5.12) into (5.15) and (5.16) yields the numerical solution. The coefficient, σ, varies with the tolerable level, V_{tol}. The value of the coefficient, σ, is zero when the tolerable level is set to zero.

5.1.1.3 Computational Dynamics

Figures 5.3–5.6 show the simulated responses for a flight distance of 3 km. The vehicle mass, m_v, moment of Inertia, I, distance, a, distance, b, hook mass, m_h, rigging length, l_g, and vehicle nominal velocity are 6000 kg, 17450 kg m^2, 3.5 m, 5 m, 50 kg, 7 m, and 46 km/h, respectively. The pilot command arises when the pilot pushes the control stick to move the aerial vehicle forward. This results in a rise in velocity until the nominal velocity is reached. The vehicle then flies at its nominal velocity for a while until the pilot pulls the control stick. The pilot commands decrease the velocity to zero in order to keep the vehicle near hover.

There are two stages in the system response. The transient stage of the response is defined as the time frame when the vehicle flies forward. The deflection during the transient stage is referred to as the transient deflection. The residual stage is

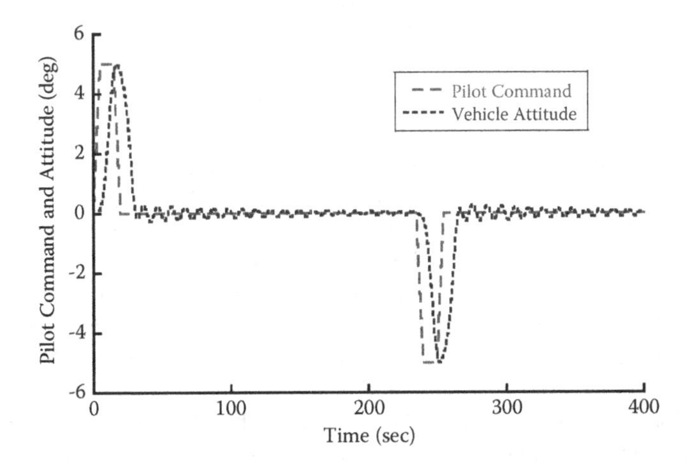

FIGURE 5.3 Simulated response of vehicle attitude to a pilot command.

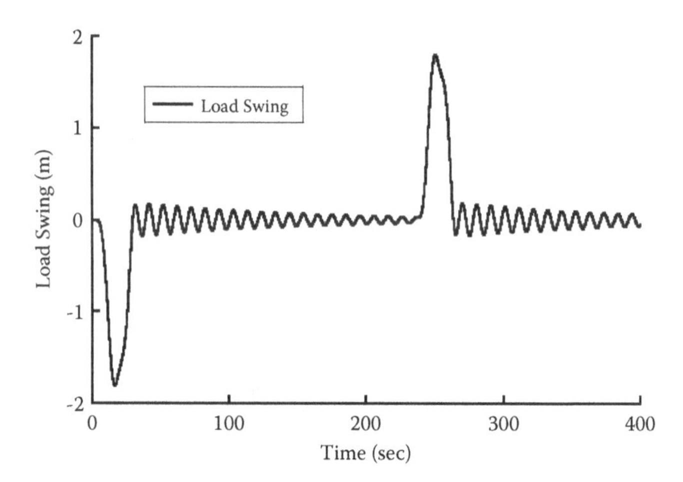

FIGURE 5.4 Simulated response of load swing to a pilot command.

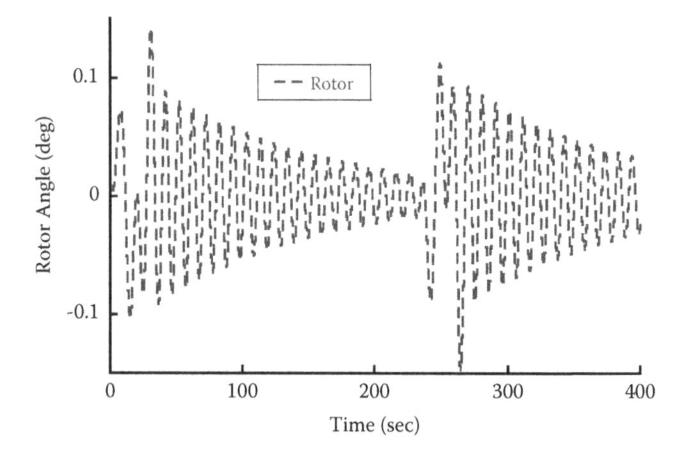

FIGURE 5.5 Simulated response of rotor angle to a pilot command.

defined as the time frame when the vehicle attempts to stop (hovering). The maximum deflection during this residual stage is defined as the residual amplitude. The residual amplitude of the vehicle attitude and load swing shown in Figures 5.3 and 5.4 are 0.27° and 0.18 m, respectively. The settling time of the vehicle attitudes is defined as the time required for the response to settle within 0.5°. Meanwhile, the settling time of the payload swing is defined as the time required for the vibrations to settle within 1% of the sum of the suspension length, l_s, and second cable equivalent length, l_y. The settling time of the vehicle attitude and load swing in Figures 5.3 and 5.4 are 9.3 s and 5.9 s, respectively.

The rotor angle for piloting the vehicle and the flight distance of the vehicle are shown in Figures 5.5 and 5.6. The maximum deflection of the rotor angle is approximately 0.15°, which is a very small value for controlling the aerial vehicle.

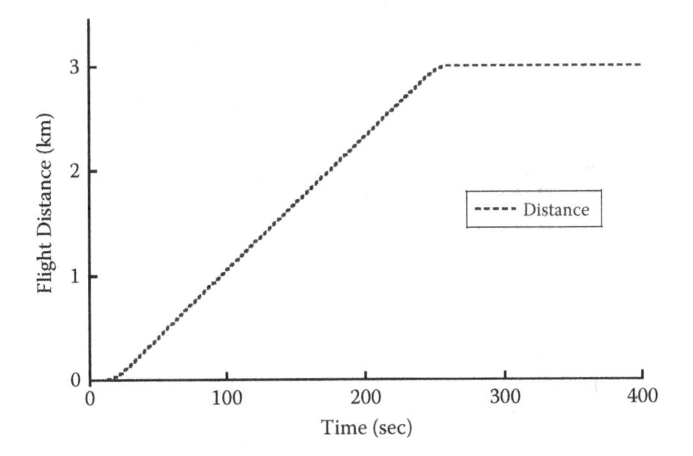

FIGURE 5.6 Simulated response of flight distance to a pilot command.

The rotor period, which is 10.3 seconds in Figure 5.5, is very similar to the oscillation period of the vehicle attitude. This is because the oscillations of the vehicle attitude are damped by adjusting the rotor angle.

The residual oscillations of the vehicle attitude and load deflection induced by various flight distances are shown in Figures 5.7 and 5.8. The suspension cable length, load length, and load mass were fixed at 15 m, 10 m, and 2000 kg, respectively. Peaks and troughs in the residual amplitude occur because the interference between the oscillations caused by the acceleration and deceleration are in phase or out of phase. The peak magnitudes decrease with increasing flight distances because of the damping effect of the MFC controller. When the combined controller is utilized, the coupling effect between the vehicle and load is greatly reduced, hence the vehicle attitudes are <0.53° and the load deflection are <0.35 m

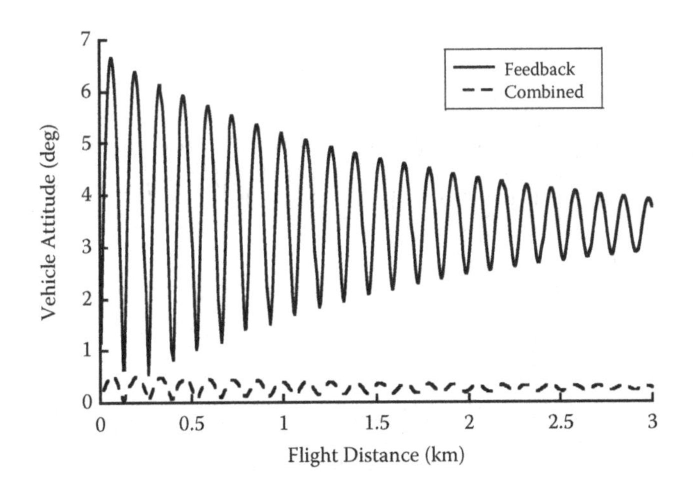

FIGURE 5.7 Residual amplitude of vehicle attitude vs. flight distance.

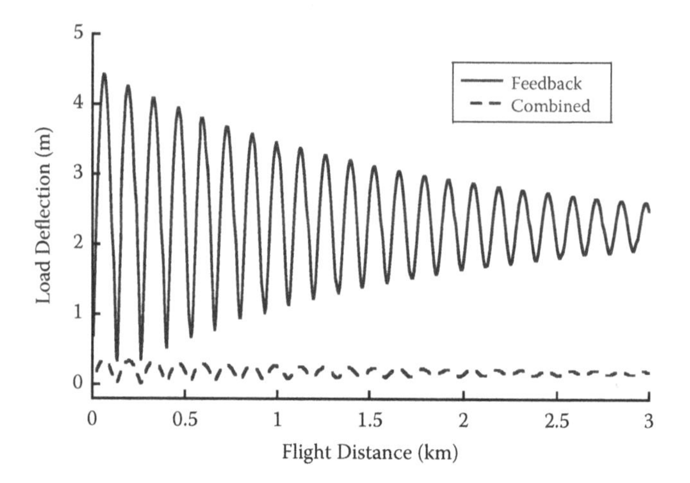

FIGURE 5.8 Residual amplitude of load deflection vs. flight distance.

for all the flight distances simulated. They satisfy the requirements of ADS-33E. The frequency in the residual amplitude is approximately the first-mode frequency of the load swing, which can be estimated from Equation (5.9).

In addition, the settling time of the vehicle attitude and load deflection for various flight distances are shown in Figures 5.9 and 5.10. The simulated results for the settling time are similar to that in the residual amplitude observed in Figures 5.7 and 5.8. The settling times in the vehicle attitude and load deflection converge to a constant value as the flight distance increases as the maneuver is executed with the feedback controller. The values given by Equation (5.9) can also be used to estimate the frequency in the settling time. The majority of the settling time of the helicopter attitude and the load deflection with the combined controller are under 9.8 seconds and 6.5 seconds, respectively. On the basis of the aforementioned analysis, the

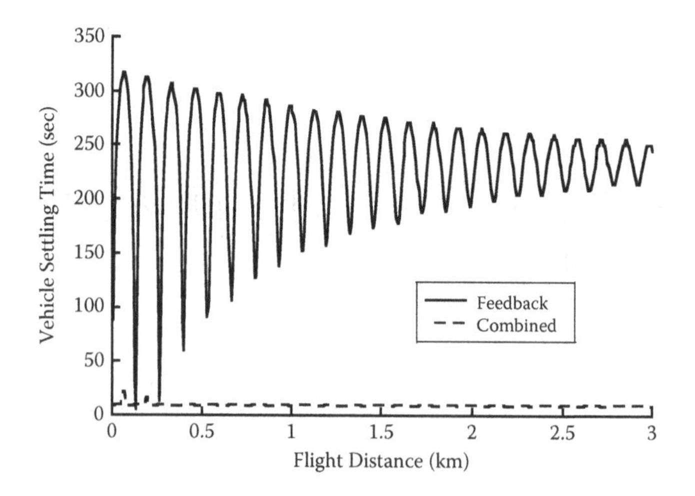

FIGURE 5.9 Settling time of vehicle attitude vs. flight distance.

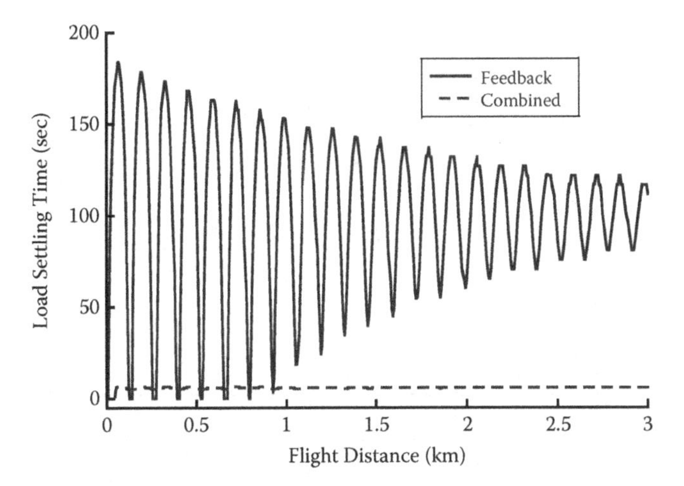

FIGURE 5.10 Settling time of load deflection vs. flight distance.

combined controller has made a significant achievement in reducing the settling time for the purpose of rapidly stabilizing the aerial vehicle and suppressing the swing load.

System parameters including suspension length, load mass, and length will change over time. Changing parameters have a great influence on the system dynamics. The robustness to variations in the system parameters for the new controller is important. The simulation parameters of flight distance, load length, and load mass were set to 2 km, 10 m, and 2000 kg, respectively. Note that the modeled suspension length was fixed at 15 m for the combined controller. Figures 5.11 and 5.12 show the residual amplitudes of the helicopter attitude and load oscillations with increasing suspension length. Peaks and troughs in the residual amplitude exist. At least two primary contributors account for such variations. First, different

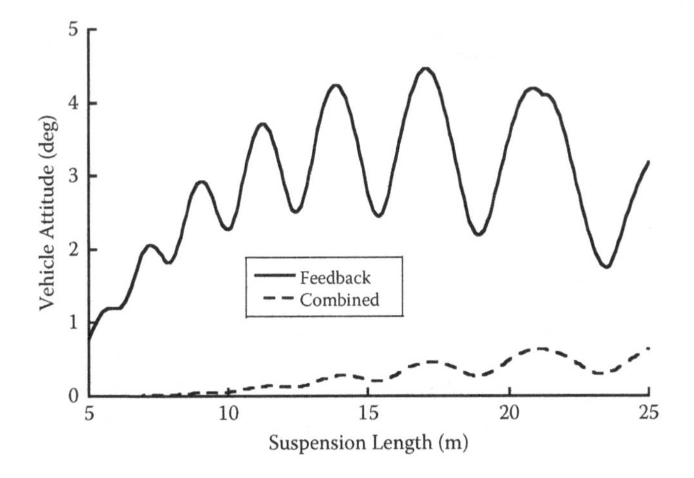

FIGURE 5.11 Residual amplitude vs. suspension cable length.

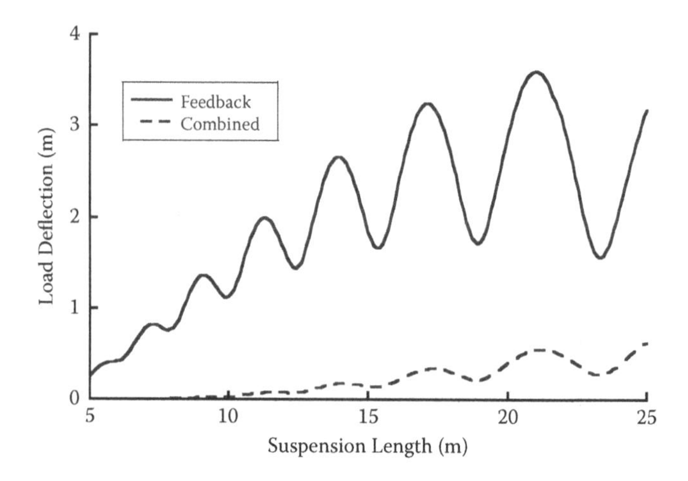

FIGURE 5.12 Residual amplitude vs. suspension cable length.

suspension lengths result in different swing frequencies. Secondly, stabilizing the vehicle attitude is more difficult for a longer suspension cable length. The combined controller limited the residual amplitudes of the helicopter attitude to <0.48° for all the cable lengths, and limited the residual amplitudes of the distributed-mass load to <0.40 m. That is a small value for aerial cranes. Decreasing the suspension length decreases the residual amplitudes with the combined controller, while increasing the suspension length increases the residual amplitudes. However, the combined controller dramatically reduced the residual amplitudes, even with a large modeling error in the suspension length.

Meanwhile, the different cable lengths causing the different settling time of the vehicle and load are presented in Figures 5.13 and 5.14. The settling time of the vehicle and load with the feedback controller are similar to those in the residual

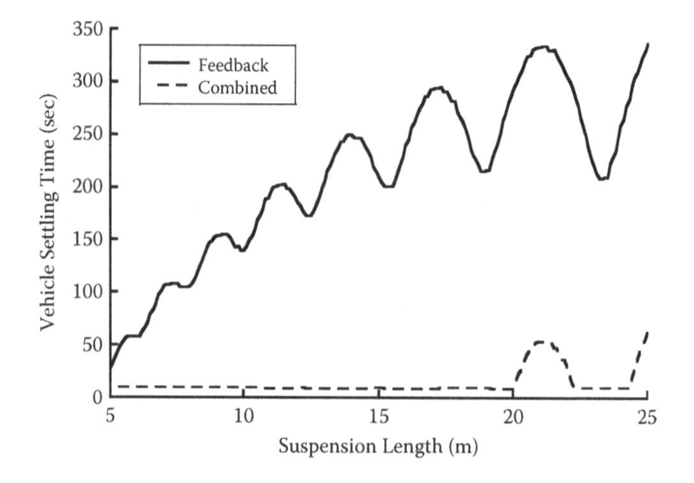

FIGURE 5.13 Settling time of vehicle attitude vs. suspension cable length.

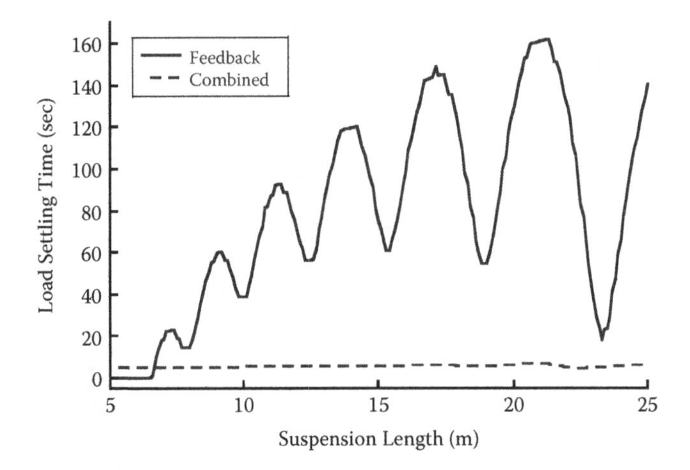

FIGURE 5.14 Settling time of load deflection vs. suspension cable length.

amplitude. The combined controller shortens the settling time prominently, and settling time of the vehicle and load are approximately 9.8 seconds and 6.3 seconds, respectively. The modeled suspension length for the combined controller was also fixed at 15 m in this case. Hence, the combined controller can tolerate large error in the suspension length.

The coupling effect between the oscillations of the distributed-mass loads and aerial vehicles is greatly influenced by the load length. The residual amplitude results are shown in Figures 5.15 and 5.16. The simulation parameters of flight distance, suspension length, and load mass were set to 2 km, 15 m, and 2000 kg, respectively. Note that the modeled load length was fixed at 10 m for the combined controller. The residual amplitudes of the vehicle and load with the feedback controller decrease with increasing load length shorter than 9 m. Both the vehicle

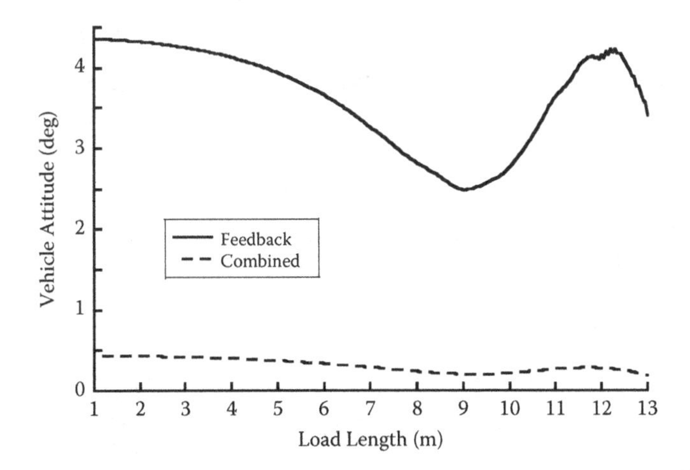

FIGURE 5.15 Residual amplitude of vehicle attitude vs. load length.

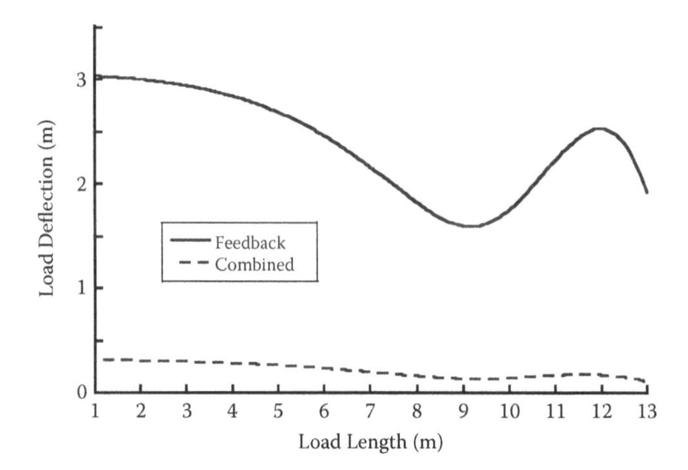

FIGURE 5.16 Residual amplitude of load deflection vs. load length.

and load have peaks and troughs in their residual amplitudes between 9 m and 13 m because the changing load lengths have some effects on the system frequency. Furthermore, the load length has small effects on the residual amplitudes with the combined controller. The combined controller suppresses the residual amplitudes of the load to <0.32 m, and the load oscillations do not significantly disturb the vehicle. The vehicle attitude was always <0.44° at the end of maneuver.

Figures 5.17 and 5.18 indicate that the settling time variation with the feedback controller is similar to the complex behavior observed in the residual amplitude. Before 9 m, the settling time decreases as the load length increases. After this point, the settling time increases to a peak value with increasing the load length. The peak in the residual amplitude and settling time occurs around a load length of 12 m. The load length has little influence on the settling time, which is

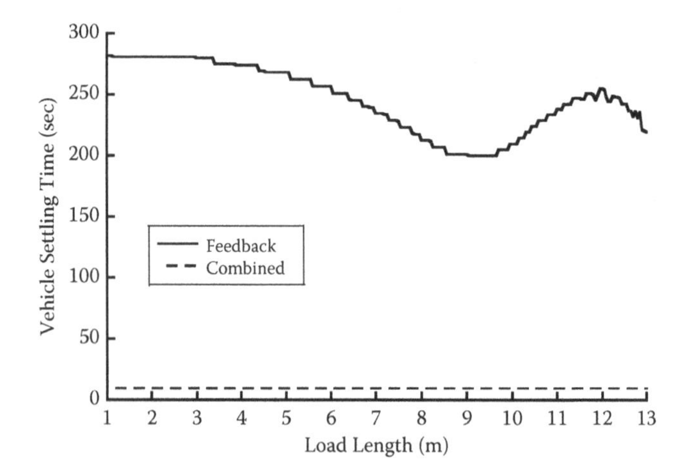

FIGURE 5.17 Settling time of vehicle attitude vs. load length.

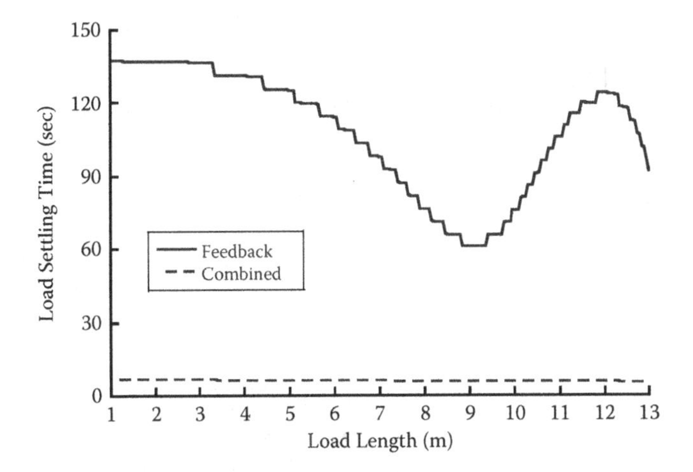

FIGURE 5.18 Settling time of load deflection vs. load length.

attributed to the robustness of the combined controller. With the combined controller, the majority of the settling times of the vehicle and load are <9.9 s and <6.6 s, respectively.

Figures 5.19 and 5.20 demonstrate clearly that some changes in the residual amplitude occur when the load mass changes. The simulation parameters of flight distance, load length, and suspension length were set to 2 km, 10 m, and 15 m, respectively. Note that the modeled load mass was fixed at 2000 kg for the combined controller. With the feedback controller, there was a rise and fall in the residual amplitude. The primary reason for such phenomenon is the load mass that affects the oscillation frequency. Nevertheless, the residual amplitudes of vehicle and load with the combined controller remained stable under 0.3° and 0.2 m, respectively.

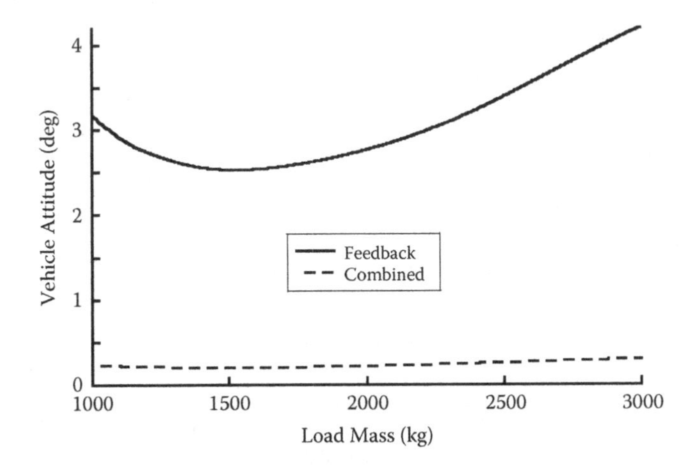

FIGURE 5.19 Residual amplitude of vehicle attitude vs. load mass.

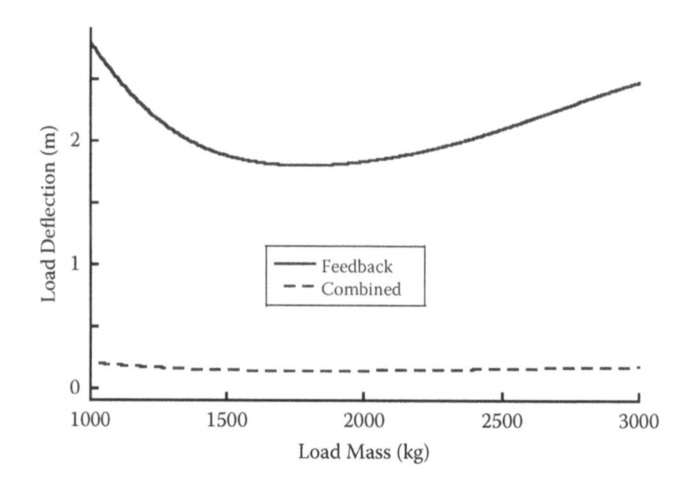

FIGURE 5.20 Residual amplitude of load deflection vs. load mass.

The corresponding settling times of the vehicle and load are presented in Figures 5.21 and 5.22. The simulated results in the settling time with the feedback controller are similar to that in the residual amplitude observed in Figures 5.19 and 5.20. By contrast, the settling times of vehicle and load with the combined controller both remain steady under 9.66 seconds and 5.91 seconds, respectively. They are safely limited within the requirements of ADS-33E [24].

When considered, in total, it is clear that the combined controller is effective in suppressing the oscillations of the vehicle attitude and the load. The controller not only helps transfer the load to the desired position precisely but also gives the pilot a safer environment to command the aerial vehicle. On the other hand, the controller shortening of the settling time is also of great importance. The short settling time certainly improves the transfer efficiency.

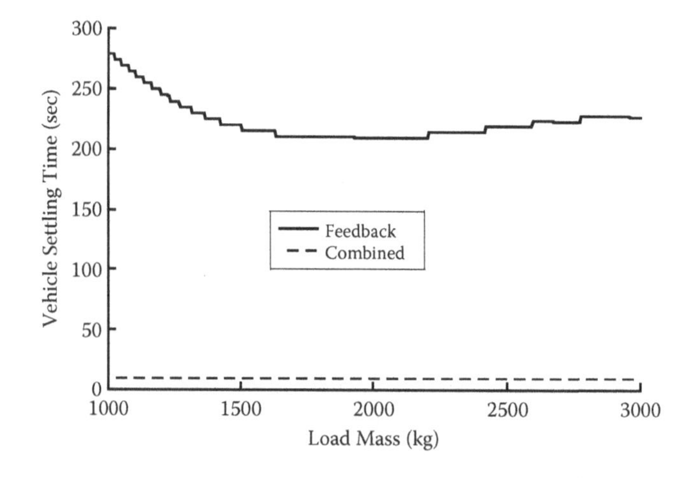

FIGURE 5.21 Settling time of vehicle attitude vs. load mass.

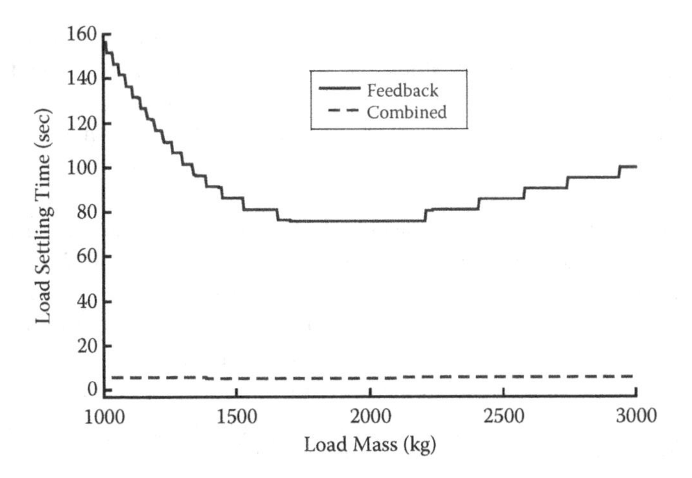

FIGURE 5.22 Settling time of load deflection vs. load mass.

5.1.2 Three-Dimensional Dynamics

5.1.2.1 Modeling

Figure 5.23 shows a schematic representation of a helicopter transporting a distributed-mass beam. The helicopter uses a gyrostat H, meaning a rigid body with three fixed-axis reaction control rotors **H**1, **H**2, and **H**3 in the model. The helicopter flies in near-hover operation along Newtonian axes $\mathbf{N}_1$ and $\mathbf{N}_2$. The mass of the helicopter is m_v, and the centroidal principal axis moments of inertia about the $\mathbf{H}_1$, $\mathbf{H}_2$, and $\mathbf{H}_3$ axes are I_{xx}, I_{yy}, and I_{zz}, respectively. The basic

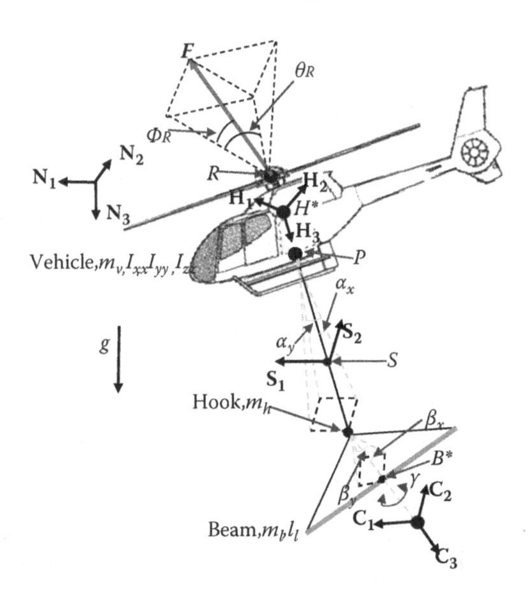

FIGURE 5.23 Model of a helicopter transporting a slender beam.

transformations are Newtonian $\mathbf{N}_1$, $\mathbf{N}_2$, and $\mathbf{N}_3$. Motion of the helicopter is divided into the displacement, x, along the $\mathbf{N}_1$ direction, the displacement, y, along the $\mathbf{N}_2$ direction, the pitch attitude, θ_v, and the roll attitude, Φ_v. Helicopters carrying distributed-mass loads often maneuver in the near-hover state, thus vertical motion and the yaw angle are uncoupled from the displacement and the attitude. In this case, this chapter assumes that the helicopter has a separate model of yaw attitude and vertical displacement. The distance between the helicopter mass center (H^*) and the main rotor hub R is a. The thrust force, F, produced by the main rotor is along a direction which has two angles θ_R and Φ_R between $\mathbf{H}_3$. The magnitude of the component force of F along the $\mathbf{N}_3$ direction balances gravity in order to maintain the helicopter flight altitude.

The distance between the helicopter center of mass, H^*, and the load suspension point, P, is d. A massless suspension cable S (length, l_s), which is assumed to be inelastic, hangs below the helicopter with two angles α_x and α_y to orient the pendulum. The suspension cable supports a hook of mass, m_h. A uniformly distributed-mass beam (B) of mass, m_l, length, l_l, and centroidal moment of inertia, I_l, is attached to the hook with three angles β_x, β_y, and γ to orient the double pendulum using two rigging cables of length, l_g. The centroid of the beam is located at its center, B^*.

Inputs to the near-hover model are the thrust angles, θ_R and Φ_R. Outputs are the helicopter displacements, x and y, the attitudes, θ_v and Φ_v, the cable swing angles, α_x, α_y, β_x, and β_y, and the twist angle, γ. The model assumes that the hook is a point mass. The aerodynamic effects on the suspended load are negligible due to near-hover operation, heavy load, and long cables.

Restricting ourselves to planar motion, the generalized speeds u_i are chosen as:

$$u_1 = \dot{x},\ u_2 = \dot{y},\ u_3 = \dot{\theta}_v,\ u_4 = \dot{\phi}_v,\ u_5 = \dot{\alpha}_x,\ u_6 = \dot{\alpha}_y,\ u_7 = \dot{\beta}_x,\ u_8 = \dot{\beta}_y,\ u_9 = \dot{\gamma}. \tag{5.17}$$

where Equation (5.17) are the generalized speeds of helicopter, first pendulum, and second pendulum. The basis transformation matrix between the three bodies H (Helicopter), S (Suspension Cable), and C (Rigging Cable) are:

$$\begin{Bmatrix} \mathbf{N}_1 \\ \mathbf{N}_2 \\ \mathbf{N}_3 \end{Bmatrix} = \begin{bmatrix} \cos\theta_v & 0 & \sin\theta_v \\ -\sin\phi_v \sin\theta_v & \cos\phi_v & \sin\phi_v \cos\theta_v \\ -\sin\theta_v \cos\phi_v & \sin\phi_v & \cos\phi_v \cos\theta_v \end{bmatrix} \begin{Bmatrix} \mathbf{H}_1 \\ \mathbf{H}_2 \\ \mathbf{H}_3 \end{Bmatrix} \tag{5.18}$$

$$\begin{Bmatrix} \mathbf{H}_1 \\ \mathbf{H}_2 \\ \mathbf{H}_3 \end{Bmatrix} = \begin{bmatrix} \cos\alpha_y & 0 & \sin\alpha_y \\ -\sin\alpha_x \sin\alpha_y & \cos\alpha_x & \sin\alpha_x \cos\alpha_y \\ -\sin\alpha_y \cos\alpha_x & \sin\alpha_x & \cos\alpha_x \cos\alpha_y \end{bmatrix} \begin{Bmatrix} \mathbf{S}_1 \\ \mathbf{S}_2 \\ \mathbf{S}_3 \end{Bmatrix} \tag{5.19}$$

$$\begin{Bmatrix} \mathbf{S}_1 \\ \mathbf{S}_2 \\ \mathbf{S}_3 \end{Bmatrix} = \begin{bmatrix} \cos\beta_y \cos\gamma - \sin\beta_x \sin\beta_y \sin\gamma & \sin\beta_x \cos\gamma & \sin\beta_y \cos\gamma - \sin\beta_x \sin\gamma \cos\beta_y \\ -\sin\gamma \cos\beta_y - \sin\beta_x \sin\beta_y \cos\gamma & \cos\beta_x \cos\gamma & \sin\beta_x \cos\beta_y \cos\gamma - \sin\beta_y \sin\gamma \\ -\sin\beta_y \cos\beta_x & \sin\beta_x & \cos\beta_x \cos\beta_y \end{bmatrix} \begin{Bmatrix} \mathbf{C}_1 \\ \mathbf{C}_2 \\ \mathbf{C}_3 \end{Bmatrix} \tag{5.20}$$

The velocity of helicopter center of mass H^* in the Newtonian frame N is:

$$^{N}\mathbf{v}^{H^*} = u_1\mathbf{N}_1 + u_2\mathbf{N}_2 \tag{5.21}$$

The angular velocity of helicopter H is:

$$^{N}\boldsymbol{\omega}^{H} = u_4\mathbf{H}_1 + h_1 u_3\mathbf{H}_2 - h_2 u_3\mathbf{H}_3 \tag{5.22}$$

The coefficients h_i are quite massive to write, and are provided in the supplementary downloadable material. The angular velocity of suspension cable S is:

$$^{N}\boldsymbol{\omega}^{S} = {}^{N}\boldsymbol{\omega}^{H} + {}^{H}\boldsymbol{\omega}^{S} = u_4\mathbf{H}_1 + h_1 u_3\mathbf{H}_2 - h_2 u_3\mathbf{H}_3 + u_5\mathbf{S}_1 - h_9 u_6\mathbf{S}_2 - h_6 u_6\mathbf{S}_3 \tag{5.23}$$

The position vector of point P from the helicopter center (H^*) is $\mathbf{P}^{H^*P}$ defined as:

$$\mathbf{P}^{H^*P} = d\mathbf{H}_3 \tag{5.24}$$

The velocity of point P is:

$$^{N}\mathbf{v}^{P} = {}^{N}\mathbf{v}^{H^*} + {}^{N}\boldsymbol{\omega}^{H} \times \mathbf{P}^{H^*P} = u_1\mathbf{N}_1 + u_2\mathbf{N}_2 - du_4\mathbf{H}_2 + h_4 u_3\mathbf{H}_1 \tag{5.25}$$

The position vector of hook h from the suspension point P is:

$$\mathbf{P}^{Ph} = l_s\mathbf{S}_3 \tag{5.26}$$

The velocity of hook h is:

$$\begin{aligned} ^{N}\mathbf{v}^{h} &= {}^{N}\mathbf{v}^{P} + {}^{N}\boldsymbol{\omega}^{S} \times \mathbf{P}^{Ph} \\ &= u_1\mathbf{N}_1 + u_2\mathbf{N}_2 + h_4 u_3\mathbf{H}_1 - du_4\mathbf{H}_2 + (h_{13}u_3 - h_{14}u_4 - h_{15}u_6)\mathbf{S}_1 \\ &\quad + (h_{17}u_3 - h_{16}u_4 - l_s u_5)\mathbf{S}_2 \end{aligned} \tag{5.27}$$

The angular velocity of distributed beam B is:

$$\begin{aligned} {}^{N}\boldsymbol{\omega}^{B} &= {}^{N}\boldsymbol{\omega}^{H} + {}^{H}\boldsymbol{\omega}^{S} + {}^{S}\boldsymbol{\omega}^{B} \\ &= u_4\mathbf{H}_1 + h_1u_3\mathbf{H}_2 - h_2u_3\mathbf{H}_3 + u_5\mathbf{S}_1 - h_9u_6\mathbf{S}_2 + h_6u_6\mathbf{S}_3 \\ &\quad + (h_{19}u_7 + h_{21}u_8)\mathbf{C}_1 + (h_{21}u_7 - h_{29}u_8)\mathbf{C}_2 + (u_9 - h_{32}u_8)\mathbf{C}_3 \end{aligned} \tag{5.28}$$

The position vector of B^* from the hook h is:

$$\mathbf{P}^{hB^*} = l_y\mathbf{C}_3 \tag{5.29}$$

A distance l_y is defined as $l_y = \sqrt{l_g^2 - 0.25l_l^2}$. The velocity of the beam center of mass B^* is:

$$\begin{aligned} {}^{N}\mathbf{v}^{B^*} &= {}^{N}\mathbf{v}^{h} + {}^{N}\boldsymbol{\omega}^{B} \times \mathbf{P}^{hB^*} \\ &= u_1\mathbf{N}_1 + u_2\mathbf{N}_2 + h_4u_3\mathbf{H}_1 - du_4\mathbf{H}_2 \\ &\quad + (h_{13}u_3 - h_{14}u_4 - h_{15}u_6)\mathbf{S}_1 + (h_{17}u_3 - h_{16}u_4 - l_su_5)\mathbf{S}_2 \\ &\quad + (h_{104}u_3 + h_{103}u_4 + h_{102}u_5 - h_{105}u_6 + l_yh_{21}u_7 - h_{19}h_{110}u_8)\mathbf{C}_1 \\ &\quad + (-h_{108}u_3 - h_{107}u_4 - h_{106}u_5 + h_{109}u_6 - h_{19}l_yu_7 - h_{88}h_{110}u_8)\mathbf{C}_2 \\ &\quad - h_{89}h_{110}u_8\mathbf{C}_3 \end{aligned} \tag{5.30}$$

At this stage, we extract the rth partial velocities $v_r^{H^*}$, the rth partial angular velocities ω_r^H of the helicopter as coefficients of u_r from Equations (5.21) and (5.22), rth partial velocities of the hook, v_r^h from Equation (5.27), the rth partial angular velocities of beam ω_r^B from Equation (5.28), and rth partial velocities $v_r^{B^*}$ of the beam center of mass B* from Equation (5.30). These partial velocities are shown in Table 5.1.

The acceleration of the helicopter, hook, and beam are:

$${}^{N}\mathbf{a}^{H^*} = \frac{{}^{N}d\,{}^{N}\mathbf{v}^{H^*}}{dt} = \dot{u}_1\mathbf{N}_1 + \dot{u}_2\mathbf{N}_2 \tag{5.31}$$

$${}^{N}\mathbf{a}^{h} = \frac{{}^{N}d\,{}^{N}\mathbf{v}^{h}}{dt} = \dot{u}_1\mathbf{N}_1 + \dot{u}_2\mathbf{N}_2 + {}^{N}\dot{\boldsymbol{\omega}}^{H} \times \mathbf{P}^{H^*P} + {}^{N}\dot{\boldsymbol{\omega}}^{S} \times \mathbf{P}^{Ph} \tag{5.32}$$

$${}^{N}\mathbf{a}^{B^*} = \frac{{}^{N}d\,{}^{N}\mathbf{v}^{B^*}}{dt} = \dot{u}_1\mathbf{N}_1 + \dot{u}_2\mathbf{N}_2 + {}^{N}\dot{\boldsymbol{\omega}}^{H} \times \mathbf{P}^{H^*P} + {}^{N}\dot{\boldsymbol{\omega}}^{S} \times \mathbf{P}^{Ph} + {}^{N}\dot{\boldsymbol{\omega}}^{B} \times \mathbf{P}^{hB^*} \tag{5.33}$$

The angular acceleration of the helicopter and beam are:

$${}^{N}\boldsymbol{\alpha}^{H} = \frac{{}^{N}d\,{}^{N}\boldsymbol{\omega}^{H}}{dt} = \dot{u}_4\mathbf{H}_1 + (h_1\dot{u}_3 - h_{189})\mathbf{H}_2 - (h_{190} + h_2\dot{u}_3)\mathbf{H}_3 \tag{5.34}$$

TABLE 5.1
Partial Velocities and Partial Angular Velocities for the System

r	$v_r^{H^*}$	ω_r^H	v_r^h	ω_r^B	$v_r^{B^*}$
1	$\mathbf{N}_1$	0	$\mathbf{N}_1$	0	$\mathbf{N}_1$
2	$\mathbf{N}_2$	0	$\mathbf{N}_2$	0	$\mathbf{N}_2$
3	0	$h_1\mathbf{H}_2 - h_2\mathbf{H}_3$	$h_4\mathbf{H}_1 + h_{13}\mathbf{S}_1 + h_{17}\mathbf{S}_2$	$h_1\mathbf{H}_2 - h_2\mathbf{H}_3$	$h_4\mathbf{H}_1 + h_{13}\mathbf{S}_1 + h_{17}\mathbf{S}_2 + h_{104}\mathbf{C}_1 - h_{108}\mathbf{C}_2$
4	0	$\mathbf{H}_1$	$-d\mathbf{H}_2 - h_{14}\mathbf{S}_1 - h_{16}\mathbf{S}_2$	$\mathbf{H}_1$	$-d\mathbf{H}_2 - h_{14}\mathbf{S}_1 - h_{16}\mathbf{S}_2 + h_{103}\mathbf{C}_1 - h_{107}\mathbf{C}_2$
5	0	0	$-l_s\mathbf{S}_2$	$\mathbf{S}_1$	$-\mathrm{l}_s\mathbf{S}_2 + h_{102}\mathbf{C}_1 - h_{106}\mathbf{C}_2$
6	0	0	$-h_{15}\mathbf{S}_1$	$-h_9\mathbf{S}_2 + h_6\mathbf{S}_3$	$-h_{15}\mathbf{S}_1 - h_{105}\mathbf{C}_1 + h_{109}\mathbf{C}_2$
7	0	0	0	$h_{19}\mathbf{C}_1 + h_{21}\mathbf{C}_2$	$l_y h_{21}\mathbf{C}_1 - h_{19} l_y \mathbf{C}_2$
8	0	0	0	$h_{21}\mathbf{C}_1 - h_{29}\mathbf{C}_2 - h_{32}\mathbf{C}_3$	$-h_{19}h_{110}\mathbf{C}_1 - h_{88}h_{110}\mathbf{C}_2 - h_{89}h_{110}\mathbf{C}_3$
9	0	0	0	$\mathbf{C}_3$	0

$$ {}^{N}\boldsymbol{\alpha}^{B} = \frac{{}^{N}d\,{}^{N}\boldsymbol{\omega}^{B}}{dt} = {}^{N}\dot{\boldsymbol{\omega}}^{H} + {}^{H}\dot{\boldsymbol{\omega}}^{S} + {}^{S}\dot{\boldsymbol{\omega}}^{B} \tag{5.35} $$

The generalized inertia forces are:

$$ (F^{*}_{r=1,2,3,4})_H = -m_v \cdot {}^{N}\mathbf{a}^{H^*} \cdot \mathbf{v}_r^{H^*} - (\mathbf{I}_H \cdot {}^{N}\boldsymbol{\alpha}^{H} + {}^{N}\boldsymbol{\omega}^{H} \cdot \mathbf{I}_H \cdot {}^{N}\boldsymbol{\omega}^{H}) \cdot \boldsymbol{\omega}_r^{H} \tag{5.36} $$

$$ (F^{*}_{4+r})_h = -m_h \cdot {}^{N}\mathbf{a}^{h} \cdot \mathbf{v}_r^{h} \tag{5.37} $$

$$ (F^{*}_{6+r=7,8,9})_B = -m_v \cdot {}^{N}\mathbf{a}^{B^*} \cdot \mathbf{v}_r^{B^*} - (\mathbf{I}_B \cdot {}^{N}\boldsymbol{\alpha}^{B} + {}^{N}\boldsymbol{\omega}^{B} \cdot \mathbf{I}_B \cdot {}^{N}\boldsymbol{\omega}^{B}) \cdot \boldsymbol{\omega}_r^{B} \tag{5.38} $$

where $\mathbf{I}_H$ and $\mathbf{I}_B$ are the inertia matrix of Helicopter (H) and Beam (B):

$$ \mathbf{I}_H = \begin{bmatrix} I_{xx} & 0 & 0 \\ 0 & I_{yy} & 0 \\ 0 & 0 & I_{zz} \end{bmatrix} \tag{5.39} $$

$$ \mathbf{I}_B = \begin{bmatrix} 0 & 0 & 0 \\ 0 & I_l & 0 \\ 0 & 0 & I_l \end{bmatrix} \tag{5.40} $$

A part of generalized active force F_r^{g} is due to the force of gravity on the helicopter, hook, and the distributed load. Another part of the generalized active force is due to the thrust force, F, produced by the main rotor. Thus, the generalized active force is:

$$ F_r = F_r^{g} + F = m_v g \mathbf{N}_3 \cdot \mathbf{N}_3 + m_h g \mathbf{N}_3 \cdot \mathbf{N}_3 + m_l g \mathbf{N}_3 \cdot \mathbf{N}_3 + F \mathbf{N}_3 \cdot \mathbf{N}_3 \tag{5.41} $$

Here dot-multiplications are performed with the partial velocities of the points of application of the forces, and F is the magnitude of the component force of F along the $\mathbf{N}_3$ direction in order to balance gravity:

$$ F = -\frac{m_v g + m_l g + m_h g}{[\cos\phi_R \cos\theta_R \cos\phi_v \cos\theta_v - \sin\phi_R \sin\phi_v \cos\theta_v - \cos\phi_R \sin\theta_R \sin\theta_v]} \tag{5.42} $$

The coordinate transformation matrix between the reference frames for the helicopter (H) and the rotor hub (R) is:

$$ \begin{Bmatrix} \mathbf{H}_1 \\ \mathbf{H}_2 \\ \mathbf{H}_3 \end{Bmatrix} = \begin{bmatrix} \cos\theta_R & -\sin\phi_R \sin\theta_R & \sin\theta_R \cos\phi_R \\ 0 & \cos\phi_R & \sin\phi_R \\ \sin\theta_R & \sin\phi_R \cos\theta_R & \cos\phi_R \cos\theta_R \end{bmatrix} \begin{Bmatrix} \mathbf{R}_1 \\ \mathbf{R}_2 \\ \mathbf{R}_3 \end{Bmatrix} \tag{5.43} $$

Kane's equation is:

$$ F_r + F_r^{*} = 0, \quad r = 1, ..9. \tag{5.44} $$

Substitution of Equations (5.36)–(5.38) and (5.41) into (5.44) yields the dynamical equations:

$$M\cdot\begin{pmatrix}\ddot{x}\\ \ddot{y}\\ \ddot{\theta}_v\\ \ddot{\varphi}_v\\ \ddot{\alpha}_x\\ \ddot{\alpha}_y\\ \ddot{\beta}_x\\ \ddot{\beta}_y\\ \ddot{\gamma}\end{pmatrix}+f\begin{pmatrix}\dot{x},\ \dot{y},\ \dot{\theta}_v,\ \dot{\varphi}_v,\ \dot{\alpha}_x,\ \dot{\alpha}_y,\ \dot{\beta}_x,\ \dot{\beta}_y,\ \dot{\gamma},\\ x,\ y,\ \theta_v,\ \varphi_v,\ \alpha_x,\ \alpha_y,\ \beta_x,\ \beta_y,\ \gamma\\ \ddot{\theta}_R,\ \dot{\theta}_R,\ \theta_R,\ \ddot{\varphi}_R,\ \dot{\varphi}_R,\ \varphi_R\end{pmatrix}=0 \quad (5.45)$$

where M is the mass matrix and f is the column matrix of gravity terms, centrifugal and Coriolis terms, and control input terms. The matrices M and f are also given in the supplementary downloadable material.

The dynamical model in Equation (5.45) includes nine system states, x, y, θ_v, Φ_v, α_x, α_y, β_x, β_y, and γ, and two inputs, θ_R and Φ_R. Therefore, it is an under-actuated system that requires a sophisticated control system. The control objectives consist of two parts: (i) attitude control of the helicopter, and (ii) swing suppression of the distributed-mass load. To achieve the objectives, a combination of feedback and command-smoothing control is presented in this chapter. A feedback controller regulates the vehicle attitude by following the states of a prescribed model and reduces the tracking errors, while a command-smoothing controller attenuates the distributed-mass load swing by smoothing the pilot commands.

5.1.2.2 Control

This section presents a controller including a feedback controller and open-loop controller. It is assumed that there is a separate controller to hold the heading and the altitude. A model-following controller (MFC) regulates the helicopter's attitude by following the states of a prescribed model and attenuating the tracking errors, while a command smoother reduces both the load swing and twisting by smoothing the pilot's commands. The configuration of the combined feedback and command-smoothing control is shown in Figure 5.24. The pilot generates a pilot command, via the control

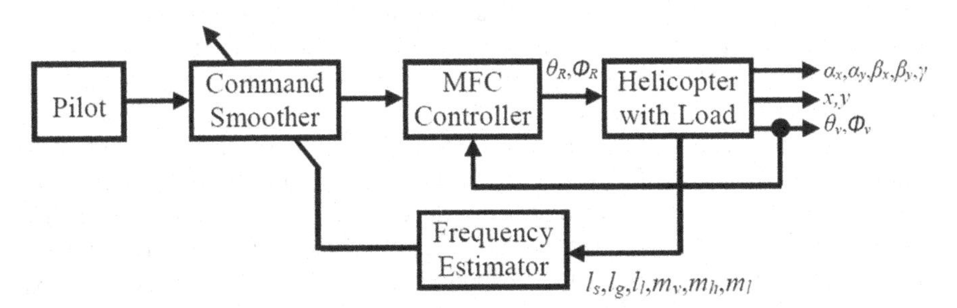

FIGURE 5.24 Combined control architecture.

stick. A command-smoothing controller modifies the pilot command to produce a smoothed command. The measurement of the vehicle attitude, θ_v and Φ_v, is used to adjust the thrust angle, θ_R and Φ_R, in a feedback control loop. Then, the attitude, θ_v and Φ_v, is forced to track the smoothed command, by the feedback controller. The system frequency is estimated by system parameters and is used to design the command-shaping controller. The smoothed command drives the aerial vehicle toward the desired position without inducing oscillations of the distributed-mass load.

5.1.2.2.1 MFC Controller for Attitude Regulation

The dynamics of the helicopter with distributed-mass load in Equation (5.45) is too complicated to analytically design a MFC controller. Therefore, an MFC controller for a simplified model without the sling load was created. Then, the result was refined as needed for the aerial crane. The controller will be used on the full nonlinear dynamics shown in Equation (5.45) to test its performance.

Ignoring the swing angles, α_x, α_y, β_x, and β_y, and the twisting angle, γ, in the dynamics model (5.45) yields a simplified model without the slung load dynamics:

$$\begin{cases} I_{xx}\ddot{\varphi}_v \cos(\phi_R + \phi_x) = a\cdot g\cdot m_v\cdot \sin(\phi_R) \\ I_{yy}\ddot{\theta}_v \cos(\theta_R + \theta_v) = a\cdot g\cdot m_v\cdot \sin(\theta_R) \\ \ddot{y} = g\cdot \tan(\phi_R + \phi_v) \\ \ddot{x} = g\cdot \tan(\theta_R + \theta_R) \end{cases} \tag{5.46}$$

Both the vehicle attitudes, θ_v and Φ_v, and the thrust angles, θ_R and Φ_R, are assumed to be small so that Equation (5.46) yields a linearized model. From the linearized model, an explicit MFC controller can be derived:

$$\begin{cases} \ddot{\phi}_{vm} + 2\zeta_p\omega_p\dot{\phi}_{vm} + \omega_p^2\cdot\Phi_{vm} = \omega_p^2\cdot c_x \\ \ddot{\theta}_{vm} + 2\zeta_p\omega_p\dot{\theta}_{vm} + \omega_p^2\cdot\theta_{vm} = \omega_p^2\cdot c_y \\ \phi_R = \dfrac{I_{xx}\ddot{\varphi}_{vm}}{a\cdot g\cdot m_v} + k_{xd}\cdot(\dot{\phi}_{vm} - \dot{\phi}_v) + k_{xp}\cdot(\phi_{vm} - \phi_v) \\ \theta_R = \dfrac{I_{yy}\ddot{\theta}_{vm}}{a\cdot g\cdot m_v} + k_{yd}\cdot(\dot{\theta}_{vm} - \dot{\theta}_v) + k_{yp}\cdot(\theta_{vm} - \theta_v) \end{cases} \tag{5.47}$$

where ζ_p is the damping ratio of the prescribed model, ω_p is the frequency of the prescribed model, c_x and c_y are smoothed pilot's commands along the $\mathbf{N}_1$ and $\mathbf{N}_2$ directions, Φ_{vm} and θ_{vm} are model outputs along the $\mathbf{N}_1$ and $\mathbf{N}_2$ directions, and k_{xp}, k_{xd}, k_{yp}, and k_{yd} are control gains.

The first two equations in Equation (5.47) are the prescribed model. The damping ratio, ζ_p, and the frequency, ω_p, of the prescribed model were designed by using the pole placement method. A prescribed model with reasonable damping ratio (0.707) and reasonable settling time (≤2 seconds) is utilized. The desired closed-loop poles were calculated to be approximately −2 ± 2i. Thus, the frequency, ω_p, was 2.83 rad/s.

The last two equations in Equation (5.47) are the asymptotic tracking control law. Suitable tracking errors can be achieved by designing control gains, k_{xp}, k_{yp}, k_{xd}, and k_{yd}. Referencing the eigenvalues discussed in the literature, we know that the coupling effect between the vehicle and the slung load decreases the real part of the eigenvalues and increases the system damping. Thus, the prescribed eigenvalues of the tracking controller are selected to be $-18 \pm 8.7i$. Given that the vehicle mass, m_v, moment of inertia, I_{xx}, I_{yy}, I_{zz}, and distance, a, in the simulations are 6000 kg, 17450 kg m^2, 20500 kg m^2, 3.5 m, respectively. Then, the control gains k_{xp}, k_{yp}, k_{xd}, and k_{yd} were calculated by the pole placement technique to be 33.916, 3.053, 39.845, and 3.586, respectively.

5.1.2.2.2 *Smoothing Pilot Commands for Oscillation Suppression*

In addition to modifying the vehicle's attitude with the MFC controller, it is useful to design a command-smoothing controller from a simplified model. Ignoring the vehicle attitude, θ_v and Φ_v, in the dynamics model (5.45) yields a simplified dynamic equation of the double-pendulum load swing. The simplified model also assumes the beam alignment coincides with the flight direction. The simplified model of double-pendulum dynamics in the longitudinal direction is given by:

$$\begin{cases} (c\cdot L^2 + e\cdot l_x^2 + c\cdot l_x^2 + c\cdot l_y^2 + 2c\cdot l_x\cdot l_y\cdot \cos\beta_x)\cdot\ddot{\alpha}_x + c\cdot(L^2 + l_y^2 + l_x\cdot l_y\cdot \cos\beta_x)\cdot\ddot{\beta}_x \\ -\, c\cdot l_x\cdot l_y\cdot \sin\beta_x\cdot(\dot{\beta}_x^2 + 2\dot{\alpha}_x\dot{\beta}_x) + [e\cdot l_x\cdot \cos\alpha_x + c\cdot l_x\cdot \cos\alpha_x + c\cdot l_y\cdot \cos(\alpha_x + \beta_x)] \\ \quad \cdot\ddot{x} \\ +\, g\cdot c\cdot l_y\cdot \sin(\alpha_x + \beta_x) + g\cdot(e + c)\cdot l_x\cdot \sin\alpha_x = 0 \\ (L^2 + l_y^2 + l_x\cdot l_y\cdot \cos\beta_x)\cdot\ddot{\alpha}_x + (L^2 + l_y^2)\cdot\ddot{\beta}_x + l_y\cdot \cos(\alpha_x + \beta_x)\cdot\ddot{x} \\ +\, l_x\cdot l_y\cdot \sin\beta_x\cdot\dot{\alpha}_x^2 + g\cdot l_y\cdot \sin(\alpha_x + \beta_x) = 0 \\ [e\cdot l_x\cdot \cos\alpha_x + c\cdot l_x\cdot \cos\alpha_x + c\cdot l_y\cdot \cos(\alpha_x + \beta_x)]\cdot\ddot{\alpha}_x + c\cdot l_y\cdot \cos(\alpha_x + \beta_x)\cdot\ddot{\beta}_x \\ +\, (1 + e + c)\cdot\ddot{y} - e\cdot l_x\cdot \sin\alpha_x\cdot\dot{\alpha}_x^2 - c\cdot l_y\cdot \sin(\alpha_x + \beta_x)\cdot[\dot{\alpha}_x + \dot{\beta}_x]^2 - c\cdot l_x\cdot \sin\alpha_x \\ \quad \cdot\dot{\alpha}_x^2 = 0 \end{cases} \tag{5.48}$$

where

$$L = l_p/(2\sqrt{3}) \tag{5.49}$$

$$l_x = l_s + a + d \tag{5.50}$$

The ratio of the hook mass to the helicopter mass is e, while the ratio of the load mass to the helicopter mass is c. The load swing angles are assumed to be small around the equilibrium position. The linearized frequencies of the double-pendulum swing resulting from Equation (5.48) are given by:

$$\omega_{1,2} = \sqrt{\frac{g\cdot(w \mp v)}{2l_x}} \tag{5.51}$$

where

$$w = \frac{1}{e \cdot l_y^2 + (c + e) \cdot L^2} \begin{pmatrix} (e^2 + e + c + e \cdot c) \cdot l_y^2 + (e + c) \cdot l_y \cdot l_x \\ + (e^2 + c^2 + 2e \cdot c + c + e) \cdot L^2 \end{pmatrix} \tag{5.52}$$

$$v = \sqrt{w^2 - \frac{4(e^2 + c^2 + 2e \cdot c + e + c) \cdot l_y \cdot l_x}{e \cdot l_y^2 + (e + c) \cdot R^2}} \tag{5.53}$$

A smoothing function inherent in the limited motion of the helicopter results in a limited response to oscillations of the suspended load. Thus, the command-smoothing controller is a smoothing function, which is a piecewise continuous function.

The model (5.48) can be approximated near the equilibrium angles as two uncoupled second-order systems with the frequencies shown in Equation (5.51). The oscillations may be suppressed by a four-pieces smoother:

$$u(\tau) = \begin{cases} M\tau e^{-2\zeta\omega\tau}, & 0 \le \tau \le 0.5T_m \\ M[(1 - K)T_m + (2K - 1)\tau]e^{-2\zeta\omega\tau}, & 0.5T_m < \tau \le T_m \\ M[(3K - K^2)T_m + (K^2 - 2K)\tau]e^{-2\zeta\omega\tau}, & T_m < \tau \le 1.5T_m \\ M[2K^2T_m - K^2\tau]e^{-2\zeta\omega\tau}, & 1.5T_m < \tau \le 2T_m \end{cases} \tag{5.54}$$

where

$$M = \zeta^2\omega^2/(1 - K^{-1})^2 \tag{5.55}$$

$$K = e^{2\pi\zeta/\sqrt{1-\zeta^2}} \tag{5.56}$$

$$T_m = 2\pi/(\omega\sqrt{1 - \zeta^2}) \tag{5.57}$$

The damping of the load swing can be assumed to be zero for the design of the smoother. The smoother can be designed by the estimates of the first-mode frequency of the load swing. The smoother has a low-pass filtering effect that can suppress the high-mode swing. Therefore, the MFC controller regulates the helicopter's attitudes, and the four-pieces smoother attenuates the oscillations of the load.

5.1.2.3 Computational Dynamics

The hook mass, m_h, suspension cable length, l_s, rigging cable length, l_g, distances, a and d, were set to 200 kg, 10 m, 10 m, 5 m, and 1.5 m, respectively. The initial twist angle of the load was fixed at 45°. The nonlinear simulation was performed using an acceleration to move the helicopter flying forward along the $\mathbf{N}_1$ direction, and then flight speed was kept at a constant value. At the end of travel, the helicopter was decelerated to stop at a desired distance. This process simulates normal straight-line flight with suspended loads.

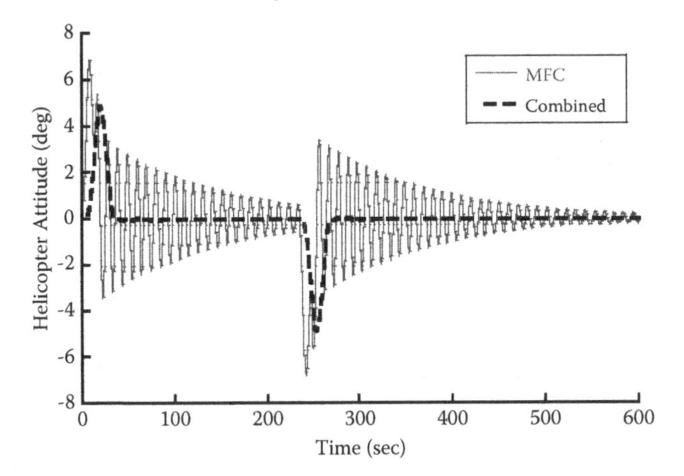

FIGURE 5.25 Simulated response of helicopter attitude.

The simulated response for a flight distance of 3 km is shown in Figures 5.25–5.28 when the load mass, m_p, and load length, l_p, were set to 2000 kg, 10 m, respectively. There are two stages in the flight system response. The transient stage of the response is defined as the period when the helicopter flies forward. The residual stage is defined as the period when the helicopter attempts to stop and hovers in the sky. The maximum defections during the transient and residual stages are referred to as the transient deflection and residual amplitude, respectively.

Figure 5.25 shows the simulated results of the helicopter's attitude. The transient deflection and residual amplitude of helicopter's attitude with the MFC controller were 6.9° and 3.5°, respectively, whereas those with the combined controller were reduced to 4.8° and 0.036°, respectively. The settling time with the MFC controller required for the residual response of the helicopter's attitude to settle within 0.5°

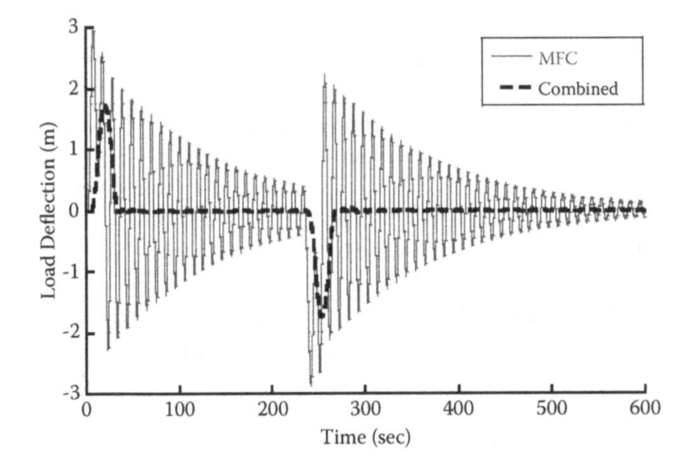

FIGURE 5.26 Simulated response of load deflection.

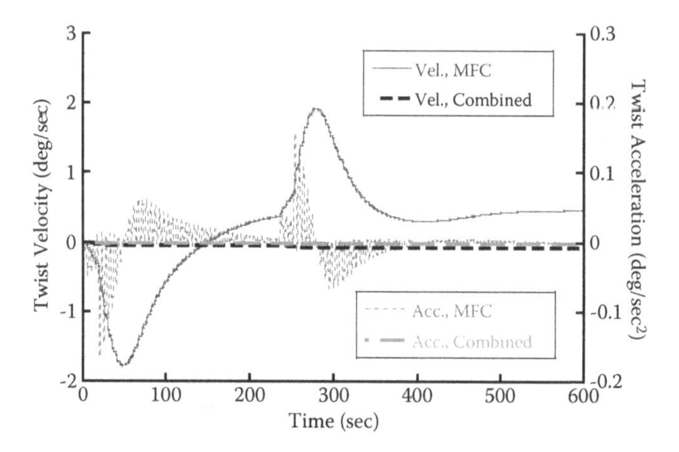

FIGURE 5.27 Simulated response of twist velocity and twist acceleration.

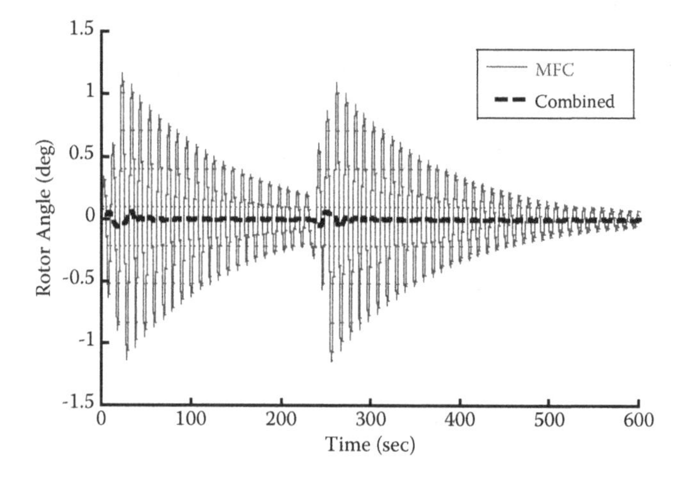

FIGURE 5.28 Simulated response of rotor angle.

was 223.6 seconds, while the settling time with the combined controller was only 9.3 seconds. Thus, the combined controller obviously reduced the oscillations of the helicopter's attitude. The oscillation frequency of helicopter's attitude in Figure 5.25 is predicted by the first-mode frequency in Equation (5.34) to be approximately 0.61 rad/s. Note that the heavy-load swing has a great impact on the helicopter's attitude.

Figure 5.26 shows the simulated results of the deflection of the load center. With the MFC controller, the transient deflection and residual amplitude of the load swing are 2.9 m and 2.3 m, respectively. The settling time required for the residual oscillation of the load to settle within 1% of the sum of the suspension length and second cable equivalent length was 95.7 seconds. With the combined controller, the transient deflection and residual amplitude were only 1.72 m and 0.024 m, respectively, while the settling time was only 5.9 seconds. The combined controller also markedly suppressed the oscillations of the load swing. The oscillation frequency of the load, which Figure 5.26 shows is approximately 0.61 rad/s, is similar to the frequency of helicopter's attitude in Figure 5.25. This is explained by the coupling effect between the helicopter and the load.

Figure 5.27 shows the simulated results of load twisting. The angular acceleration of the load twisting with the MFC controller has maximum and minimum values when the corresponding extreme values of both the attitude and load swing occurs. Additionally, the angular velocity of the load twisting with the MFC controller has a trough at approximately 50 seconds and a peak near 300 seconds. After the trough and peak, the angular acceleration and velocity approach zero and a constant, respectively. This is because the load swing results in the load twisting. The settling time with the MFC controller required for the residual oscillation of the angular acceleration to settle within $0.001\,°/s^2$ was 312.8 seconds, and the corresponding angular velocity was 0.283°/s. This valve can be considered as the steady-state velocity of the load swing. By using the combined controller, the load twisting dynamics were kept steady, and there was no significant settling process. The steady-state velocity was reduced to 0.056°/s. Thus, the combined controller reduced the load twist speed by 70%. The combined controller limited the angular velocity and acceleration of the load twisting to low values, thereby allowing for safer operation.

Figure 5.28 shows simulated results of the rotor angle. The maximum rotor angle with the MFC controller was 1.16°, while that with the combined controller was only 0.062°. The combined controller requires only a small rotor angle for controlling the helicopter. The rotor frequency in Figure 5.28 is as constant as the oscillation frequency of helicopter's attitude shown in Figure 5.25 and load deflection shown in Figure 5.26. These results can be attributed to the coupling effect between the helicopter and the load, and the damping effect of the attitude oscillations by adjusting the rotor angles.

A battery of simulations was performed to investigate the effectiveness of the combined controller for varying flight distances. The load mass, m_p, and load length, l_p, were set to 2000 kg and 10 m, respectively. The transient deflection, residual amplitude, and settling time of the helicopter's attitude are presented in Figure 5.29 as a function of flight distance. Due to in phase or out of phase interactions between the oscillations caused by the acceleration and deceleration, the transient deflection, residual amplitude, and settling time of the helicopter's attitude with the MFC controller has a periodic amplitude. Additionally, frequencies in the transient deflection, residual amplitude, and settling time are restricted by the first-mode frequency resulting from Equation (5.34). Both the magnitude of peaks and troughs in the transient deflection, residual amplitude, and settling time decrease with the increasing flying distance. This is mainly because the MFC controller provides a damping effect. Finally, the transient deflection, residual amplitude and settling time of the helicopter's attitude stabilize at approximately 6.84°, 3.35°, and 23.2 seconds. When the combined controller is utilized, the transient deflection, residual amplitude, and the settling time comply with the requirements of ADS-33E, falling below 4.9°, 0.07°, and 12.1 seconds. Thus, the combined controller reduces the transient deflection, residual amplitude, and settling time of the helicopter attitude to very low levels for a wide range of flight motions.

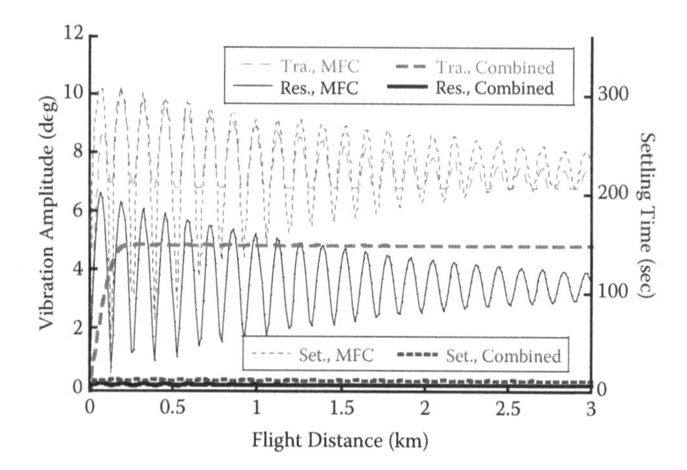

FIGURE 5.29 Transient deflection, residual amplitude, and settling time of helicopter attitude against flight distance.

Figure 5.30 shows the transient deflection, residual amplitude, and settling time of the load swing. The simulated results of the load motion are similar to that of the helicopter attitude observed in Figure 5.29. Moreover, the frequency is also restricted by the first-mode frequency of the load swing linked to the value of Equation (5.34). As the flight distance increases, the transient deflection, residual amplitude, and settling time with the MFC controller settle to approximately 2.94 m, 2.3 m, and 10.1 seconds. However, with the use of the combined controller, the transient deflection, residual amplitude, and settling time are limited to less than 1.75°, 0.05 m, and 6.8 seconds, respectively. Thus, the combined controller also suppresses the transient deflection, residual amplitude, and settling time of the load deflection to low levels over a wide range of flight motions.

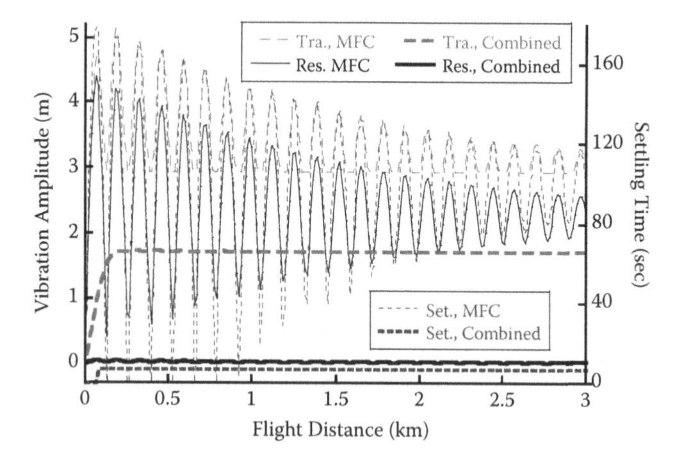

FIGURE 5.30 Transient deflection, residual amplitude, and settling time of load deflection against flight distance.

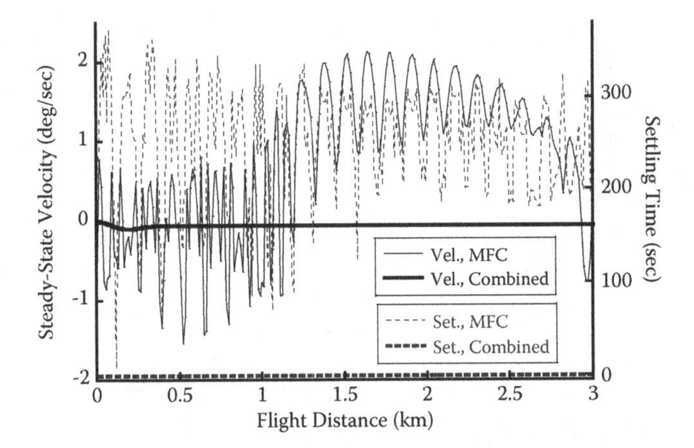

FIGURE 5.31 Transient deflection, residual amplitude, and settling time of load twisting against flight distance.

The results for load twisting are shown in Figure 5.31. For short flying distances, the twisting dynamics effects on the steady state velocity and settling time are very complicated. After this initial range, the frequency of the steady-state velocity is also similar to the first-mode frequency in Equation (5.34). Around the flight distance of 1.5 km, the steady-state velocity is at the maximum value. Then, the peak magnitude decreases with increasing flight distances because load twisting is coupled with both the helicopter's attitude and the load swing. The combined controller limits the steady-state velocity of the load twisting to below 0.12°/s for all the cases shown in Figure 5.31. The combined controller attenuates an average of 94% more steady-state velocity than the MFC controller and suppresses total settling time.

An additional set of simulations was performed to assess the robustness of the combined controller for various modeling errors in the system parameters. The flight distance was set to 3 km. The load length was varied from 7 m to 13 m in the simulations. The load mass increased as the load length increased because of constant load density. However, the modeled load length and mass for the combined controller were fixed at 10 m and 2000 kg, respectively. The corresponding design frequency for the combined controller was 0.68 rad/s.

The transient deflection, residual amplitude, and settling time of the helicopter attitude for the fixed design frequency are shown in Figure 5.32. By using the MFC controller, the transient deflection of helicopter's attitude in the transient stage shows no significant change between 7 m and 10 m. The transient deflection does rise from 10 m, and reaches the local maximum at 11.3 m. After this peak point, the transient deflection falls until 12.6 m, and then stays constant again. The residual amplitude contains peaks and troughs as the load length is varied. A trough occurs at 8.7 m, while a peak occurs at 11.3 m. Both the peak and trough are the results of the first-mode frequency in Equation (5.34). At the same time, the settling time is related closely to the residual amplitude. This is also because the changing load suspension length affects the system frequency in Equation (5.34). Under the same

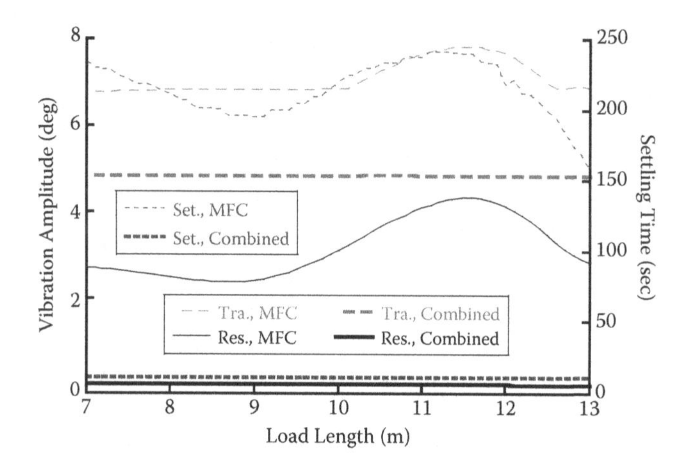

FIGURE 5.32 Transient deflection, residual amplitude, and settling time of helicopter attitude against load length and mass.

conditions, the combined controller can suppress the transient deflection, residual amplitude, and settling time better than the MFC controller by an average of 30.8%, 99%, and 94.6%, respectively.

Figure 5.33 shows the transient deflection, residual amplitude, and settling time of the load deflection. Because the dynamics of the helicopter's attitude couples with that of the load swing, the trends in the load deflection are similar to the helicopter's attitude. Before 10 m, the transient deflection decreases slightly. Then, transient deflection has a peak value when the load length reaches 11.3 m. Meanwhile, both the residual deflection and the settling time have trough values at the load length of 8.7 m. They climb to the local maximum at 11.3 m. The combined controller limits the transient deflection, residual deflection, and settling time to under 1.87 m, 0.032 m, and 7.13 s, respectively, for the parameter ranges shown.

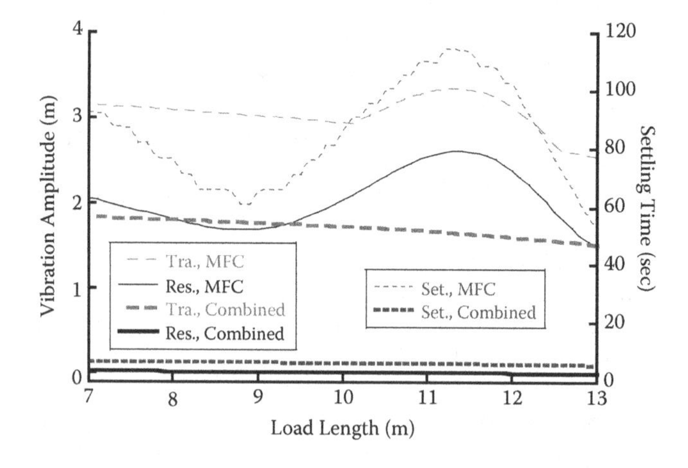

FIGURE 5.33 Transient deflection, residual amplitude, and settling time of load deflection against load length and mass.

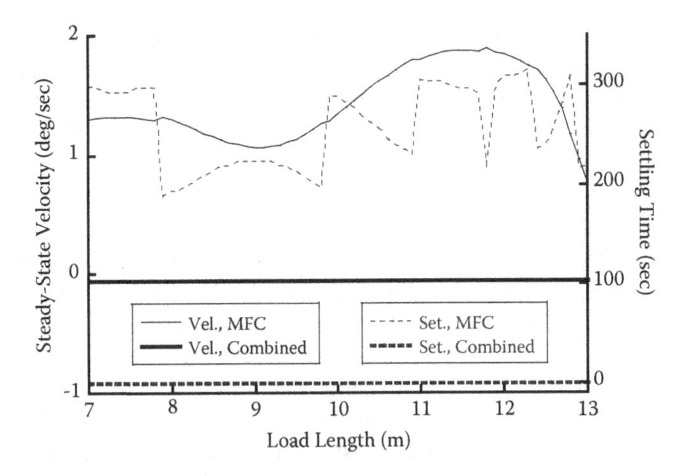

FIGURE 5.34 Transient deflection, residual amplitude, and settling time of load twisting against load length and mass.

The simulated load twisting is shown in Figure 5.34. The dynamics of the steady-state velocity is similar to that of the helicopter's attitude and the load deflection. The steady-state speed is reduced to the lowest valve at the load suspension length of 8.7 m, and rises to a local maximum at 11.3 m. When using the MFC controller, the dynamics of the settling time is very complicated. The combined controller restricts the steady-state velocity to nearly –0.06°/s and virtually eliminates the settling time. Therefore, the combined controller reduces steady-state speed of the load swing over the MFC controller by an average of 95.6%.

The simulations demonstrate that the combined controller is beneficial for reducing the oscillations of the helicopter's attitude, load swing, and load twisting. Therefore, the combined controller provides the pilot a safer environment to operate the helicopter, and a higher efficiency for transferring loads to the desired position precisely.

5.2 QUADCOPTERS SLUNG LOADS

Quadcopters suspending bulky and heavy loads may provide material-handling services in transportation, construction, and rescue missions [25–29]. Unfortunately, a cable-suspended load attached to the quadcopter constitutes a flexible structure. The unwanted oscillations of the flexible structure caused from operator-commanded motions and external disturbances are the major limitation for safe and efficient transportations. Therefore, it is essential to study the dynamics and control of quadcopters slung loads.

Prior literatures have reported the oscillation suppression of quadcopters slung objects. To reduce the oscillations, many examples show that feedback control systems are effective by using measurement and estimation of the suspended-load oscillations [30–38]. However, accurately sensing the suspended load is difficult, and the feedback controller might conflict with the operator's action. Open-loop control techniques have been used to decrease oscillations by modifying commands of the operator to produce a low-swing motion, such as optimal trajectory planning [39–48], flatness-based control [49], and input shaper [50]. In practice, the open-loop controller can reduce oscillations from operator-commanded motions, but cannot control oscillations from external disturbances. Meanwhile, significant efforts have concentrated on modeling and dynamics of the quadcopters slung loads [51–55]. However, all abovementioned works focus on the single-hoist dynamics, which is suitable for small-size loads.

Mounting numbers of examples indicate that the dual-hoist mechanism should be used to transport bulky loads for safe operations. Under the dual-hoist mechanism, the large-size objects, which should be modeled as distributed masses, are suspended by two cables below the quadcopters. This mechanism is suitable for transporting large-size loads. Additionally, the forces acting on the suspension cables under dual-hoist mechanisms are less than that under single-hoist configurations. Hence, the dual-hoist mechanism can provide a safer condition to manipulate the quadrotor slung load. However, the distributed-mass load oscillations under dual-hoist configuration corrupt the quadcopter attitude seriously because of strong coupling effects. Therefore, manipulation tasks may be very challenging.

In the recent literature, some effects have been placed on helicopters or tiltrotors slung large-size loads [56–59], in which loads were also attached to aircrafts by single-hoist configuration. Additionally, some papers [60,61] reported multiple quadcopters cooperate to transport large loads, in which cables were attached to the mass center of quadcopters. Therefore, coupling effect between the quadcopter attitude and load swing was ignored in those papers.

5.2.1 Planar Dynamics

5.2.1.1 Modeling

Figure 5.35 shows a schematic representation of a quadcopter carrying a distributed-mass beam by two independent cables undergoing planar motions. The quadcopter **D**

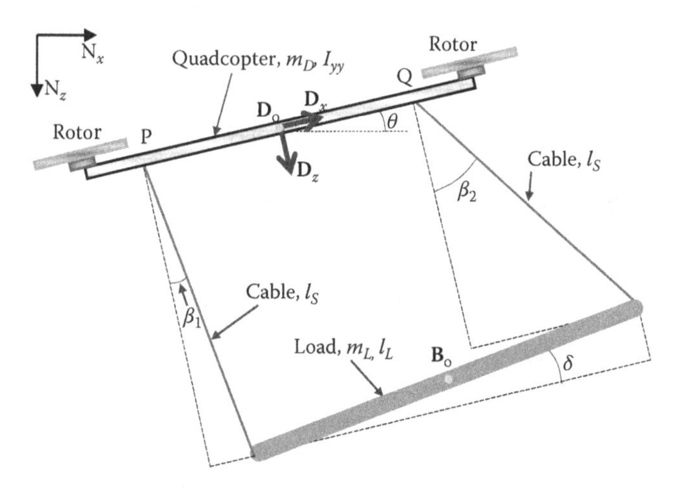

FIGURE 5.35 Planar model of a quadcopter carrying a load by dual-hoist mechanism.

flies along the $\mathbf{N}_x$ direction in near-hover operations. The mass of the quadcopter is m_D, and the moment of inertia about the $\mathbf{D}_y$ axis is I_{yy}. Motions of the quadcopter are divided into the displacements, x and z, along the $\mathbf{N}_x$ and $\mathbf{N}_z$ directions, and the attitude angle, θ, about the $\mathbf{N}_y$ axis in the inertial Newtonian frame. Let $\mathbf{D}_x\mathbf{D}_y\mathbf{D}_z$ be the moving Cartesian coordinates fixed to the quadcopter such that the $\mathbf{N}_y$ axis is parallel to the $\mathbf{D}_y$ axis. The inertial coordinates $\mathbf{N}_x\mathbf{N}_y\mathbf{N}_z$ can be converted to the moving coordinates $\mathbf{D}_x\mathbf{D}_y\mathbf{D}_z$ by rotating the attitude, θ, about the $\mathbf{N}_y$ axis.

Two cables, $\mathbf{C}_1$ and $\mathbf{C}_2$, of length, l_S, which are assumed to be massless and inelastic, hang below the quadcopter at the suspension points, P and Q. The quadcopter center of mass, $\mathbf{D}_o$, locates at the middle point between the load suspension points, P and Q. The distance between the load suspension points, P and Q, is $2l_A$. A uniformly distributed-mass rigid beam, $\mathbf{B}$, of mass, m_L, and length, l_L, serves as the large load. The mass center, $\mathbf{B}_o$, of the load locates at the middle point of the beam. The beam is attached to the quadcopter by two suspension cables. The swing angles of the suspension cables $\mathbf{C}_1$ and $\mathbf{C}_2$ are β_1 and β_2, respectively. The angle between beam direction and the $\mathbf{D}_x$ direction is defined as the slope angle, δ, of the load, through which the beam rotates about the vertical direction. Therefore, the quadcopter coordinates $\mathbf{D}_x\mathbf{D}_y\mathbf{D}_z$ can be converted to the load coordinates $\mathbf{B}_x\mathbf{B}_y\mathbf{B}_z$ by rotating the slope angle, δ, about the $\mathbf{D}_y$ axis.

Inputs to the planar near-hover model are the thrust force, F_z, along the $\mathbf{D}_z$ direction, and the pitching moment, M_θ, about the $\mathbf{D}_y$ direction. Outputs are the quadcopter displacements, x and z, the quadcopter attitude, θ, the swing angles, β_1 and β_2, and the beam slope angle, δ. The model assumes that aerodynamic effects on the suspended load are negligible due to near-hover operation, long cables, and heavy loads. The initial direction of the beam length coincides with $\mathbf{D}_x\mathbf{D}_z$ plane so that the thrust force, F_z, and pitching moment, M_θ, cannot cause beam twisting about the suspension cables.

The Kane's equations will be applied to derive the dynamical model in Figure 5.35. The generalized speeds, $u_{k=1-6}$, are chosen as the quadcopter velocities, $\dot{x}$ and $\dot{z}$, the angular velocity of the quadcopter attitude, $\dot{\theta}$, the angular velocities of the cable swing, $\dot{\beta}_1$ and $\dot{\beta}_2$, and angular velocity of the beam slope, $\dot{\delta}$, respectively. Then the velocity of the quadcopter center, $\mathbf{D}_o$, in the Newtonian reference frame $\mathbf{N}_x\mathbf{N}_y\mathbf{N}_z$ can be written as:

$$^{N}\mathbf{v}^{D_o} = u_1\mathbf{N}_x + u_2\mathbf{N}_z \tag{5.58}$$

Meanwhile, the angular velocity of the quadcopter is given by:

$$^{N}\boldsymbol{\omega}^{D} = u_3\mathbf{D}_y \tag{5.59}$$

The velocity of the beam center, $\mathbf{B}_o$, in the Newtonian reference frame can be written as:

$$\begin{aligned} ^{N}\mathbf{v}^{B_o} = {} & [u_1 - u_3 l_A \sin\theta + (u_3 + u_5) l_S \cos(\theta + \beta_2) + 0.5(u_3 + u_6) l_L \sin(\theta + \delta)]\mathbf{N}_x \\ & + [u_2 - u_3 l_A \cos\theta - (u_3 + u_5) l_S \sin(\theta + \beta_2) + 0.5(u_3 + u_6) l_L \cos(\theta + \delta)]\mathbf{N}_z \end{aligned} \tag{5.60}$$

The angular velocity of the beam, **B**, is given by:

$$ {}^{N}\boldsymbol{\omega}^{B} = u_3\mathbf{D}_y + u_6\mathbf{B}_y \tag{5.61} $$

The quadcopter, beam, and two cables constitute a four-bar linkage. Then the displacement constraint of the four-bar linkage can be written as:

$$ \begin{cases} 2l_A + l_S \sin\beta_2 - l_L \cos\delta - l_S \sin\beta_1 = 0 \\ l_S \cos\beta_2 + l_L \sin\delta - l_S \cos\beta_1 = 0 \end{cases} \tag{5.62} $$

The corresponding velocity constraint and acceleration constraint of the four-bar linkage are:

$$ \begin{cases} \dot{\beta}_2 \cos(\beta_2 - \delta) - \dot{\beta}_1 \cos(\beta_1 - \delta) = 0 \\ \dot{\delta} l_L \cos(\beta_2 - \delta) + \dot{\beta}_1 l_S \sin(\beta_1 - \beta_2) = 0 \end{cases} \tag{5.63} $$

and

$$ \begin{cases} \ddot{\beta}_2 = \tan(\beta_2 - \delta)\dot{\beta}_2^2 - \dfrac{l_L\dot{\delta}^2}{l_S\cos(\beta_2 - \delta)} - \dfrac{[\sin(\beta_1 - \delta)\dot{\beta}_1^2 - \cos(\beta_1 - \delta)\dot{u}_4]}{\cos(\beta_2 - \delta)} \\ \ddot{\delta} = \dfrac{l_S[\dot{\beta}_2^2 - \cos(\beta_1 - \beta_2)\dot{\beta}_1^2 - \sin(\beta_1 - \beta_2)\ddot{\beta}_1]}{l_L\cos(\beta_2 - \delta)} - \tan(\beta_2 - \delta)\dot{\delta}^2 \end{cases} \tag{5.64} $$

The constraints (5.62)–(5.64) indicate that only one angle among the cable swing, β_1 and β_2, and beam slope, δ, is independent while the other two are dependent. The independent motion variable is defined as β_1, and the dependent motion variables are β_2 and δ. Then the cable swing, β_1, can exhibit all dynamic behavior of cable swing and beam slope. Additionally, the beam slope, δ, is always equal to zero when the beam length, l_L, equals the distance, $2l_A$, between the two suspension points, P and Q.

A part of generalized active force is the gravity of quadcopter and beam, and another part of generalized active force is the thrust force, F_z, and pitching moment, M_θ, produced by the rotors. Therefore, the generalized active force, $F_{k=1\text{–}4}$, can be written as:

$$ F_1 = -F_z \sin\theta \tag{5.65} $$

$$ F_2 = m_L g + m_D g - F_z \cos\theta \tag{5.66} $$

$$ F_3 = M_\theta + m_L g[0.5l_L \cos(\theta + \delta) - l_A \cos\theta - l_S \sin(\theta + \beta_2)] \tag{5.67} $$

$$ F_4 = -m_L g l_S \sin(\theta + \beta_2)\frac{\cos(\beta_1 - \delta)}{\cos(\beta_2 - \delta)} - 0.5 m_L g l_L \cos(\theta + \delta)\frac{l_S \sin(\beta_1 - \beta_2)}{l_L \cos(\beta_2 - \delta)} \tag{5.68} $$

where g is the gravitational acceleration. Meanwhile, the generalized inertia force, $F_{k=1\text{–}4}^{*}$, is:

$$F_1^* = -m_D\dot{u}_1 - m_L W_1 \tag{5.69}$$

$$F_2^* = -m_D\dot{u}_2 - m_L W_2 \tag{5.70}$$

$$F_3^* = -\dot{u}_3 I_{yy} - \frac{(\dot{u}_3 + \dot{u}_6)m_L l_L^2}{12} - m_L W_1 \begin{bmatrix} -l_A \sin\theta \\ +l_S \cos(\theta + \beta_2) \\ +0.5 l_L \sin(\theta + \delta) \end{bmatrix} - m_L W_2 \begin{bmatrix} -l_A \cos\theta \\ -l_S \sin(\theta + \beta_2) \\ +0.5 l_L \cos(\theta + \delta) \end{bmatrix} \tag{5.71}$$

$$\begin{aligned} F_4^* &= \frac{\cos(\beta_1 - \delta)}{\cos(\beta_2 - \delta)}[-l_S \cos(\theta + \beta_2)m_L W_1 + l_S \sin(\theta + \beta_2)m_L W_2] \\ &- \frac{l_S \sin(\beta_1 - \beta_2)}{l_L \cos(\beta_2 - \delta)}\begin{bmatrix} -\frac{1}{12}(\dot{u}_3 + \dot{u}_6)m_L l_L^2 - 0.5 l_L \sin(\theta + \delta)m_L W_1 \\ -0.5 l_L \cos(\theta + \delta)m_L W_2 \end{bmatrix} \end{aligned} \tag{5.72}$$

where

$$\begin{aligned} W_1 &= \dot{u}_1 - (u_3^2 l_A \cos\theta + \dot{u}_3 l_A \sin\theta) + (\dot{u}_3 + \dot{u}_5)l_S \cos(\theta + \beta_2) \\ &- (u_3 + u_5)^2 l_S \sin(\theta + \beta_2) + 0.5(u_3 + u_6)^2 l_L \cos(\theta + \delta) \\ &+ 0.5(\dot{u}_3 + \dot{u}_6)l_L \sin(\theta + \delta) \end{aligned} \tag{5.73}$$

$$\begin{aligned} W_2 &= \dot{u}_2 - (-u_3^2 l_A \sin\theta + \dot{u}_3 l_A \cos\theta) - (\dot{u}_3 + \dot{u}_5)l_S \sin(\theta + \beta_2) \\ &- (u_3 + u_5)^2 l_S \cos(\theta + \beta_2) - 0.5(u_3 + u_6)^2 l_L \sin(\theta + \delta) \\ &+ 0.5(\dot{u}_3 + \dot{u}_6)l_L \cos(\theta + \delta) \end{aligned} \tag{5.74}$$

By using the Kane's equation, the sum of the generalized active force described in Equations (5.65)–(5.68) and generalized inertia force given in Equations (5.69)–(5.72) is equal to zero. Then the nonlinear equations of the motion for the model shown in Figure 5.35 yield:

$$m_D\dot{u}_1 + m_L W_1 + F_z \sin\theta = 0 \tag{5.75}$$

$$m_D\dot{u}_2 + m_L W_2 - m_L g - m_D g + F_z \cos\theta = 0 \tag{5.76}$$

$$\begin{aligned} &\dot{u}_3 I_{yy} + \frac{(\dot{u}_3 + \dot{u}_6)m_L l_L^2}{12} - M_\theta - m_L g[0.5 l_L \cos(\theta + \delta) - l_A \cos\theta - l_S \sin(\theta + \beta_2)] \\ &+ m_L W_1 \begin{bmatrix} -l_A \sin\theta \\ +l_S \cos(\theta + \beta_2) \\ +0.5 l_L \sin(\theta + \delta) \end{bmatrix} + m_L W_2 \begin{bmatrix} -l_A \cos\theta \\ -l_S \sin(\theta + \beta_2) \\ +0.5 l_L \cos(\theta + \delta) \end{bmatrix} = 0 \end{aligned} \tag{5.77}$$

$$
\begin{aligned}
& m_L g l_S \sin(\theta + \beta_2)\frac{\cos(\beta_1 - \delta)}{\cos(\beta_2 - \delta)} + 0.5 m_L g l_L \cos(\theta + \delta)\frac{l_S \sin(\beta_1 - \beta_2)}{l_L \cos(\beta_2 - \delta)} \\
& + \frac{\cos(\beta_1 - \delta)}{\cos(\beta_2 - \delta)}\left[l_S \cos(\theta + \beta_2) m_L W_1 - l_S \sin(\theta + \beta_2) m_L W_2 \right] \\
& - \frac{l_S \sin(\beta_1 - \beta_2)}{l_L \cos(\beta_2 - \delta)}\left[\begin{array}{l} \frac{1}{12}(\dot{u}_3 + \dot{u}_6) m_L l_L^2 + 0.5 l_L \sin(\theta + \delta) m_L W_1 \\ +0.5 l_L \cos(\theta + \delta) m_L W_2 \end{array} \right] = 0
\end{aligned}
\tag{5.78}
$$

Note that the combination of nonlinear equations (5.75)–(5.78) and constraints (5.62)–(5.64) will exhibit overall system dynamical behavior shown in Figure 5.35.

Resulting from the nonlinear equations (5.75)–(5.78) and constraints (5.62)–(5.64), equilibrium points of beam slope angle, δ, and quadcopter attitude, θ, are both zero, and equilibrium points of swing angles, β_1 and β_2, are given by:

$$
\begin{cases}
\beta_{10} = -\arcsin\left(\frac{l_L - 2l_A}{2l_S}\right) \\
\beta_{20} = \arcsin\left(\frac{l_L - 2l_A}{2l_S}\right)
\end{cases}
\tag{5.79}
$$

where β_{10} and β_{20} are equilibrium angles of swing angles β_1 and β_2, respectively.

The nonlinear equations (5.75)–(5.78) and constraints (5.62)–(5.64) are an under-actuated system that requires a sophisticated control system. The control objective consists of two parts, attitude control of the quadcopter and swing suppression of the distributed-mass load. To achieve the objective, a combination of feedback controller and input shaper are presented in this chapter. A feedback controller regulates the quadcopter attitude by following the state of a prescribed model and reducing the tracking error, while an input shaper attenuates the oscillations of the distributed-mass load by shaping the operator's commands.

5.2.1.2 Control

Figure 5.36 shows the configuration of the combined feedback control and input shaper. The operator generates a driving command, θ_r, via the control interface. An input shaper is convolved with the driving command, θ_r, to create a shaped

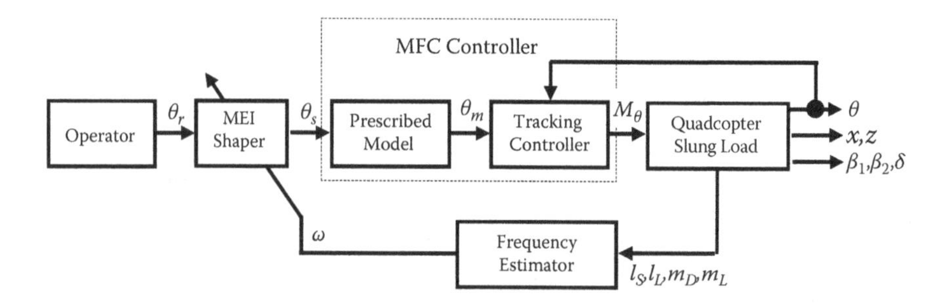

FIGURE 5.36 Combined feedback controller and input shaper architecture.

command, θ_s. The system frequency, ω, is estimated by system parameters and is used to design the input shaper.

The shaped command, θ_s, is sent to a prescribed model to generate a prescribed output of the vehicle attitude, θ_m. The measurement of the quadcopter attitude, θ, is used to adjust the pitching moment, M_θ, in a feedback control loop. The input of the feedback loop is the prescribed output, θ_m, such that the quadcopter attitude, θ, asymptotically tracks the prescribed output, θ_m. Then, the feedback controller stabilizes the quadcopter attitude. Meanwhile, it will also reject load oscillations resulting from the external disturbances due to the coupling effect between the quadcopter and load. In addition, the combined feedback control and input shaper method can suppress oscillations of the quadcopter and load without performing heavy computational tasks.

5.2.1.2.1 MFC for Attitude Regulation

The MFC controller has been widely used in attitude stabilization for aerial vehicle because of good control performance. The MFC control architecture includes a prescribed model and a tracking controller. The prescribed model provides a linear dynamic. The tracking controller cancels the error dynamics between the nonlinear quadcopter slung loads and the prescribed model.

An explicit MFC controller will be used to regulate quadcopter's attitude. The nonlinear equations (5.75)–(5.78) and constraints (5.62)–(5.64) are too complex to analytically design the MFC controller. Therefore, the MFC controller for a simplified model would be created. Then the design result would be refined as needed for nonlinear dynamics of quadcopters slung loads.

The oscillatory deflections of the quadcopter attitude, θ, swing angles, β_1 and β_2, and beam slope, δ, around the equilibrium points are denoted by θ_t, β_{1t}, β_{2t}, and δ_t, respectively. From the nonlinear equations (5.75)–(5.78) and acceleration constraint (5.64), a linearized model can be derived by assuming small oscillations around the equilibrium points:

$$\begin{cases} B_{1,4}\ddot{x} + [B_{1,5}\theta_t + B_{1,7}\beta_{1t} + B_{1,9}\beta_{2t} + B_{1,11}\delta_t]F_z + B_{1,14}M_\theta = 0 \\ B_{2,4}\ddot{z} + [B_{2,6}\theta_t + B_{2,8}\beta_{1t} + B_{2,11}\beta_{2t} + B_{2,14}\delta_t]M_\theta + B_{2,16}F_z + B_{2,18}g = 0 \\ B_{3,4}\ddot{\theta}_t + [B_{3,5}\beta_{1t} + B_{3,7}\beta_{2t} + B_{3,9}\delta_t]F_z + B_{3,12}M_\theta = 0 \\ B_{4,4}\ddot{\beta}_{1t} + [B_{4,5}\beta_{1t} + B_{4,7}\beta_{2t} + B_{4,9}\delta_t]F_z + [B_{4,6}\beta_{1t} + B_{4,8}\beta_{2t} + B_{4,10}\delta_t \\ \qquad + B_{4,12}]M_\theta = 0 \\ B_{5,4}\ddot{\beta}_{2t} + [B_{5,5}\beta_{1t} + B_{5,7}\beta_{2t} + B_{5,9}\delta_t]F_z + [B_{5,6}\beta_{1t} + B_{5,8}\beta_{2t} + B_{5,10}\delta_t \\ \qquad + B_{5,12}]M_\theta = 0 \\ B_{6,4}\ddot{\delta}_t + [B_{6,5}\beta_{1t} + B_{6,7}\beta_{2t} + B_{6,9}\delta_t]F_z + [B_{6,6}\beta_{1t} + B_{6,8}\beta_{2t} + B_{6,10}\delta_t + B_{6,12}]M_\theta \\ \qquad = 0 \end{cases} \tag{5.80}$$

where $B_{k,j}$ are coefficients. Because the swing angle is a very small factor, ignoring the effect of load swing and slope on the motion of the quadcopter yields a simplified model of the attitude:

$$B_{3,4}\ddot{\theta}_t + B_{3,12}M_\theta = 0 \tag{5.81}$$

From the simplified model (5.80), an explicit MFC controller can be derived:

$$\begin{cases} \ddot{\theta}_m + 2\zeta_m\omega_m\dot{\theta}_m + \omega_m{}^2\theta_m = \omega_m{}^2\theta_s \\ M_\theta = -\frac{B_{3,4}}{B_{3,12}}\ddot{\theta}_m + k_p(\theta_m - \theta) + k_d(\dot{\theta}_m - \dot{\theta}) \end{cases} \tag{5.82}$$

where ω_m and ζ_m are the frequency and damping ratio of the prescribed model, θ_m is the output of the prescribed model, and k_p and k_d are control gains.

The first equation in Equation (5.82) is the prescribed model. The pole placement method was used to design the frequency, ω_m, and damping ratio, ζ_m, of the prescribed model. The desired poles were chosen to be $-2 \pm i2$ for creating the reasonable damping ratio and short settling time. Therefore, the frequency, ω_m, and damping ratio, ζ_m, in the prescribed model were calculated to be 2.83 rad/s and 0.707, respectively.

The second equation in Equation (5.82) describes the asymptotic tracking control law. The tracking controller includes a feedforward control part and a feedback control part. The first term on the right-hand side of the second equation in Equation (5.82) is the feedforward control part, which improves the transient-response characteristics by using the model-inversion technique. The second and third terms on the right-hand side of the second equation in Equation (5.82) are the feedback control part, which cancels the error dynamics of the model tracking. The feedforward and feedback control methods inherent in the tracking controller exhibit the good robustness to reject the unwanted nonlinear dynamics.

Substituting Equation (5.82) into (5.81) yields the equation of motions for the closed-loop system:

$$-\frac{B_{3,4}}{B_{3,12}}(\ddot{\theta}_m - \ddot{\theta}) + k_d(\dot{\theta}_m - \dot{\theta}) + k_p(\theta_m - \theta) = 0 \tag{5.83}$$

Careful selection of the gains, k_p and k_d, in Equation (5.82) will achieve stability and suitable tracking errors. The negative real parts in the closed-loop poles will result in the stability of the control system (5.82). Meanwhile, the coupling effect between the quadcopter and the suspended load increases the system damping, and decreases the real part of the poles. Therefore, the closed-loop poles are designed to be $-6 \pm i2$. When the quadcopter mass, m_D, moment of inertia, I_{yy}, suspension distance, l_A, and cable length, l_S, were fixed at 85 kg, 4.5 kg m^2, 1 m, and 5 m, the gains k_p and k_d by the pole placement technique were calculated to be 442.17 and 132.65, respectively.

It is assumed that the vertical motion of the quadcopter is unaffected by motion of the load swing and quadcopter attitude. Then a vertical motion controller is designed for near-hover control:

$$F_z = (m_D + m_L)g \tag{5.84}$$

From the linearized model (5.81) and controller (5.82) and (5.84), the natural frequency and damping ratio of the quadcopter slung load can be derived:

$$T_1\lambda^4 + T_2\lambda^3 + T_3\lambda^2 + T_4\lambda + T_5 = 0 \tag{5.85}$$

where $T_{k=1-5}$ are coefficients and

$$\lambda = -\zeta\omega \pm i\omega\sqrt{1-\zeta^2} \tag{5.86}$$

ω is the natural frequency, ζ is the corresponding damping ratio, and i is the imaginary unit.

The natural frequency and damping ratio are dependent on the cable length, load length, quadcopter mass, and load mass. As the cable length increases, the frequency decreases sharply. Figures 5.37 and 5.38 show the natural frequency and damping ratio for various load length when the quadcopter mass, m_D, moment of inertia, I_{yy}, suspension distance, l_A, and cable length, l_S, were fixed at 85 kg, 4.5 kg m^2, 1 m, and 5 m, respectively. The linear mass density of the beam was set to 5 kg/m, then the load mass increases as the load length increases. As the load length increases, the first-mode frequency changes slightly, meanwhile the second-mode frequency increases slowly. The second-mode damping ratio reaches very high values for all load lengths. Therefore, second-mode vibrations will decrease sharply because of large damping and frequency. Instead, the first-mode vibrations cannot be ignored because the corresponding frequency and damping are both very low. Therefore, there is a need to design another controller to attenuate the first-mode oscillations.

As an initial test of the performance of the MFC controller, Figures 5.39 and 5.40 show simulated response of quadcopter attitude and load oscillations resulting from non-zero initial conditions. The quadcopter mass, m_D, moment of inertia, I_{yy}, suspension distance, l_A, cable length, l_S, load mass, m_L, and load length, l_L, in the

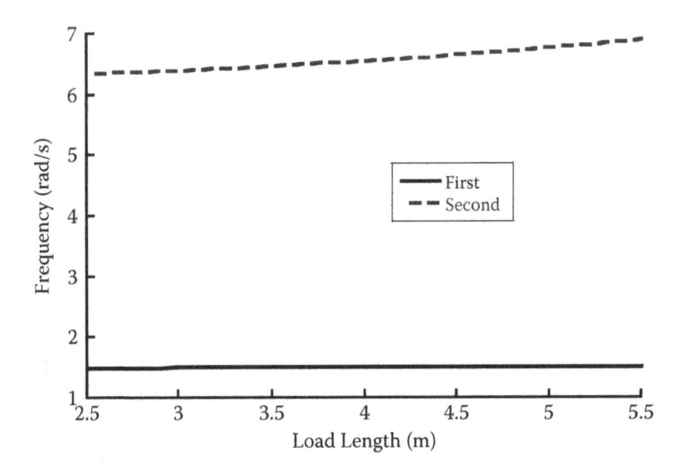

FIGURE 5.37 Frequency vs. load length.

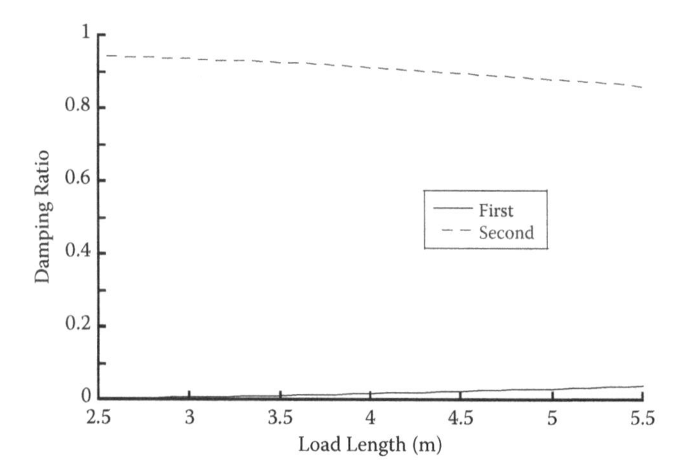

FIGURE 5.38 Damping ratio vs. load length.

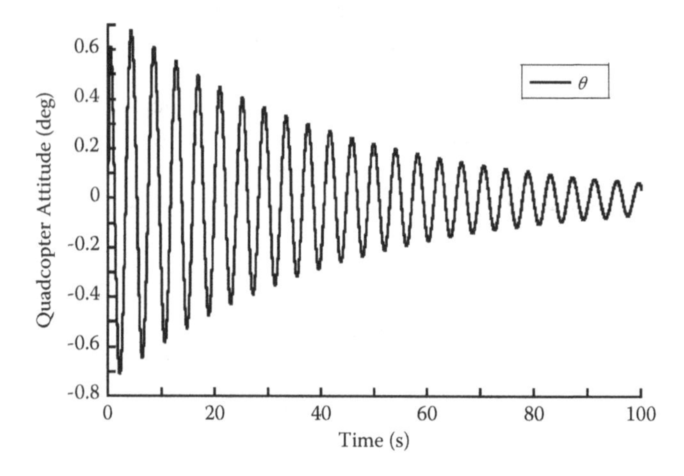

FIGURE 5.39 Responses of quadcopter attitude resulting from non-zero initial conditions.

simulations were fixed at 85 kg, 4.5 kg m^2, 1 m, 5 m, 20 kg, and 4 m, respectively. The initial value of swing angles, β_1 and β_2, and the slope angle, δ, were set to −1.537°, 21.72°, and 5.065°, respectively. The initial swing in this case might be considered as the oscillations induced by external disturbances including aerodynamic effects from rotors, wind gusts, and severe weather. Initial oscillations of the load would cause oscillations of the quadcopter attitude because of the coupling effect between the quadcopter and load. The frequency and damping ratio in Figures 5.37 and 5.38 can be estimated from Equation (5.85).

The load oscillations resulting from external disturbances may cause a tension moment acting on the quadcopter via the two suspension cables. The damping effect designed in the MFC controller reduced oscillations of quadcopter attitude. Then

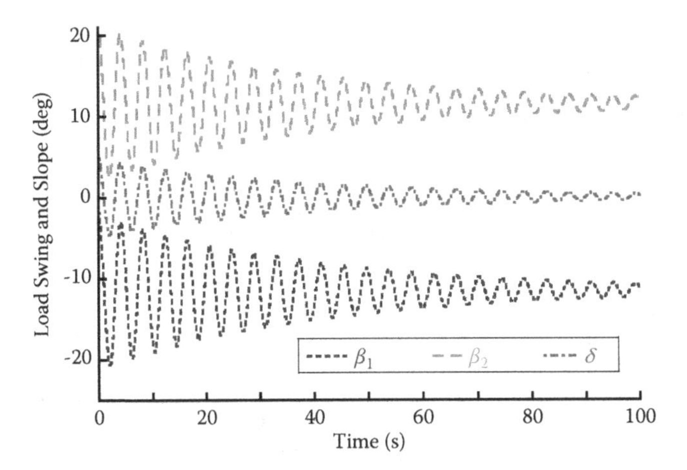

FIGURE 5.40 Responses of load swing and slope resulting from non-zero initial conditions.

the load oscillations were also attenuated due to the coupling effect between aerial vehicles and loads. Therefore, all the quadcopter attitude, load swing, and slope approached their corresponding equilibrium points. Note the equilibria of attitude angle, θ, swing angles, β_1, β_2, and the slope angle, δ, were 0°, −11.539°, 11.539°, and 0°, respectively. With the MFC controller, the settling time required for the response of the quadcopter attitude to settle within 0.1° was 81.08 seconds, that of the swing angles, β_1 and β_2, and slope angle, δ, to settle within 0.5° were 118.03 seconds, 118.03 seconds, and 91.07 seconds, respectively. Therefore, the MFC controller can control the quadcopter attitude, and reject oscillations resulting from external disturbances.

5.2.1.2.2 Modified Extra-Insensitive (MEI) Input Shaper

Although the MFC controller can stabilize quadcopter's attitude and reject disturbances, it cannot be used to reduce oscillations induced by pilot commands because of a large rise-time penalty. The pilot-induced oscillations should be controlled by a MEI input shaper. The MEI shaper is given by:

$$\begin{aligned} &A_1 = \frac{1}{(1+K)^2}, \quad A_2 = \frac{K}{(1+K)^2}, \quad A_3 = \frac{K}{(1+K)^2}, \quad A_4 = \frac{K^2}{(1+K)^2}, \\ &\tau_1 = 0, \quad \tau_2 = \frac{\pi}{q\omega\sqrt{1-\zeta^2}}, \quad \tau_3 = \frac{\pi}{p\omega\sqrt{1-\zeta^2}}, \quad \tau_4 = \left(\frac{1}{q} + \frac{1}{p}\right)\frac{\pi}{\omega\sqrt{1-\zeta^2}}, \end{aligned} \tag{5.87}$$

where δ is the impulse function and

$$K = e^{-\pi\zeta/\sqrt{1-\zeta^2}} \tag{5.88}$$

Figures 5.41 and 5.42 show two modified coefficients for varying tolerable level and damping ratio. The tolerable vibration level ranges from 0% to 10%. In the case of zero tolerable level, two modified coefficients both are limited to one. As the

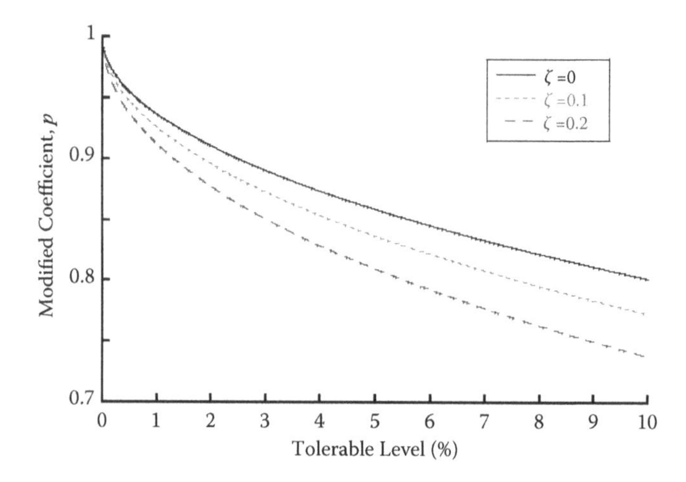

FIGURE 5.41 Modified coefficient p for various tolerable vibration levels and damping ratio.

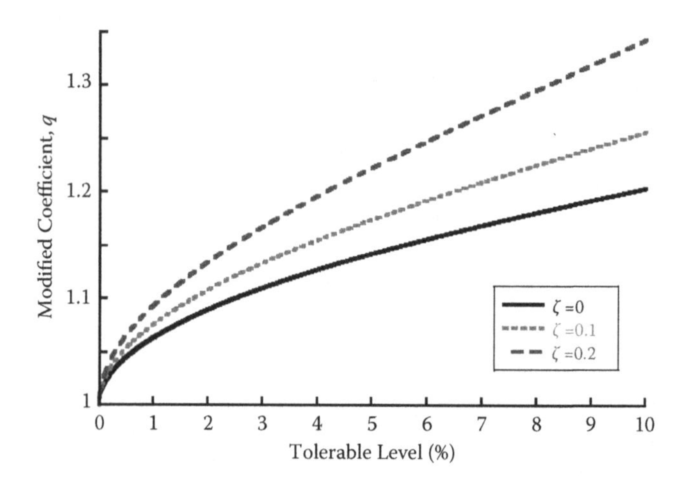

FIGURE 5.42 Modified coefficient q for various tolerable vibration levels and damping ratio.

tolerable level and damping ratio increase, the modified coefficient p decreases and modified coefficient q increases.

The frequency sensitivity curves are shown in Figure 5.43 designed for zero damping. The normalized frequency is defined as the ratio of the real frequency to the design frequency. The MEI shaper limits the residual vibrations to the tolerable level at the design frequency. Meanwhile, the residual vibrations at two modified frequencies are suppressed to zero by the MEI shaper. The insensitivity, which provides a quantitative measure of robustness, is defined as the range of each curve that lies below the tolerable vibration level. The 5% insensitivity ranges from 0.798 to 1.205, and the 10% insensitivity ranges from 0.715 to 1.298. Increasing the tolerable level increases distance between the modified frequencies, and then

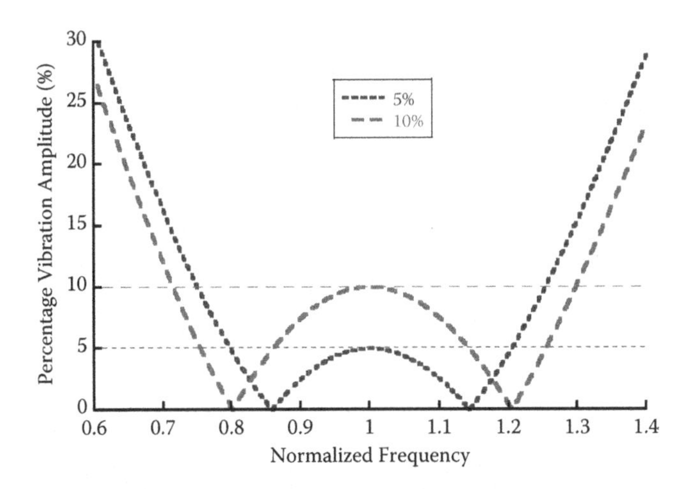

FIGURE 5.43 Frequency sensitivity curves for MEI shaper.

widens the range of the frequency insensitivity. The MEI shaper exhibits a notch filtering effect, and suppresses a wide range of system frequencies.

5.2.1.3 Computational Dynamics

The MFC+MEI combined controller is designed from the linearization of the nonlinear equations (5.75)–(5.78) and constraints (5.62)–(5.64) by assuming small oscillations around equilibria. However, the robustness to modeling error and couple effect between the quadcopter and load have been considered in the design of control gains. Therefore, the MFC+MEI combined controller will be effective in the case of large oscillations. The effectiveness and robustness of the controller will be used on the nonlinear equations (5.75)–(5.78) and constraints (5.62)–(5.64) to test the performance in this section.

The quadcopter mass, m_D, moment of inertia, I_{yy}, suspension distance, l_A, cable length, l_S, and linear mass density of the beam were set to 85 kg, 4.5 kg m^2, 1 m, 5 m, and 5 kg/m, respectively. The tolerable level of the MEI shaper in the simulations is set to 5%. There are two stages in the flight system response. The transient stage of the response is defined as the period when the quadcopter flies forward. The residual stage is defined as the period when the quadcopter attempts to stop and hovers in the sky. The peak-to-peak deflections during the transient and residual stages are referred to as the transient deflection and residual amplitude, respectively.

A simulated response to a flight distance of 68 m is shown in Figures 5.44–5.47 when the beam length, l_L, and beam mass, m_L, were set to 4 m and 20 kg, respectively. The simulation first gave an acceleration to move the quadcopter flying along $\mathbf{N}_x$ direction, and then kept flight speed at a constant value of 3 m/s. At the end of travel, the quadcopter was decelerated at a desired distance. This process simulates normal straight-line flight with suspended loads. The corresponding flight command, θ_r, is shown in Figure 5.44. The command, θ_r, is modified by the MEI shaper to produce the shaped command, θ_s, also shown in Figure 5.44. Without the MEI shaper, the input to the MFC controller is the original command, θ_r. When the

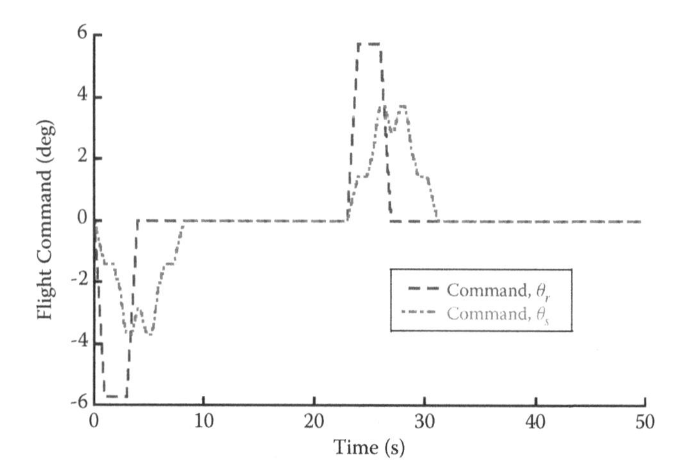

FIGURE 5.44 Time response of flight commands to a flight distance of 68 m.

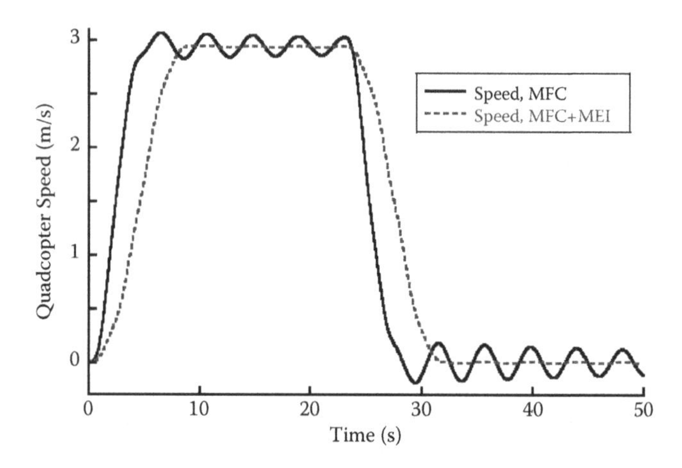

FIGURE 5.45 Time response of quadcopter speed to a flight distance of 68 m.

MFC + MEI combined controller is implemented, the input to the MFC controller is shaped command, θ_s. The design frequency and damping ratio for the MEI shaper are 1.52 rad/s and 0.016, which can be calculated from Equation (5.85) by system parameters, and can also be found in Figure 5.45.

Figure 5.46 shows the simulated response of the quadcopter attitude, θ. The transient deflection, residual amplitude, and settling time of the quadcopter attitude with the MFC controller were 12.4°, 4.4°, and 86.8 seconds, respectively. Meanwhile, those with the MFC+MEI combined controller were 7.3°, 0.77°, and 0.59 seconds, respectively. The settling time of the quadcopter attitude is defined as the time required for the residual response to settle within 0.1°.

The simulated results of the load deflection are also shown in Figure 5.47. The load deflection is defined as the peak-to-peak deflection of the load center. With the MFC

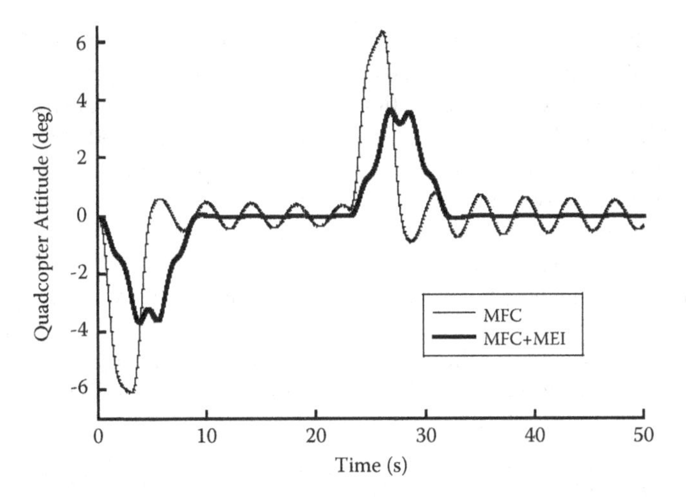

FIGURE 5.46 Time response of quadcopter attitude to a flight distance of 68 m.

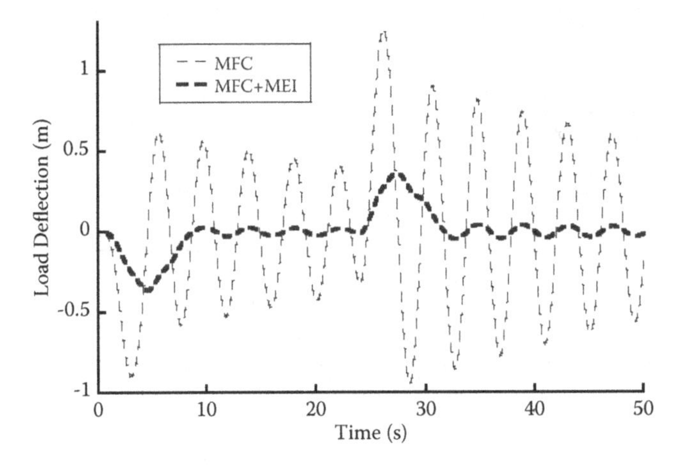

FIGURE 5.47 Time response of load deflection to a flight distance of 68 m.

controller, transient deflection, residual amplitude, and settling time of the load deflection were 2.15 m, 1.85 m, and 129.9 seconds, respectively. The settling time of the load deflection is referred to as the time required for the residual response to settle within 0.8% of the cable length. With the MFC + MEI combined controller, transient deflection, residual amplitude, and settling time of the load deflection were 0.72 m, 0.09 m, and 5.9 seconds, respectively. Therefore, the combined controller markedly suppressed the oscillations of quadcopter attitude and load swing. Additionally, the oscillation frequencies of the quadcopter attitude and load swing with the MFC controller were predicted by the frequency estimating Equation (5.85) about 1.52 rad/s. The MFC controller can attenuate oscillations of the quadcopter and load, but is weak to reduce oscillations quickly. The MEI shaper benefits short of settling time for the oscillation suppression of the quadcopter and load.

The simulated response of the quadcopter speed is shown in Figure 5.45. Oscillations also occurred in the response of the quadcopter speed with the MFC controller. Meanwhile, the oscillation frequency of the quadcopter speed in Figure 5.45, and the frequency of quadcopter attitude and load swing in Figures 5.46 and 5.47 are approximately equal. Those are explained by the fact that the coupling effect between the quadcopter and the load. No oscillations exist in the response with the MFC+MEI combined controller because the combined controller can suppress total oscillations.

Another set of simulation serves to investigate the effectiveness of the combined controller for varying flight distances. The load mass and load length were confined to 20 kg and 4 m, respectively. The transient deflection, residual amplitude, and settling time of the quadcopter attitude and load swing are presented in Figures 5.48–5.53.

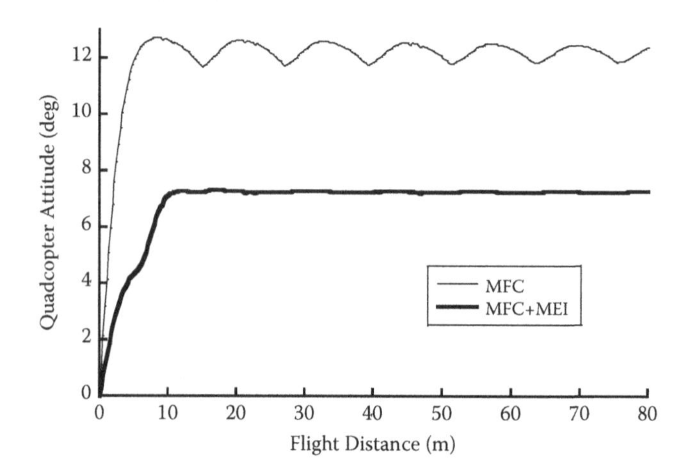

FIGURE 5.48 Transient deflection of quadcopter attitude against flight distance.

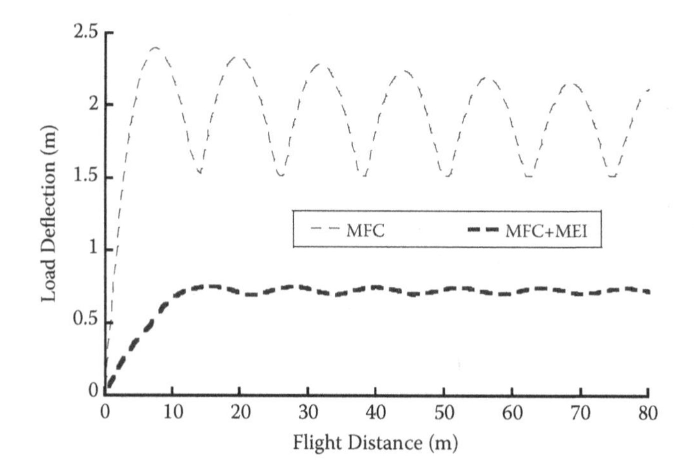

FIGURE 5.49 Transient deflection of quadcopter attitude and load deflection against flight distance.

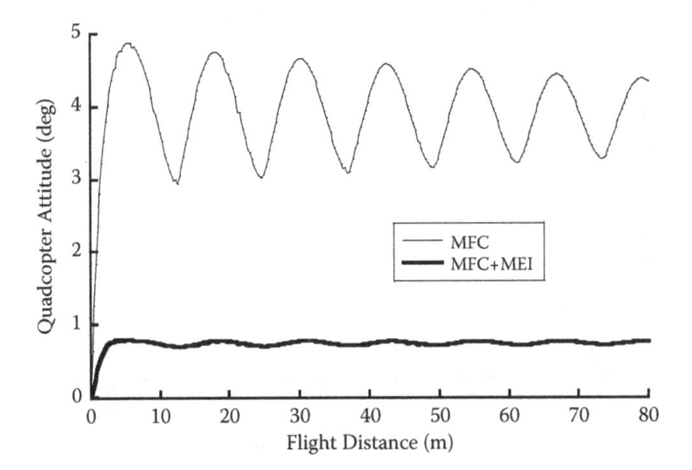

FIGURE 5.50 Residual amplitude of quadcopter attitude against flight distance.

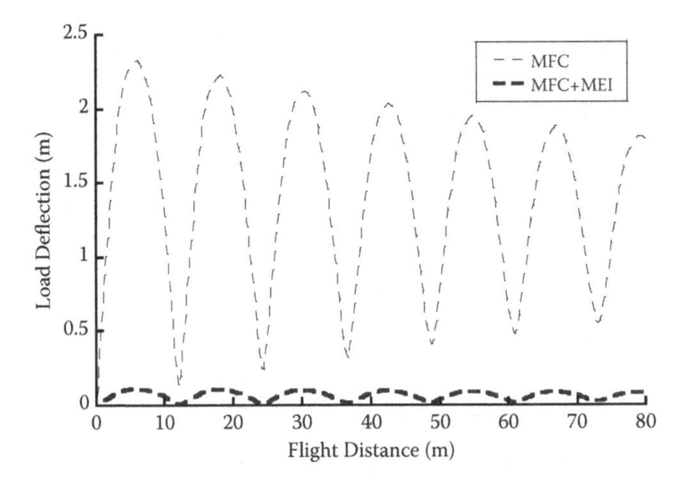

FIGURE 5.51 Residual amplitude of load deflection against flight distance.

The transient deflection, residual amplitude, and settling time change periodically because of in phase or out of phase between the oscillations caused by the acceleration and deceleration. Meanwhile, frequencies in the transient deflection, residual amplitude, and settling time were predicted by the values from Equation (5.85). Increasing flight distance decreases the magnitude of peaks and troughs due to the damping effect created by the MFC controller. Lastly, the transient deflection, residual amplitude, and settling time of the quadcopter attitude and load swing stabilized at constant values.

The design frequency and damping ratio for the MEI shaper are 1.52 rad/s and 0.016, respectively. With the MFC + MEI combined controller, the transient deflection, residual amplitude, and settling time of the quadcopter attitude were reduced by an average of 42.1%, 81.0%, and 99.2%, and the transient deflection,

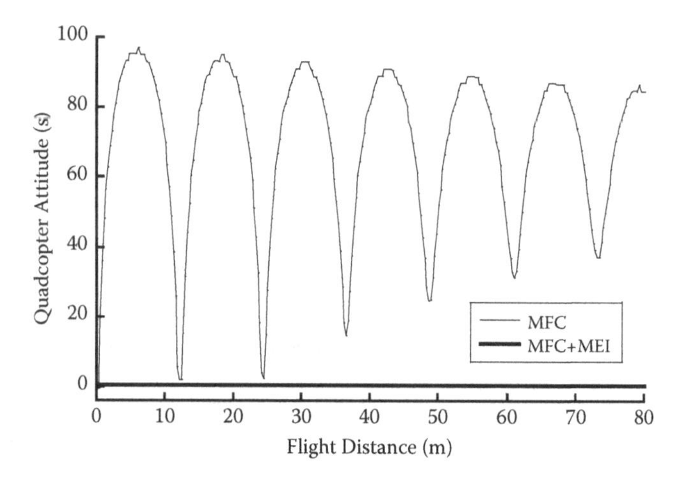

FIGURE 5.52 Settling time of quadcopter attitude against flight distance.

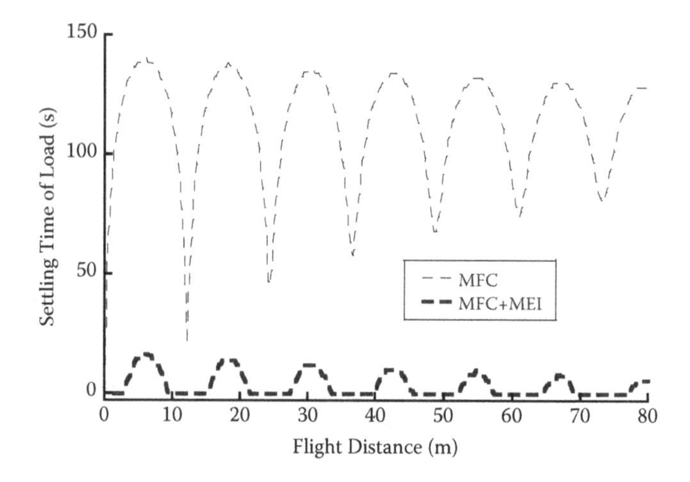

FIGURE 5.53 Settling time of load deflection against flight distance.

residual amplitude, and settling time of the load swing were reduced by an average of 63.9%, 95.0%, and 97.0%. Thus, the combined controller limited the oscillations of quadcopter attitude and load swing to low levels for various flight motions.

The last set of simulations is aimed at examining the robustness of the combined controller for various modeling errors in the system parameter. The simulated results of the quadcopter attitude and load swing for the fixed design frequency are shown in Figures 5.54–5.59. The load length ranged from 2.5 m to 5.5 m, and increasing load length increased the load mass. However, the modeled value of the load length and mass were fixed at 4 m and 20 kg for the design of the MFC + MEI combined controller. The corresponding design frequency and damping ratio for the MEI shaper were 1.52 rad/s and 0.016, respectively.

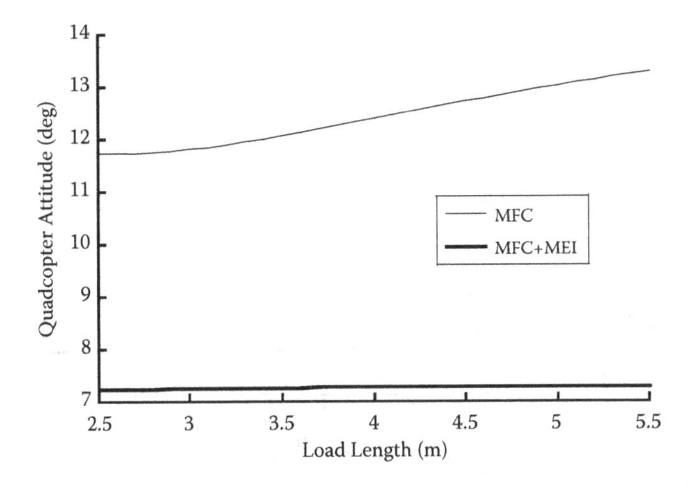

FIGURE 5.54 Transient deflection of quadcopter attitude for various load length.

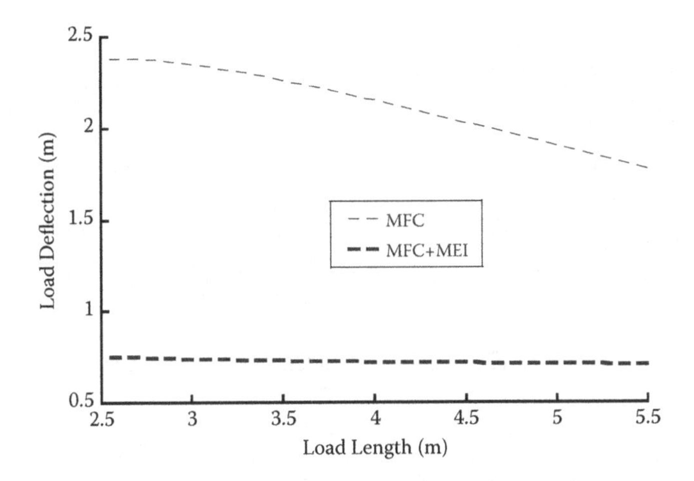

FIGURE 5.55 Transient deflection of load deflection for various load length.

With the MFC controller, the transient deflection and residual amplitude of the quadcopter attitude increased as the load length increased, and the transient deflection and residual amplitude of the load swing decreased with increasing load length. A local maximum in the settling time of the quadcopter attitude arose at the load length of 2.8 m. The settling time of load deflection decreased sharply as the load length increased. Under the same conditions, the MFC+MEI combined controller suppressed the transient deflection, residual amplitude, and settling time of the quadcopter attitude by an average of 41.4%, 82.8%, and 99.4%, respectively. The MFC+MEI combined controller reduced the transient deflection, residual amplitude, and settling time of the load deflection by an average of 65.8%, 95.1%, and 89.3%, respectively.

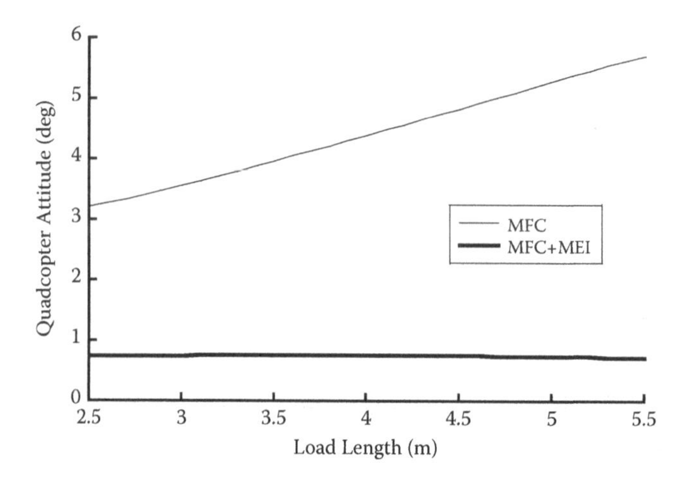

FIGURE 5.56 Residual amplitude of quadcopter attitude for various load length.

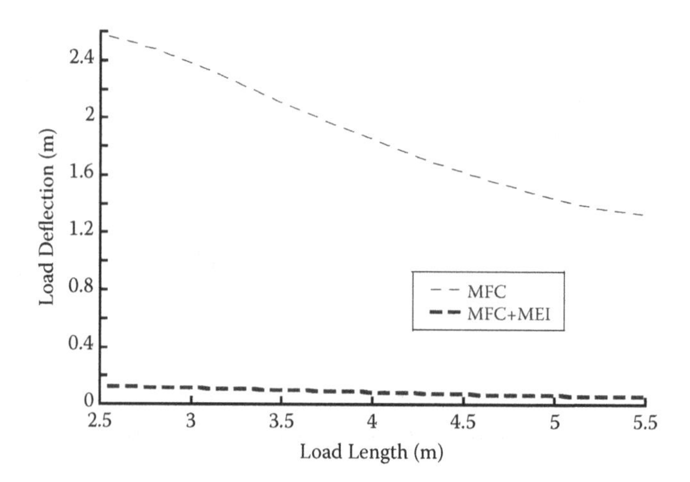

FIGURE 5.57 Residual amplitude of load deflection for various load length.

All of the transient deflection, residual amplitude, and settling time with the MFC + MEI controller were nearly independent of changes in the load length and mass. This is because the MEI shaper can limit vibrations below the tolerable level of 5% when the normalized frequency varied from 0.798 to 1.205. In this case, the first-mode frequency ranged between 1.49 rad/s and 1.52 rad/s. The frequency of 1.49 rad/s corresponds to the normalized frequency of 0.98, while the 1.52 rad/s is corresponding to the normalized frequency of one.

Simulations demonstrated that the MFC + MEI combined controller is beneficial to reduce the transient and residual oscillations of the quadcopter attitude and load swing in near hover operations. Therefore, the combined controller provides a safer environment to operator the quadcopter, and a higher efficiency to transfer the loads to the desired position precisely.

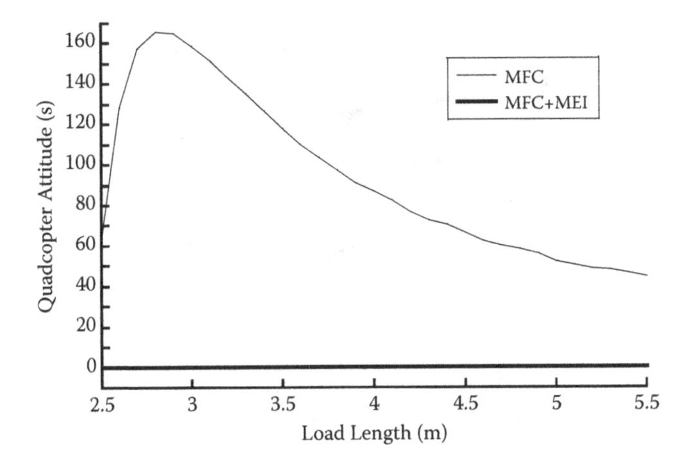

FIGURE 5.58 Settling time of quadcopter attitude for various load length.

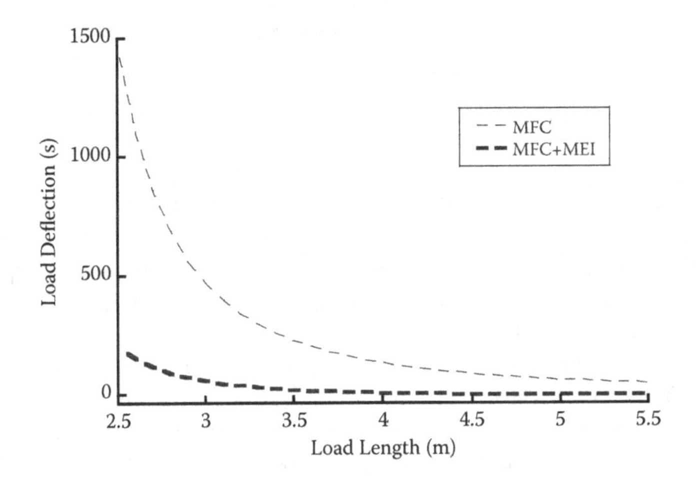

FIGURE 5.59 Settling time of load deflection for various load length.

5.2.2 Three-Dimensional Dynamics

5.2.2.1 Modeling

Figure 5.60 shows a schematic representation of a quadcopter transporting a load. The quadcopter uses a gyrostat **D**, meaning a rigid body with three fixed-axis $\mathbf{D}_x$, $\mathbf{D}_y$, $\mathbf{D}_z$ in the model. The mass of the quadcopter is m_D, and the centroidal principal axis moments of inertia about the $\mathbf{D}_x$, $\mathbf{D}_y$, $\mathbf{D}_z$ axis are I_{xx}, I_{yy} and I_{zz}, respectively. The basic transformations are Newtonian $\mathbf{N}_x$, $\mathbf{N}_y$, and $\mathbf{N}_z$. Motion of the quadcopter is divided into the displacement, x, along the $\mathbf{N}_x$ direction, the displacement, y, along the $\mathbf{N}_y$ direction, the displacement, z, along the $\mathbf{N}_z$ direction, yaw attitude, ψ, pitch attitude, θ, and the roll attitude, φ. The inertial coordinates $\mathbf{N}x\mathbf{N}y\mathbf{N}z$ can be converted to the moving Cartesian coordinates $\mathbf{D}x\mathbf{D}y\mathbf{D}z$ of the quadcopter by

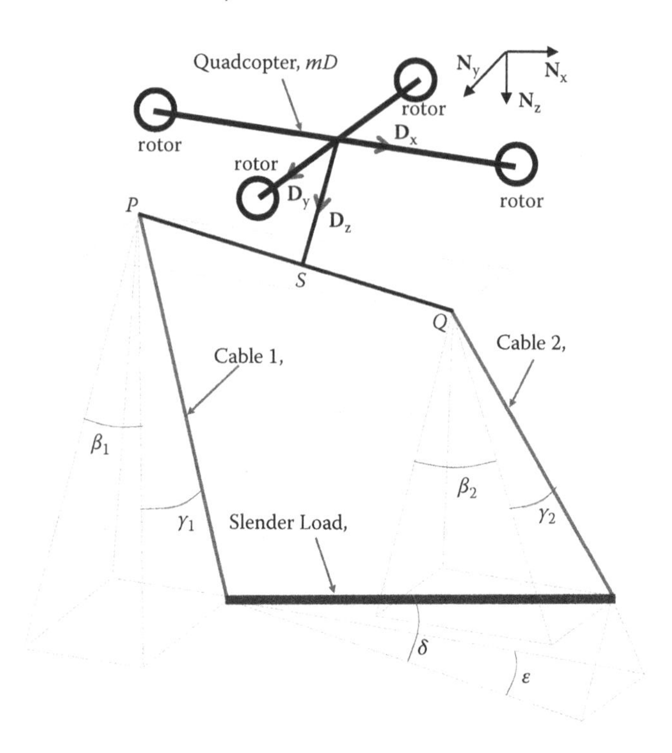

FIGURE 5.60 Model of a quadcopter transporting a load by double-hoist mechanisms.

rotating the yaw attitude, ψ, pitch attitude, θ, and the roll attitude, φ, respectively. The quadrotors produce thrust force, F_z, along $\mathbf{D}_z$ direction and thrust moments, M_φ, M_θ, and M_ψ, about $\mathbf{D}_x$, $\mathbf{D}_y$, and $\mathbf{D}_z$ directions.

A rigid link attaches a connection point, S, to the quadcopter center of mass, D^*, along the $\mathbf{D}_z$ direction. The distance between the quadcopter center of mass, D^*, and the point, S, is $l_C\mathbf{D}_z$. Another rigid link, which is parallel to the $\mathbf{D}_x\mathbf{D}_y$ plane, connects two suspension points, P and Q, to the point, S. The distance between quadcopter center, D^*, and suspension point, Q, is $l_A\mathbf{D}_x + l_B\mathbf{D}_y + l_C\mathbf{D}_z$, while that between the quadcopter center, D^*, and suspension point, P, is $-l_A\mathbf{D}_x - l_B\mathbf{D}_y + l_C\mathbf{D}_z$. The coupling effects between quadcopter attitudes and load oscillations may vary by changing link lengths, l_A, l_B, and l_C.

Two massless and inelastic cables, $\mathbf{C}_1$ and $\mathbf{C}_2$, of same length, l_S, suspend below the quadcopter with swing angles, β_1, γ_1, β_2, and γ_2, to orient the pendulum. A uniformly distributed-mass load ($\mathbf{B}$) of mass, m_L, length, l_L, and centroidal moment of inertia, I_L, is attached to the cables. The load is slender such that the cross-sectional area is assumed to be small. The centroid of the load is located at its center, B^*. The slope angle, δ, of the load is the angle between the load direction and $\mathbf{D}_x\mathbf{D}_y$ plane, while the twist angle, ε, is the angle between the load direction and $\mathbf{D}_x\mathbf{D}_z$ plane. Therefore, the coordinates $\mathbf{D}_x\mathbf{D}_y\mathbf{D}_z$ can also be converted to the moving Cartesian coordinates $\mathbf{B}_x\mathbf{B}_y\mathbf{B}_z$ of the load by rotating the twist angle, ε, and the slope angle, δ, respectively.

Inputs to the near-hover model are the thrust force, F_z, and thrust moments, M_ψ, M_θ, and M_φ. Outputs are the quadcopter displacements, x, y, and z, the attitudes, ψ, θ, and φ, the swing angles, β_1, γ_1, β_2, and γ_2, the twist angle, ε, and the slope angle, δ. The generalized speeds, u_r, (r = 1, 2, ..., 12) are chosen as:

$$u_1 = \dot{x},\ u_2 = \dot{y},\ u_3 = \dot{z},\ u_4 = \dot{\varphi},\ u_5 = \dot{\theta},\ u_6 = \dot{\psi}. \tag{5.89}$$

$$u_7 = \dot{\beta}_1,\ u_8 = \dot{\gamma}_1,\ u_9 = \dot{\beta}_2,\ u_{10} = \dot{\gamma}_2,\ u_{11} = \dot{\varepsilon},\ u_{12} = \dot{\delta}. \tag{5.90}$$

where Equation (5.89) is the generalized speed of quadcopter, and Equation (5.90) is the generalized speed of the load. The basis transformation matrix among the bodies **D** (Quadcopter), $\mathbf{C}_1$ (Cable 1), $\mathbf{C}_2$ (Cable 2), and **B** (Load) are:

$$\begin{Bmatrix} \mathbf{D}_x \\ \mathbf{D}_y \\ \mathbf{D}_z \end{Bmatrix} = \begin{bmatrix} \cos\theta\cos\psi & \cos\theta\sin\psi & -\sin\theta \\ \sin\varphi\sin\theta\cos\psi - \cos\varphi\sin\psi & \sin\varphi\sin\theta\sin\psi + \cos\varphi\cos\psi & \sin\varphi\cos\theta \\ \cos\varphi\sin\theta\cos\psi + \sin\varphi\sin\psi & \cos\varphi\sin\theta\sin\psi - \sin\varphi\cos\psi & \cos\varphi\cos\theta \end{bmatrix} \begin{Bmatrix} \mathbf{N}_x \\ \mathbf{N}_y \\ \mathbf{N}_z \end{Bmatrix} \tag{5.91}$$

$$\begin{Bmatrix} \mathbf{C}_{1x} \\ \mathbf{C}_{1y} \\ \mathbf{C}_{1z} \end{Bmatrix} = \begin{bmatrix} \cos\beta_1 & 0 & -\sin\beta_1 \\ \sin\beta_1\sin\gamma_1 & \cos\gamma_1 & \cos\beta_1\sin\gamma_1 \\ \sin\beta_1\cos\gamma_1 & -\sin\gamma_1 & \cos\beta_1\cos\gamma_1 \end{bmatrix} \begin{Bmatrix} \mathbf{D}_x \\ \mathbf{D}_y \\ \mathbf{D}_z \end{Bmatrix} \tag{5.92}$$

$$\begin{Bmatrix} \mathbf{C}_{2x} \\ \mathbf{C}_{2y} \\ \mathbf{C}_{2z} \end{Bmatrix} = \begin{bmatrix} \cos\beta_2 & 0 & -\sin\beta_2 \\ \sin\beta_2\sin\gamma_2 & \cos\gamma_2 & \cos\beta_2\sin\gamma_2 \\ \sin\beta_2\cos\gamma_2 & -\sin\gamma_2 & \cos\beta_2\cos\gamma_2 \end{bmatrix} \begin{Bmatrix} \mathbf{D}_x \\ \mathbf{D}_y \\ \mathbf{D}_z \end{Bmatrix} \tag{5.93}$$

$$\begin{Bmatrix} \mathbf{B}_x \\ \mathbf{B}_y \\ \mathbf{B}_z \end{Bmatrix} = \begin{bmatrix} \cos\varepsilon\cos\delta & \sin\varepsilon\cos\delta & -\sin\delta \\ -\sin\varepsilon & \cos\varepsilon & 0 \\ \cos\varepsilon\sin\delta & \sin\varepsilon\sin\delta & \cos\delta \end{bmatrix} \begin{Bmatrix} \mathbf{D}_x \\ \mathbf{D}_y \\ \mathbf{D}_z \end{Bmatrix} \tag{5.94}$$

The angular velocity of quadcopter in the Newtonian frame **N** is:

$$^N\boldsymbol{\omega}^D = (u_4 - u_6\sin\theta)\mathbf{D}_x + (u_5\cos\varphi + u_6\sin\varphi\cos\theta)\mathbf{D}_y + (u_6\cos\varphi\cos\theta - u_5\sin\varphi)\mathbf{D}_z \tag{5.95}$$

The velocity of quadcopter center of mass D^* in the Newtonian frame **N** is:

$$^N\mathbf{v}^{D^*} = u_1\mathbf{N}_x + u_2\mathbf{N}_y + u_3\mathbf{N}_z \tag{5.96}$$

The angular velocity of the load is:

$$ {}^N\boldsymbol{\omega}^B = {}^N\boldsymbol{\omega}^D + {}^D\boldsymbol{\omega}^B = {}^N\boldsymbol{\omega}^D - u_{11}\sin\delta\mathbf{B}_x + u_{12}\mathbf{B}_y + u_{11}\cos\delta\mathbf{B}_z \quad (5.97) $$

The angular velocity of cables $\mathbf{C}_1$, $\mathbf{C}_2$ are:

$$ {}^N\boldsymbol{\omega}^{C_1} = {}^N\boldsymbol{\omega}^D + {}^D\boldsymbol{\omega}^{C_1} = {}^N\boldsymbol{\omega}^D + u_8\mathbf{C}_{1x} + u_7\cos\gamma_1\mathbf{C}_{1y} - u_7\sin\gamma_1\mathbf{C}_{1z} \quad (5.98) $$

$$ {}^N\boldsymbol{\omega}^{C_2} = {}^N\boldsymbol{\omega}^D + {}^D\boldsymbol{\omega}^{C_2} = {}^N\boldsymbol{\omega}^D + u_{10}\mathbf{C}_{2x} + u_9\cos\gamma_2\mathbf{C}_{2y} - u_9\sin\gamma_2\mathbf{C}_{2z} \quad (5.99) $$

The velocity of the load center of mass B^* is:

$$ {}^N\mathbf{v}^{B^*} = {}^N\mathbf{v}^{D^*} + {}^N\boldsymbol{\omega}^D \times (l_A\mathbf{D}_x + l_B\mathbf{D}_y + l_C\mathbf{D}_z) + {}^N\boldsymbol{\omega}^{C_2} \times l_S\mathbf{C}_{2z} - {}^N\boldsymbol{\omega}^B \times 0.5l_L\mathbf{B}_x \quad (5.100) $$

The derivative of Equations (5.95) and (5.97) with respect to time derives the angular acceleration, ${}^N\boldsymbol{\alpha}^D$, of quadcopter, and angular acceleration, ${}^N\boldsymbol{\alpha}^B$, of load, respectively. The derivative of Equations (5.96) and (5.100) with respect to time yields the acceleration, ${}^N\mathbf{a}^{D^*}$, of quadcopter center of mass, and acceleration, ${}^N\mathbf{a}^{B^*}$, of load center of mass, respectively. Meanwhile, we extract the rth partial velocities $\mathbf{v}_r^{D^*}$, the rth partial angular velocities $\boldsymbol{\omega}_r^D$ of the quadcopter as coefficients of u_r from Equations (5.96) and (5.95), the rth partial angular velocities of load $\boldsymbol{\omega}_r^B$ from Equation (5.97), and rth partial velocities $\mathbf{v}_r^{B^*}$ of the load center of mass B^* from Equations (5.100).

The generalized inertia forces are:

$$ \begin{aligned} F_r^* &= -m_D \cdot {}^N\mathbf{a}^{D^*} \cdot \mathbf{v}_r^{D^*} - (\mathbf{I}_D \cdot {}^N\boldsymbol{\alpha}^D + {}^N\boldsymbol{\omega}^D \times \mathbf{I}_D \cdot {}^N\boldsymbol{\omega}^D) \cdot \boldsymbol{\omega}_r^D \\ &\quad - m_L \cdot {}^N\mathbf{a}^{B^*} \cdot \mathbf{v}_r^{B^*} - (\mathbf{I}_B \cdot {}^N\boldsymbol{\alpha}^B + {}^N\boldsymbol{\omega}^B \times \mathbf{I}_B \cdot {}^N\boldsymbol{\omega}^B) \cdot \boldsymbol{\omega}_r^B \end{aligned} \quad (5.101) $$

where $\mathbf{I}_D$ and $\mathbf{I}_B$ are the inertia matrix of quadcopter (**D**) and load (**B**), respectively. They satisfy:

$$ \mathbf{I}_D = \begin{bmatrix} I_{xx} & 0 & 0 \\ 0 & I_{yy} & 0 \\ 0 & 0 & I_{zz} \end{bmatrix} \quad (5.102) $$

$$ \mathbf{I}_B = \begin{bmatrix} 0 & 0 & 0 \\ 0 & I_L & 0 \\ 0 & 0 & I_L \end{bmatrix} \quad (5.103) $$

A part of generalized active forces is due to the force of gravity on the quadcopter and load. Another part of the generalized active forces is due to the thrust force and moments produced by the rotors. Thus, the generalized active force is:

$$F_r = (m_D g\mathbf{N}_z - F_z\mathbf{D}_z)\cdot\mathbf{v}_r^{D^*} + (M_\varphi\mathbf{D}_x + M_\theta\mathbf{D}_y + M_\psi\mathbf{D}_z)\cdot\boldsymbol{\omega}_r^D + m_L g\mathbf{N}_z\cdot\mathbf{v}_r^{B^*} \tag{5.104}$$

where $F_z\mathbf{D}_z$ is the thrust force along the $\mathbf{D}_z$ direction. In order to maintain the quadcopter flight altitude, the magnitude of the thrust force along the $\mathbf{N}_z$ direction is $F_z\mathbf{D}_z\cdot\mathbf{N}_z$ for balancing gravity of the quadcopter and load.

The quadcopter, cables, and load generate a three-dimensional four-bar linkage, in which contains three velocity constraints:

$$u_7 l_S \cos\beta_1 \cos\gamma_1 - u_8 l_S \sin\beta_1 \sin\gamma_1 - u_9 l_S \cos\beta_2 \cos\gamma_2 + u_{10} l_S \sin\beta_2 \sin\gamma_2 - u_{11} l_L \sin\varepsilon \cos\delta - u_{12} l_L \cos\varepsilon \sin\delta = 0 \tag{5.105}$$

$$u_8 l_S \cos\gamma_1 - u_{10} l_S \cos\gamma_2 - u_{11} l_L \cos\varepsilon \cos\delta + u_{12} l_L \sin\varepsilon \sin\delta = 0 \tag{5.106}$$

$$u_7 l_S \sin\beta_1 \cos\gamma_1 + u_8 l_S \cos\beta_1 \sin\gamma_1 - u_9 l_S \sin\beta_2 \cos\gamma_2 - u_{10} l_S \cos\beta_2 \sin\gamma_2 + u_{12} l_L \cos\delta = 0 \tag{5.107}$$

The Kane's equation describes that the sum of generalized inertia forces in Equation (5.101) and generalized active forces in Equation (5.104) should be limited to zero. Then the nonlinear dynamic equations of the motion yield:

$$[\mathrm{M}]\{\dot{\mathrm{U}}\} + \{\mathrm{Z}\} = 0 \tag{5.108}$$

where M is the mass matrix, $\dot{\mathrm{U}}$ is the column matrix of derivatives of generalized speed u_r with respect to time, and Z is the column matrix of gravity terms, centrifugal and Coriolis terms, and input terms.

The dynamical model in Equation (5.108) includes 12 system states, x, y, z, φ, θ, ψ, β_1, γ_1, β_2, γ_2, ε, and δ, and four inputs, F_z, M_φ, M_θ, and M_ψ. Therefore, it is an under-actuated system that requires a sophisticated control system. The control objectives consist of three parts: (i) attitude control of the quadcopter, (ii) swing and twisting suppression of the distributed-mass load, and (iii) external-disturbance rejection. To achieve the objectives, a combination of feedback control and hybrid filter will be presented in this chapter. A feedback controller regulates the quadcopter attitude and rejects external disturbances, while a hybrid filter attenuates the load swing and twisting caused by pilot commands.

5.2.2.2 Control

This section presents a combined control scheme including a feedback controller and three prefilters. An MFC regulates the quadcopter's attitude by following the states of a prescribed model and attenuating the tracking errors, while three prefilters reduce both the load swing and twisting by modifying the pilot commands.

The configuration of the combined feedback and prefilter control is shown in Figure 5.61. The operator generates pilot commands, ψ_r, θ_r, and φ_r, via the remote-control transmitter. Three prefilters (hybrid filter) modify the pilot

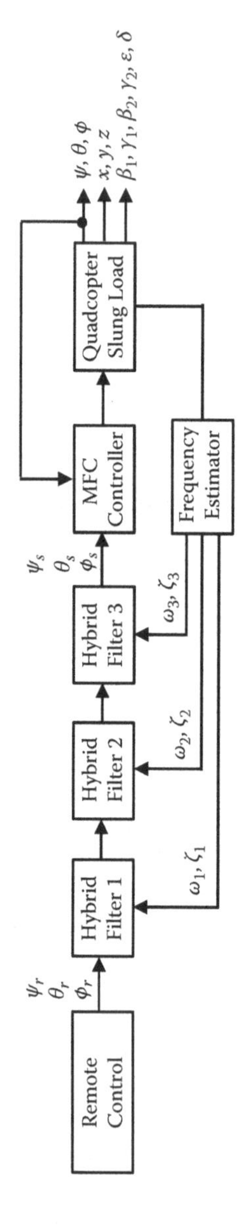

FIGURE 5.61 Combined control architecture.

commands to produce modified commands, ψ_s, θ_s, and φ_s. The measurement of the quadcopter attitude, ψ, θ, and φ, is used to adjust the thrust moments M_ψ, M_θ, and M_φ, in a feedback control loop. Then, the attitudes are forced to track the modified commands by the MFC controller. The frequencies and corresponding damping ratios are estimated by system parameters and are applied to design the three filters. The combined feedback controller and prefilter drive the aerial cranes toward the desired position with minimal oscillations of the quadcopter attitude, load swing, and twisting.

5.2.2.2.1 MFC Controller for Attitude Regulation

The dynamics of a quadcopter slung load in Equation (5.108) is too complicated to analytically design an MFC controller. Oscillations of the quadcopter and load are assumed to be small, and the model is also assumed to undergo planar motions. Then three linearized models can be derived in the $\mathbf{D}_x\mathbf{D}_y$, $\mathbf{D}_x\mathbf{D}_z$, and $\mathbf{D}_y\mathbf{D}_z$ planes. Three explicit MFC controllers are designed to control the corresponding attitude of the quadcopter from the three linearized models. The prescribed model of the MFC controller is described as:

$$\begin{cases} \ddot{\psi}_m + 2\zeta_m\omega_m\dot{\psi}_m + \omega_m^2\psi_m = \omega_m^2\psi_s \\ \ddot{\theta}_m + 2\zeta_m\omega_m\dot{\theta}_m + \omega_m^2\theta_m = \omega_m^2\theta_s \\ \ddot{\varphi}_m + 2\zeta_m\omega_m\dot{\varphi}_m + \omega_m^2\varphi_m = \omega_m^2\varphi_s \end{cases} \tag{5.109}$$

where ζ_m is the damping ratio of the prescribed model, ω_m is the frequency of the prescribed model, ψ_s, θ_s, and φ_s, are modified commands, ψ_m, θ_m, and φ_m, are model outputs. A prescribed model with reasonable damping ratio (0.707) and reasonable settling time (≤4 seconds) is utilized. The desired closed-loop poles were designed to be approximately $-1 \pm j1$. Thus, the frequency, ω_m, was 1.41 rad/s by using the pole placement method. In addition, the asymptotic tracking control law of the MFC controller is expressed as:

$$\begin{cases} M_\psi = B_\psi\cdot\ddot{\psi}_m + k_{\psi d}\cdot(\dot{\psi}_m - \dot{\psi}) + k_{\psi p}\cdot(\psi_m - \psi) \\ M_\theta = B_\theta\cdot\ddot{\theta}_m + k_{\theta d}\cdot(\dot{\theta}_m - \dot{\theta}) + k_{\theta p}\cdot(\theta_m - \theta) \\ M_\varphi = B_\varphi\cdot\ddot{\varphi}_m + k_{\varphi d}\cdot(\dot{\varphi}_m - \dot{\varphi}) + k_{\varphi p}\cdot(\varphi_m - \varphi) \end{cases} \tag{5.110}$$

where $k_{\psi p}$, $k_{\psi d}$, $k_{\theta p}$, $k_{\theta d}$, $k_{\varphi p}$, and $k_{\varphi d}$ are control gains, and B_ψ, B_θ, and B_φ are coefficients. Suitable tracking errors can be achieved by designing control gains. The coupling effect between the aerial vehicle and suspended load decreases the real part of the eigenvalues in Equation (5.110) and increases the system damping. Given that the quadcopter mass, m_D, and moment of inertia, I_{xx}, I_{yy}, and I_{zz} are 85 kg, 4.5 kg m^2, 4.5 kg m^2, and 6 kg m^2, respectively. Thus, the prescribed eigenvalues of attitudes ψ, θ, and φ in Equation (5.110) are selected to be $-12 \pm j4$, $-6 \pm j2$, and $-6 \pm j2$, respectively. Therefore, the control gains $k_{\psi p}$, $k_{\psi d}$, $k_{\theta p}$, $k_{\theta d}$, $k_{\varphi p}$, and $k_{\varphi d}$ were calculated by the pole placement technique to be 960, 144, 180, 54, 431.6, and 129.5, respectively.

Resulting from the three linearized models and the MFC controller, each plane includes two linearized frequencies and corresponding damping ratios. The first-mode damping ratio is near zero, while the second-mode damping ratio is near one. The second-mode oscillations can be neglected because corresponding amplitudes may damp quickly. Actually, the three-dimensional four-bar linkage shown in Figure 5.60 includes three natural frequencies with near-zero damping, and the MFC controller adds another three frequencies with high damping. Therefore, only three first-mode frequencies and damping ratios in three planar motions should be considered because they exhibit low frequency and near zero damping.

The linearized frequencies are dependent on the system parameters, such as quadcopter mass, m_D, cable length, l_S, load length, l_L, load mass, m_L, and suspension distances, l_A, l_B, and l_C. The cable length, l_S, and suspension distance, l_C, have fundamental influence on linearized frequencies. Increasing cable length, l_S, and suspension distances, l_C, decrease three linearized frequencies sharply. The load length and mass might also have some impacts on the linearized frequencies and damping. The three frequencies and damping in the $\mathbf{D}_x\mathbf{D}_y$, $\mathbf{D}_x\mathbf{D}_z$, and $\mathbf{D}_y\mathbf{D}_z$ planes induced by various load length and mass are shown in Figures 5.62 and 5.63. The cable length, l_S, and suspension distances, l_A, l_B, and l_C, were set to 5 m, 0 m, 1 m, and 0.1 m, respectively. The load length ranged from 2.5 m to 5.5 m. While the linear mass density of the load was fixed at 5 kg/m in this chapter, increasing load length increases load mass. Both the frequency and damping ratio in the $\mathbf{D}_y\mathbf{D}_z$ plane increase as the load length and mass increase. The frequency in the $\mathbf{D}_x\mathbf{D}_z$ plane changes slightly, while the corresponding damping ratio increases sharply with increasing load length. In the $\mathbf{D}_x\mathbf{D}_y$ plane, increasing load length decreases the frequency significantly and increases the damping ratio.

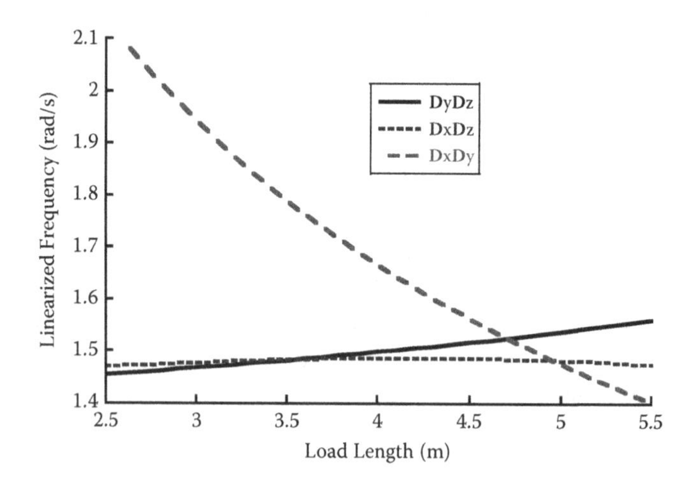

FIGURE 5.62 Linearized frequencies for various load length and mass.

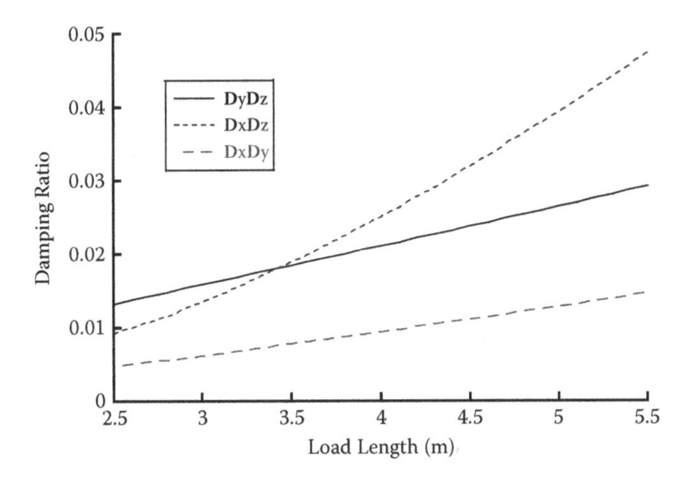

FIGURE 5.63 Damping ratio for various load length and mass.

The frequency and damping in the $\mathbf{D}_y\mathbf{D}_z$ plane are more related with the load swing, β_1 and β_2, and pitch attitude, θ, that in the $\mathbf{D}_x\mathbf{D}_z$ plane are more related with the load swing, γ_1 and γ_2, load slope, δ, and roll attitude, φ, and that in the $\mathbf{D}_x\mathbf{D}_y$ plane are more related with the load twisting, ε, and yaw attitude, ψ. The frequency in the $\mathbf{D}_y\mathbf{D}_z$ plane approaches the frequency in the $\mathbf{D}_x\mathbf{D}_z$ plane near the load length of 3.57 m, where the internal resonance might arise. The same phenomenon will occur near load length of 4.71 m and 4.96 m.

5.2.2.2.2 Hybrid Filter for Suppressing Load Oscillations

The load swing and twisting caused by external disturbance might also induce oscillations of quadcopter attitude because of coupling effect between quadcopter and load. The damping effect in the MFC controller can attenuate the oscillations of the quadcopter attitude. Then the load oscillations can also be reduced by MFC controller because of the coupling effect. However, the MFC controller should not be used to suppress the load swing and twisting caused by pilot commands. This is because the settling time for attenuating load oscillations by MFC controller would be very long.

In addition to regulating the quadcopter's attitude with the MFC controller, it is useful to design the prefilter to suppress load oscillations induced by pilot commands. The prefilter is a combination of discrete- and continuous-time function, which inherent in the limited motion of the quadcopter results in a limited response to oscillations of the suspended load.

The nonlinear dynamics of the quadcopters slung loads in Equation (5.108) with the MFC controller in Equations (5.109) and (5.110) can be approximated near the equilibria as three second-order systems with linearized frequencies and damping ratios shown in Figures 5.62 and 5.63. A hybrid filter, $h(\tau)$, of impulse and continuous function is given by:

$$h(\tau) = \begin{cases} A_1, & \tau = 0 \\ c_1 \omega_k \cdot e^{-\zeta_k \omega_k \tau}, & \frac{(1-r)\pi}{\omega_k \sqrt{1-\zeta_k^2}} \leq \tau \leq \frac{(1+r)\pi}{\omega_k \sqrt{1-\zeta_k^2}} \\ 0, & \text{others} \end{cases} \tag{5.111}$$

where r is the modified factor of the continuous function. The magnitudes c_1 and A_1 are given by:

$$c_1 = \frac{1}{\left(\frac{2}{\sqrt{1-\zeta_k^2}} \sin(r\pi) + \frac{e^{-\pi\zeta_k/\sqrt{1-\zeta_k^2}}}{\zeta_k} \left(e^{r\pi\zeta_k/\sqrt{1-\zeta_k^2}} - e^{-r\pi\zeta_k/\sqrt{1-\zeta_k^2}} \right) \right)} \tag{5.112}$$

$$A_1 = \frac{2c_1}{\sqrt{1-\zeta_k^2}} \sin(r\pi) \tag{5.113}$$

The hybrid filter (5.111) includes one impulse and one exponential function. The duration of the hybrid filter (5.111) is:

$$T_h = \frac{(1+r)\pi}{\omega_k \sqrt{1-\zeta_k^2}} \tag{5.114}$$

The hybrid filter (5.111) is different from the input shaper (a series of impulses) and command smoother (continuous function). The duration of hybrid filter can be designed by changing the modified factor, r. The modified factor, r, has a large effect on the frequency insensitivity and duration of the hybrid filter. Increasing modified factor, r, increases frequency insensitivity and duration of the hybrid function. The 5% frequency insensitivity for the hybrid filter ranges from 0.9680 to 1.0321 when modified factor, r, is set to 0.1.

Dynamic analyses indicate that oscillations with three linearized frequencies and damping ratios in Figures 5.62 and 5.63 should be suppressed. Therefore, three hybrid filters with three linearized frequencies and damping ratios are placed in series to attenuate swing and twisting of the load. The MFC controller shown in Equations (5.109) and (5.110) regulates the quadcopter's attitudes and rejects external disturbances, and three hybrid filters shown in Equation (5.111) attenuates the load oscillations caused by pilot commands.

By double-hoist mechanisms and two suspension links, the swing angles, β_1, γ_1, β_2, and γ_2, the slope angle, δ, and the twist angle, ε, are fully coupled with yaw attitude, ψ, pitch attitude, θ, and the roll attitude, φ. The coupling effects benefit disturbance rejection. This is because the MFC controller stabilizes the attitudes, and will also reduce the load oscillations by the coupling impact. Therefore, the load oscillations caused by external disturbances, which do not need to measure on-the-fly, can be suppressed by using the double-hoist mechanisms and two suspension links. Single-hoist configuration and cable connection directly to aerial vehicles cannot cause fully coupling effect between the load oscillations and quadcopter attitudes.

5.2.2.3 Computational Dynamics

In the mid-2000s, a prefilter on pilot commands has been experimentally verified on large-scale aerial vehicles, and then is used by Sikorsky. Meanwhile, it was demonstrated that the prefilter is difficult to achieve good experimental results on small-scale aerial vehicles. The work of experimental verification on large-scale quadcopter is massive. Therefore, this article only presented the numerical verification of disturbance rejection, effectiveness, and robustness of the control method. The combined MFC controller and prefilter will be applied to the dynamic model in Equation (5.19) to test the dynamic behavior. The quadcopter mass, m_D, and moment of inertia, I_{xx}, I_{yy}, and I_{zz}, suspension distances, l_A, l_B, and l_C, cable length, l_S, and linear mass density of the load in this section are fixed at 85 kg, 4.5 kg· m^2, 4.5 kg· m^2, 6 kg m^2, 0 m, 1 m, 0.1 m, 5 m, and 5 kg/m, respectively.

5.2.2.3.1 Disturbance Rejection

External disturbances might cause load oscillations in many cases. The first group of simulations is conducted to investigate the control performance of rejecting external disturbances. Figures 5.64–5.66 show a simulated response to a non-zero initial condition caused by external disturbances when the load length is 4 m. The initial values of load swing, β_1, γ_1, β_2, and γ_2, load slope, δ, load twisting, ε, were set to –6.8182°, 15.6532°, 2.9851°, –6.4229°, –2.6012°, and 78.0025°, respectively. Meanwhile, the initial values of three quadcopter attitudes were set to be zero. Then load oscillations would also induce oscillations of quadcopter attitudes by coupling effect between the quadcopter and load.

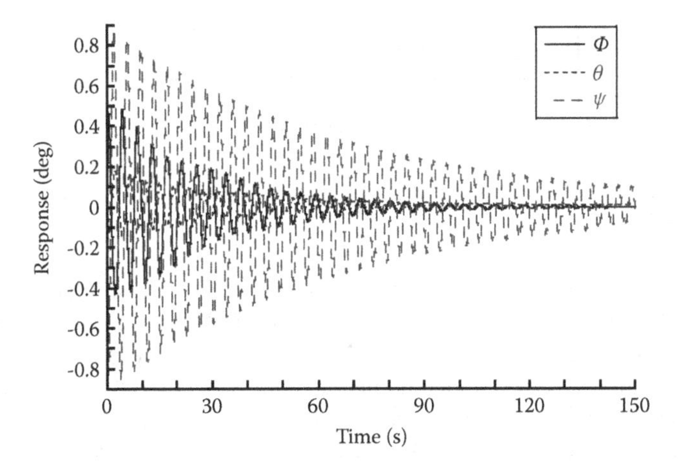

FIGURE 5.64 Simulated response of quadcopter attitude, φ, θ, and ψ, to a non-zero initial condition.

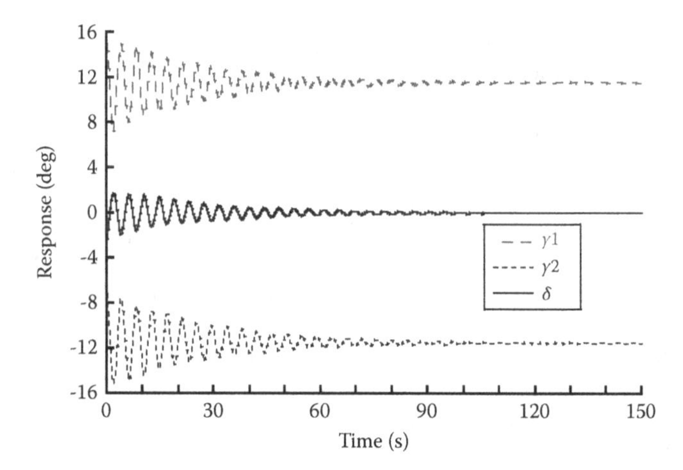

FIGURE 5.65 Simulated response of load swing and slope, γ_1, γ_2, and δ, to a non-zero initial condition.

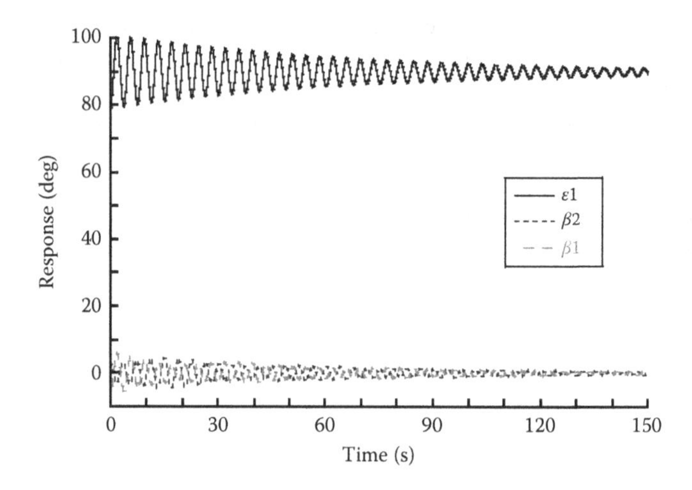

FIGURE 5.66 Simulated response of load swing and twisting, β_1, β_2, and ε, to a non-zero initial condition.

The frequencies of attitude, ψ, and load twisting, ε, are similar because they are related with the linearized frequency in the $\mathbf{D}_x\mathbf{D}_y$ plane. Other frequencies are similar because frequencies in the $\mathbf{D}_y\mathbf{D}_z$ and $\mathbf{D}_x\mathbf{D}_z$ planes are similar in this case. The damping effect inherent in the proposed control scheme attenuated oscillations of quadcopter attitude, load swing, and twisting, which approach their corresponding equilibria. Double-hoist mechanisms create fully coupling effect between the quadcopter and load. The presented controller reduced oscillations of quadcopter and load caused by external disturbances using the coupling impact.

5.2.2.3.2 *Effectiveness of Oscillation Suppression*

A nonlinear simulation was performed to move the quadcopter flying about the $\mathbf{D}_z$ direction, and then flying forward along the $\mathbf{D}_y$ direction. At the beginning of flight, the acceleration drives the quadcopter to reach the desirable flight direction in the $\mathbf{D}_z$ direction, and then move at a constant flight speed along the $\mathbf{D}_y$ direction. At the end of travel, the quadcopter was decelerated to stop at a desired distance. This process simulates normal straight-line flight along a required direction.

There are two stages in the flight system response. The transient stage of the response is defined as the period when the quadcopter flies forward. The residual stage is defined as the period when the quadcopter attempts to stop and hovers in the sky. The maximum peak-to-peak defections during the transient and residual stages are referred to as the transient deflection and residual amplitude, respectively.

The simulated response for a yaw angle of 6° about the $\mathbf{D}_z$ direction and a flight distance of 60 m along the $\mathbf{D}_y$ direction is shown in Figures 5.67–5.69 when load length, l_L, was set to 4 m. The transient deflection, residual amplitude, and settling time of quadcopter's attitude, φ, with the MFC controller were 12.53°, 1.05°, and 46.42 s, respectively, whereas those with the combined MFC controller and three prefilters were reduced to 6.76°, 0.26°, and 0.37 s, respectively. The settling time of quadcopter attitudes is defined as the time required for the residual response of the quadcopter's attitude to settle within 0.1°. Thus, the combined MFC controller and three prefilters obviously reduced the oscillations of the quadcopter's attitude. The oscillation frequency of quadcopter's attitude is predicted accurately by the corresponding result in Figure 5.62.

With the MFC controller, the transient deflection, residual amplitude, and settling time of the load swing, γ_1, were 10.93°, 8.09°, and 58.82 s, and those of the load swing, β_1, were 2.20°, 1.28°, and 11.84 seconds. The settling time of load oscillations is defined as time required for the residual oscillations of the load to settle within 0.5°. With the combined MFC controller and three prefilters, the transient deflection and residual amplitude of the load swing, γ_1, were only 0.97° and 0.20°, while those of the

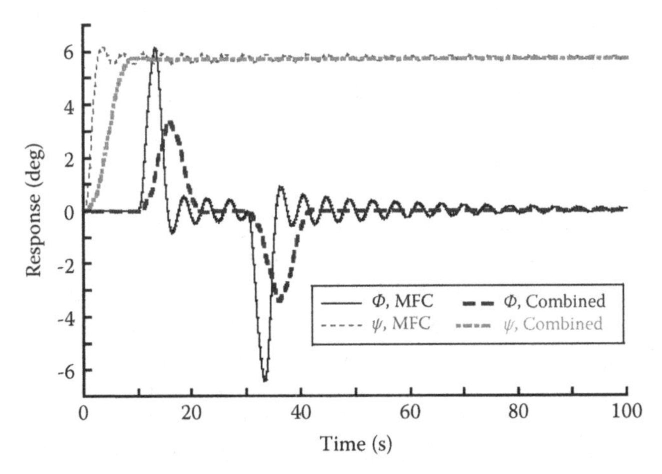

FIGURE 5.67 Simulated response of quadcopter attitude, φ and ψ, to a flight distance.

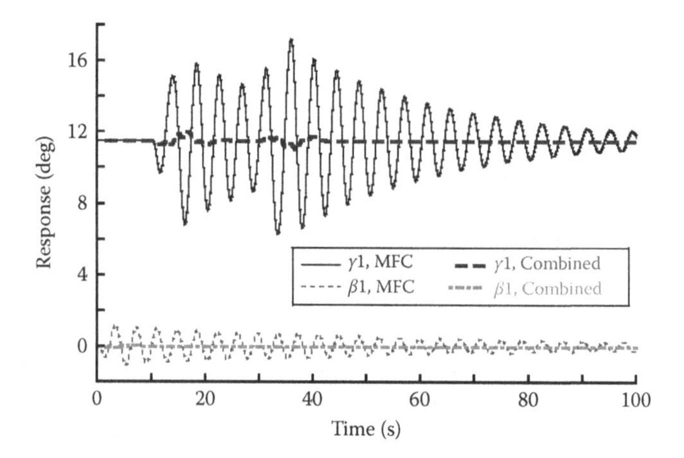

FIGURE 5.68 Simulated response of load swing, γ_1 and β_1, to a flight distance.

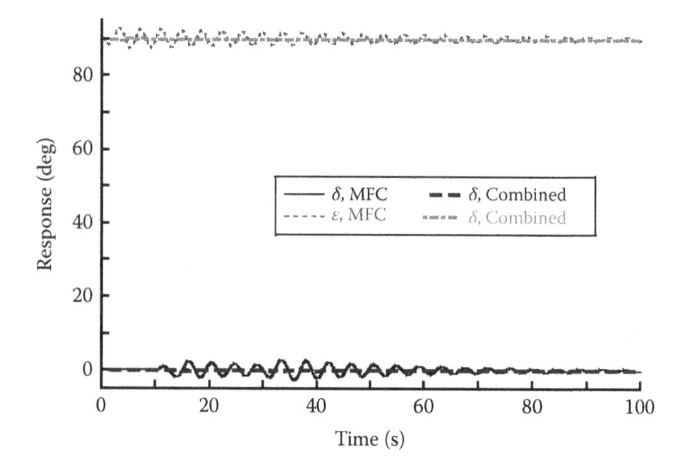

FIGURE 5.69 Simulated response of load slope and twisting, δ and ε, to a flight distance.

load swing, β_1, were 0.19° and 0.0008°. The combined MFC controller and three prefilters totally suppressed the settling time of the load swing, γ_1 and β_1. The oscillation frequencies of the load are similar to the frequencies of quadcopter's attitude. This is explained by the coupling effect between the quadcopter and the load.

The transient deflection, residual amplitude, settling time of the load twisting, ε, with the MFC controller were 5.39°, 2.94°, and 72.12 seconds, and those with the combined MFC controller and three prefilters were 0.46°, 0.002°, and 0.0 seconds. Meanwhile, the transient deflection, residual amplitude, and settling time of the load slope, δ, with the MFC controller were 5.46°, 4.05°, and 39.82 seconds, and those with the combined MFC controller and three prefilters were 0.49°, 0.10°, and 0.0 seconds. Thus, the combined control method limited the load twisting and slope to low values, thereby allowing for safer operation.

A battery of simulations was performed to investigate the effectiveness of the combined controller for varying flight distances. The load mass, m_L, and load length, l_L, were set to 20 kg and 4 m, respectively. The residual amplitude and settling time of the quadcopter's attitude, φ, are presented in Figures 5.70 and 5.71 as a function of flight distance. Due to in phase or out of phase interactions between the oscillations caused by the acceleration and deceleration, the residual amplitude, and settling time of the quadcopter's attitude, φ, with the MFC controller has a periodic amplitude. Additionally, frequency of the quadcopter's attitude, φ, is restricted by the frequency in Figure 5.62 because of coupling effect among attitude, φ, load swing, γ_1 and γ_2, and load slope, δ. Both the magnitude of peaks and troughs in the residual amplitude and settling time decrease with the increasing flying

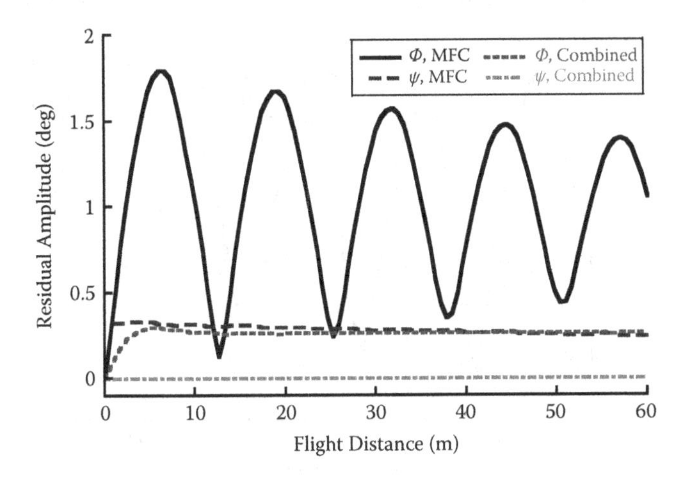

FIGURE 5.70 Residual amplitude of quadcopter attitude, φ and ψ, against flight distance.

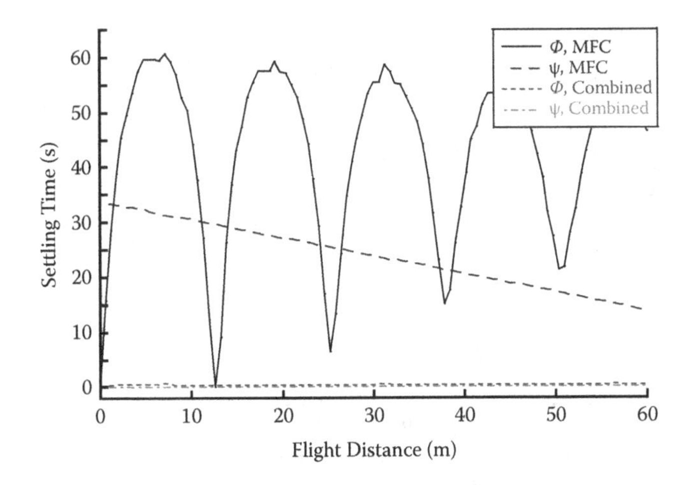

FIGURE 5.71 Settling time of quadcopter attitude, φ and ψ, against flight distance.

distance. This is mainly because the MFC controller provides a damping effect. Finally, the residual amplitude and settling time of the quadcopter's attitude, φ, stabilized at approximately 0.93° and 42.5 s. Both the residual amplitude and settling time of the quadcopter's attitude, ψ, decrease slowly as the flight distances increase. When the combined MFC controller and three prefilters is utilized, the residual amplitude and the settling time of attitude, φ, fell below 0.3° and 0.7 seconds. Meanwhile, the combined control method eliminated all of residual amplitude and settling time of attitude, ψ. Thus, the combined MFC controller and three prefilters reduces the residual amplitude and settling time of the quadcopter attitudes to very low levels for a wide range of flight motions.

Figures 5.72 and 5.73 show the residual amplitude and settling time of the load swing, γ_1 and β_1. The simulated results of the load swing, γ_1, are similar to that of

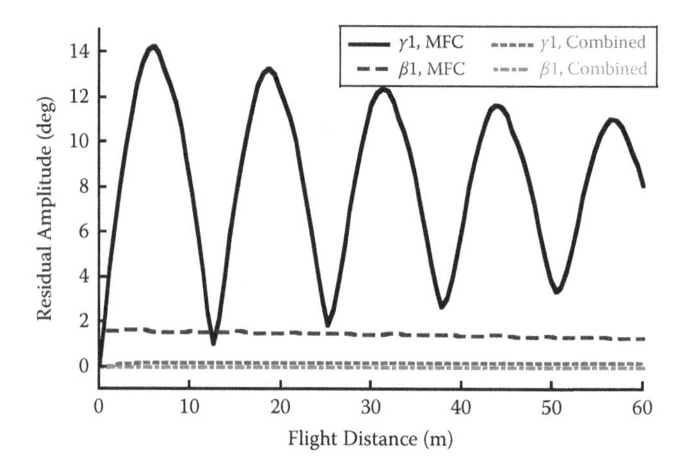

FIGURE 5.72 Residual amplitude of load swing, γ_1 and β_1, against flight distance.

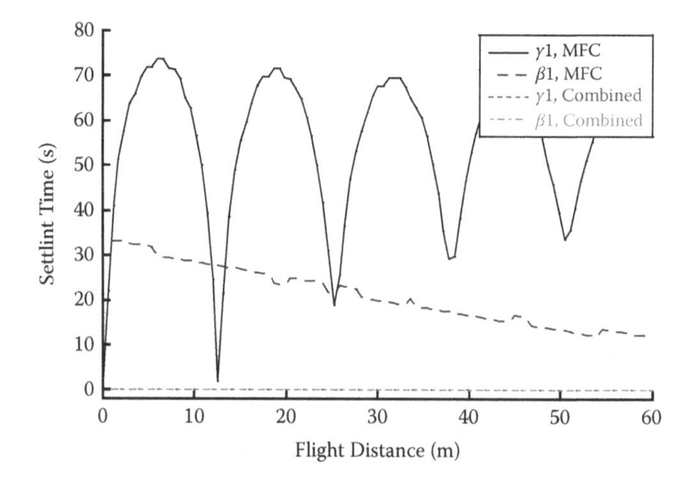

FIGURE 5.73 Settling time of load swing, γ_1 and β_1, against flight distance.

the quadcopter attitude, φ, observed in Figures 5.70 and 5.71. Moreover, the frequency, γ_1, is also restricted by the frequency linked to the value in Figure 5.62. As the flight distance increases, the residual amplitude and settling time of the load swing, γ_1, with the MFC controller settled to approximately 7.3° and 54.9 seconds. Both the residual amplitude and settling time of the load swing, β_1, decrease slightly as the flight distances increase. However, with the use of the combined MFC controller and three prefilters, the residual amplitude of the load swing, γ_1 and β_1, were limited to less than 0.2° and 0.001°, respectively. The combined control method also suppresses all of the settling time of the load swing, γ_1 and β_1, to very low levels over a wide range of flight motions.

The results for load twisting, ε, and slope, δ, are shown in Figures 5.74 and 5.75. The dynamics of the load slope are also similar to that of the quadcopter

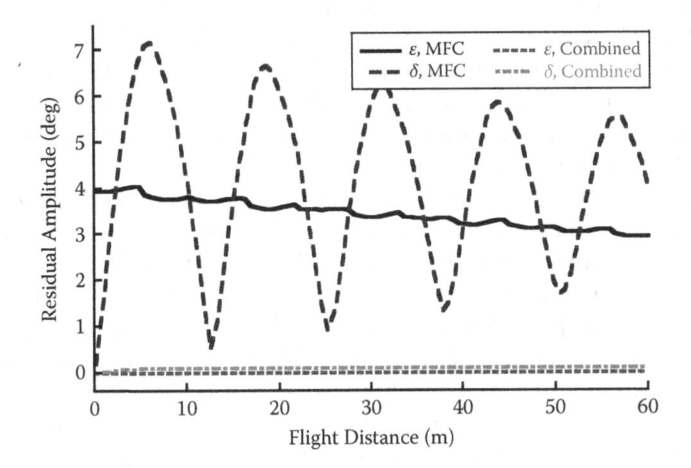

FIGURE 5.74 Residual amplitude of load slope and twisting, δ and ε, against flight distance.

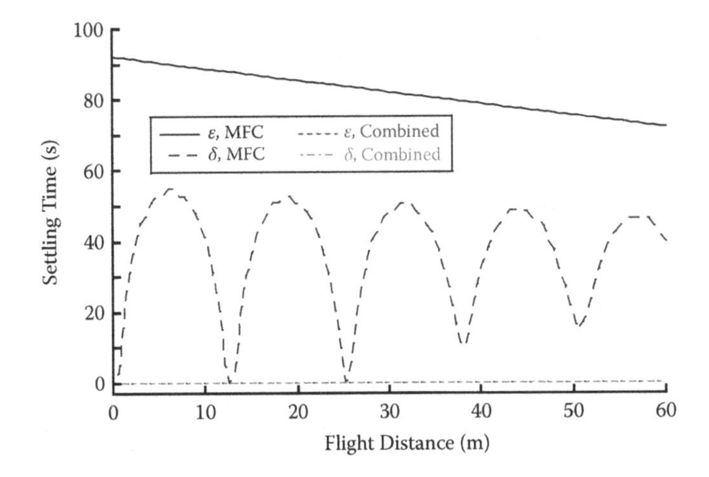

FIGURE 5.75 Settling time of load slope and twisting, δ and ε, against flight distance.

attitude, φ, and the load swing, γ_1. The residual amplitude and settling time of the load twisting changed slightly for load length shown. The combined MFC controller and three prefilters limited the residual amplitude of the load twisting and slope to below 0.01° and 0.1° for all the cases shown in Figures 5.74 and 5.75. The combined control method attenuated an average of 99.9% and 97.7% more residual amplitude of the load twisting and slope than the MFC controller and suppresses total settling time.

5.2.2.3.3 Robustness to Modeling Errors in System Parameter

An additional set of simulations was performed to assess the robustness of the combined control scheme for various modeling errors in the system parameters. The flight distance was set to 60 m along the $\mathbf{D}_y$ direction after a yaw angle of 6° about the $\mathbf{D}_z$ direction. The load length was varied from 2.5 m to 5.5 m in the simulations. The load mass increased as the load length increased because of constant load density. However, the modeled load length and mass for the combined controller were fixed at 4 m and 20 kg, respectively. The corresponding design frequencies and damping ratios for the combined controller were 1.4880 rad/s, 0.025, 1.5001 rad/s, 0.021, 1.6667 rad/s, 0.009.

The residual amplitude and settling time of the quadcopter attitude for the fixed design frequency are shown in Figures 5.76 and 5.77. By using the MFC controller, increasing load length increases the residual amplitude of quadcopter attitude, φ, and decreases corresponding settling time. Meanwhile, the residual amplitude of quadcopter attitude, ψ, increases slowly as the load length increases. The settling time of quadcopter attitude, ψ, shows no significant change before 3.7 m. The settling time does rise from 3.7 m and reaches the local maximum at 4.7 m. After this peak point, the settling time falls. Under the same conditions, the combined control method suppressed the residual amplitude and settling time of quadcopter attitude, φ, better than the MFC controller by an average of

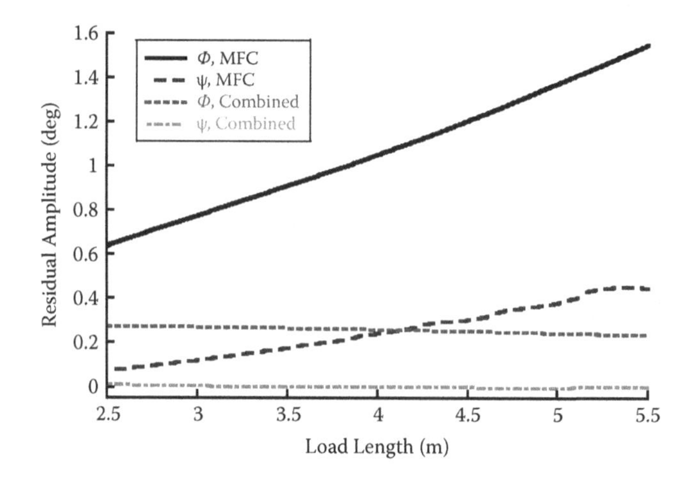

FIGURE 5.76 Residual amplitude quadcopter attitude, φ and ψ, against load length and mass.

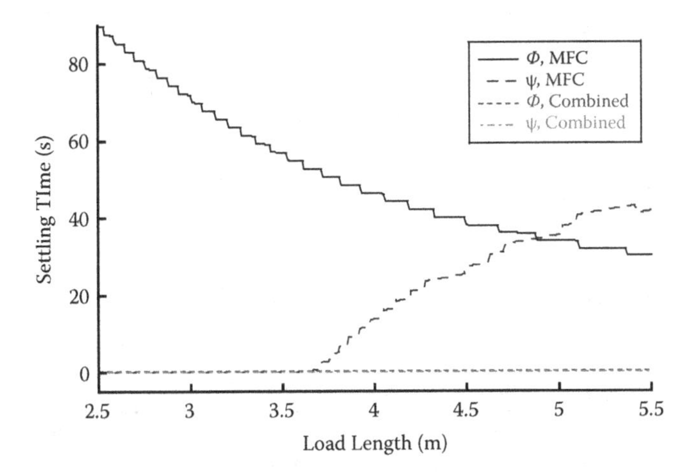

FIGURE 5.77 Settling time of quadcopter attitude, φ and ψ, against load length and mass.

75.7% and 99.3%, and those of quadcopter attitude, ψ, by an average of 99.0% and 100%.

Figures 5.78 and 5.79 shows the residual amplitude and settling time of the load swing, γ_1 and β_1. With the MFC controller, as the load length increases, the residual amplitude and settling time of the load swing, γ_1, decreases. Increasing load length increases the residual amplitude of the load swing, β_1, slightly. Meanwhile, the settling time of the load swing, β_1, keeps zero until 3.5 m, and then increases slowly. The combined MFC controller and three prefilters limited the residual deflection and settling time of the load swing, γ_1 and β_1, to under 0.2°, 0.0 seconds, 0.1°, 0.0 seconds, respectively, for the parameter ranges shown.

The simulated load twisting and slope are shown in Figures 5.80 and 5.81. When using the MFC controller, the residual amplitude of the load twisting changes

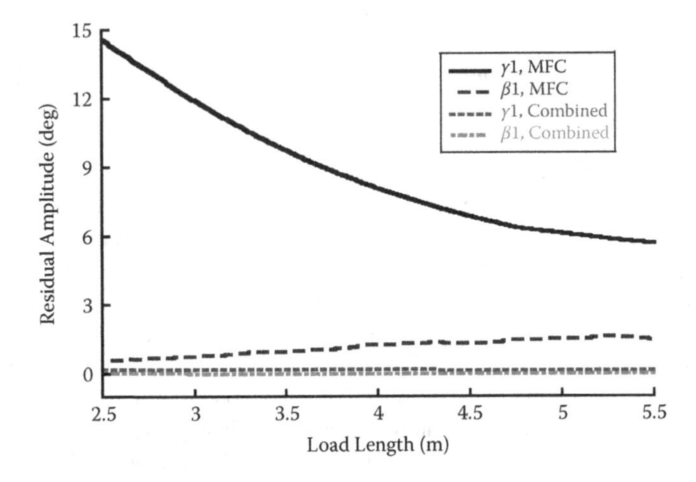

FIGURE 5.78 Residual amplitude of load swing, γ_1 and β_1, against load length and mass.

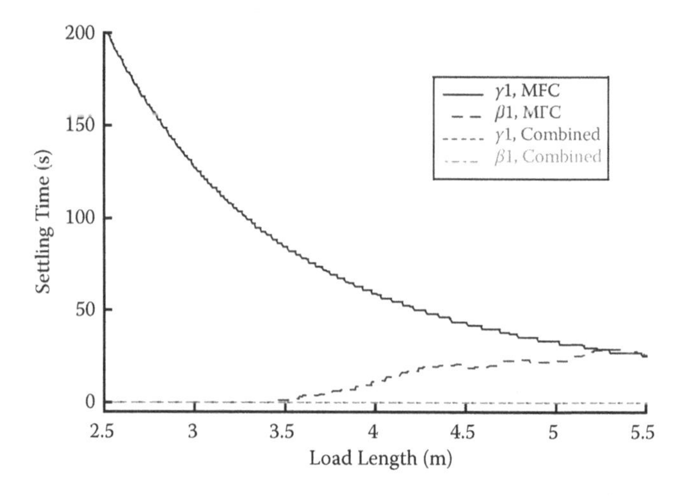

FIGURE 5.79 Settling time of load swing, γ_1 and β_1, against load length and mass.

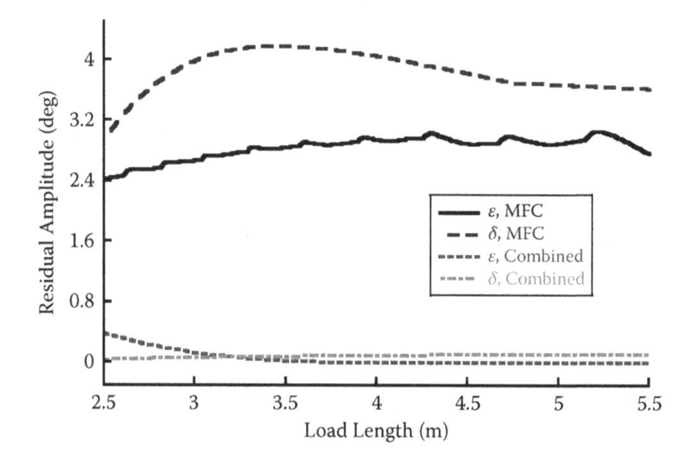

FIGURE 5.80 Residual amplitude of load slope and twisting, δ and ε, against load length and mass.

slightly and the settling time of the load twisting decreases with increasing load length. The residual amplitude of load slope contains a peak as the load length is varied. The settling time of load slope decreases with increasing load length. Therefore, the combined control scheme reduced residual amplitude and settling time of the load twisting over the MFC controller by an average of 98.0% and 100%, and those of the load slope by an average of 97.6% and 100%.

Simulations demonstrate that the combined MFC controller and three prefilters is beneficial for reducing the oscillations of the quadcopter's attitude, load swing, and load twisting. Therefore, the combined control method provides a safer environment to operate the quadcopter slung load, and a higher efficiency for transferring loads to the desired position precisely.

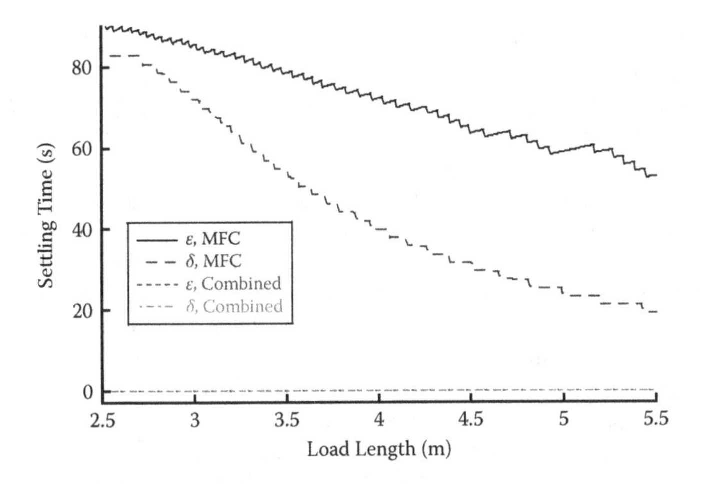

FIGURE 5.81 Settling time of load slope and twisting, δ and ε, against load length and mass.

5.3 QUADCOPTERS SLUNG LIQUID CONTAINERS

Water containers suspended below large quadrotors may be applied for fire-fighting services. Undoubtedly, flight motions of the aerial vehicle and external disturbances may cause undesirable oscillations of the quadrotor attitude, container swing, and liquid sloshing. The quadrotor attitude and cable-suspended container exhibit compound-pendulum dynamics, while motions of the free liquid surface in the tank display liquid-sloshing dynamics. Moreover, the interaction between the fluid sloshing and container may also corrupt the load swing and quadrotor attitude seriously. Therefore, the coupled attitude-pendulum-sloshing dynamics degrade effectiveness and safety of the flight operation. Operation for this type of system is very challenging because of coupling dynamic behavior among quadrotor's attitude, load swing, and liquid sloshing. Therefore, the study of attitude-pendulum-sloshing dynamics is essential.

Broad attentions have been directed at modeling and dynamics of quadrotors with external suspended rigid loads. Mounting numbers of the example have provided the solutions to the oscillation problems caused by the suspended-load swing for quadrotors slung rigid loads. The feedback control schemes use estimation of the suspended-load oscillations to reduce swing in closed-loop systems. Unfortunately, accurately measuring the suspended-load oscillations is challenging. In order to produce a minimal swing, optimal trajectory planning, flatness-based control, and command shaping techniques have been used to reduce the oscillations by modifying the pilot commands. However, external disturbances cannot be rejected. All abovementioned works are focused on the suspended rigid loads, which only exhibits multiple pendulum dynamics. The theoretical findings and experimental results cannot apply on quadrotors slung liquid tanks because of coupled attitude-pendulum-sloshing dynamics.

Meanwhile, many efforts have been devoted to sloshing dynamics in the liquid tank. Baffles and absorbers are widely used to reduce fluid energy. However, weight and complexity of the overall system increase. A further solution is the active control method. But the active control approaches are very difficult to carry out in the real transport. Feedback control schemes measure motions of liquid surface in a feedback loop for oscillations suppression. Nevertheless, accurately sensing the sloshing states is challenging. Open-loop control techniques modify the driving commands to generate a prescribed motion with minimal sloshing. However, the previous open-loop controller cannot reject the oscillations caused by external disturbances. All above-mentioned works are only focused on sloshing dynamics. Additionally, some of works described sloshing dynamics using the equivalent mechanical model, such as pendulum model. However, the equivalent pendulum model for fluid sloshing dynamics is significantly different from attitude-pendulum-sloshing dynamics.

Dynamics and control of multiple quadrotors slung a liquid tank have been reported in the previous literature. However, the quadrotor attitude was ignored, and the complex sloshing dynamics were simplified as the equivalent mass-spring model. The dynamics of vehicle attitude is critical for safe flight. Moreover, the equivalent mass-spring model of sloshing dynamics cannot describe the complex attitude-pendulum-sloshing dynamics. Therefore, no attention has been directed at the coupled dynamics among the vehicle attitude, container swing, and liquid sloshing because of complex dynamics in the quadrotor slung liquid tank.

5.3.1 Modeling

Figure 5.82 shows a planar model of a quadrotor carrying a liquid container. The quadrotor, D, flies along the N_x direction in near-hover operation. The mass of

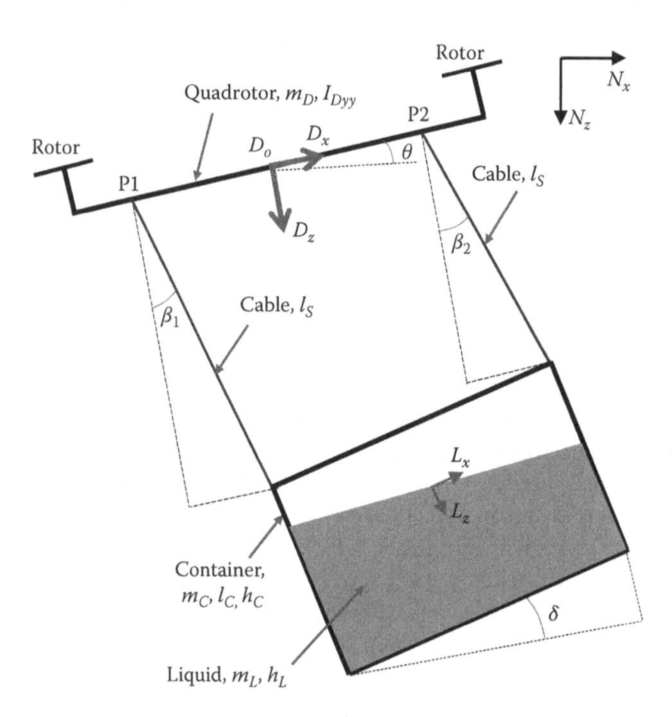

FIGURE 5.82 Planar model of a quadrotor transporting a liquid tank.

the quadrotor is m_D, and the moment of inertia about the D_y axis is I_{Dyy}. Motions of the quadrotor are divided into the displacements, x_D and z_D, along the N_x and N_z directions, and the attitude angle, θ, about the N_y axis in the inertial Newtonian frame. Let $D_xD_yD_z$ be the moving Cartesian coordinates fixed to the quadrotor such that the D_y axis is parallel to the N_y axis. Thus, the inertial coordinates $N_xN_yN_z$ can be converted to the moving coordinates $D_xD_yD_z$ by rotating the attitude, θ, about the N_y axis.

Two cables, C_1 and C_2, of length, l_S, which are assumed to be massless and inelastic, hang below the quadrotor at the suspension points, P_1 and P_2. The quadrotor center of gravity, D_o, locates at the middle point between the suspension points, P_1 and P_2. The distance between the suspension points, P_1 and P_2, are $2l_A$. A liquid rectangular tank, T, of mass, m_C, length, l_C, and height, h_C, is attached to the quadrotor by two suspension cables. The swing angles of the suspension cables C_1 and C_2 are β_1 and β_2, respectively. The angle between the length direction of the container and the D_x direction is defined as the slope angle, δ, of the container, through which the container rotates about the D_y direction.

The rigid tank contains a liquid whose surface at rest is at a height, h_L, from the bottom of the tank. The mass of the liquid is m_L. The quadrotor coordinates $D_xD_yD_z$ can be converted to the liquid coordinates $L_xL_yL_z$ by rotating the slope angle, δ, about the D_y axis. The origin of the liquid coordinates locates at the middle point of the undisturbed free surface. The fluid surface elevation measured from the undisturbed free surface in the L_z direction denotes η.

Inputs to the planar near-hover model shown in Figure 5.82 are the thrust force, F_z, along the D_z direction, and the pitching moment, M_θ, about the D_y direction. Outputs are the quadrotor displacements, x_D and z_D, quadrotor attitude, θ, swing angles, β_1 and β_2, slope angle, δ, and surface elevation, η. The model does not include the aerodynamic effects on the suspended tank due to near-hover operation, long cables, and heavy loads. The aerodynamic effects on the tank are considered as the external disturbances in this article. The disturbances will be rejected by designing the oscillation control methods. The initial direction of the tank length coincides with D_xD_z plane so that the thrust force, F_z, and pitching moment, M_θ, cannot cause container twisting about the suspension cables. The model also assumes that the cable length does not change during the motions. The following assumptions are employed to simplify the fluid dynamics: (1) the flow is assumed to be irrotational; (2) the fluid is nonviscous homogeneous and incompressible; and (3) the displacement and velocity of the liquid free surface are small.

5.3.1.1 Compound-Pendulum Dynamics

The generalized speeds, $u_{k=1\text{–}6}$, are chosen as:

$$u_1 = \dot{x}_D,\ u_2 = \dot{z}_D,\ u_3 = \dot{\theta},\ u_4 = \dot{\beta}_1,\ u_5 = \dot{\beta}_2,\ u_6 = \dot{\delta} \tag{5.115}$$

where the first three equations in Equation (5.115) are generalized speeds of quadrotor, the last three are generalized speeds of suspended container. The basis transformation matrix among the bodies D (quadrotor), C_1 (cable 1), C_2 (cable 2), T (tank), and L (liquid) are:

$$\begin{Bmatrix} \mathbf{D}_x \\ \mathbf{D}_z \end{Bmatrix} = \begin{bmatrix} \cos\theta & -\sin\theta \\ \sin\theta & \cos\theta \end{bmatrix} \begin{Bmatrix} \mathbf{N}_x \\ \mathbf{N}_z \end{Bmatrix} \tag{5.116}$$

$$\begin{Bmatrix} \mathbf{C}_{1x} \\ \mathbf{C}_{1z} \end{Bmatrix} = \begin{bmatrix} \cos\beta_1 & -\sin\beta_1 \\ \sin\beta_1 & \cos\beta_1 \end{bmatrix} \begin{Bmatrix} \mathbf{D}_x \\ \mathbf{D}_z \end{Bmatrix} \tag{5.117}$$

$$\begin{Bmatrix} \mathbf{C}_{2x} \\ \mathbf{C}_{2z} \end{Bmatrix} = \begin{bmatrix} \cos\beta_2 & -\sin\beta_2 \\ \sin\beta_2 & \cos\beta_2 \end{bmatrix} \begin{Bmatrix} \mathbf{D}_x \\ \mathbf{D}_z \end{Bmatrix} \tag{5.118}$$

$$\begin{Bmatrix} \mathbf{T}_x \\ \mathbf{T}_z \end{Bmatrix} = \begin{bmatrix} \cos\delta & -\sin\delta \\ \sin\delta & \cos\delta \end{bmatrix} \begin{Bmatrix} \mathbf{D}_x \\ \mathbf{D}_z \end{Bmatrix} \tag{5.119}$$

$$\begin{Bmatrix} \mathbf{L}_x \\ \mathbf{L}_z \end{Bmatrix} = \begin{bmatrix} \cos\delta & -\sin\delta \\ \sin\delta & \cos\delta \end{bmatrix} \begin{Bmatrix} \mathbf{D}_x \\ \mathbf{D}_z \end{Bmatrix} \tag{5.120}$$

Then the velocity of the quadrotor center, D_o, in the Newtonian reference frame can be written as:

$$^{N}\mathbf{v}^{D_o} = u_1\mathbf{N}_x + u_2\mathbf{N}_z \tag{5.121}$$

Meanwhile, the angular velocity of the quadrotor is given by:

$$^{N}\boldsymbol{\omega}^{D} = u_3\mathbf{D}_y \tag{5.122}$$

The velocity of the tank center, T_o, in the Newtonian reference frame can be written as:

$$^{N}\mathbf{v}^{T_o} = {}^{N}\mathbf{v}^{D_o} - l_A u_3\mathbf{D}_z + l_S(u_3 + u_5)\mathbf{C}_{2x} + 0.5l_C(u_3 + u_6)\mathbf{T}_z + 0.5h_C(u_3 + u_6)\mathbf{T}_x \tag{5.123}$$

The angular velocity of the tank, T, is given by:

$$^{N}\boldsymbol{\omega}^{T} = {}^{N}\boldsymbol{\omega}^{D} + u_6\mathbf{T}_y \tag{5.124}$$

A part of generalized active force is the gravity of quadrotor and tank, and another part of generalized active force is the thrust force, F_z, and pitching moment, M_θ, produced by the rotors, and the fluid force, F_{fx} and F_{fz}, acting on the tank sidewall along the L_x and L_z directions, respectively. Therefore, the generalized active force, $F_{k=1\text{–}6}$, can be written as:

$$F_1 = F_{fx}\cos(\theta + \delta) + F_{fz}\sin(\theta + \delta) - F_z\sin\theta \tag{5.125}$$

$$F_2 = (m_D + m_C)g - F_{fx}\sin(\theta + \delta) + F_{fz}\cos(\theta + \delta) - F_z\cos\theta \tag{5.126}$$

$$F_3 = M_\theta + 0.5F_{fx}(h_C - h_L) + \begin{bmatrix} F_{fx}cos(\theta+\delta) \\ +F_{fz}sin(\theta+\delta) \end{bmatrix} \begin{bmatrix} 0.5h_C cos(\theta+\delta) + l_S cos(\theta+\beta_2) \\ +0.5l_C sin(\theta+\delta) - l_A sin\theta \end{bmatrix}$$
$$- \begin{bmatrix} m_C g + F_{fz}\cos(\theta+\delta) \\ -F_{fx}\sin(\theta+\delta) \end{bmatrix} \begin{bmatrix} 0.5h_C \sin(\theta+\delta) - 0.5l_C\cos(\theta+\delta) \\ + l_S \sin(\theta+\beta_2) + l_A\cos\theta \end{bmatrix} \quad (5.127)$$

$$F_4 = 0 \quad (5.128)$$

$$F_5 = l_S\cos(\theta+\beta_2)\begin{bmatrix} F_{fx}\cos(\theta+\delta) \\ + F_{fz}\sin(\theta+\delta) \end{bmatrix} - l_S\sin(\theta+\beta_2)\begin{bmatrix} m_C g + F_{fz}\cos(\theta+\delta) \\ - F_{fx}\sin(\theta+\delta) \end{bmatrix} \quad (5.129)$$

$$F_6 = 0.5F_{fx}(h_C - l_C) + 0.5\begin{bmatrix} h_C\cos(\theta+\delta) \\ +l_C\sin(\theta+\delta) \end{bmatrix}\begin{bmatrix} F_{fx}\cos(\theta+\delta) \\ +F_{fz}\sin(\theta+\delta) \end{bmatrix}$$
$$+ 0.5[l_C\cos(\theta+\delta) - h_C\sin(\theta+\delta)]\begin{bmatrix} m_C g + F_{fz}\cos(\theta+\delta) \\ -F_{fx}\sin(\theta+\delta) \end{bmatrix} \quad (5.130)$$

where g is the gravitational acceleration. Meanwhile, the generalized inertia force, $F_{k=1-6}^{*}$, is:

$$F_1^* = -m_D\dot{u}_1 + m_C\begin{pmatrix} \sin(\theta+\delta)[0.5h_C(u_3+u_6)^2 - 0.5l_C(\dot{u}_3+\dot{u}_6)] \\ -\cos(\theta+\delta)[0.5l_C(u_3+u_6)^2 + 0.5h_C(\dot{u}_3+\dot{u}_6)] \\ -\dot{u}_1 + l_A u_3^2\cos\theta - l_S\cos(\theta+\beta_2)(\dot{u}_3+\dot{u}_5) \\ + l_S sin(\theta+\beta_2)(u_3+u_5)^2 + l_A\dot{u}_3\sin\theta \end{pmatrix} \quad (5.131)$$

$$F_2^* = -m_D\dot{u}_2 + m_C\begin{pmatrix} \cos(\theta+\delta)[0.5h_C(u_3+u_6)^2 - 0.5l_C(\dot{u}_3+\dot{u}_6)] \\ -\dot{u}_2 + \sin(\theta+\delta)[0.5l_C(u_3+u_6)^2 + 0.5h_C(\dot{u}_3+\dot{u}_6)] \\ -l_A u_3^2\sin\theta + l_S\sin(\theta+\beta_2)(\dot{u}_3+\dot{u}_5) \\ + l_S\cos(\theta+\beta_2)(u_3+u_5)^2 + l_A\dot{u}_3\cos\theta \end{pmatrix} \quad (5.132)$$

$$F_3^* = -I_{Dyy}\dot{u}_3 - I_{Tyy}(\dot{u}_3+\dot{u}_6)$$
$$- m_C\begin{bmatrix} 0.5h_C\sin(\theta+\delta) \\ -0.5l_C\cos(\theta+\delta) \\ + l_S\sin(\theta+\beta_2) \\ + l_A\cos\theta \end{bmatrix}\begin{bmatrix} \cos(\theta+\delta)(0.5h_C(u_3+u_6)^2 - 0.5l_C(\dot{u}_3+\dot{u}_6)) \\ -\dot{u}_2 + \sin(\theta+\delta)(0.5l_C(u_3+u_6)^2 + 0.5h_C(\dot{u}_3+\dot{u}_6)) \\ -l_A u_3^2\sin\theta + l_S\sin(\theta+\beta_2)(\dot{u}_3+\dot{u}_5) \\ + l_S\cos(\theta+\beta_2)(u_3+u_5)^2 + l_A\dot{u}_3\cos\theta \end{bmatrix}$$
$$+ m_C\begin{bmatrix} 0.5h_C\cos(\theta+\delta) \\ + l_S\cos(\theta+\beta_2) \\ + 0.5l_C\sin(\theta+\delta) \\ - l_A\sin\theta \end{bmatrix}\begin{bmatrix} \sin(\theta+\delta)(0.5h_C(u_3+u_6)^2 - 0.5l_C(\dot{u}_3+\dot{u}_6)) \\ -\cos(\theta+\delta)(0.5l_C(u_3+u_6)^2 + 0.5h_C(\dot{u}_3+\dot{u}_6)) \\ -\dot{u}_1 + l_A u_3^2\cos\theta - l_S\cos(\theta+\beta_2)(\dot{u}_3+\dot{u}_5) \\ + l_S\sin(\theta+\beta_2)(u_3+u_5)^2 + l_A\dot{u}_3\sin\theta \end{bmatrix} \quad (5.133)$$

$$F_4^* = 0 \tag{5.134}$$

$$F_5^* = m_C \left(\begin{aligned} & l_S \cos(\theta + \beta_2) \begin{bmatrix} \sin(\theta + \delta)(0.5h_C(u_3 + u_6)^2 - 0.5l_C(\dot{u}_3 + \dot{u}_6)) \\ - \cos(\theta + \delta)(0.5l_C(u_3 + u_6)^2 + 0.5h_C(\dot{u}_3 + \dot{u}_6)) \\ - \dot{u}_1 + l_A u_3^2 \cos\theta - l_S \cos(\theta + \beta_2)(\dot{u}_3 + \dot{u}_5) \\ + l_S \sin(\theta + \beta_2)(u_3 + u_5)^2 + l_A \dot{u}_3 \sin\theta \end{bmatrix} \\ & - l_S \sin(\theta + \beta_2) \begin{bmatrix} \cos(\theta + \delta)(0.5h_C(u_3 + u_6)^2 - 0.5l_C(\dot{u}_3 + \dot{u}_6)) \\ - \dot{u}_2 + \sin(\theta + \delta)(0.5l_C(u_3 + u_6)^2 + 0.5h_C(\dot{u}_3 + \dot{u}_6)) \\ - l_A u_3^2 \sin\theta + l_S \sin(\theta + \beta_2)(\dot{u}_3 + \dot{u}_5) \\ + l_S \cos(\theta + \beta_2)(u_3 + u_5)^2 + l_A \dot{u}_3 \cos\theta \end{bmatrix} \end{aligned} \right) \tag{5.135}$$

$$\begin{aligned} F_6^* = & -I_{Tyy}(\dot{u}_3 + \dot{u}_6) \\ & + m_C \left(\begin{aligned} & \begin{bmatrix} 0.5h_C cos(\theta + \delta) \\ + 0.5l_C sin(\theta + \delta) \end{bmatrix} \begin{bmatrix} \sin(\theta + \delta)(0.5h_C(u_3 + u_6)^2 - 0.5l_C(\dot{u}_3 + \dot{u}_6)) \\ - \cos(\theta + \delta)(0.5l_C(u_3 + u_6)^2 + 0.5h_C(\dot{u}_3 + \dot{u}_6)) \\ - \dot{u}_1 + l_A u_3^2 \cos\theta - l_S \cos(\theta + \beta_2)(\dot{u}_3 + \dot{u}_5) \\ + l_S \sin(\theta + \beta_2)(u_3 + u_5)^2 + l_A \dot{u}_3 \sin\theta \end{bmatrix} \\ & + \begin{bmatrix} 0.5l_C cos(\theta + \delta) \\ - 0.5h_C sin(\theta + \delta) \end{bmatrix} \begin{bmatrix} \cos(\theta + \delta)(0.5h_C(u_3 + u_6)^2 - 0.5l_C(\dot{u}_3 + \dot{u}_6)) \\ - \dot{u}_2 + \sin(\theta + \delta)(0.5l_C(u_3 + u_6)^2 + 0.5h_C(\dot{u}_3 + \dot{u}_6)) \\ - l_A u_3^2 \sin\theta + l_S \sin(\theta + \beta_2)(\dot{u}_3 + \dot{u}_5) \\ + l_S \cos(\theta + \beta_2)(u_3 + u_5)^2 + l_A \dot{u}_3 \cos\theta \end{bmatrix} \end{aligned} \right) \end{aligned} \tag{5.136}$$

where the tank moment of inertia about the T_y axis is I_{Tyy}.

By using the Kane method, the sum of the generalized active forces described in Equations (5.125)–(5.130) and generalized inertia forces given in Equations (5.131)–(5.136) should be equal to zero. Then the nonlinear equations of the motion for the model shown in Figure 5.82 yield:

$$[W]\cdot\{\dot{U}\} + \{Q\} + \{Z\} = \mathbf{0} \tag{5.137}$$

where W is the mass matrix, and $\dot{U}$ is the column matrix of derivatives of generalized speeds, $u_{k=1\text{–}6}$, with respect to time, Q is the column matrix of centrifugal and Coriolis terms, and Z is the column matrix of gravity terms and control input terms. The first two terms in the left side of Equation (5.137) are the generalized inertia forces described in Equations (5.131)–(5.136), and the third term in the left side of Equation (5.137) is the generalized active forces given in Equations (5.125)–(5.130).

The quadrotor, tank, and two cables constitute a four-bar linkage. Then the displacement constraint of the four-bar linkage can be written as:

$$\begin{cases} 2l_A + l_S \sin\beta_2 - l_C \cos\delta - l_S \sin\beta_1 = 0 \\ l_S \cos\beta_2 + l_C \sin\delta - l_S \cos\beta_1 = 0 \end{cases} \tag{5.138}$$

The geometric constraint given in Equation (5.138) shows that determining one of the three angles, β_1, β_2, and δ, is enough to know the other two. While the independent motion variable is defined as u_4, the dependent motion variables are defined as u_5 and u_6. Then the corresponding velocity constraint and acceleration constraint of the four-bar linkage are:

$$\begin{cases} \dot{\beta}_2 \cos(\beta_2 - \delta) - \dot{\beta}_1 \cos(\beta_1 - \delta) = 0 \\ \dot{\delta} l_C \cos(\beta_2 - \delta) + \dot{\beta}_1 l_S \sin(\beta_1 - \beta_2) = 0 \end{cases} \tag{5.139}$$

and

$$\begin{cases} \ddot{\beta}_2 = \tan(\beta_2 - \delta)\dot{\beta}_2^{\,2} - \dfrac{l_C \dot{\delta}^2}{l_S \cos(\beta_2 - \delta)} - \dfrac{[\sin(\beta_1 - \delta)\dot{\beta}_1^{\,2} - \cos(\beta_1 - \delta)\ddot{\beta}_1]}{\cos(\beta_2 - \delta)} \\ \ddot{\delta} = \dfrac{l_S [\dot{\beta}_2^{\,2} - \cos(\beta_1 - \beta_2)\dot{\beta}_1^{\,2} - \sin(\beta_1 - \beta_2)\ddot{\beta}_1]}{l_C \cos(\beta_2 - \delta)} - \tan(\beta_2 - \delta)\dot{\delta}^2 \end{cases} \tag{5.140}$$

5.3.1.2 Sloshing Dynamics

The velocity of the fluid for the irrotational flow in the container is given by:

$$\mathbf{v} = v_x \mathbf{L}_x + v_z \mathbf{L}_z + \frac{\partial \varphi}{\partial x}\mathbf{L}_x + \frac{\partial \varphi}{\partial z}\mathbf{L}_z \tag{5.141}$$

where the perturbed velocity potential function is φ, the displacements in the L_x and L_z directions are denoted as x and z, respectively. Velocities v_x and v_z in Equation (5.141) are given by:

$$\begin{aligned} v_x = {} & u_1 \cos(\theta + \delta) - u_2 \sin(\theta + \delta) + l_A u_3 \sin\delta \\ & + l_S (u_3 + u_5)\cos(\beta_2 - \delta) + (h_C - 0.5h_L)(u_3 + u_6) \end{aligned} \tag{5.142}$$

$$\begin{aligned} v_z = {} & u_1 \sin(\theta + \delta) + u_2 \cos(\theta + \delta) - l_A u_3 \cos\delta - l_S (u_3 + u_5)\sin(\beta_2 - \delta) \\ & + 0.5 l_C (u_3 + u_6) \end{aligned} \tag{5.143}$$

The corresponding acceleration of the fluid in the container is given by:

$$\begin{aligned} \mathbf{a} = {} & a_x \mathbf{L}_x + a_z \mathbf{L}_z + \nabla \frac{\partial \varphi}{\partial t} + \frac{1}{2}\nabla\left[\left(\frac{\partial \varphi}{\partial x}\right)^2 + \left(\frac{\partial \varphi}{\partial z}\right)^2\right] + 2(u_3 + u_6)\frac{\partial \varphi}{\partial z}\mathbf{L}_x - 2(u_3 \\ & + u_6)\frac{\partial \varphi}{\partial x}\mathbf{L}_z \end{aligned} \tag{5.144}$$

where

$$\begin{aligned} a_x = {} & \dot{u}_1 \cos(\theta + \delta) - \dot{u}_2 \sin(\theta + \delta) - l_A u_3^2 \cos\delta + l_A \dot{u}_3 \sin\delta + l_S (\dot{u}_3 \\ & + \dot{u}_5) \cos(\beta_2 - \delta) - l_S (u_3 + u_5)^2 \sin(\beta_2 - \delta) + 0.5 l_C (u_3 + u_6)^2 \\ & + (h_C - 0.5 h_L)(\dot{u}_3 + \dot{u}_6) \end{aligned} \tag{5.145}$$

$$\begin{aligned} a_z = {} & \dot{u}_1 \sin(\theta + \delta) + \dot{u}_2 \cos(\theta + \delta) - l_A u_3^2 \sin\delta - l_A \dot{u}_3 \cos\delta - l_S (\dot{u}_3 \\ & + \dot{u}_5) \sin(\beta_2 - \delta) - l_S (u_3 + u_5)^2 \cos(\beta_2 - \delta) + 0.5 l_C (\dot{u}_3 + \dot{u}_6) \\ & - (h_C - 0.5 h_L)(u_3 + u_6)^2 \end{aligned} \tag{5.146}$$

Resulting from Equation (5.144), employing Euler's fluid dynamics equation, and ignoring the high-order terms yield:

$$\frac{p}{\rho} + \frac{\partial \varphi}{\partial t} + x[a_x + g \sin(\theta + \delta)] + z[a_z - g\cos(\theta + \delta)] = 0 \tag{5.147}$$

where ρ is the area mass density of the liquid. The fluid pressure is zero at the fluid free surface. Employing Bernoulli's equation yields:

$$\frac{\partial \varphi}{\partial t}\Big|_{z=\eta} + x[a_x + g \sin(\theta + \delta)] + \eta[a_z - g\cos(\theta + \delta)] = 0 \tag{5.148}$$

Resulting from the abovementioned assumptions, the boundary values in terms of the perturbed velocity potential is summarized as follows:

$$\frac{\partial^2 \varphi}{\partial x^2} + \frac{\partial^2 \varphi}{\partial z^2} = 0 \tag{5.149}$$

$$\frac{\partial \varphi}{\partial x}\Big|_{x=\pm 0.5 l_C} = 0 \tag{5.150}$$

$$\frac{\partial \varphi}{\partial z}\Big|_{z=h_L} = 0 \tag{5.151}$$

$$\frac{\partial \varphi}{\partial z}\Big|_{z=\eta} = \frac{\partial \eta}{\partial t} \tag{5.152}$$

The perturbed velocity potential and surface elevation can be assumed to be:

$$\varphi(x, z, t) = \sum_k \phi_k(x, z) \cdot \dot{q}_k(t) \tag{5.153}$$

$$\eta(x, z, t) = \sum_k \sigma_k(x, z) \cdot q_k(t) \tag{5.154}$$

where k is the positive integer, q_k is the time-dependent function of the kth vibrational mode of the sloshing, and ϕ_k and σ_k are the corresponding spatial functions. Substituting Equations (5.153) and (5.154) into (5.149)–(5.152) yields:

$$\frac{\partial^2 \phi_k}{\partial x^2} + \frac{\partial^2 \phi_k}{\partial z^2} = 0 \tag{5.155}$$

$$\frac{\partial \phi_k}{\partial x}|_{x=\pm 0.5 l_C} = 0 \tag{5.156}$$

$$\frac{\partial \phi_k}{\partial z}|_{z=h_L} = 0 \tag{5.157}$$

$$\frac{\partial \phi_k}{\partial z}|_{z=\eta} = \sigma_k = \frac{{\omega_k}^2 \phi_k}{[a_z - g\cos(\theta + \delta)]}|_{z=\eta} \tag{5.158}$$

The spatial function, ϕ_k, and frequency function, ω_k, of the sloshing can be obtained by solving Equations (5.155)–(5.158):

$$\phi_k(x, z) = \cos\left[k\pi\left(\frac{x}{l_C} + \frac{1}{2}\right)\right]\cosh\left[\frac{\pi k}{l_C}(z - h_L)\right] \tag{5.159}$$

$$\omega_k = \sqrt{\frac{k\pi}{l_C}[g\cos(\theta + \delta) - a_z]\tanh\left(k\pi\frac{h_L}{l_C}\right)} \tag{5.160}$$

The sloshing frequencies in Equation (5.160) depend on the system parameters, quadrotor attitude, container swing, container slope, and quadrotor acceleration. The sloshing frequencies exhibit complex nonlinear behavior because nonlinear motions of quadrotor and container also have large impact on the sloshing frequency. Substituting Equations (5.153) and (5.154) into (5.148), then multiplying by spatial function, ϕ_k, and integrating over the free surface ($-0.5l_C \le x \le 0.5l_C$) result in:

$$\ddot{q}_k(t) + \omega_k^2 q_k(t) + \gamma_k[a_x + g\sin(\theta + \delta)] = 0 \tag{5.161}$$

where

$$\gamma_k = \frac{\int_{-0.5l_C}^{0.5l_C} x\cdot\phi_k dx}{\int_{-0.5l_C}^{0.5l_C} \phi_k\cdot\phi_k dx} = \frac{2l_C\cos(k\pi - 1)}{k^2\pi^2\cosh\left[\frac{k\pi}{l_C}(z - h_L)\right]} \tag{5.162}$$

The coefficient, γ_k, in Equation (5.162) is limited to zero when the k is set to even. Resulting from Equations (5.153), (5.154), and (5.161), both the perturbed velocity potential, φ, and surface elevation, η, are a sum of response for each of an infinite number of sloshing modes. The surface elevation at the fluid free surface is described as:

$$\eta(x, 0, t) = \sum_k -\frac{k\pi}{l_C}\cos\left[k\pi\left(\frac{x}{l_C} + \frac{1}{2}\right)\right]\sinh\left(k\pi\frac{h_L}{l_C}\right)q_k(t) \tag{5.163}$$

The fluid forces, F_{fx} and F_{fz}, along the L_x and L_z directions can be written in the form:

$$F_{fx} = \int_{\eta(0.5l_C,t)}^{h_L} p(0.5l_C, z)dz - \int_{\eta(-0.5l_C,t)}^{h_L} p(-0.5l_C, z)dz \tag{5.164}$$

$$F_{fz} = \int_{-0.5l_C}^{0.5l_C} p(x, h_L)dx \tag{5.165}$$

where

$$p(x, z) = -\rho\left(\frac{\partial\varphi}{\partial t} + x[a_x + g\sin(\theta + \delta)] + z[a_z - g\cos(\theta + \delta)]\right) \tag{5.166}$$

where ρ is the area mass density of the liquid. Then the liquid mass satisfies $m_L = \rho l_C h_L$. Substituting Equation (5.166) into (5.164) and (5.165) yields the fluid force on the container sidewall:

$$\begin{aligned} F_{fx} &= \tfrac{\rho}{2}[a_z - g\cos(\theta + \delta)][\eta^2(0.5l_C, t) - \eta^2(-0.5l_C, t)] \\ &\quad - m_L[a_x + g\sin(\theta + \delta)] - \Sigma_k[\cos(k\pi) - 1]\tfrac{\rho l_C}{k\pi}\sinh\left(k\pi\tfrac{h_L}{l_C}\right)\ddot{q}_k(t) \end{aligned} \tag{5.167}$$

$$F_{fz} = -m_L[a_z - g\cos(\theta + \delta)] \tag{5.168}$$

Therefore, the nonlinear equation (5.137), constraints (5.138)–(5.140), and fluid forces (5.167) and (5.168) describe compound-pendulum dynamics, while the sloshing frequency (5.160), mode shape (5.159), and corresponding time function (5.161) report sloshing dynamics. Additionally, the coupling effects between pendulum and sloshing dynamics are also included in the dynamic model by the quadrotor attitude, cable swing, container slope, and fluid forces on the tank sidewall.

5.3.2 Control

5.3.2.1 Control of Quadrotor's Attitude

Resulting from Equations (5.137) and (5.161), equilibrium points of quadrotor attitude, θ, tank slope angle, δ, and surface elevation, η, are zero, and equilibrium points of swing angles, β_1 and β_2, are given by:

$$\begin{cases} \beta_{10} = -\arcsin\left(\frac{l_C - 2l_A}{2l_S}\right) \\ \beta_{20} = \arcsin\left(\frac{l_C - 2l_A}{2l_S}\right) \end{cases} \tag{5.169}$$

The system dynamics (5.137) and (5.161) are too complicated to analytically design controller for stabilize attitude. Therefore, a model-following (MF) controller will be obtained from a simplified model. The control gains will be refined to account for the nonlinear dynamics. Finally, the resulting controller will be applied on the nonlinear dynamic equations (5.137) and (5.161) to test its performance.

The oscillatory deflections of the vehicle attitude, θ, swing angles, β_1 and β_2, and slope angle, δ, around the equilibrium points are donated by, θ_t, β_{1t}, β_{2t}, and δ_t, respectively. From Equations (5.137) and (5.161), a simplified model can be derived by assuming small oscillations around the equilibrium points and neglecting high modes in sloshing dynamics:

$$\begin{cases} H_{1,5}\ddot{x}_D + H_{1,6}\theta_t F_z + H_{1,8}\beta_{1_t}F_z + H_{1,10}\beta_{2_t}F_z + H_{1,12}\delta_t F_z + H_{1,14}q_1 F_z + H_{1,17}M_\theta \\ \quad = 0 \\ H_{2,5}\ddot{z}_D + H_{2,7}\theta_t M_\theta + H_{2,9}\beta_{1_t}M_\theta + H_{2,12}\beta_{2_t}M_\theta + H_{2,15}\delta_t M_\theta + H_{2,19}F_z + H_{2,21} \\ \quad = 0 \\ H_{3,5}\ddot{\theta}_t + H_{3,6}\beta_{1_t}F_z + H_{3,8}\beta_{2_t}F_z + H_{3,10}\delta_t F_z + H_{3,12}q_1 F_z + H_{3,14}M_\theta = 0 \\ H_{4,5}\ddot{\beta}_{1_t} + H_{4,6}\beta_{1_t}F_z + H_{4,7}\beta_{1_t}M_\theta + H_{4,8}\beta_{2_t}F_z + H_{4,9}\beta_{2_t}M_\theta + H_{4,10}\delta_t F_z \\ \quad + H_{4,11}\delta_t M_\theta + H_{4,12}q_1 F_z + H_{4,13}q_1 M_\theta + H_{4,15}M_\theta = 0 \\ H_{5,5}\ddot{q}_1 + H_{5,6}\beta_{1_t}F_z + H_{5,7}\beta_{1_t}M_\theta + H_{5,8}\beta_{2_t}F_z + H_{5,9}\beta_{2_t}M_\theta + H_{5,10}\delta_t F_z \\ \quad + H_{5,11}\delta_t M_\theta + H_{5,12}q_1 F_z + H_{5,13}q_1 M_\theta + H_{5,15}M_\theta = 0 \\ H_{6,5}\ddot{\beta}_{2_t} + H_{6,6}\beta_{1_t}F_z + H_{6,7}\beta_{1_t}M_\theta + H_{6,8}\beta_{2_t}F_z + H_{6,9}\beta_{2_t}M_\theta + H_{6,10}\delta_t F_z \\ \quad + H_{6,11}\delta_t M_\theta + H_{6,12}q_1 F_z + H_{6,13}q_1 M_\theta + H_{6,15}M_\theta = 0 \\ H_{7,5}\ddot{\delta}_t + H_{7,6}\beta_{1_t}F_z + H_{7,7}\beta_{1_t}M_\theta + H_{7,8}\beta_{2_t}F_z + H_{7,9}\beta_{2_t}M_\theta + H_{7,10}\delta_t F_z \\ \quad + H_{7,11}\delta_t M_\theta + H_{7,12}q_1 F_z + H_{7,13}q_1 M_\theta + H_{7,15}M_\theta = 0 \end{cases} \tag{5.170}$$

where q_1 is the time function of the first sloshing mode, and $H_{k,j}$ are the coefficients.

The simplified model (5.170) yields a linearized model of the quadrotor's attitude by ignoring the effect of load swing and fluid sloshing on the motion of the quadrotor:

$$H_{3,5}\ddot{\theta} + H_{3,14}M_\theta = 0 \tag{5.171}$$

From the linearized model (5.171), an explicit MF controller can be derived:

$$\begin{cases} \ddot{\theta}_m + 2\zeta_m\omega_m\dot{\theta}_m + \omega_m{}^2\theta_m = \omega_m{}^2\theta_s \\ M_\theta = -\frac{H_{3,5}}{H_{3,14}}\ddot{\theta}_m + k_p(\theta_m - \theta) + k_d(\dot{\theta}_m - \dot{\theta}) \end{cases} \tag{5.172}$$

where ω_m and ζ_m are the frequency and damping ratio of the prescribed model, θ_m is the model output, and k_p and k_d are control gains. The first equation in Equation (5.172) is the prescribed model, while the second one describes the asymptotic tracking control law.

The pole placement method was used to design the frequency, ω_m, and damping ratio, ζ_m. The desired closed-loop poles were $-1 \pm i1$ for the reasonable damping ratio and settling time in the prescribed model. Therefore, the frequency, ω_m, in the prescribed model was calculated to be approximately 1.41 rad/s, and the corresponding damping ratio is 0.707.

Careful selection of the gains, k_p and k_d, in the tracking controller will achieve stability of the closed-loop system and suitable tracking errors. Resulting from Equations (5.171) and (5.172), the negative real parts in the design eigenvalues causes the stability of the control system. In addition, the coupling effect between the quadrotor and the suspended load increases the system damping, and decreases the real part of the eigenvalues. Therefore, the eigenvalues are designed to be $-8 \pm i2$. Given that the quadcopter mass, m_D, moment of inertia, I_{Dyy}, suspension distance, l_A, cable length, l_S, tank length, l_C, tank height, h_C, tank mass, m_C, moment of inertia, I_{Tyy}, liquid height, h_L, liquid mass, m_L, and area mass density, ρ, of the liquid were fixed at 85 kg, 4.5 kg m^2, 1 m, 5 m, 0.6 m, 0.5 m, 4 kg, 0.5 kg m^2, 0.35 m, 21 kg, and 100 kg/m^2, the gains, k_p, and k_d, by the pole placement technique were calculated to be 839.61 and 197.56, respectively.

The thrust force, F_z, in the D_z direction should be designed to balance the gravity of quadrotor mass and load mass. Then substituting the controller (5.172) into the simplified model (5.170) derives the natural frequency and damping ratio of the quadrotor slung load:

$$\mu_1\lambda^6 + \mu_2\lambda^5 + \mu_3\lambda^4 + \mu_4\lambda^3 + \mu_5\lambda^2 + \mu_6\lambda + \mu_7 = 0 \tag{5.173}$$

where

$$\lambda = -\zeta\Omega \pm i\Omega\sqrt{1 - \zeta^2} \tag{5.174}$$

Ω is the natural frequency of quadrotor slung liquid tank, ζ is the corresponding damping ratio, i is the imaginary unit, and $\mu_{k=1-7}$ are coefficients.

The natural frequency and damping ratio in Equation (5.173) are dependent on the system parameters. Figures 5.83 and 5.84 show the natural frequency and damping ratio for various cable length when the quadcopter mass, m_D, moment of inertia, I_{Dyy}, suspension distance, l_A, tank length, l_C, tank height, h_C, tank mass, m_C, liquid height, h_L, liquid mass, m_L, and area mass density, ρ, of the liquid were fixed at 85 kg, 4.5 kg m^2, 1 m, 0.6 m, 0.5 m, 4 kg, 0.35 m, 21 kg, and 100 kg/m^2, respectively. Increasing cable length decreases first-mode frequency and damping ratio. The cable length has little impact on the second-mode frequency and damping ratio. As the cable length increases, the third-mode frequency decreases slightly and corresponding damping ratio increases. In this case, the first-mode oscillations are more related with the pendulum motion, and the second-mode oscillations are more related with the sloshing motion. The third mode oscillations are created by the MF controller. Therefore, the third-mode damping ratio reaches a very high value.

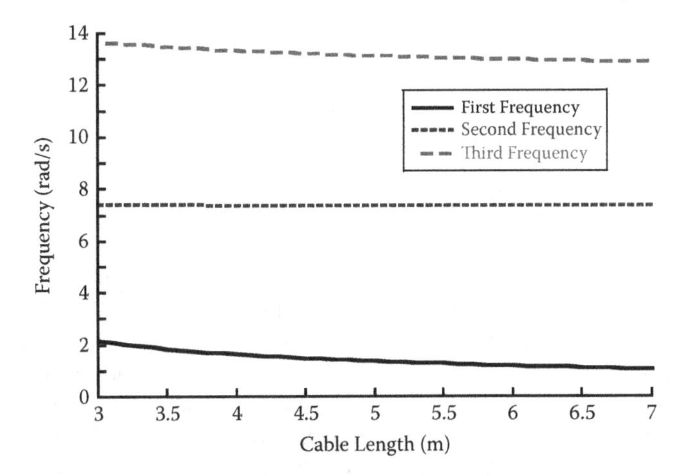

FIGURE 5.83 Natural frequency against cable length.

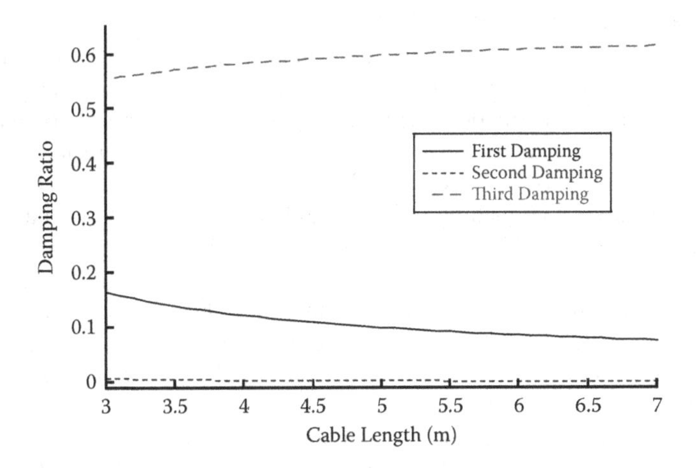

FIGURE 5.84 Damping ratio against cable length.

5.3.2.2 Control of Swing-Sloshing Dynamics

The four-pieces (FP) smoother is utilized to attenuate the coupled oscillations in the swing-sloshing dynamics. The FP smoother is given by:

$$\mathrm{FP}(\tau) = \begin{cases} \tau R e^{-2\zeta_1 \Omega_1 \tau}, & 0 \le \tau \le 0.5T_1 \\ (T_1 - K^{-1}T_1 - \tau + 2K^{-1}\tau) R e^{-2\zeta_1 \Omega_1 \tau}, & 0.5T_1 \le \tau \le T_1 \\ (3K^{-1}T_1 - K^{-2}T_1 - 2K^{-1}\tau + K^{-2}\tau) R e^{-2\zeta_1 \Omega_1 \tau}, & T_1 \le \tau \le 1.5T_1 \\ (2K^{-2}T_1 - K^{-2}\tau) R e^{-2\zeta_1 \Omega_1 \tau}, & 1.5T_1 < \tau \le 2T_1 \end{cases} \tag{5.175}$$

where Ω_1 is the first-mode natural frequency and ζ_1 is the corresponding damping ratio.

$$K = e^{(-\pi\zeta_1/\sqrt{1-\zeta_1^2})} \tag{5.176}$$

$$R = 4\zeta_1^2\Omega_1^2/(1 + K - K^2 - K^3)^2 \tag{5.177}$$

$$T_1 = 2\pi/(\Omega_1\sqrt{1 - \zeta_1^2}) \tag{5.178}$$

The smoother (5.175) is a combination of notch filter and low passing filter. The notch filter effect targets the first-mode oscillations shown in Figure 5.83, while the low passing filtering effect attenuates the high-mode oscillations. Thus, the smoother (5.175) is designed by using the first-mode frequency and damping ratio, which can be estimated from Equation (5.173) and can also be found in Figure 5.83. It is very challenging to estimate the high-mode frequency and damping ratio. Therefore, it is fortunate the smoother (5.175) may only use the first-mode frequency and damping ratio.

5.3.2.3 Control System Architecture

Figure 5.85 shows the control architecture for quadrotors slung liquid tanks. The operators generate the original commands, θ_r, via control interface. The commands, θ_r, filter through the FP smoother (5.175) to produce the smoothed commands, θ_s. The smoothed commands drive the MF controller (5.172) to create the pitching moment, M_θ, in a closed loop by measuring the quadrotor attitude, θ. The pitching moment, M_θ, moves the quadrotor slung liquid container to the desired positions. The system parameters will be used to estimate the first-mode natural frequency, Ω_1, and damping ratio, ζ_1, from Equation (5.173) for the FP smoother. The MF controller is applied for attitude stabilization, while the FP smoother is used for suppressing oscillations caused by operator's commands. Note that the thrust force, F_z, does not exist in Figure 5.85 because it is only applied to balance the gravity for hover flight.

Comparing with the single-hoist configuration, the dual-hoist mechanism creates fully coupling effect among quadrotor attitude, load swing, and liquid sloshing. The coupling impact benefits the disturbance rejection. The external disturbances resulting from the aerodynamic properties of rotors and wind gusts may cause container swing about the cable. The swing might also result in oscillations of liquid

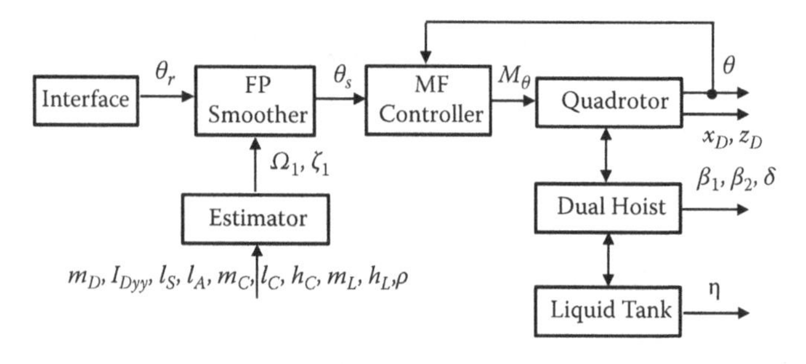

FIGURE 5.85 Control architecture.

sloshing and quadrotor attitude. The MF controller will mitigate oscillations of the vehicle attitude in a closed loop. The oscillation suppression of the quadrotor attitude will also produce the oscillation reduction of container swing and liquid sloshing because of the fully coupling effect. Hence, it is unnecessary to measure the oscillations of container swing and liquid sloshing under dual-hoist mechanisms. It is very challenging to measure the container swing and liquid sloshing in a real condition. Therefore, dual-hoist configurations help to reject the oscillations caused by external disturbances including aerodynamic effect and wind gusts.

5.3.3 Computational Dynamics

The combined FP smoother and MF controller method will be applied on the nonlinear model (5.137), constraints (5.138)–(5.140), surface elevation (5.154), and corresponding time function (5.161) to test the control performance. The quadcopter mass m_D, moment of inertia I_{Tyy}, moment of inertia I_{Dyy}, suspension distance l_A, tank length l_C, tank height h_C, tank mass m_C, liquid height h_L, liquid mass m_L, flight speed, liquid mass density ρ, sloshing measuring point is 85 kg, 0.5 kg m^2, 4.5 kg m^2, 1 m, 0.6 m, 0.5 m, 4 kg, 0.35 m, 21 kg, 3 m/s, 100 kg/m^2, $x = 0.5l_C$; $z = 0$, respectively. Three group of simulations will be conducted to test the disturbance rejection, effectiveness, and robustness of the control systems.

5.3.3.1 Disturbance Rejection

When the cable length, l_S, was fixed at 5 m, the corresponding equilibria of quadrotor attitude, θ, swing angles, β_1 and β_2, slope angle, δ, and surface elevation, η, were 0°, 8.048°, −8.048°, 0°, and 0 mm, respectively. External disturbances such as wind gusts and aerodynamics from rotors may induce a non-zero initial condition around the equilibria. Figures 5.86–5.89 show a simulated response of the quadrotor attitude, θ, swing angle, β_2, surface elevation, η, and fluid force, F_{fx}, to a non-zero initial condition. The initial values of quadrotor attitude, θ, swing angles, β_1 and β_2,

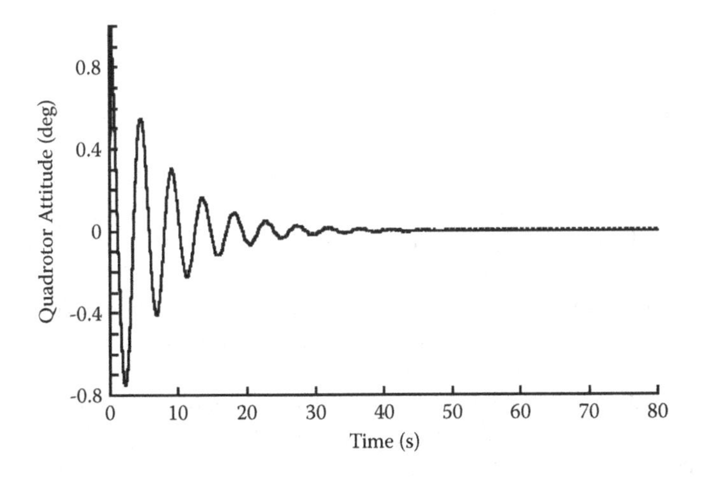

FIGURE 5.86 Time response of quadrotor attitude to a non-zero initial condition.

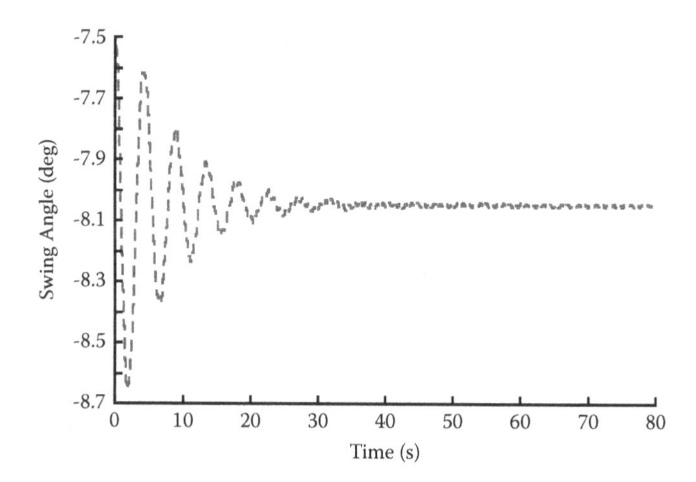

FIGURE 5.87 Time response of swing angle to a non-zero initial condition.

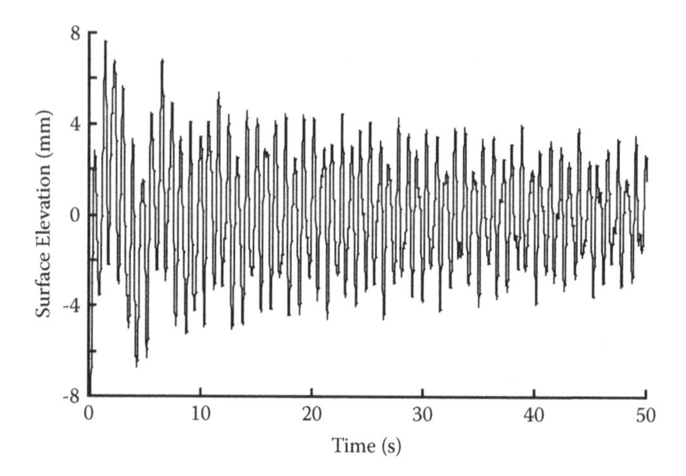

FIGURE 5.88 Time response of surface elevation to a non-zero initial condition.

slope angle, δ, and surface elevation, η, were set to 0°, 10.037°, 6.091°, −4.617°, and −7.980 mm, respectively. The container oscillations and fluid sloshing will cause the oscillations of the quadrotor attitude because of the interaction among quadrotor, container, and liquid.

The motions of the quadrotor attitude, container swing and liquid sloshing are related with the first- and second-mode frequencies resulting from Equation (5.173) and shown in Figure 5.83. Moreover, the damping effect inherent in MF controller reduced the oscillational amplitudes in Figures 5.86–5.89 because of the fully coupling impact among quadrotor, container, and liquid produced by dual-hoisting mechanisms. Finally, the responses of quadrotor attitude, container swing, container slope, and surface elevation stabilized at the corresponding equilibria. The settling time required

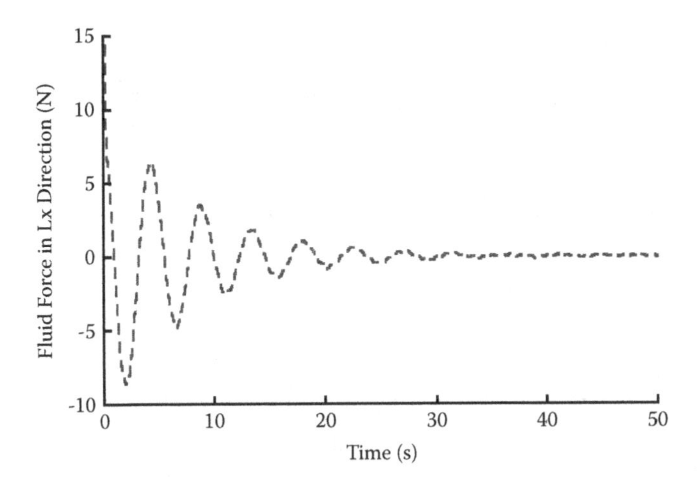

FIGURE 5.89 Time response of fluid force to a non-zero initial condition.

for the response of the quadrotor attitude to settle within 0.1° was 16.39 seconds. The settling time of the swing angle, β_2, is defined as the time required for the response curve to reach and stay within 0.2° about the equilibrium point. Thus, the settling time of swing angle, β_2, was 9.25 seconds. The settling time of the surface elevation is referred to as the time required for the response to settle within 2 mm. Therefore, the settling time of the surface elevation was 143.49 seconds.

5.3.3.2 Effectiveness of the Control System

When the cable length, l_S, was fixed at 5 m, the first-mode frequency and damping ratio were 1.383 rad/s and 0.0995, respectively. Meanwhile, the second-mode frequency and damping ratio were 7.387 rad/s and 0.002, respectively. Therefore, the FP smoother was implemented by using the first-mode frequency of 1.383 rad/s and damping ratio of 0.0995.

Figures 5.90–5.95 show the simulated responses to a flight distance of 60 m. The operator first gave an acceleration to drive the aerial crane flying along N_x direction. The speed of the aerial crane would increase from zero to the nominal speed of 3 m/s. At the end of flight, the operator gave a deceleration to decrease the quadrotor speed to zero, and the aerial vehicle hovered at the desired position. The operated commands, θ_r, shown in Figure 5.90 filtered through the FP smoother to create smoothed commands, θ_s, in Figure 5.90 which were the input to the MF controller for stabilizing the quadcopter attitude, θ. The maximum peak-to-peak deflection of the quadrotor attitude, swing angle, and surface elevation during the flight of aerial crane is defined as transient deflection, while the maximum peak-to-peak deflection during the hover of aerial crane is denoted as residual amplitude.

Figures 5.91 and 5.92 show the responses of the quadrotor attitude, θ, and the responses of the swing angle, β_2, and surface elevation, η, are shown in Figure 5.93. While the equilibria of the quadrotor attitude and surface elevation were both zero, the response of the swing angle, β_2, stabilized at the equilibrium

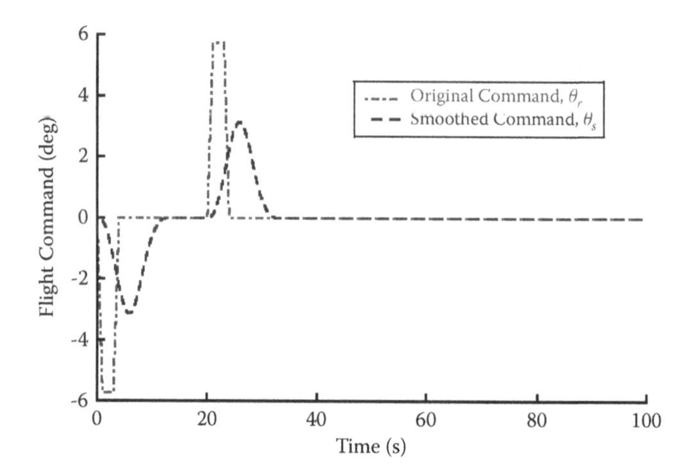

FIGURE 5.90 Time response of flight command to a flight distance of 60 m.

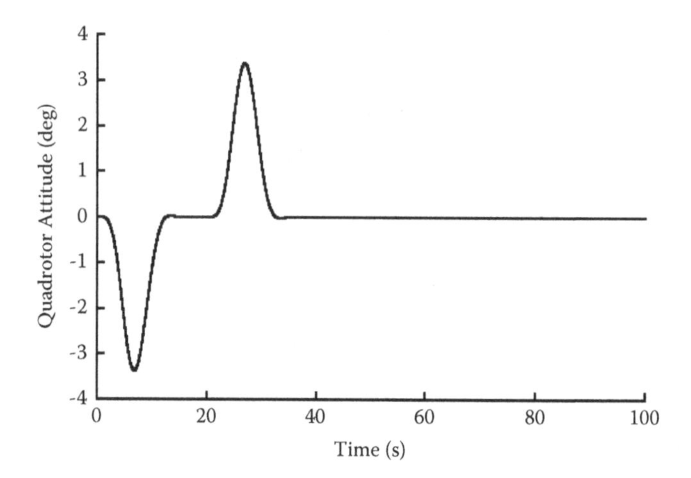

FIGURE 5.91 Time response of quadrotor attitude to a flight distance of 60 m.

point of −8.048°. The oscillations of quadrotor attitude, cable swing, and liquid sloshing in Figures 5.91–5.93 disappeared. This is because the notch filtering effect inherent in the FP smoother completely suppressed the first-mode oscillations. Meanwhile, the low-passing effect of the FP smoother reduced significantly the amplitudes of second- and third-mode oscillations of quadrotor attitude, cable swing, and liquid sloshing.

When the combined FP smoother and MF controller scheme was utilized, the transient deflection and residual amplitude of the quadrotor attitude in Figure 5.91 were 6.735° and 0.025°, respectively. Meanwhile, the transient deflection and residual amplitude of the swing angle, β_2, in Figure 5.92 were 0.408° and 0.016°, respectively. Additionally, the transient deflection and residual amplitude of the

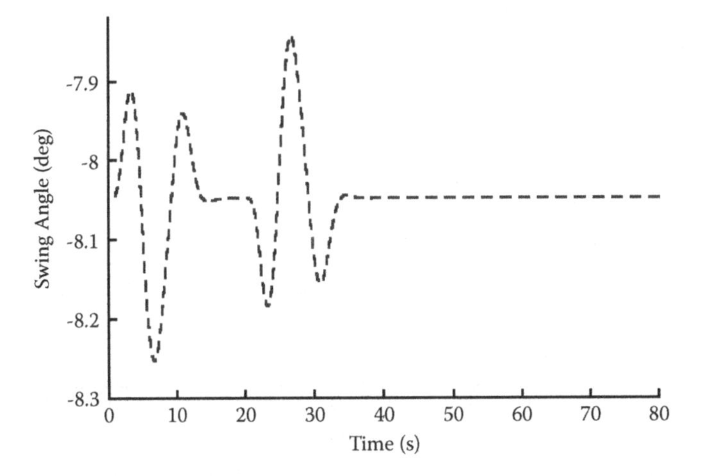

FIGURE 5.92 Time response of swing angle to a flight distance of 60 m.

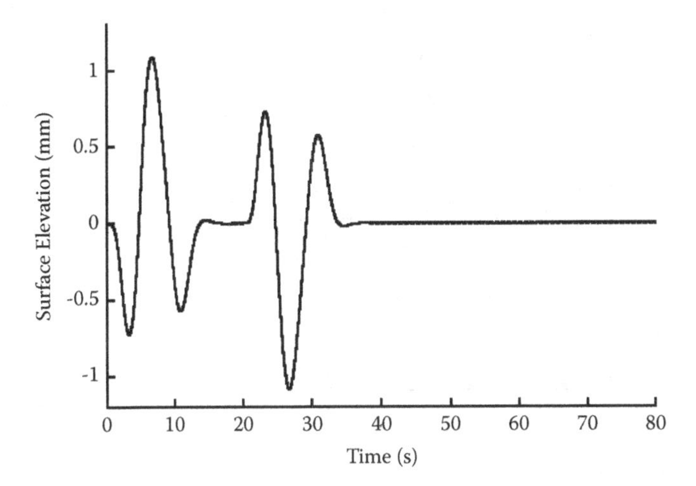

FIGURE 5.93 Time response of surface elevation to a flight distance of 60 m.

surface elevation, η, in Figure 5.93 were 2.171 mm and 0.085 mm, respectively. Therefore, the combined controller suppressed oscillations of the vehicle attitude, cable pendulum and liquid sloshing to a very low level.

Figures 5.94 and 5.95 show the responses of the pitching moment, M_θ, and fluid force, F_{fx}, in the L_x direction. The oscillations of the pitching moment, M_θ, and fluid force, F_{fx}, is similar to the oscillations of the quadrotor attitude, swing angle, and surface elevation in Figures 5.91–5.93. These results can be attributed to the coupling effect among the quadrotor, suspended container, and liquid sloshing. During the residual stage, the combined control method requires only a small pitching moment for controlling the quadrotor slung liquid tank, and produces a small fluid force for acting on the container sidewall.

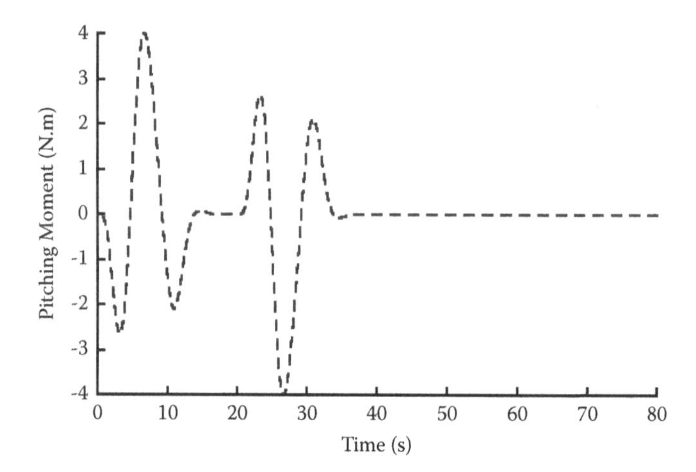

FIGURE 5.94 Time response of pitching moment to a flight distance of 60 m.

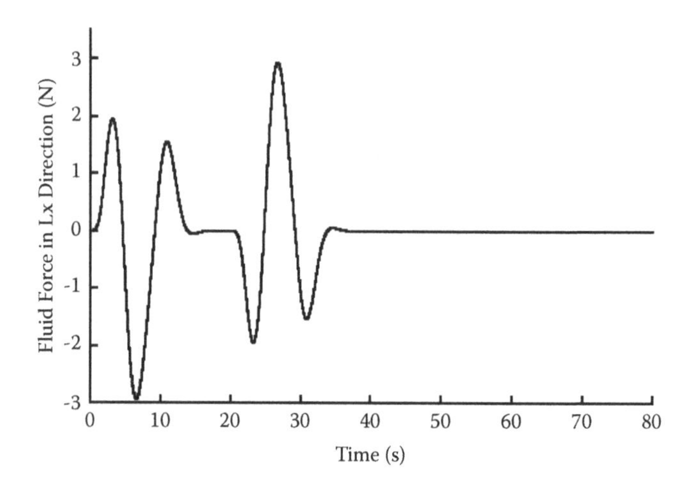

FIGURE 5.95 Time response of fluid force to a flight distance of 60 m.

5.3.3.3 Robustness of the Control System

Theoretical analyses indicate that the cable length has a large influence on the system dynamics. However, cable length will certainly vary over time in a real condition. Therefore, a robust control system is required in the presence of variation in the cable length. The cable length, l_S, varied from 3.5 m to 6.5 m, while both the FP smoother and MF controller were designed for the constant length of 5 m. Figure 5.96 shows the transient deflection and residual amplitude of quadrotor attitude, θ, and swing angle, β_2, from those tests, while Figure 5.97 shows the transient deflection and residual amplitude of surface elevation, η.

The transient deflection of quadrotor attitude, θ, and swing angle, β_2, and surface elevation, η, increased with increasing cable length. The residual amplitudes of

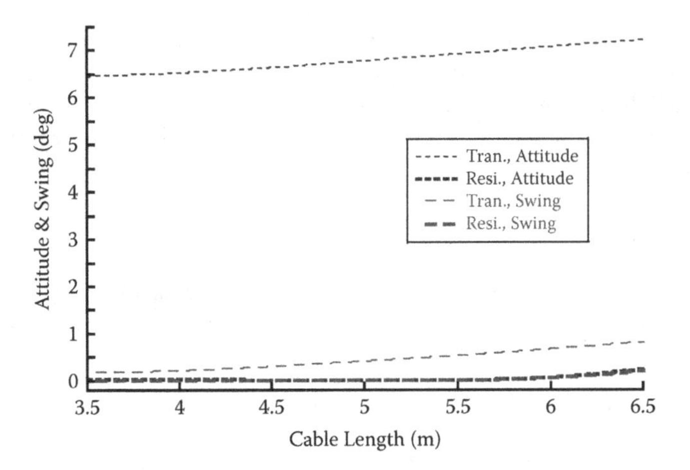

FIGURE 5.96 Transient deflection and residual amplitude of quadrotor attitude and swing angle for various cable length.

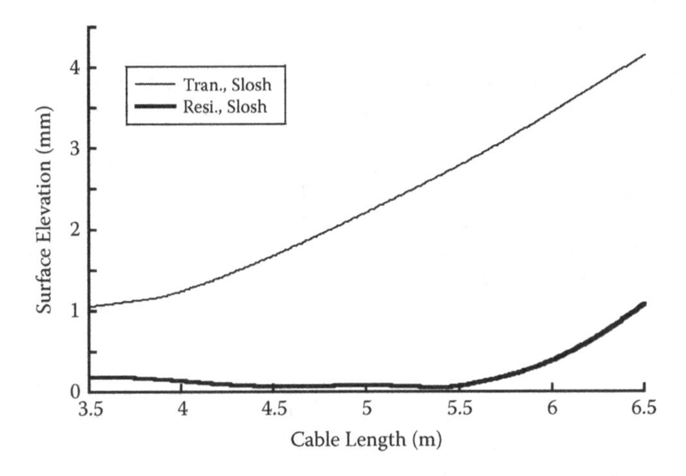

FIGURE 5.97 Transient deflection and residual amplitude of surface elevation for various cable length.

quadrotor attitude, θ, and swing angle, β_2, are independent of the cable length. A minimum in the residual amplitude of surface elevation, η, arose near the cable length of 5 m. This is because the FP smoother and MF controller are both designed for the cable length of 5 m. As the cable length increased from cable length of 5 m, the residual amplitude of surface elevation, η, increased rapidly. When the cable length decreased from cable length of 5 m, the residual amplitude of surface elevation, η, increased slowly. This is because the short length of the cable is corresponding to the high frequency, while the long length of the cable corresponds to the low frequency. The low-passing filtering effect inherent in the combined FP smoother and MF controller technique can reduce more oscillations at the high frequency. However, the combined controller suppressed oscillations of quadrotor attitude, load swing, and liquid sloshing to low level for a large error in the cable length. Therefore, simulated results in Figures 5.96 and 5.97 indicate that the combined controller exhibits a good robustness to changes in the system parameters and working conditions.

REFERENCES

[1] S. Asseo, R. Whitbec, Control requirements for sling-load stabilization in heavy lift helicopters, *Journal of the American Helicopter Society*, 18 (3) (1973) 23–31.

[2] T. Dukes, Maneuvering heavy sling loads near hover part I: damping the pendulous mode, *Journal of the American Helicopter Society*, 18 (2) (1973) 2–11.

[3] R. Raz, A. Rosen, T. Ronen, Active aerodynamic stabilization of a helicopter/sling-load system, *Journal of Aircraft*, 26 (1989) 822–828.

[4] N. Gupta, A. Bryson, Near-hover control of a helicopter with a hanging load, *Journal of Aircraft*, 13 (1973) 217–222.

[5] A. Hutto, Flight-test report on the heavy-lift helicopter flight-control system, *Journal of the American Helicopter Society*, 21 (1) (1976) 32–40.

[6] S. Oh, J. Ryu, S. Agrawal, Dynamics and control of a helicopter carrying a payload using a cable-suspended robot, *Journal of Mechanical Design*, 128 (5) (2006) 1113–1121.

[7] H. Omar, Designing anti-swing fuzzy controller for helicopter slung-load system near hover by particle swarms, *Aerospace Science and Technology*, 29 (1) (2013) 223–234.

[8] T. Oktay, C. Sultan, Modeling and control of a helicopter slung-load system, *Aerospace Science and Technology*, 29 (1) (2013) 206–222.

[9] J. Trachte, F. Gonzalez, A. Mcfadyen, Nonlinear model predictive control for a multi-rotor with heavy slung load, *International Conference on Unmanned Aircraft Systems*, Orlando, USA, 2014, pp. 1105–1110.

[10] S. El Ferik, G. Ahmed, H. Omar, Load swing control for an unmanned aerial vehicle with a slung load, *11th International Multi-Conference on Systems, Signals & Devices*, Barcelona, Spain, 2014, pp. 1–9.

[11] C. Ivler, J. Powell, M. Tischler, Design and flight test of a cable angle feedback flight control system for the RASCAL JUH-60 helicopter, *Journal of the American Helicopter Society*, 59 (4) (2014) 1–15.

[12] J. Krishnamurthi, J. F. Horn, Helicopter slung load control using lagged cable angle feedback, *Journal of the American Helicopter Society*, 60 (2) (2015) 1–12.

[13] C. Ivler, Constrained state-space coupling numerator solution and helicopter external load control design application, *Journal of Guidance, Control, and Dynamics*, 38 (10) (2015) 2004–2010.

[14] M. Bisgaard, A. La Cour-Harbo, J. Bendtsen, Adaptive control system for autonomous helicopter slung load operations, *Control Engineering Practice*, 18 (7) (2010) 800–811.

[15] C. Ivler, J. Powell, M. Tischler, J. Fletcher, M. Ott, Design and flight test of a cable angle/rate feedback flight control system for the RASCAL JUH-60 helicopter, *68th American Helicopter Society International Annual Forum*, Fort Worth, USA, 2012, pp. 1405–1429.

[16] C. Adams, J. Potter, W. Singhose, Input-shaping and model-following control of a helicopter carrying a suspended load, *Journal of Guidance, Control, and Dynamics*, 38 (1) (2015) 94–105.

[17] W. Hall, A. Bryson, Inclusion of rotor dynamics in controller design for helicopters, *Journal of Aircraft*, 10 (4) (1973) 200–206.
[18] M. Bernard, J. Bendtsen, A. La Cour-Harbo, Modeling of generic slung load system, *Journal of Guidance, Control, and Dynamics*, 32 (2) (2009) 573–585.
[19] M. Bernard, K. Kondak, G. Hommel, Load transportation system based on autonomous small size helicopters, *Aeronautical Journal*, 114 (2010) 191–198.
[20] K. Enciu, A. Rosen, Nonlinear dynamical characteristics of fin-stabilized underslung loads, *AIAA Journal*, 53 (3) (2015) 723–738.
[21] G. Guglieri, P. Marguerettaz, Dynamic stability of a helicopter with an external suspended load, *Journal of the American Helicopter Society*, 59 (4) (2014) 1–12.
[22] Y. Cao, Z. Wang, Equilibrium characteristics and stability analysis of helicopter slung-load system, *Proceedings of the Institution of Mechanical Engineers, Part G: Journal of Aerospace Engineering*, 231 (6) (2016) 1056–1064.
[23] J.J. Potter, C.J. Adams, W. Singhose, A planar experimental remote-controlled helicopter with a suspended load, *IEEE/ASME Transactions on Mechatronics*, 20 (5) (2015) 2496–2503.
[24] Aeronautical design standard 33E: handling qualities requirements for military rotorcraft, U.S. Army Aviation, and Missile Command, Aviation Engineering Directorate, ADS-33E-PRF, Redstone Arsenal, AL, 2000.
[25] H. Glida, L. Abdou, A. Chelihi, et al., Optimal model-free backstepping control for a quadrotor helicopter, *Nonlinear Dynamics*, 100 (4) (2020) 3449–3468.
[26] A. Eskandarpour, I. Sharf, A constrained error-based MPC for path following of quadrotor with stability analysis, *Nonlinear Dynamics*, 99 (2) (2020) 899–918.
[27] S. Harshavarthini, R. Sakthivel, C. Ahn, Finite-time reliable attitude tracking control design for nonlinear quadrotor model with actuator faults, *Nonlinear Dynamics*, 96 (4) (2019) 2681–2692.
[28] U. Ansari, A. Bajodah, B. Kada, Development and experimental investigation of a Quadrotor's robust generalized dynamic inversion control system, *Nonlinear Dynamics*, 96 (2) (2019) 1541–1557.
[29] Q. Xu, Z. Wang, Z. Zhen, Adaptive neural network finite time control for quadrotor UAV with unknown input saturation, *Nonlinear Dynamics*, 98 (3) (2019) 1973–1998.
[30] S. Tang, V. Wüest, V. Kumar, Aggressive flight with suspended payloads using vision-based control, *IEEE Robotics and Automation Letters*, 3 (2) (2018) 1152–1159.
[31] M. Vahdanipour, M. Khodabandeh, Adaptive fractional order sliding mode control for a quadcopter with a varying load, *Aerospace Science and Technology*, 86 (2019) 737–747.
[32] W. Dong, Y. Ding, L. Yang, et al., An efficient approach for stability analysis and parameter tuning in delayed feedback control of a flying robot carrying a suspended load, *Journal of Dynamic Systems, Measurement, and Control*, 141 (8) (2019) 081015.
[33] X. Liang, Y. Fang, N. Sun, et al., A novel energy-coupling-based hierarchical control approach for unmanned quadcopter transportation systems, *IEEE/ASME Transactions on Mechatronics*, 24 (1) (2019) 248–259.
[34] E. De Angelis, F. Giulietti, G. Pipeleers, Two-time-scale control of a multirotor aircraft for suspended load transportation, *Aerospace Science and Technology*, 84 (2019) 193–203.
[35] S. Yang, B. Xian, Energy-based nonlinear adaptive control design for the Quadcopter UAV system with a suspended payload, *IEEE Transactions on Industrial Electronics*, 67 (3) (2019) 2054–2064.
[36] A.R. Godbole, K. Subbarao, Nonlinear control of unmanned aerial vehicles with cable suspended payloads, *Aerospace Science and Technology*, 93 (2019) 105299.

[37] G. Yu, D. Cabecinhas, R. Cunha, et al., Nonlinear backstepping control of a quadrotor-slung load system, *IEEE/ASME Transactions on Mechatronics*, 24 (5) (2019) 2304–2315.

[38] X. Bin, S. Wang, S. Yang, Nonlinear adaptive control for an unmanned aerial payload transportation system: theory and experimental validation, *Nonlinear Dynamics*, 98 (3) (2019) 1745–1760.

[39] P. Cruz, R. Fierro, Cable-suspended load lifting by a quadcopter UAV: hybrid model, trajectory generation, and control, *Autonomous Robots*, 41 (8) (2017) 1629–1643.

[40] H. Sayyaadi, A. Soltani, Modeling and control for cooperative transport of a slung fluid container using quadcopter, *Chinese Journal of Aeronautics*, 31 (2) (2018) 262–272.

[41] B. Barikbin, A. Fakharian, Trajectory tracking for quadcopter UAV transporting cable-suspended payload in wind presence, *Transactions of the Institute of Measurement and Control*, 41 (5) (2019) 1243–1255.

[42] A. Mohammadi, E. Abbasi, M. Ghayour, et al., Formation control and path tracking for a group of quadcopter to carry out a suspended load, *Modares Mechanical Engineering*, 19 (4) (2019) 887–899.

[43] D. Cabecinhas, R. Cunha, C. Silvestre, A trajectory tracking control law for a Quadcopter with slung load, *Automatica*, 106 (2019) 384–389.

[44] M. Guo, D. Gu, W. Zha, et al., Controlling a quadcopter carrying a cable-suspended load to pass through a window, *Journal of Intelligent & Robotic Systems*, 98 (2020) 387–401.

[45] L. Qian, H. Liu, Path following control of a quadcopter UAV with a cable suspended payload under wind disturbances, *IEEE Transactions on Industrial Electronics*, 67(3) (2019) 2021–2029.

[46] D. Cabecinhas, R. Cunha, C. Silvestre, A trajectory tracking control law for a quadrotor with slung load, *Automatica*, 106 (2019) 384–389.

[47] L. Qian, H. Liu, Path following control of a quadrotor UAV with a cable suspended payload under wind disturbances, *IEEE Transactions on Industrial Electronics*, 67 (3) (2019) 2021–2029.

[48] D. Hashemi, H. Heidari, Trajectory planning of quadrotor UAV with maximum payload and minimum oscillation of suspended load using optimal control, *Journal of Intelligent & Robotic Systems*, 100 (2020) 1369–1381.

[49] O. Ogunbodede, R. Yoshinaga, T. Singh, Vibration control of unmanned aerial vehicle with suspended load using the concept of differential flatness, *American Control Conference*, Philadelphia, PA, USA, 2019, pp. 4268–4273.

[50] P. Homolka, M. Hromčík, T. Vyhlídal, Input shaping solutions for drones with suspended load: first results, *International Conference on Process Control*, Strbske Pleso, Slovakia, 2017, pp. 30–35.

[51] I. Palunko, P. Cruz, R. Fierro, Agile load transportation: safe and efficient load manipulation with aerial robots, *IEEE Robotics & Automation Magazine*, 19(3) (2012) 69–79.

[52] S. Sadr, A. Moosavian, P. Zarafshan, Dynamics modeling and control of a quadcopter with swing load, *Journal of Robotics*, 2014 (2014) 265897.

[53] P. Cruz, M. Oishi, R. Fierro, Lift of a cable-suspended load by a quadcopter: a hybrid system approach, *American Control Conference*, Chicago, IL, USA, 2015, pp. 1887–1892.

[54] N. Johnson, W. Singhose, Dynamics and modeling of a quadcopter with a suspended payload. *Applied Aerodynamics Conference*, Atlanta, Georgia, USA, 2018, p. 4213.

[55] E. de Angelis, Swing angle estimation for multicopter slung load applications, *Aerospace Science and Technology*, 89 (2019) 264–274.

[56] N. Gupta, A. Bryson, Near-hover control of a helicopter with a hanging load, *Journal of Aircraft*, 13 (3) (1976) 217–222.
[57] K. Enciu, A. Rosen, Nonlinear dynamical characteristics of fin-stabilized underslung loads, *AIAA Journal*, 53 (3), (2015) 723–738.
[58] Y. Cao, Z. Wang, Equilibrium characteristics and stability analysis of helicopter slung-load system, *Proceedings of the Institution of Mechanical Engineers, Part G: Journal of Aerospace Engineering*, 231 (6) (2016) 1056–1064.
[59] B.S. Rego, G.V. Raffo, Suspended load path tracking control using a tilt-rotor UAV based on zonotopic state estimation, *Journal of the Franklin Institute*, 356 (4) (2019) 1695–1729.
[60] J. Erskine, A. Chriette, S. Caro, Wrench analysis of cable-suspended parallel robots actuated by quadcopter unmanned aerial vehicles, *Journal of Mechanisms and Robotics*, 11 (2) (2019) 020909.
[61] B. Shirani, N. Majdeddin, I. Izadi, Cooperative load transportation using multiple UAVs, *Aerospace Science and Technology*, 84 (2019) 158–169.

6 Dual Cranes

Multiple cranes carrying a bulky payload are commonly applied in many engineering applications. Suspension cables attach a large payload to multiple cranes. Undouble, oscillations of payload have a large influence on the operating speed and safety of cranes. Many scientists have been focused on a single crane transporting a payload. However, less attentions have been devoted to multiple cranes moving a large payload.

6.1 PLANAR MOTIONS

Prior literature [1–9] has discussed the dynamic modeling of multiple cranes carrying large payloads. The dynamics of this system is more complicated than typically found on the single crane having distributed-mass payload dynamics and double-pendulum dynamics [10–22]. The estimation and analysis of natural frequencies in multiple cranes moving distributed-mass payloads are very challenging. Averaging the two frequencies of each cable independently was used to forecast the system frequency in the literature [4–6]. However, the average frequency is only correct when the length of two cables are same. Additionally, the literature [9] estimated the equilibrium point of the dual cranes by assuming the same length of the two cables. However, the cable lengths are always different under real working conditions.

Mounting numbers of examples have given the solutions to the oscillation suppression of multiple cranes moving large-size payloads. The developed control methods have concentrated on the open-loop control, which modifies the operated commands to drive the multiple cranes undergoing low-swing motions. Leban et al. presented an inverse kinematic controller to limit the payload swing of dual cranes. Experiments on a small-scale dual crane demonstrated the effectiveness of the inverse kinematic controller [1]. Perig et al. proposed an optimal control method to move two cranes with minimum swing of the payload. They demonstrated the effectiveness by simulating crossbeam transportation [2]. Unfortunately, the computational load on-the-fly is a barrier in the application of the inverse kinematic and optimal control methods discussed in [3].

Literature [3–8] demonstrated the effectiveness of the input shaping technique for dual cranes. In practice, the input shaper depends on the accurate estimates of the natural frequency, at which little attention has been directed. Therefore, the frequency estimation is an obstacle to the input shaping techniques. Lu et al. presented a feedback controller for dual cranes [9]. However, the research findings are only effective in the case of same suspension cable lengths.

DOI: 10.1201/9781003247210-6

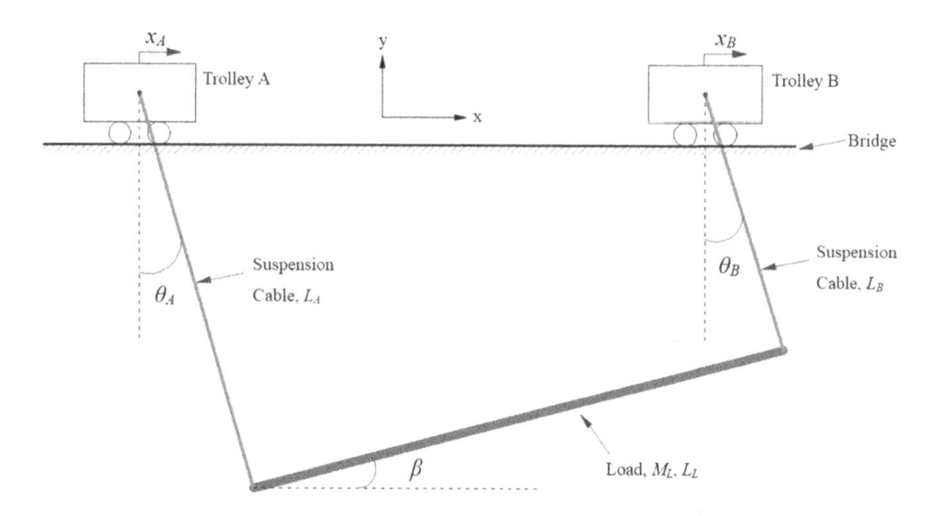

FIGURE 6.1 Model of dual planar cranes moving a distributed-mass beam.

6.1.1 Modeling

Figure 6.1 shows a schematic representation of two cranes transporting a distributed-mass beam undergoing planar motions. Two trolleys of mass, M_A and M_B, slide independently along a bridge in the x direction. Thus, the displacement of the two trolleys along the x direction in the Inertial Newtonian frame are x_A and x_B. Two massless suspension cables of length, L_A and L_B, which are assumed to be inelastic, hang below the two trolleys independently and support a uniformly distributed-mass beam of mass, M_L, and length, L_L. The swing angles of the suspension cables are θ_A and θ_B, respectively. The angle between beam direction and the x direction is defined as the pitch angle, β, through which the beam rotates about the vertical direction.

The inputs to the model in Figure 6.1 are the acceleration, $\ddot{x}_A$ and $\ddot{x}_B$, of the two trolleys. The outputs are the swing angles, θ_A and θ_B, and the pitch angle, β, of the payload. The initial direction of the beam length coincides with the oxy plane such that the motion of the dual cranes in Figure 6.1 cannot cause beam twisting about the suspension cables. This simplifies the system to a planar dynamic model. The kinetic energy of the dual planar cranes in Figure 6.1 is:

$$T = \frac{M_L}{2}\begin{bmatrix} \dot{x}_A^2 + L_A^2\dot{\theta}_A^2 + \frac{1}{3}L_L^2\dot{\beta}^2 + 2L_A\dot{x}_A\dot{\theta}_A\cos\theta_A \\ - L_L\dot{x}_A\dot{\beta}\sin\beta + L_AL_L\dot{\theta}_A\dot{\beta}\sin(\theta_A - \beta) \end{bmatrix} + \frac{M_A\dot{x}_A^2}{2} + \frac{M_B\dot{x}_B^2}{2} \tag{6.1}$$

While the trolley centroid is assumed as zero potential energy surface, the potential energy of the dual planar cranes is:

$$V = \frac{1}{2}gM_LL_L\sin\beta - gM_LL_A\cos\theta_A \tag{6.2}$$

Then by using the generalized Lagrange method, the nonlinear equations of motion for the model are derived as:

$$
\begin{aligned}
&2L_A\begin{bmatrix}3\cos^2(\beta-\theta_B)+\sin^2(\theta_A-\theta_B)\\+3\sin(\beta-\theta_A)\sin(\theta_A-\theta_B)\cos(\beta-\theta_B)\end{bmatrix}\ddot{\theta}_A\\
&-L_A\begin{bmatrix}3\sin(2\theta_A-\theta_B-\beta)\cos(\beta-\theta_B)\\-2\sin(\theta_A-\theta_B)\cos(\theta_A-\theta_B)\end{bmatrix}\dot{\theta}_A^2\\
&-L_B[3\sin(\beta-\theta_A)\cos(\beta-\theta_B)+2\sin(\theta_A-\theta_B)]\dot{\theta}_B^2\\
&-L_L\begin{bmatrix}3\cos(\beta-\theta_A)\cos^2(\beta-\theta_B)\\+3\sin(\beta-\theta_A)\sin(\beta-\theta_B)\cos(\beta-\theta_B)\\+2\sin(\theta_A-\theta_B)\sin(\beta-\theta_B)\end{bmatrix}\dot{\beta}^2\\
&-\begin{bmatrix}3\sin\theta_B\sin(\beta-\theta_A)\cos(\beta-\theta_B)\\+2\sin(\theta_A-\theta_B)\sin\theta_B\end{bmatrix}(\ddot{x}_A-\ddot{x}_B)\\
&+3\begin{bmatrix}\sin\beta\sin(\theta_A-\theta_B)\cos(\beta-\theta_B)\\+2\cos\theta_A\cos^2(\beta-\theta_B)\end{bmatrix}\ddot{x}_A+3\begin{bmatrix}2\sin\theta_A\cos^2(\beta-\theta_B)\\-\cos\beta\sin(\theta_A-\theta_B)\cos(\beta-\theta_B)\end{bmatrix}g=0
\end{aligned}
\tag{6.3}
$$

$$
\begin{aligned}
&L_A\cos(\beta-\theta_A)\ddot{\theta}_A+L_A\sin(\beta-\theta_A)\dot{\theta}_A^2-L_B\cos(\beta-\theta_B)\ddot{\theta}_B\\
&-L_B\sin(\beta-\theta_B)\dot{\theta}_B^2-L_L\dot{\beta}^2+\cos\beta(\ddot{x}_A-\ddot{x}_B)=0
\end{aligned}
\tag{6.4}
$$

$$
\begin{aligned}
&L_A\sin(\theta_A-\theta_B)\ddot{\theta}_A+L_A\cos(\theta_A-\theta_B)\dot{\theta}_A^2-L_B\dot{\theta}_B^2+L_L\cos(\beta-\theta_B)\ddot{\beta}\\
&-L_L\sin(\beta-\theta_B)\dot{\beta}^2-\sin\theta_B(\ddot{x}_A-\ddot{x}_B)=0
\end{aligned}
\tag{6.5}
$$

where g is the gravitational constant. Resulting from the dynamic model (6.3)–(6.5), the constraints for calculating equilibrium angles are given by:

$$
\begin{cases}2\cdot\sin\theta_{A0}\cdot\cos(\beta_0-\theta_{B0})-\cos\beta_0\cdot\sin(\theta_{A0}-\theta_{B0})=0\\L_A\cdot\sin\theta_{A0}-L_B\cdot\sin\theta_{B0}+L_L\cdot\cos\beta_0+x_A-x_B=0\\L_A\cdot\cos\theta_{A0}-L_B\cdot\cos\theta_{B0}-L_L\cdot\sin\beta_0=0\end{cases}
\tag{6.6}
$$

where θ_{A0}, θ_{B0}, and β_0 are equilibrium angle of the swing angle, θ_A, swing angle, θ_B, and pitch angle, β, respectively. A linearized model can be derived from the dynamic model (6.3)–(6.5) by assuming small oscillations around the equilibrium angles (6.6). Then resulting from the linearized model, the equation of natural frequency of the dual planar cranes are:

$$
R_1\cdot\omega^6+R_2\cdot\omega^4+R_3\cdot\omega^2+R_4=0
\tag{6.7}
$$

where ω is the natural frequency of the dual planar cranes, and R_1, R_2, R_3, and R_4 are

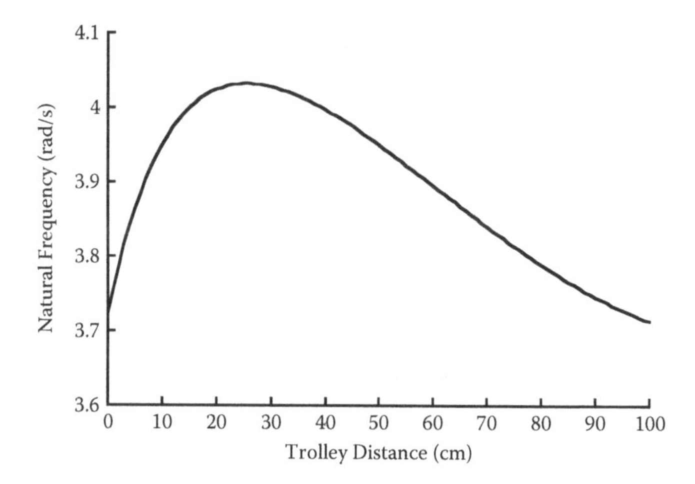

FIGURE 6.2 Natural frequency vs. trolley distance.

the coefficients. The closed-form of the natural frequency is challenging. Instead, solving Equation (6.7) yields the numerical solutions of the natural frequency. Equation (6.7) has three solutions including one frequency and the other two zero. The two zero solutions occur because of zero stiffness in both the subsystems (6.4) and (6.5).

The natural frequency resulting from Equation (6.7) depends on the trolley distance, suspension cable lengths, and payload length. Figure 6.2 shows the natural frequency as a function of trolley distance when the suspension length, L_A, suspension length, L_B, and beam length, L_L, were fixed at 82 cm, 64 cm, and 120 cm, respectively. The natural frequency increases as the trolley distance increases before 25 cm. After this point, the natural frequency decreases.

Figure 6.3 shows the natural frequency for various suspension length, L_A, when the trolley distance, suspension length, L_B, and beam length, L_L, were fixed at

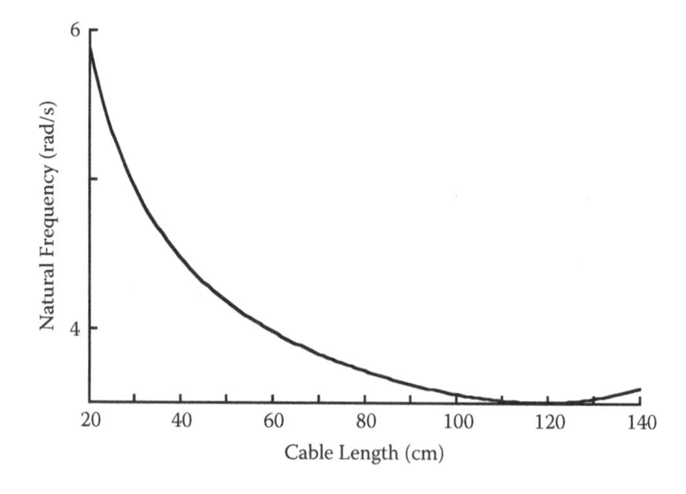

FIGURE 6.3 Natural frequency vs. suspension length.

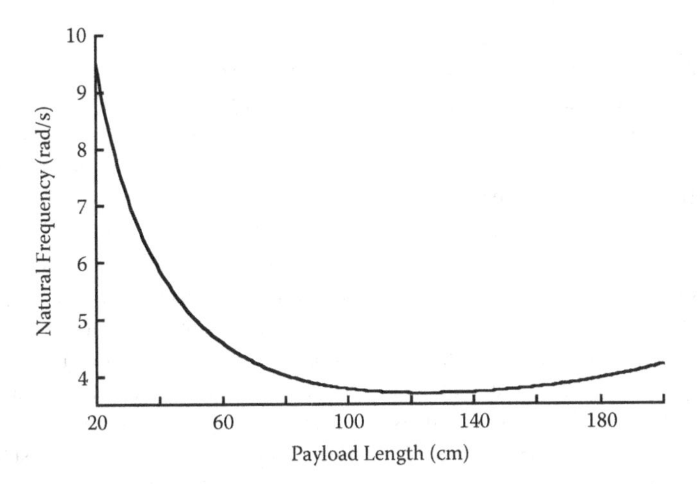

FIGURE 6.4 Natural frequency vs. payload length.

120 cm, 64 cm, and 120 cm, respectively. The natural frequency decreases sharply as the suspension length increases before 120 cm, and then increases slightly.

Figure 6.4 shows the natural frequency induced by varying beam length, L_L, when the trolley distance, suspension length, L_A, and suspension length, L_B, were fixed at 120 cm, 82 cm, and 64 cm, respectively. A local minimum occurs at the payload length of 125 cm. Small changes in the payload length result in relatively large changes in the natural frequency. Figures 6.2–6.4 indicate that the natural frequency varies slightly as the trolley distance changes, and varies sharply as the payload length and suspension cable length change.

The command used to drive the model (6.3)–(6.5) is a trapezoidal velocity profile. The initial distance between the two trolleys, maximum driving velocity,

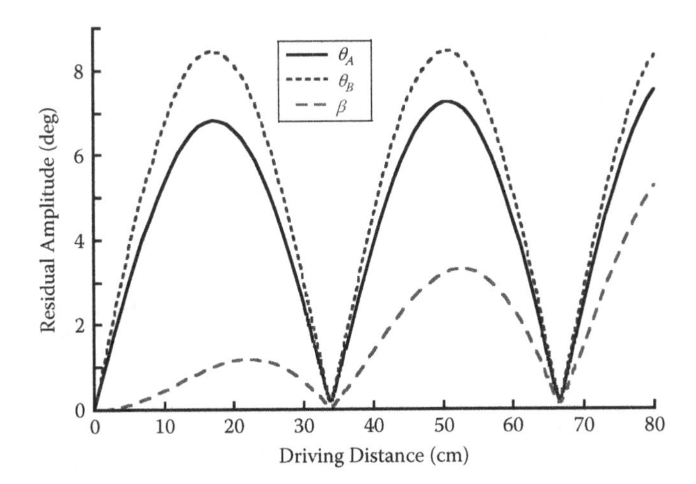

FIGURE 6.5 Simulated residual amplitudes of the payload swing and pitch against driving distance.

and maximum driving acceleration in this chapter were fixed at 120 cm, 20 cm/s, and 2 m/s^2, respectively. Figure 6.5 shows the peak-to-peak residual amplitudes induced by various driving distances of trolley A when the suspension length, L_A, suspension length, L_B, and beam length, L_L, were fixed at 82 cm, 64 cm, and 120 cm, respectively. The peaks in the swing angles and pitch angle are caused by the constructive interference between the oscillations resulting from the acceleration and deceleration of the trolley. Those two oscillations are sometimes in phase and sometimes out-of-phase, thereby causing peaks and troughs, respectively. The distance between the peaks and troughs is corresponding to the natural frequency of the system, which can be estimated by using the numerical solution of Equation (6.7). Moreover, as the driving distances increase, the magnitudes of the peak in the swing angles, θ_A and θ_B, and the pitch angle, β, increase. Hence, decreasing the payload swing results in the small oscillations of the payload pitch.

Figure 6.6 shows the simulated residual amplitudes for varying suspension length, L_A, when the driving distance, suspension length, L_B, and beam length, L_L, were fixed at 20 cm, 64 cm, and 120 cm, respectively. The residual amplitude in the swing angle, θ_A, increases as the suspension length, L_A, increases before 29 cm. After this point, it decreases as the suspension length increases. The residual amplitude in the swing angle, θ_B, increases with increasing suspension length, L_A, before 60 cm, and then decreases. A local maximum in the pitch angle, β, occurs at the suspension length of 50 cm, while a near-zero residual oscillation in the pitch angle arises at the suspension length of 130 cm. The near-zero oscillation in the payload pitch results from the zero equilibrium angles, θ_{A0} and θ_{B0}, of the swing in this specified case. The zero equilibrium angles, θ_{A0} and θ_{B0}, result in an uncoupled effect between the swing of the two suspension cables.

Figure 6.7 shows the simulated residual amplitudes as a function of payload length, L_L, when the driving distance, suspension length, L_A, and suspension length, L_B, were fixed at 20 cm, 82 cm, and 64 cm, respectively. A trough in the swing pitch

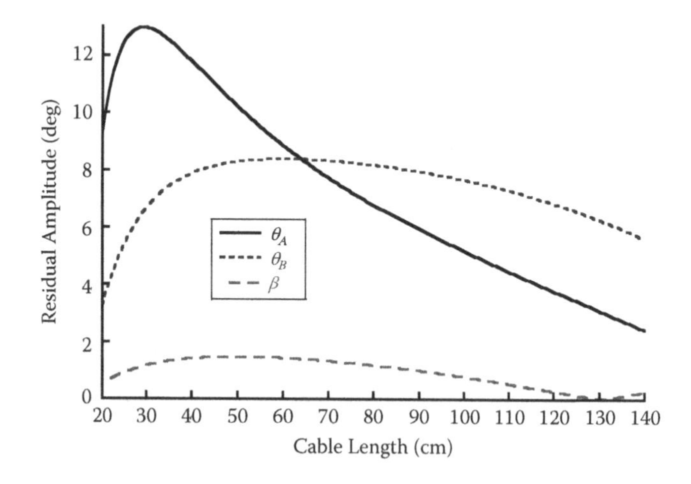

FIGURE 6.6 Simulated residual amplitudes of the payload swing and pitch against suspension length.

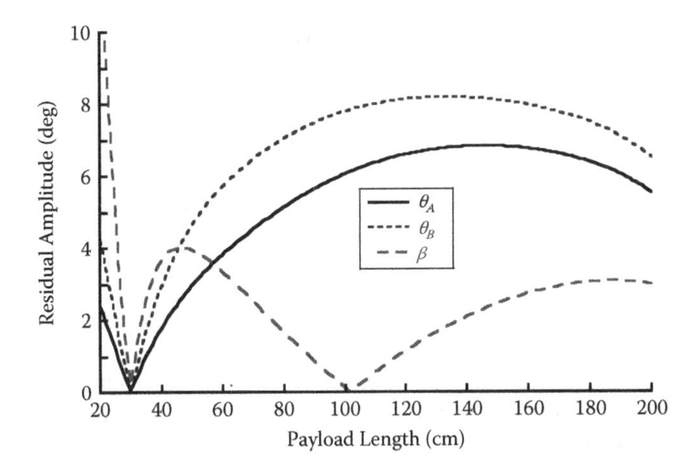

FIGURE 6.7 Simulated residual amplitudes of the payload swing and pitch against payload length.

occurs at the beam length of 30 cm because of zero swing in this case. Another zero oscillations in the payload pitch arise at the beam length of 101 cm. This is also because the equilibrium angles, θ_{A0} and θ_{B0}, of the swing are both limited to zero at this point. Therefore, the system dynamic behavior is the interference results of the equilibrium angles given in Equation (6.6) and natural frequency given in Equation (6.7).

6.1.2 Vibration Control

The original driving commands filter through the MEI shaper to create shaped commands, which move the dual planar cranes with minimum oscillations of the payload. The MEI shaper is given by:

$$\text{MEI} = \begin{bmatrix} A_k \\ \tau_k \end{bmatrix} = \begin{bmatrix} \frac{1}{[1+K]^2} & \frac{K}{[1+K]^2} & \frac{K}{[1+K]^2} & \frac{K^2}{[1+K]^2} \\ 0 & \frac{T_m}{2q} & \frac{T_m}{2p} & \left(\frac{T_m}{2q} + \frac{T_m}{2p}\right) \end{bmatrix} \tag{6.8}$$

where

$$T_m = 2\pi/(\omega\sqrt{1-\zeta^2}) \tag{6.9}$$

$$K = e^{(-\pi\zeta/\sqrt{1-\zeta^2})} \tag{6.10}$$

The MEI shaper is a four-impulse function. The duration of the MEI shaper is:

$$\tau_n = \frac{T_m}{2q} + \frac{T_m}{2p} \tag{6.11}$$

Note that the duration is approximately a damped vibration period. The residual amplitudes at the design frequency could be constrained below a tolerable level, V_{tol}:

$$e^{-\zeta\omega\tau_n}\cdot\sqrt{S^2(\omega,\zeta) + C^2(\omega,\zeta)} \leq V_{tol} \tag{6.12}$$

where τ_n is the duration of the MEI shaper. Forcing the derivative of the residual vibrations with respect to frequency to zero should increase the robustness to modeling errors in the frequency:

$$S(\omega,\zeta)\cdot\frac{\partial S(\omega,\zeta)}{\partial\omega} + C(\omega,\zeta)\cdot\frac{\partial C(\omega,\zeta)}{\partial\omega} - \zeta\tau_n\cdot[S^2(\omega,\zeta) + C^2(\omega,\zeta)] = 0 \tag{6.13}$$

Substituting Equation (6.8) into (6.12) and (6.13) yields the numerical solution of the modified coefficients p and q. While the tolerable level of the vibrations is set to 5%, the solution of the modified coefficients for the undamped system are p = 0.8584 and q = 1.1429.

6.1.3 Computational Dynamics

The numerical verification of the effectiveness and robustness of the MEI shaper for controlling the dual cranes carrying distributed-mass beams will be conducted in this section. The MEI shaper should estimate the natural frequency for vibration reduction by using the system parameters. System parameters include trolley distance, suspension cable length, and payload length. However, it is very challenging to know the system parameters accurately because they will change over time. Variations in those parameters have a large effect on the system dynamics. Thus, the robustness to changes in the system parameters for the MEI shaper is important.

The residual oscillations induced by various driving distances are shown in Figure 6.8. The suspension length, L_A, suspension length, L_B, and beam length, L_L, were fixed at 82 cm, 64 cm, and 120 cm, respectively. The modeled trolley distance for the MEI shaper was fixed at 120 cm. Peaks and troughs occur also because the interference between the oscillations caused by the acceleration and deceleration are in phase or out of phase. The MEI shaper suppresses the payload swing, θ_A, payload swing, θ_B, and payload pitch, β, by an average of 94.9%, 94.9%, and 95.0%, respectively. The simulations indicate that both the MEI shaper can limit the residual oscillations of the payload swing and pitch to the tolerable level of 5%, approximately.

Figure 6.9 shows the variations of simulated residual amplitudes with changing suspension length, L_A. The driving distance, suspension length, L_B, and beam length, L_L, were set to 20 cm, 64 cm, and 120 cm, respectively. The modeled suspension

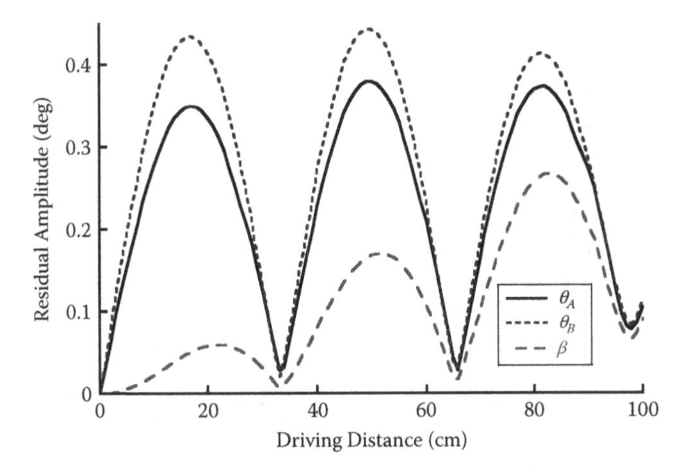

FIGURE 6.8 Simulated residual amplitudes against driving distance.

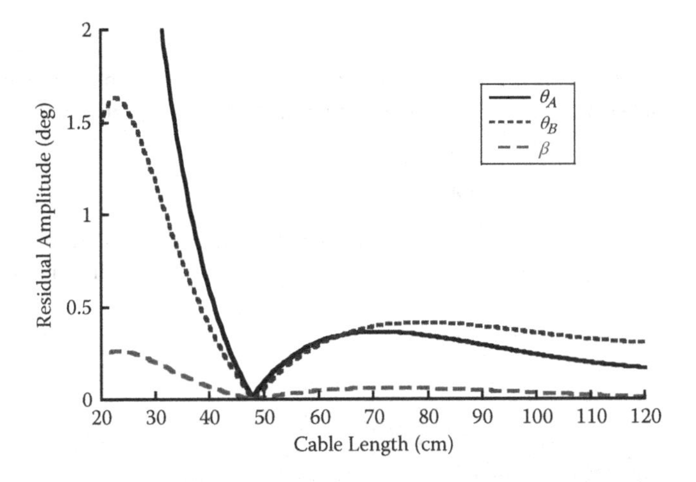

FIGURE 6.9 Simulated residual amplitudes against suspension length.

length, L_A, was fixed at 82 cm for the MEI shaper. Near-zero oscillations arise at the suspension length of 49 cm, which corresponds to the nearby frequency, $q{\cdot}\omega$. Zero-vibration constraints result in the near-zero oscillations at the suspension length of 49 cm. The payload swing and pitch increase sharply as the suspension length decreases from 49 cm because the sensitivity of the corresponding frequency changes drastically.

Figure 6.10 demonstrates clearly that some changes in the residual amplitude occur when the payload length varies. The simulated parameters of driving distance, suspension length, L_A, and suspension length, L_B, were set to 20 cm, 82 cm, and 64 cm, respectively. The modeled beam length, L_L, for the MEI shaper was fixed at 120 cm. Near-zero oscillations occur at the beam length of 62 cm and 192 cm, which correspond to the two nearby frequencies. The beam lengths between 62 cm and 192 cm are corresponding to the natural frequencies ranged from $p{\cdot}\omega$ to $q{\cdot}\omega$. Both the tolerable-level constraint and zero-derivative constraint limit the beam lengths between 62 cm

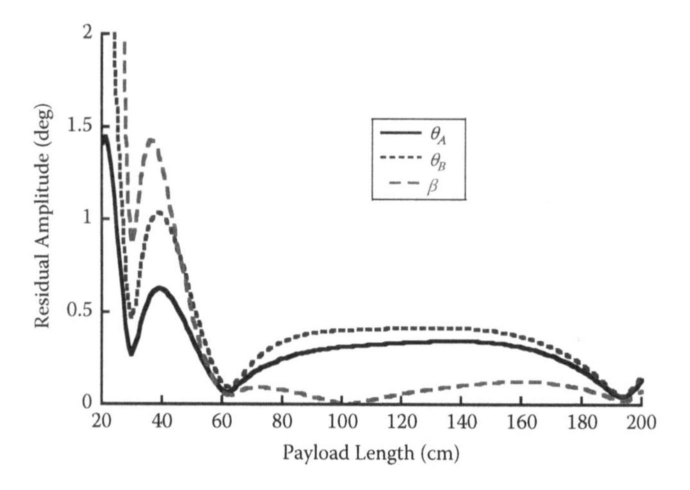

FIGURE 6.10 Simulated residual amplitudes against payload length.

and 192 cm below the tolerable level. As the beam length decreases from 62 cm or increases from 192 cm, the payload swing and pitch increase sharply. This is also because the sensitivity of the corresponding frequencies varies significantly.

6.1.4 Experimental Verification

Experiments are conducted on two small-scale bridge cranes carrying a slender beam shown in Figure 6.11. The Panasonic AC servomotors with encoders drive

FIGURE 6.11 Test bench of dual cranes carrying a beam.

two trolleys. The control hardware consists of two host computers for program development and user interface, and two DSP-based motion control cards connecting the computers to servo amplifiers. A MATLAB® script applies the control algorithm to filter through the operator commands (baseline trapezoidal-velocity commands), and generates the shaped commands for the drives. A slender beam was served as the distributed-mass payload, and was attached to the two trolleys by using two suspension cables. The wooden beam has a length of 120 cm and mass of 489 g. The initial distance between the two trolleys was 120 cm. The cable lengths were set to 82 cm and 64 cm. Two CMOS cameras were mounted to the two trolleys independently to record the horizontal swing displacements of the two black markers on the beam. The markers A and B were fixed nearby the connection points between the beam and the suspension cables.

The effect of varying driving distances is investigated in the first set of experiments. Figures 6.12 and 6.13 show the simulated and experimental amplitudes of residual amplitudes without the controller. Trolley B was still and trolley A was driven for a distance ranged from 3 cm to 54 cm. Peaks and troughs are found to arise in the uncontrolled residual amplitudes as the driving distance is changed. The distance between the peaks corresponds to the natural frequency, which can be forecasted by Equation (6.7). The uncontrolled experimental values match the corresponding simulated curves very well. The uncontrolled experimental results are lower than the simulated curves. This is because the model in Equations (6.3)–(6.5) is undamped, while the experimental system had some small amount of damping. The damping decreased the amplitudes of the experimental oscillations.

The simulated and experimental results of residual amplitudes with the MEI shaper are also shown in Figures 6.12 and 6.13 for comparison purpose. The modeled frequency for the MEI shaper was fixed at 3.6952 rad/s, which corresponds to the trolley distance of 120 cm. The MEI shaper suppressed the experimental deflection of marker A by an average of 94.2%. Meanwhile, the MEI shaper reduced the experimental deflection of marker B by an average of 94.6%. The controlled results show a

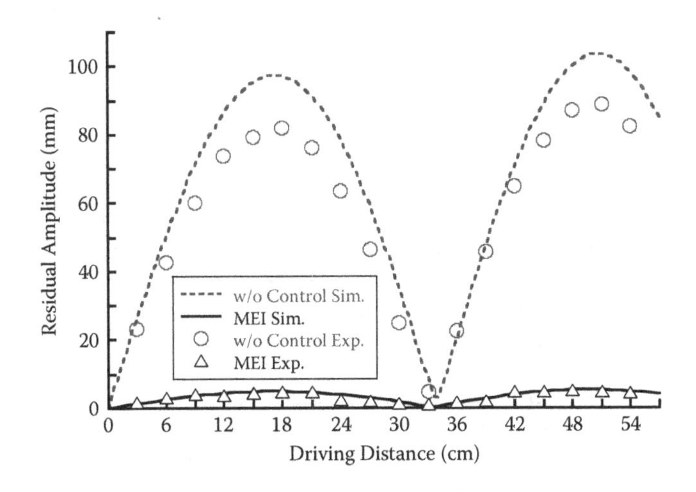

FIGURE 6.12 Experimental residual amplitudes in Marker A for various trolley distances.

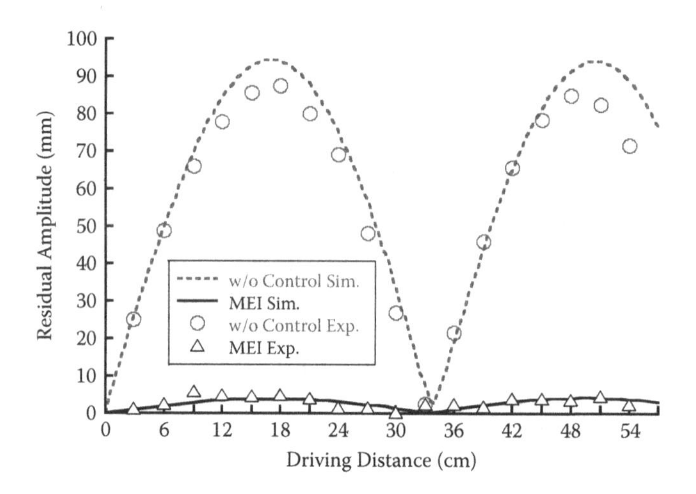

FIGURE 6.13 Experimental residual amplitudes in Marker B for various trolley distances.

significant suppression of the payload swing for all cases. The experimental values with the MEI shaper are slightly different from the simulated results because there certainly exists the modeling error in the design frequency and uncertainty in the overall system. However, the experimental results still follow the general shape as the simulated curve. The results in Figures 6.12 and 6.13 verify the simulated dynamical behavior and the effectiveness of the MEI shaper.

The second set of experiments is aimed at examining the effect of various modeling errors in the design frequency. Trolley A was driven to a distance of 24 cm from the initial position of 0 cm such that the distance between two trolleys was 96 cm after the experiment. Figure 6.14 shows the experimental response to a small modeling error in the design frequency. During the experiment, the design frequency was held constant at 3.6952 rad/s for the MEI shaper. The uncontrolled

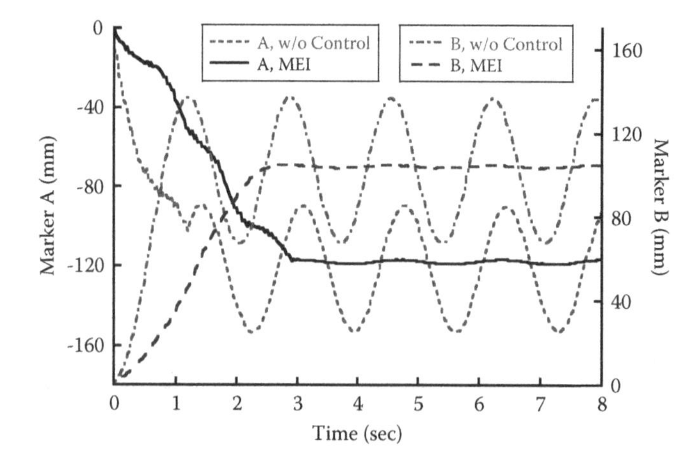

FIGURE 6.14 Experimental response to a small error in the frequency.

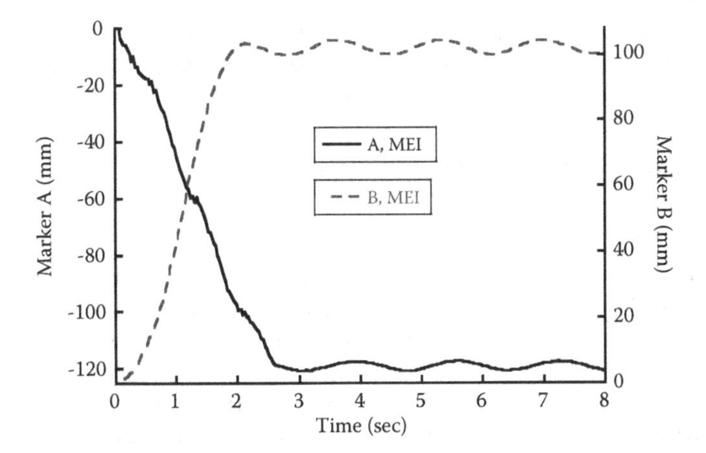

FIGURE 6.15 Experimental response to a positive error in the frequency.

residual oscillations in the deflection of markers A and B are found to have amplitudes of 63.5 mm and 69.2 mm, respectively. The residual amplitudes in the deflection of markers A and B with the MEI shaper are reduced to 2.3 mm and 1.5 mm, respectively. The MEI shaper has dramatically suppressed the payload swing in the case of small modeling error.

Figure 6.15 shows the experimental response for the modeled frequency of 4.8038 rad/s (corresponding to +30% error). With the MEI shaper, the residual amplitude of markers A and B are 3.2 mm and 4.7 mm, respectively. Figure 6.16 shows the experimental response when the modeled frequency is fixed at 2.5866 rad/s (corresponding to −30% error). The residual amplitude of markers A and B with the MEI shaper are 23.7 mm and 23.4 mm, respectively. The experimental findings demonstrate that the MEI shaper is effective in suppressing the oscillations of the dual cranes, and provide good robustness to modeling errors in the frequency.

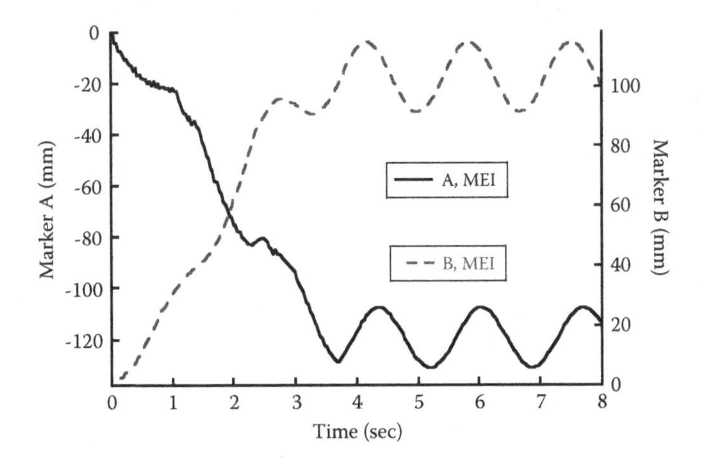

FIGURE 6.16 Experimental response to a negative error in the frequency.

Furthermore, the jerk-limitation feature inherent in the motion caused by the MEI-shaped commands is also of great importance. The jerk limitation certainly improves the performance in oscillation suppression.

6.2 THREE-DIMENSIONAL MOTIONS

6.2.1 MODELING

Figure 6.17 shows a schematic representation of dual cranes carrying a distributed-mass load. Trolleys A and B slide independently along bridges and runways in the $\mathbf{N}_x$ and $\mathbf{N}_z$ directions. x_A and x_B are the displacements of the two trolleys along the $\mathbf{N}_x$ direction in the Inertial Newtonian frame, while z_A and z_B are the displacements of the two trolleys along the $\mathbf{N}_z$ direction in the Inertial Newtonian frame.

Two massless and inelastic cables, $\mathbf{C_A}$ and $\mathbf{C_B}$, connect a uniformly distributed-mass payload, **P**, to the two trolleys independently. The length of cables, $\mathbf{C_A}$ and $\mathbf{C_B}$, are L_A and L_B. The mass and length of the payload are m_L and L_L, respectively. φ_A and θ_A are defined as the swing angles of the cable $\mathbf{C_A}$ about the $\mathbf{N}_x$ and $\mathbf{N}_z$ directions. φ_B and θ_B are denoted as the swing angles of the cable $\mathbf{C_B}$ about the $\mathbf{N}_x$ and $\mathbf{N}_z$ directions.

Rotating the swing angles θ_A and φ_A respectively can convert the inertial coordinates $\mathbf{N}_x\mathbf{N}_y\mathbf{N}_z$ to the moving Cartesian coordinates $\mathbf{C}_{\mathbf{A}x}\mathbf{C}_{\mathbf{A}y}\mathbf{C}_{\mathbf{A}z}$ of the cable $\mathbf{C_A}$. Rotating the pitch angle β and twist angle γ respectively can convert the inertial coordinates $\mathbf{N}_x\mathbf{N}_y\mathbf{N}_z$ to the moving Cartesian coordinates $\mathbf{P}_x\mathbf{P}_y\mathbf{P}_z$ of the payload **P.** The acceleration of the trolleys **A** and **B** is the input to the model. The swing angles, θ_A and φ_A, of the cable $\mathbf{C_A}$, the swing angles, θ_B and φ_B, of the cable $\mathbf{C_B}$, the pitch angle, β, and the twist angle, γ, of the payload **P** are the outputs. The motion of the trolleys is also assumed to be unaffected by motion of the payload due to the large mechanical impedance in the drive system.

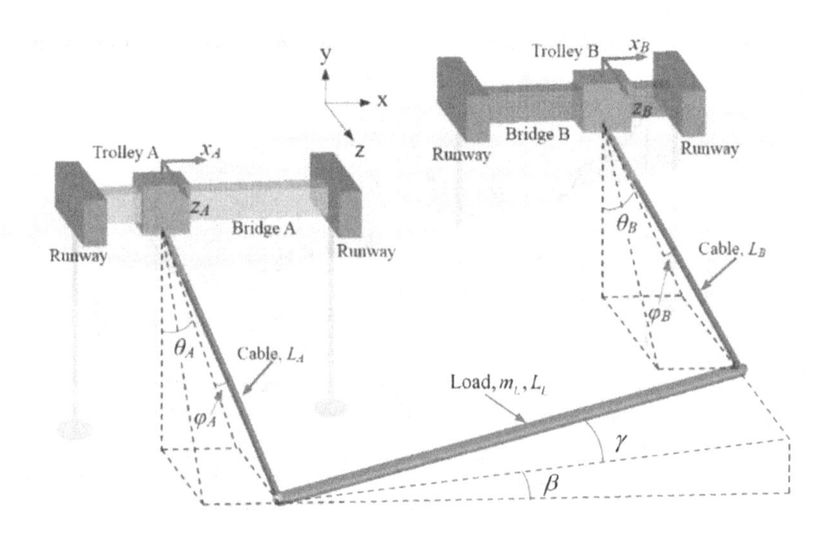

FIGURE 6.17 Model of dual cranes transporting a slender load.

The generalized speeds, u_i, in the Kane equation are:

$$u_1 = \dot{\theta}_A, \quad u_2 = \dot{\varphi}_A, \quad u_3 = \dot{\theta}_B, \quad u_4 = \dot{\varphi}_B, \quad u_5 = \dot{\beta}, \quad u_6 = \dot{\gamma} \tag{6.14}$$

Then the velocity of the payload center in the Newtonian reference frame is:

$$\begin{aligned} {}^{N}v^{P_0} = {} & u_1 L_A \cos\varphi_A \cdot \mathbf{C}_{\mathbf{A}x} - u_2 L_A \cdot \mathbf{C}_{\mathbf{A}z} + 0.5 u_5 L_L \cos\gamma \cdot \mathbf{P}_y \\ & - 0.5 u_6 L_L \cdot \mathbf{P}_z + \dot{x}_A \cdot \mathbf{N}_x + \dot{z}_A \cdot \mathbf{N}_z \end{aligned} \tag{6.15}$$

The angular velocity of the payload can be written as:

$${}^{N}\omega^{P} = -u_5 \sin\gamma \cdot \mathbf{P}_x + u_6 \cdot \mathbf{P}_y + u_5 \cos\gamma \cdot \mathbf{P}_z \tag{6.16}$$

The dependent motion variables are chosen as φ_B, β, and γ. Therefore, the independent motion variables are θ_A, φ_A, and θ_B. Then the velocity constraints among those variables are:

$$\begin{aligned} & L_A(u_1 \cos\varphi_A \cos\theta_A - u_2 \sin\varphi_A \sin\theta_A) \\ & + L_B(u_4 \sin\varphi_B \sin\theta_B - u_3 \cos\varphi_B \cos\theta_B) \\ & - L_L(u_5 \sin\beta \cos\gamma + u_6 \cos\beta \sin\gamma) + (\dot{x}_A - \dot{x}_B) = 0 \end{aligned} \tag{6.17}$$

$$\begin{aligned} & L_A(u_2 \sin\varphi_A \cos\theta_A + u_1 \sin\theta_A \cos\varphi_A) - L_B(u_4 \sin\varphi_B \cos\theta_B + u_3 \cos\varphi_B \sin\theta_B) \\ & \quad + L_L(u_5 \cos\beta \cos\gamma - u_6 \sin\beta \sin\gamma) = 0 \end{aligned} \tag{6.18}$$

$$u_4 L_B \cos\varphi_B - u_2 L_A \cos\varphi_A - u_6 L_L \cos\gamma + (\dot{z}_A - \dot{z}_B) = 0 \tag{6.19}$$

The generalized active force defined in the Kane equation can be described as:

$$F_1 = -g L_A m_L \cos\varphi_A \sin\theta_A + \frac{g L_A m_L \cos\varphi_A \left[\begin{array}{c} \cos\beta \sin\theta_A \left(\begin{array}{l} \sin\gamma \cos\beta \cos\varphi_B \\ - \sin\varphi_B \sin\theta_B \cos\gamma \end{array} \right) \\ + \sin\beta \sin\gamma \cos\varphi_B \cos(\beta - \theta_A) \\ - \cos\beta \cos\theta_A \left(\begin{array}{l} \sin\beta \sin\gamma \cos\varphi_B \\ + \sin\varphi_B \cos\gamma \cos\theta_B \end{array} \right) \end{array} \right]}{2[\sin\gamma \cos\varphi_B + \sin\varphi_B \cos\gamma \sin(\beta - \theta_B)]} \tag{6.20}$$

$$F_2 = -gL_A m_L \sin\varphi_A \cos\theta_A + \frac{gL_A m_L \begin{bmatrix} \cos\beta \sin\varphi_A \sin\theta_A \begin{pmatrix} \sin\beta \sin\gamma \cos\varphi_B \\ + \sin\varphi_B \cos\gamma \cos\theta_B \end{pmatrix} \\ + \sin\gamma \begin{pmatrix} \sin\beta \sin\varphi_A \cos\varphi_B \sin(\beta - \theta_A) \\ - \sin\beta \sin\varphi_B \cos\varphi_A \sin(\beta - \theta_B) \\ - \cos\beta \sin\varphi_B \cos\varphi_A \cos(\beta - \theta_B) \end{pmatrix} \\ + \cos\beta \sin\varphi_A \cos\theta_A \begin{pmatrix} \sin\gamma \cos\beta \cos\varphi_B \\ - \sin\varphi_B \sin\theta_B \cos\gamma \end{pmatrix} \end{bmatrix}}{2[\sin\gamma \cos\varphi_B + \sin\varphi_B \cos\gamma \sin(\beta - \theta_B)]} \tag{6.21}$$

$$F_3 = \frac{-gL_B m_L \cos\varphi_B \begin{bmatrix} \sin\beta \sin\gamma \cos\varphi_B \cos(\beta - \theta_B) \\ - \cos\beta \sin\varphi_B \cos\gamma \\ - \cos\beta \sin\gamma \cos\varphi_B \sin(\beta - \theta_B) \end{bmatrix}}{2[\sin\gamma \cos\varphi_B + \sin\varphi_B \cos\gamma \sin(\beta - \theta_B)]} \tag{6.22}$$

where g is the gravitational constant. The generalized inertia force defined in the Kane equation can be written as:

$$F_i^* = -m_L \cdot {}^N a^{P_0} \cdot {}^N v_i^{P_0} - \left(I^{P/P_0} \cdot {}^N\alpha^P + {}^N\omega^P \times I^{P/P_0} \cdot {}^N\omega^P\right) \cdot {}^N\omega_i{}^P \tag{6.23}$$

where the moment of inertia of the payload about the payload center P_0 is $I^{P/P0}$, the acceleration of the payload center in the Newtonian reference frame is ${}^N a^{P_0}$, angular acceleration of the payload in the Newtonian reference frame is ${}^N\alpha^P$, the ith partial velocity of the payload center is ${}^N v_i^{P_0}$, and the ith partial angular velocities of the payload is ${}^N\omega_i^P$.

By using the Kane equation, the nonlinear equations of the motion for the model yield:

$$[M]\{\dot{U}\} + \{f\} = 0 \tag{6.24}$$

where M is the mass matrix, $\dot{U}$ is the column matrix of derivatives of generalized speed with respect to time, and f is the column matrix of gravity terms, centrifugal and Coriolis terms, and control input terms.

6.2.2 Dynamics

Limiting the trolley acceleration, velocity and acceleration of the payload swing, pitch, and twisting to zero in Equation (6.24) yield the equilibrium points:

$$\cos\beta_0 \cos\theta_{A0} \begin{bmatrix} \cos\gamma_0 \sin\varphi_{B0} \cos\theta_{B0} \\ + \sin\beta_0 \sin\gamma_0 \cos\varphi_{B0} \end{bmatrix} + \cos\beta_0 \sin\theta_{A0} \begin{bmatrix} \cos\gamma_0 \sin\varphi_{B0} \sin\theta_{B0} \\ - \cos\beta_0 \sin\gamma_0 \cos\varphi_{B0} \end{bmatrix}$$
$$+ 2\sin\theta_{A0} \begin{bmatrix} \sin\gamma_0 \cos\varphi_{B0} \\ + \cos\gamma_0 \sin\varphi_{B0} \sin(\beta_0 - \theta_{B0}) \end{bmatrix} - \sin\beta_0 \sin\gamma_0 \cos\varphi_{B0} \cos(\beta_0 - \theta_{A0}) = 0 \tag{6.25}$$

$$\begin{aligned}
&\cos\beta_0 \cos\varphi_{A0} \sin\varphi_{B0} \sin\gamma_0 \cos(\beta_0 - \theta_{B0}) \\
&+ \sin\beta_0 \sin\gamma_0 \begin{bmatrix} \cos\varphi_{A0} \sin\varphi_{B0} \sin(\beta_0 - \theta_{B0}) \\ - \cos\varphi_{B0} \sin\varphi_{A0} \sin(\beta_0 - \theta_{A0}) \end{bmatrix} \\
&- \cos\beta_0 \sin\varphi_{A0} \sin\theta_{A0} \begin{bmatrix} \cos\gamma_0 \sin\varphi_{B0} \cos\theta_{B0} \\ + \sin\beta_0 \sin\gamma_0 \cos\varphi_{B0} \end{bmatrix} \\
&+ \cos\beta_0 \sin\varphi_{A0} \cos\theta_{A0} \begin{bmatrix} \cos\gamma_0 \sin\varphi_{B0} \sin\theta_{B0} \\ - \cos\beta_0 \sin\gamma_0 \cos\varphi_{B0} \end{bmatrix} \\
&+ 2\sin\varphi_{A0} \cos\theta_{A0} \begin{bmatrix} \sin\gamma_0 \cos\varphi_{B0} \\ + \cos\gamma_0 \sin\varphi_{B0} \sin(\beta_0 - \theta_{B0}) \end{bmatrix} = 0
\end{aligned} \tag{6.26}$$

$$\begin{aligned}
&\cos\beta_0 \cos\gamma_0 \sin\varphi_{B0} + \cos\beta_0 \sin\gamma_0 \cos\varphi_{B0} \sin(\beta_0 - \theta_{B0}) \\
&- \sin\beta_0 \sin\gamma_0 \cos\varphi_{B0} \cos(\beta_0 - \theta_{B0}) = 0
\end{aligned} \tag{6.27}$$

$$L_A \cos\varphi_{A0} \sin\theta_{A0} + L_L \cos\beta_0 \cos\gamma_0 - L_B \cos\varphi_{B0} \sin\theta_{B0} + x_A - x_B = 0 \tag{6.28}$$

$$L_A \cos\varphi_{A0} \cos\theta_{A0} - L_B \cos\varphi_{B0} \cos\theta_{B0} - L_L \sin\beta_0 \cos\gamma_0 = 0 \tag{6.29}$$

$$L_B \sin\varphi_{B0} - L_A \sin\varphi_{A0} - L_L \sin\gamma_0 + z_A - z_B = 0 \tag{6.30}$$

where θ_{A0}, θ_{B0}, φ_{A0}, φ_{B0}, β_0, and γ_0 are the equilibrium angles. Assuming small oscillations around the equilibria yields a linearized model. Then natural frequencies of the three-dimensional dual cranes can be derived from the linearized model. Because the three constraints (6.28)–(6.30) are corresponding to zero-stiffness subsystems, only three natural frequencies, ω_i, occur for the model (6.24).

The cable length, payload length, and trolley distance have some influence on the three natural frequencies. Natural frequencies decrease as the cable length increases. Three frequencies induced by various distances $(x_B - x_A)$ between the two trolleys are shown in Figure 6.18, while three frequencies induced by various load lengths are shown in Figure 6.19. The length of the cable $\mathbf{C_A}$, length of cable $\mathbf{C_B}$, distance $(z_B - z_A)$ between the two trolleys, and load length were fixed at 81 cm, 62 cm, 30 cm, and 119 cm, respectively. As the distance $(x_B - x_A)$ increases, the first two frequencies change slightly, and the third frequency increases obviously.

When the length of the cable $\mathbf{C_A}$, length of cable $\mathbf{C_B}$, distance $(z_B - z_A)$, and distance $(x_B - x_A)$ between the two trolleys were fixed at 81 cm, 62 cm, 30 cm, and 120 cm, increasing load length decreases the third frequency, ω_3, sharply. Meanwhile, the second frequency, ω_2, decreases as the load length increases before 130 cm, and approaches the first frequency, ω_1, near 130 cm. The first and second frequencies both increase slightly as the load length increases from 130 cm.

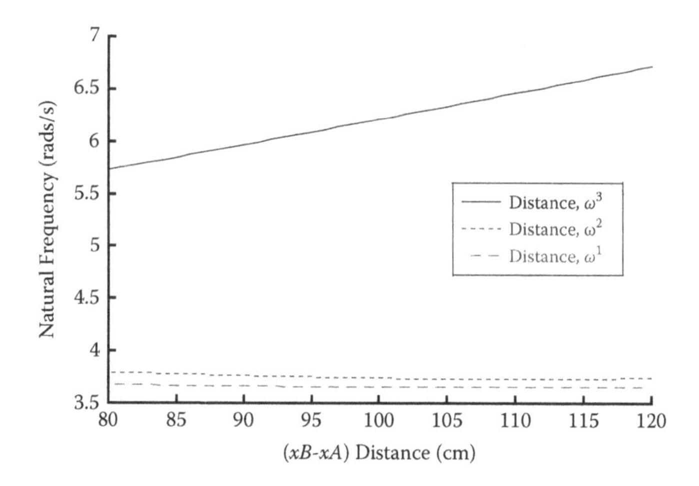

FIGURE 6.18 Natural frequencies against trolley distance and load length.

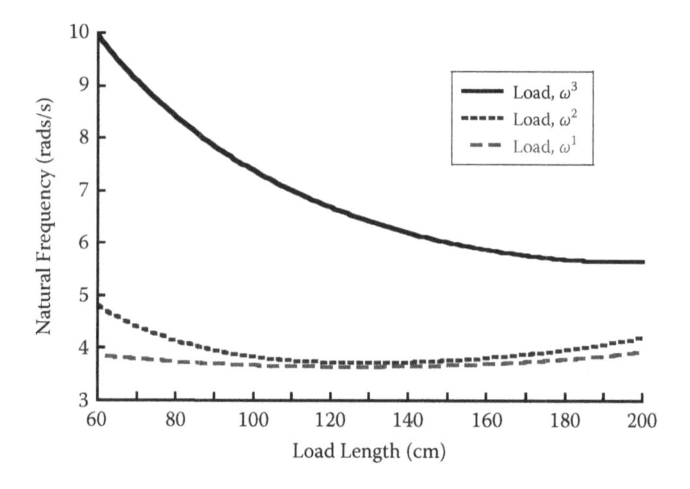

FIGURE 6.19 Natural frequencies against trolley distance and load length.

The dynamic behavior of the payload swing and pitch depends on the first two frequencies, ω_1 and ω_2. Meanwhile, the behavior of the payload twisting is dependent on the third frequency, ω_3. The system parameters have a large influence on the third frequency. Therefore, more robustness to changes in the third frequency is needed to design the corresponding control system. It is challenging to control the payload twisting in three-dimensional dual cranes.

6.2.3 Oscillation Control

A combined control architecture for dual cranes is shown in Figure 6.20. In order to control the first-mode oscillations, an MEI input shaper with the notch filtering

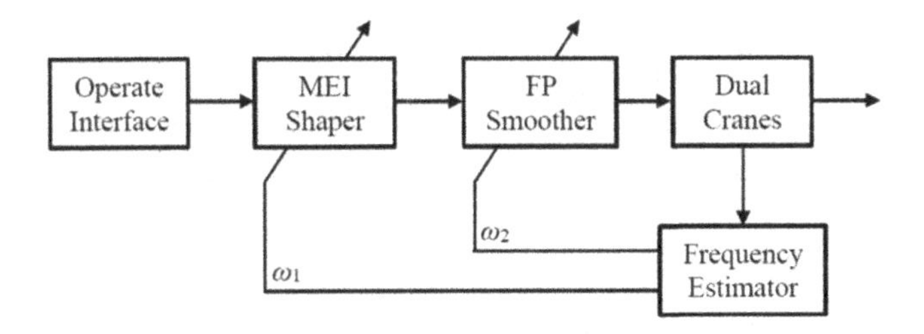

FIGURE 6.20 Combined control architecture.

effect is designed. To attenuate the oscillations of the last two modes, a four-pieces (FP) smoother is designed. Modified commands were created by filtering the original driving commands with the MEI shaper and FP smoother. The modified commands drove the dual cranes to the desired position and caused minimal oscillations of the payload. The MEI shaper and FP smoother were designed by predicting the first and second frequencies.

The nonlinear model (6.24) may be approximated as three second-order systems with three abovementioned natural frequencies near the equilibrium point. A four-impulses MEI shaper can suppress the first-mode oscillations with the frequency, ω_1, and damping ratio, ζ_1. The MEI shaper is given by:

$$\text{MEI} = \begin{bmatrix} A_k \\ \tau_k \end{bmatrix} = \begin{bmatrix} \frac{1}{[1+K_1]^2} & \frac{K_1}{[1+K_1]^2} & \frac{K_1}{[1+K_1]^2} & \frac{K_1^2}{[1+K_1]^2} \\ 0 & \frac{T_1}{q} & \frac{T_1}{p} & \left(\frac{T_1}{q} + \frac{T_1}{p}\right) \end{bmatrix} \tag{6.31}$$

where kth-impulse amplitude and time are A_k and τ_k, respectively. The coefficients K_1 and T_1 are:

$$K_1 = e^{(-\pi\zeta_1/\sqrt{1-\zeta_1^2})} \tag{6.32}$$

$$T_1 = \pi/(\omega_1\sqrt{1 - \zeta_1^2}) \tag{6.33}$$

Two modified coefficients, p and q, can be calculated from the constraints:

$$e^{-\zeta_1\omega_1\tau_n}\sqrt{S_1^2(\omega_1, \zeta_1) + C_1^2(\omega_1, \zeta_1)} \le V_{tol} \tag{6.34}$$

$$S_1(\omega_1, \zeta_1)\cdot\frac{\partial S_1(\omega_1, \zeta_1)}{\partial\omega_1} + C_1(\omega_1, \zeta_1)\cdot\frac{\partial C_1(\omega_1, \zeta_1)}{\partial\omega_1} - \zeta_1\tau_n\cdot[S_1{}^2(\omega_1, \zeta_1) + C_1{}^2(\omega_1, \zeta_1)] = 0 \tag{6.35}$$

where

$$S_1(\omega_1, \zeta_1) = \sum_k A_k \cdot e^{\zeta_1 \omega_1 \tau_k} \sin(\omega_1 \sqrt{1 - \zeta_1^2}\, \tau_k) \tag{6.36}$$

$$C_1(\omega_1, \zeta_1) = \sum_k A_k \cdot e^{\zeta_1 \omega_1 \tau_k} \cos(\omega_1 \sqrt{1 - \zeta_1^2}\, \tau_k) \tag{6.37}$$

$$\tau_n = \left(\frac{1}{q} + \frac{1}{p}\right) \cdot \frac{\pi}{(\omega_1 \sqrt{1 - \zeta_1^2})} \tag{6.38}$$

the tolerable level of the vibration is V_{tol}. When the tolerable level of the vibration and damping ratio are set to 5% and 0, the solution of the modified coefficients is $p = 0.8584$ and $q = 1.1429$.

The FP smoother is given by:

$$u(\tau) = \begin{cases} W\tau e^{-2\zeta_2 \omega_2 \tau}, & 0 \le \tau \le T_2 \\ W(2T_2 - 2T_2 K_2^{-1} - \tau + 2K_2^{-1}\tau) e^{-2\zeta_2 \omega_2 \tau}, & T_2 \le \tau \le 2T_2 \\ W(6K_2^{-1}T_2 - 2K_2^{-2}T_2 - 2K_2^{-1}\tau + K_2^{-2}\tau) e^{-2\zeta_2 \omega_2 \tau}, & 2T_2 \le \tau \le 3T_2 \\ W(4K_2^{-2}T_2 - K_2^{-2}\tau) e^{-2\zeta_2 \omega_2 \tau}, & 3T_2 < \tau \le 4T_2 \end{cases} \tag{6.39}$$

where

$$K_2 = e^{(-\pi\zeta_2/\sqrt{1-\zeta_2^2})} \tag{6.40}$$

$$W = \frac{4\zeta_2^2 \omega_2^2}{[1 + K_2 - K_2^2 - K_2^3]^2} \tag{6.41}$$

$$T_2 = \pi/(\omega_2 \sqrt{1 - \zeta_2^2}) \tag{6.42}$$

The FP smoother can control the second-mode oscillations with the frequency, ω_2, and damping ratio, ζ_2. The notch filtering effect for the first-mode frequency is provided by the MEI shaper, and the notch filtering effect for the second-mode frequency and low-pass filtering effect for the third-mode frequency are provided by the FP smoother. Then oscillations of all modes would be suppressed by the combination of MEI shaper and FP smoother.

The MEI + FP combined filter is designed by using estimates of natural frequencies, which are obtained by assuming small oscillations around the equilibrium position. In the case of large oscillations, the nonlinear frequency depends on the amplitude of the oscillation. Therefore, errors between the natural frequency and the nonlinear frequency cannot be neglected in the case of great oscillations. The wider frequency insensitivity of the MEI + FP combined filter suppresses the oscillations caused by the errors between the natural frequency and the nonlinear frequency.

6.2.4 Computational Dynamics

The nonlinear model (6.24) will be used to validate the effectiveness of the MEI+FP combined filter. The simulated residual amplitude of the radial swing angle, θ_A, tangential swing angle, φ_A, pitch angle, β, and twist angle, γ, induced by various driving distances are shown in Figure 6.21. The payload length, L_L, cable length, L_A, and cable length, L_B, were fixed at 119 cm, 81 cm, and 62 cm, respectively. The maximum driving velocity is 20 cm/s, while the acceleration is fixed at 2 m/s^2. The residual amplitude is defined as maximum peak-to-peak deflection of the payload during the trolley stopes.

Peaks and troughs contain the simulated residual amplitudes of the swing, pitch, and twisting because the oscillations caused by the acceleration and deceleration are in phase or out of phase. The first or second frequencies are corresponding to the distance between the peaks and troughs in the swing and pitch because the first two frequencies are close in this case. Meanwhile, the third frequency is corresponding to the distance between the peaks and troughs in the twisting. More peaks and troughs occurred in the twisting because the third frequency is higher than the first two frequencies. The MEI + FP combined filter suppresses oscillations of the swing, pitch, and twisting to a very low level. The controlled residual amplitudes of the radial swing angle, θ_A, tangential swing angle, φ_A, pitch angle, β, and twist angle, γ, are limited to $<0.02°$, $<0.03°$, $<0.005°$, and $<0.05°$, respectively.

Figure 6.22 shows the simulated residual amplitude of the radial swing angle, θ_A, tangential swing angle, φ_A, pitch angle, β, and twist angle, γ, when the simulated cable length, L_A, ranged from 50 cm to 110 cm. The driving distance, payload length, L_L, and cable length, L_B, were fixed at 20 cm, 119 cm, and 62 cm, respectively. As the cable length increased, residual amplitudes of the radial swing

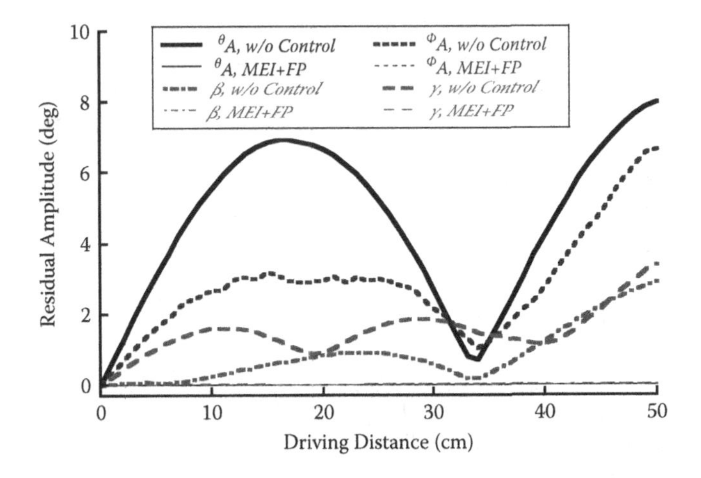

FIGURE 6.21 Simulated results for various driving distances.

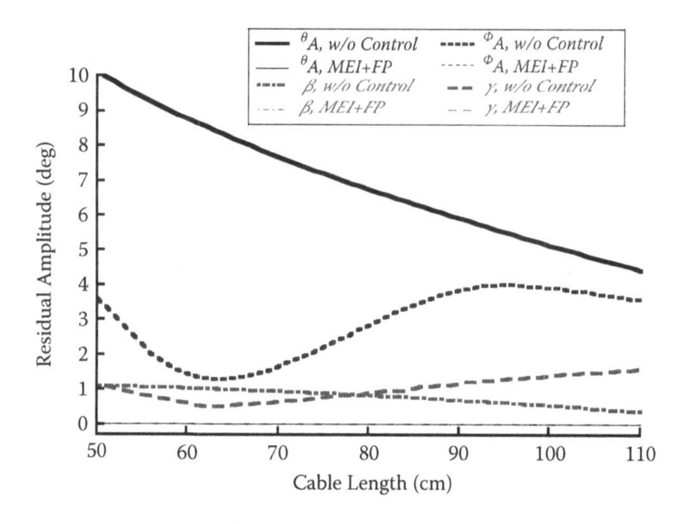

FIGURE 6.22 Simulated results for various cable lengths.

angle, θ_A, decreased. The residual amplitudes of the tangential swing angle, φ_A, and twist angle, γ, occur local minimum near the cable length of 64 cm. This is because the lengths of two suspension cables in this case were equal. The modeled cable length for the design of the combined filter was held constant at 80 cm. The residual amplitude of the radial swing angle, θ_A, tangential swing angle, φ_A, pitch angle, β, and twist angle, γ, were limited to <0.2°, <0.07°, <0.01°, and <0.01° by the MEI + FP combined filter.

The simulated residual amplitude of the radial swing angle, θ_A, tangential swing angle, φ_A, pitch angle, β, and twist angle, γ, for varying beam lengths are shown in the Figure 6.23 when the load length ranged from 60 cm to 180 cm. The driving

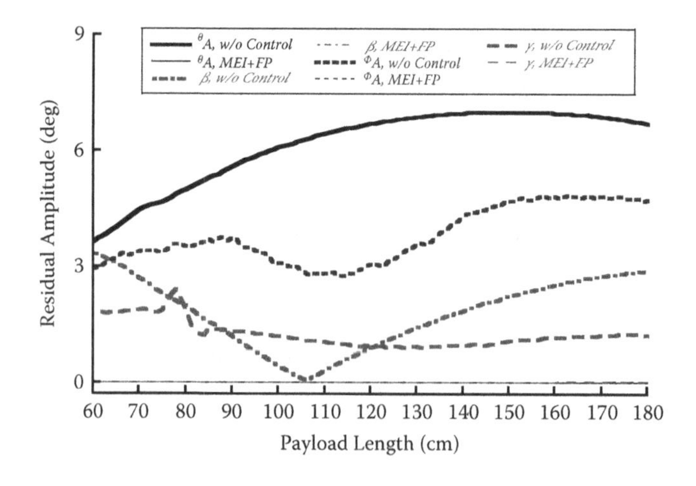

FIGURE 6.23 Simulated results for various payload lengths.

distance, cable length, L_A, cable length, L_B, were fixed at 20 cm, 81 cm, and 62 cm, respectively. Because the equilibrium angles of the payload swing were limited to zero near the payload length of 105 mm, a zero oscillation in the pitch angle, β, arose. The simulated amplitudes of twist angle, γ, and tangential swing angle, φ_A, are non-smooth due to the nonlinearity. The modeled payload length for the design of the combined filter was set to 119 cm. The MEI + FP combined filter limited the residual amplitude of the radial swing angle, θ_A, tangential swing angle, φ_A, pitch angle, β, and twist angle, γ, to <0.03°, <0.02°, <0.04°, and <0.02°, respectively.

6.2.5 Experimental Investigation

The testbed of two small-scale bridge cranes moving a slender beam was used to perform experimental validation. Two Panasonic AC servomotors with encoders drove two trolleys independently. Two DSP-based motion control cards attach host computers to servo amplifiers. The control algorithm in two MATLAB scripts was applied to filter through the baseline trapezoidal-velocity commands. The shaped commands were generated for the two drives. A slender distributed-mass wood beam served as the payload. The beam has a length of 119 cm and mass of 465 g. Two suspension cables attached the beam to the two trolleys. The initial distance, $(x_B - x_A)$, and the initial distance, $(z_B - z_A)$, between two trolleys were 120 cm and 30 cm, respectively.

The cable lengths were fixed at 81 cm and 62 cm. Two markers were fixed nearby the connection points between the beam and the cables. Two markers were applied to estimate the payload swing displacements and the payload twist angle. In order to record the displacements of the two black markers on the beam, two CMOS cameras were mounted near the suspension points between the trolleys and cables independently. The position deflections of marker A in the radial and tangential directions are denoted as PAx and PAz, respectively. The deflections, PAx and PAz, are corresponding to the radial swing angle, θ_A, and tangential swing angle, φ_A, respectively. The inverse tangent function can be applied to calculate the twist angle, γ, of the payload from the two markers. In addition, only three angles are independent variables, while the other three are dependent. Therefore, the experimental results of deflections, PAx and PAz, and twist angle, γ, exhibit all dynamic behavior of payload swing, pitch, and twisting.

The simulated and experimental residual oscillations of the deflection, PAx and PAz, and the twist angle, γ, are shown in Figures 6.24 and 6.25. Trolley **A** drove a distance of 30 cm along the $\mathbf{N}_x$ direction while trolley **B** was still. After trolley **A** stopped, the distance $(x_B - x_A)$ between the trolleys was 90 cm. The steady-state values of the deflections, PAx and PAz, and the twist angle, γ, were 119.1 mm, −39.7 mm, and 13.5°, respectively. The steady-state values are corresponding to the equilibrium points. The oscillation frequencies of the deflections, PAx and PAz, in Figure 6.24 are corresponding to the first two frequencies, and the oscillation frequency of the payload twisting in Figure 6.25 is corresponding to the third-mode frequency.

The simulated results in the natural frequency and equilibrium point estimated the experimental data very well. The uncontrolled peak-to-peak residual amplitudes

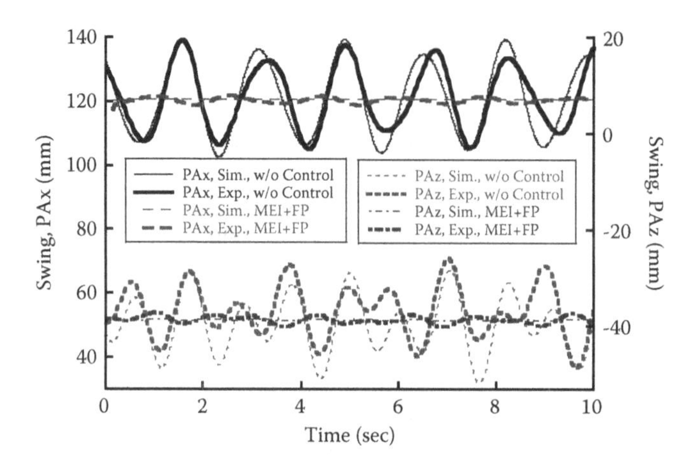

FIGURE 6.24 Deflection of marker A to an operator's command.

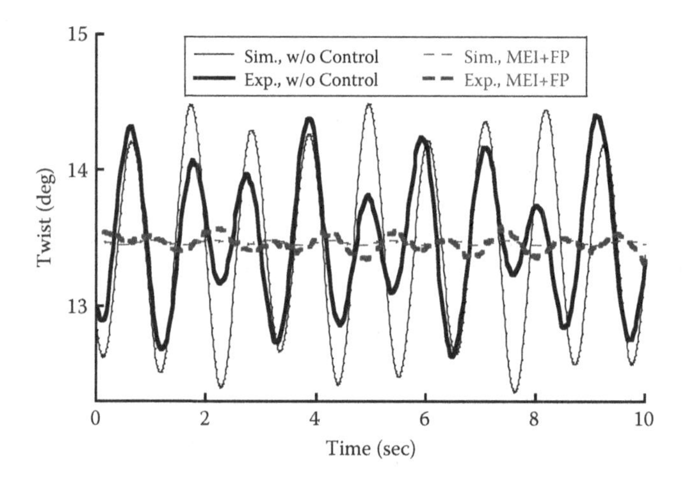

FIGURE 6.25 Payload twisting to an operator's command.

of the deflections, PAx and PAz, and the twist angle, γ, are 32.3 mm, 23.4 mm, and 1.78°, respectively. Figures 6.24 and 6.25 also show the control effect with the MEI + FP combined filter. The combined filter dramatically suppresses both the payload swing and twisting. This is because the experimental residual amplitudes of the deflections, PAx and PAz, and the twist angle, γ, were 3.2 mm, 3.2 mm, and 0.28°, respectively.

The comparison of simulated and experimental residual amplitudes for varying modeling error in the design frequency is shown in Figures 6.26 and 6.27. Trolley **A** drove 30 cm along the $\mathbf{N}_x$ direction while trolley **B** was motionless. The normalized frequency of 1.0 is corresponding to the two design frequencies of 3.66 rad/s and 3.75 rad/s, respectively. The experimental residual amplitudes of the deflections,

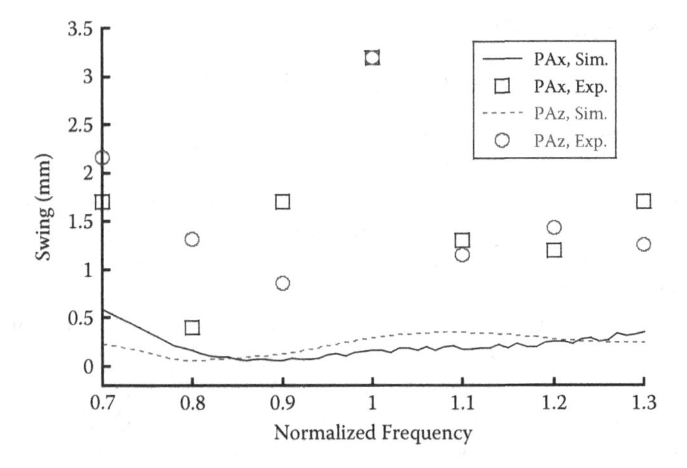

FIGURE 6.26 Experimental residual amplitudes of marker A to modeling errors in the design frequency.

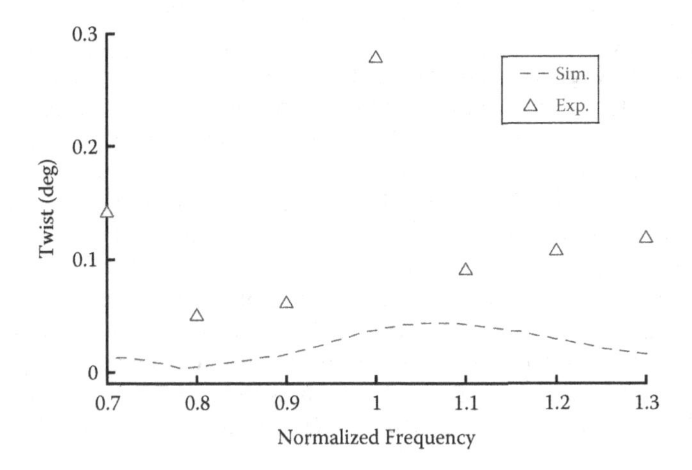

FIGURE 6.27 Experimental residual amplitudes of payload twisting to modeling errors in the design frequency.

PAx and PAz, and the twist angle, γ, with the MEI + FP filter were 3.2 mm, 3.2 mm, and 0.28°, respectively.

Both the payload swing and twisting decreased to near-zero values at the normalized frequency ranged from 1.2 to 1.3. This is because zero vibrations are designed for the MEI + FP filter at the nearby frequencies. The experimental data follow the general shape as the corresponding simulated results. The MEI + FP combined filter limited the residual oscillations of the swing deflection, PAx and PAz, and the twist angle, γ, to very low levels for all the normalized frequencies. This is because the combined filter exhibits good robustness in the frequency.

6.3 DUAL AERIAL CRANES

6.3.1 Modeling of Dual Quadrotors Slung Loads

A schematic representation of dual quadrotors carrying a slender bar is shown in Figure 6.28. The dual quadrotors use two gyrostats, $\mathbf{Q}_1$ and $\mathbf{Q}_2$, which mean rigid bodies with three fixed-axis, $\mathbf{Q}_{1x}\mathbf{Q}_{1y}\mathbf{Q}_{1z}$ and $\mathbf{Q}_{2x}\mathbf{Q}_{2y}\mathbf{Q}_{2z}$, in the quadrotors. The inertial coordinates $\mathbf{N}_x\mathbf{N}_y\mathbf{N}_z$ can be converted to the moving coordinates $\mathbf{Q}_{1x}\mathbf{Q}_{1y}\mathbf{Q}_{1z}$ by rotating the yaw attitude, ψ_1, pitch attitude, θ_1, and roll attitude, φ_1, and to the moving coordinates $\mathbf{Q}_{2x}\mathbf{Q}_{2y}\mathbf{Q}_{2z}$ by rotating the yaw attitude, ψ_2, pitch attitude, θ_2, and roll attitude, φ_2. The left quadrotor $\mathbf{Q}_1$ generates thrust force, f_1, along the $\mathbf{Q}_{1z}$ direction and thrust moments, $M_{\varphi 1}$, $M_{\theta 1}$, and $M_{\psi 1}$, about the $\mathbf{Q}_{1x}\mathbf{Q}_{1y}\mathbf{Q}_{1z}$ directions. Meanwhile, the right quadrotor $\mathbf{Q}_2$ generates thrust force, f_2, along the $\mathbf{Q}_{2z}$ direction and thrust moments, $M_{\varphi 2}$, $M_{\theta 2}$, and $M_{\psi 2}$, about the $\mathbf{Q}_{2x}\mathbf{Q}_{2y}\mathbf{Q}_{2z}$ directions. The quadrotor $\mathbf{Q}_1$ flies with displacements of x_1, y_1, and z_1, along the $\mathbf{N}_x\mathbf{N}_y\mathbf{N}_z$ directions, and the displacements of quadrotor $\mathbf{Q}_2$ are denoted by x_2, y_2, z_2, in the inertial coordinates. Both the mass of the dual quadrotors is denoted by m. The centroidal principal axis moments of inertia about the moving coordinates in the dual quadrotors are denoted by I_{xx}, I_{yy}, and I_{zz}, respectively.

Two rigid links attach two suspension points, P_1 and P_2, to the mass center, $\mathrm{Q}_{1\mathrm{o}}$ and $\mathrm{Q}_{2\mathrm{o}}$, of the quadrotors, $\mathbf{Q}_1$, and $\mathbf{Q}_2$, along the $\mathbf{Q}_{1z}$, and $\mathbf{Q}_{2z}$, directions, respectively. The lengths of two links are both L. Therefore, the distance between the mass center, $\mathrm{Q}_{1\mathrm{o}}$, of the quadrotor and the suspension point, P_1, is $L\mathbf{Q}_{1z}$, while the distance between the mass center, $\mathrm{Q}_{2\mathrm{o}}$, of the quadrotor and the suspension point, P_2, is $L\mathbf{Q}_{2z}$. Two massless and inelastic cables, $\mathbf{C}_1$ and $\mathbf{C}_2$, of the same length, l, attach two suspension points, P_1 and P_2, to a slender bar $\mathbf{B}$. The swing angles of the cable $\mathbf{C}_1$ are denoted by α_{1x} and α_{1y} to orient the pendulum, and that of the cable $\mathbf{C}_2$ are denoted by α_{2x} and α_{2y}. The mass and length of the slender bar are denoted by m_b and l_b, respectively. The cross-sectional area of the bar is small because of the slender load. The inertial coordinates $\mathbf{N}_x\mathbf{N}_y\mathbf{N}_z$ can be converted to the moving coordinates $\mathbf{B}_x\mathbf{B}_y\mathbf{B}_z$ by rotating the twist angle, β, and slope angle, γ, respectively.

Inputs to the model in Figure 6.28 are thrust forces, f_1 and f_2, and thrust moments, $M_{\varphi 1}$, $M_{\theta 1}$, $M_{\psi 1}$, $M_{\varphi 2}$, $M_{\theta 2}$, and $M_{\psi 2}$. Outputs are quadrotor displacements x_1,

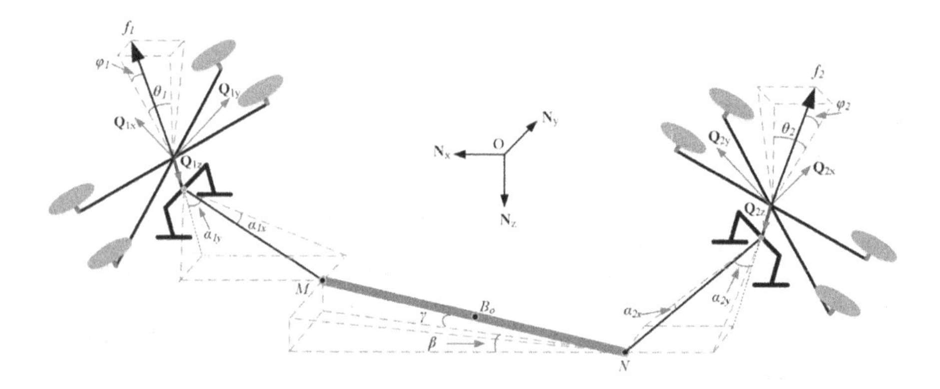

FIGURE 6.28 Model of dual quadrotors carrying a slender bar.

y_1, z_1, x_2, y_2, and z_2, attitudes, ψ_1, θ_1, φ_1, ψ_2, θ_2, and φ_2, swing angles, α_{1x}, α_{1y}, α_{2x}, and α_{2y}, twist angle, β, and slope angle, γ.

The basis transformation matrix among bodies quadrotors, $\mathbf{Q}_1$ and $\mathbf{Q}_2$, cables, $\mathbf{C}_1$ and $\mathbf{C}_2$, and bar $\mathbf{B}$ are expressed by:

$$\begin{bmatrix} \mathbf{Q}_{1x} \\ \mathbf{Q}_{1y} \\ \mathbf{Q}_{1z} \end{bmatrix} = \begin{bmatrix} 1 & 0 & 0 \\ 0 & C_{\phi_1} & S_{\phi_1} \\ 0 & -S_{\phi_1} & C_{\phi_1} \end{bmatrix} \begin{bmatrix} C_{\theta_1} & 0 & -S_{\theta_1} \\ 0 & 1 & 0 \\ S_{\theta_1} & 0 & C_{\theta_1} \end{bmatrix} \begin{bmatrix} C_{\psi_1} & S_{\psi_1} & 0 \\ -S_{\psi_1} & C_{\psi_1} & 0 \\ 0 & 0 & 1 \end{bmatrix} \begin{bmatrix} \mathbf{N}_x \\ \mathbf{N}_y \\ \mathbf{N}_z \end{bmatrix} \quad (6.43)$$

$$\begin{bmatrix} \mathbf{Q}_{2x} \\ \mathbf{Q}_{2y} \\ \mathbf{Q}_{2z} \end{bmatrix} = \begin{bmatrix} 1 & 0 & 0 \\ 0 & C_{\phi_2} & S_{\phi_2} \\ 0 & -S_{\phi_2} & C_{\phi_2} \end{bmatrix} \begin{bmatrix} C_{\theta_2} & 0 & -S_{\theta_2} \\ 0 & 1 & 0 \\ S_{\theta_2} & 0 & C_{\theta_2} \end{bmatrix} \begin{bmatrix} C_{\psi_2} & S_{\psi_2} & 0 \\ -S_{\psi_2} & C_{\psi_2} & 0 \\ 0 & 0 & 1 \end{bmatrix} \begin{bmatrix} \mathbf{N}_x \\ \mathbf{N}_y \\ \mathbf{N}_z \end{bmatrix} \quad (6.44)$$

$$\begin{bmatrix} \mathbf{C}_{1x} \\ \mathbf{C}_{1y} \\ \mathbf{C}_{1z} \end{bmatrix} = \begin{bmatrix} 1 & 0 & 0 \\ 0 & C_{\alpha_{1x}} & S_{\alpha_{1x}} \\ 0 & -S_{\alpha_{1x}} & C_{\alpha_{1x}} \end{bmatrix} \begin{bmatrix} C_{\alpha_{1y}} & 0 & -S_{\alpha_{1y}} \\ 0 & 1 & 0 \\ S_{\alpha_{1y}} & 0 & C_{\alpha_{1y}} \end{bmatrix} \begin{bmatrix} \mathbf{Q}_{1x} \\ \mathbf{Q}_{1y} \\ \mathbf{Q}_{1z} \end{bmatrix} \quad (6.45)$$

$$\begin{bmatrix} \mathbf{C}_{2x} \\ \mathbf{C}_{2y} \\ \mathbf{C}_{2z} \end{bmatrix} = \begin{bmatrix} 1 & 0 & 0 \\ 0 & C_{\alpha_{2x}} & S_{\alpha_{2x}} \\ 0 & -S_{\alpha_{2x}} & C_{\alpha_{2x}} \end{bmatrix} \begin{bmatrix} C_{\alpha_{2y}} & 0 & -S_{\alpha_{2y}} \\ 0 & 1 & 0 \\ S_{\alpha_{2y}} & 0 & C_{\alpha_{2y}} \end{bmatrix} \begin{bmatrix} \mathbf{Q}_{2x} \\ \mathbf{Q}_{2y} \\ \mathbf{Q}_{2z} \end{bmatrix} \quad (6.46)$$

$$\begin{bmatrix} \mathbf{B}_x \\ \mathbf{B}_y \\ \mathbf{B}_z \end{bmatrix} = \begin{bmatrix} C_\gamma & 0 & -S_\gamma \\ 0 & 1 & 0 \\ S_\gamma & 0 & C_\gamma \end{bmatrix} \begin{bmatrix} C_\beta & S_\beta & 0 \\ -S_\beta & C_\beta & 0 \\ 0 & 0 & 1 \end{bmatrix} \begin{bmatrix} \mathbf{N}_x \\ \mathbf{N}_y \\ \mathbf{N}_z \end{bmatrix} \quad (6.47)$$

where the following abbreviations are used:

$$\begin{cases} S_{\phi_1} \triangleq \sin(\phi_1), \quad C_{\phi_1} \triangleq \cos(\phi_1), \quad S_{\theta_1} \triangleq \sin(\theta_1), \quad C_{\theta_1} \triangleq \cos(\theta_1), \\ S_{\psi_1} \triangleq \sin(\psi_1), \\ C_{\psi_1} \triangleq \cos(\psi_1), \quad S_{\phi_2} \triangleq \sin(\phi_2), \quad C_{\phi_2} \triangleq \cos(\phi_2), \quad S_{\theta_2} \triangleq \sin(\theta_2), \\ C_{\theta_2} \triangleq \cos(\theta_2), \\ S_{\psi_2} \triangleq \sin(\psi_2), \quad C_{\psi_2} \triangleq \cos(\psi_2), \quad S_{\alpha_{1x}} \triangleq \sin(\alpha_{1x}), \quad C_{\alpha_{1x}} \triangleq \cos(\alpha_{1x}), \\ S_{\alpha_{1y}} \triangleq \sin(\alpha_{1y}), \\ C_{\alpha_{1y}} \triangleq \cos(\alpha_{1y}), \quad S_{\alpha_{2x}} \triangleq \sin(\alpha_{2x}), \quad C_{\alpha_{2x}} \triangleq \cos(\alpha_{2x}), \quad S_{\alpha_{2y}} \triangleq \sin(\alpha_{2y}), \\ C_{\alpha_{2y}} \triangleq \cos(\alpha_{2y}), \quad S_\gamma \triangleq \sin(\gamma), \quad C_\gamma \triangleq \cos(\gamma), \quad S_\beta \triangleq \sin(\beta), \\ C_\beta \triangleq \cos(\beta). \end{cases} \quad (6.48)$$

The velocities of the mass center, Q_{1o} and Q_{2o}, of the quadrotors, $\mathbf{Q}_1$ and $\mathbf{Q}_2$, in the Newtonian reference frame are described as:

$$\begin{cases} {}^N\mathbf{v}^{Q_{1o}} = \dot{x}_1\mathbf{N}_x + \dot{y}_1\mathbf{N}_y + \dot{z}_1\mathbf{N}_z \\ {}^N\mathbf{v}^{Q_{2o}} = \dot{x}_2\mathbf{N}_x + \dot{y}_2\mathbf{N}_y + \dot{z}_2\mathbf{N}_z \end{cases} \tag{6.49}$$

The angular velocities of the quadrotors, $\mathbf{Q}_1$ and $\mathbf{Q}_2$, in the Newtonian reference frame are given by:

$$\begin{cases} {}^N\boldsymbol{\omega}^{Q_1} = (\dot{\phi}_1 - \dot{\psi}_1 S_{\theta_1})\,\mathbf{Q}_{1x} + (\dot{\psi}_1 \cdot S_{\phi_1} C_{\theta_1} + \dot{\theta}_1 C_{\theta_1})\mathbf{Q}_{1y} + (\dot{\psi}_1 C_{\phi_1} C_{\theta_1} - \dot{\theta}_1 S_{\phi_1})\mathbf{Q}_{1z} \\ {}^N\boldsymbol{\omega}^{Q_2} = (\dot{\phi}_2 - \dot{\psi}_2 S_{\theta_2})\,\mathbf{Q}_{2x} + (\dot{\psi}_2 S_{\phi_2} C_{\theta_2} + \dot{\theta}_2 C_{\theta_2})\mathbf{Q}_{2y} + (\dot{\psi}_2 C_{\phi_2} C_{\theta_2} - \dot{\theta}_2 S_{\phi_2})\mathbf{Q}_{2z} \end{cases} \tag{6.50}$$

The velocity of the mass center, B_o, of the bar $\mathbf{B}$ in the Newtonian reference frame is:

$$\begin{aligned} {}^N\mathbf{v}^{B_o} = {} & \dot{x}_1\mathbf{N}_x + \dot{y}_1\mathbf{N}_y + \dot{z}_1\mathbf{N}_z + L\,(\dot{\theta}_1 C_{\theta_1} + \dot{\psi}_1 S_{\phi_1} C_{\theta_1})\,\mathbf{Q}_{1x} \\ & - L\,(\dot{\psi}_1 C_{\phi_1} C_{\theta_1} - \dot{\theta}_1 S_{\phi_1})\,\mathbf{Q}_{1y} + l\dot{\alpha}_{1y}\mathbf{C}_{1x} - L\dot{\alpha}_{1x}\mathbf{C}_{1y} - 0.5 l_b \dot{\beta} C_\gamma \mathbf{B}_y \\ & + 0.5 l_b \dot{\gamma}\mathbf{B}_z \end{aligned} \tag{6.51}$$

The angular velocity of the bar $\mathbf{B}$ in the Newtonian reference frame is:

$${}^N\boldsymbol{\omega}^B = -\dot{\beta} S_\gamma \mathbf{B}_x + \dot{\gamma}\mathbf{B}_y + \dot{\beta} C_\gamma \mathbf{B}_z \tag{6.52}$$

A three-dimensional six-bar linkage is created by two quadrotors, two rigid links, two cables, and load. The six-bar linkage contains a displacement constraint:

$$(x_2 - x_1)\mathbf{N}_x + (y_2 - y_1)\mathbf{N}_y + (z_2 - z_1)\mathbf{N}_z + L\mathbf{Q}_{1z} + l\mathbf{C}_{1z} - l_b\mathbf{B}_x - l\mathbf{C}_{2x} - L\mathbf{Q}_{2z} = \mathbf{0} \tag{6.53}$$

The generalized speeds, u_i ($i = 1, 2, \ldots, 18$), are chosen as:

$$\begin{cases} u_1 = \dot{x}_1, & u_2 = \dot{y}_1, & u_3 = \dot{z}_1, & u_4 = \dot{\psi}_1, & u_5 = \dot{\theta}_1, & u_6 = \dot{\phi}_1, \\ u_7 = \dot{x}_2, & u_8 = \dot{y}_2, & u_9 = \dot{z}_2, & u_{10} = \dot{\psi}_2, & u_{11} = \dot{\theta}_2, & u_{12} = \dot{\phi}_2, \\ u_{13} = \dot{\alpha}_{1x}, & u_{14} = \dot{\alpha}_{1y}, & u_{15} = \dot{\alpha}_{2x}, & u_{16} = \dot{\alpha}_{2y}, & u_{17} = \dot{\gamma}, & u_{18} = \dot{\beta}. \end{cases} \tag{6.54}$$

The generalized active forces include the gravity of quadrotors and load, thrust force, and thrust moment. Then the generalized active forces, F_i ($i = 1, 2, \ldots, 18$), can be written as:

$$F_i = (mg\mathbf{N}_z - f_1\mathbf{Q}_{1z}) \cdot {}^{N}\mathbf{v}_i^{Q_{1o}} + (mg\mathbf{N}_z - f_2\mathbf{Q}_{2z}) \cdot {}^{N}\mathbf{v}_i^{Q_{2o}} + m_b g\mathbf{N}_z \cdot {}^{N}\mathbf{v}_i^{B_o} + (M_{\psi 1}\mathbf{Q}_{1z} + M_{\theta 1}\mathbf{Q}_{1y} + M_{\varphi 1}\mathbf{Q}_{1x}) \cdot {}^{N}\boldsymbol{\omega}_i^{Q_1} + (M_{\psi 2}\mathbf{Q}_{2z} + M_{\theta 2}\mathbf{Q}_{2y} + M_{\varphi 2}\mathbf{Q}_{2x}) \cdot {}^{N}\boldsymbol{\omega}_i^{Q_2} \tag{6.55}$$

where $\cdot$ denotes the dot product, ${}^{N}\mathbf{v}_i^{Q_{1o}}$, ${}^{N}\mathbf{v}_i^{Q_{2o}}$, and ${}^{N}\mathbf{v}_i^{B_o}$ denote the ith partial velocities of mass center, Q_{1o} and Q_{2o}, of two quadrotors, and B_o of the load, ${}^{N}\boldsymbol{\omega}_i^{Q_1}$, ${}^{N}\boldsymbol{\omega}_i^{Q_2}$ denote the ith partial angular velocities of the quadrotors, $\mathbf{Q}_1$ and $\mathbf{Q}_2$. The derivative of Equations (6.49) and (6.51) with respect to generalized speed, u_i, give rise to ${}^{N}\mathbf{v}_i^{Q_{1o}}$, ${}^{N}\mathbf{v}_i^{Q_{2o}}$, and ${}^{N}\mathbf{v}_i^{B_o}$, respectively. ${}^{N}\boldsymbol{\omega}_i^{Q_1}$ and ${}^{N}\boldsymbol{\omega}_i^{Q_2}$ derive from the derivative of Equations (6.50) and (6.52) with respect to generalized speed, u_i.

The corresponding generalized inertia forces, F_i^* ($i = 1, 2, \ldots, 18$), can be expressed as:

$$F_i^* = -(\mathbf{I}^{Q_1/Q_{1o}} \cdot {}^{N}\boldsymbol{\alpha}^{Q_1} + {}^{N}\boldsymbol{\omega}^{Q_1} \times \mathbf{I}^{Q_1/Q_{1o}} \cdot {}^{N}\boldsymbol{\omega}^{Q_1}) \cdot {}^{N}\boldsymbol{\omega}_i^{Q_1} - m\, {}^{N}\mathbf{a}^{Q_{1o}} \cdot {}^{N}\mathbf{v}_i^{Q_{1o}} - (\mathbf{I}^{Q_2/Q_{2o}} \cdot {}^{N}\boldsymbol{\alpha}^{Q_2} + {}^{N}\boldsymbol{\omega}^{Q_2} \times \mathbf{I}^{Q_2/Q_{2o}} \cdot {}^{N}\boldsymbol{\omega}^{Q_2}) \cdot {}^{N}\boldsymbol{\omega}_i^{Q_2} - m\, {}^{N}\mathbf{a}^{Q_{2o}} \cdot {}^{N}\mathbf{v}_i^{Q_{2o}} - (\mathbf{I}^{B/B_o} \cdot {}^{N}\boldsymbol{\alpha}^{B} + {}^{N}\boldsymbol{\omega}^{B} \times \mathbf{I}^{B/B_o} \cdot {}^{N}\boldsymbol{\omega}^{B}) \cdot {}^{N}\boldsymbol{\omega}_i^{B} - m\, {}^{N}\mathbf{a}^{B_o} \cdot {}^{N}\mathbf{v}_i^{B_o} \tag{6.56}$$

where $\mathbf{I}^{Q_1/Q_{1o}}$, $\mathbf{I}^{Q_2/Q_{2o}}$, and $\mathbf{I}^{B/B_o}$ denote the inertia dyadic of the quadrotors, $\mathbf{Q}_1$ and $\mathbf{Q}_2$, and load $\mathbf{B}$, ${}^{N}\boldsymbol{\alpha}^{Q_1}$, ${}^{N}\boldsymbol{\alpha}^{Q_2}$, and ${}^{N}\boldsymbol{\alpha}^{B}$ represent the angular acceleration of the quadrotors, $\mathbf{Q}_1$ and $\mathbf{Q}_2$, and load $\mathbf{B}$, ${}^{N}\mathbf{a}^{Q_{1o}}$, ${}^{N}\mathbf{a}^{Q_{2o}}$, and ${}^{N}\mathbf{a}^{B_o}$ denote the acceleration of the mass center, Q_{1o} and Q_{2o}, of two quadrotors, and B_o, of the load, ${}^{N}\boldsymbol{\omega}_i^{B}$ denotes the ith partial angular velocities of the load, $\mathbf{B}$. The derivative of Equations (6.50) and (6.52) with respect to time results in ${}^{N}\boldsymbol{\alpha}^{Q_1}$, ${}^{N}\boldsymbol{\alpha}^{Q_2}$, and ${}^{N}\boldsymbol{\alpha}^{B}$, respectively. ${}^{N}\mathbf{a}^{Q_{1o}}$, ${}^{N}\mathbf{a}^{Q_{2o}}$, and ${}^{N}\mathbf{a}^{B_o}$ derive from the derivative of Equations (6.49) and (6.51) with respect to time. ${}^{N}\boldsymbol{\omega}_i{}^{Q_1}$ and ${}^{N}\boldsymbol{\omega}_i{}^{Q_2}$ derive from the derivative of Equation (6.50) with respect to generalized speed, u_i. ${}^{N}\boldsymbol{\omega}_i^{B}$ derives from the derivative of Equation (6.52) with respect to generalized speed, u_i. The inertia dyadic can be written as:

$$\mathbf{I}^{Q_1/Q_{1o}} = I_{xx}\mathbf{Q}_{1x}\mathbf{Q}_{1x} + I_{yy}\mathbf{Q}_{1y}\mathbf{Q}_{1y} + I_{zz}\mathbf{Q}_{1z}\mathbf{Q}_{1z} \tag{6.57}$$

$$\mathbf{I}^{Q_2/Q_{2o}} = I_{xx}\mathbf{Q}_{2x}\mathbf{Q}_{2x} + I_{yy}\mathbf{Q}_{2y}\mathbf{Q}_{2y} + I_{zz}\mathbf{Q}_{2z}\mathbf{Q}_{2z} \tag{6.58}$$

$$\mathbf{I}^{B/B_o} = \frac{1}{12}m_b l_b^2\mathbf{B}_y\mathbf{B}_y + \frac{1}{12}m_b l_b^2\mathbf{B}_z\mathbf{B}_z \tag{6.59}$$

The full dynamic model of the motion of dual quadrotors transporting a slender bar can produce by forcing the sum of the generalized active forces (6.55) and generalized inertia forces (6.56) to zero:

$$\begin{aligned}
&(mg\mathbf{N}_z - f_1\mathbf{Q}_{1z}) \cdot {}^{N}\mathbf{v}_i^{Q_{1o}} + (mg\mathbf{N}_z - f_2\mathbf{Q}_{2z}) \cdot {}^{N}\mathbf{v}_i^{Q_{2o}} + m_b g\mathbf{N}_z \cdot {}^{N}\mathbf{v}_i^{B_o} \\
&+ (M_{\psi1}\mathbf{Q}_{1z} + M_{\theta1}\mathbf{Q}_{1y} + M_{\varphi1}\mathbf{Q}_{1x}) \cdot {}^{N}\boldsymbol{\omega}_i^{Q_1} + (M_{\psi2}\mathbf{Q}_{2z} + M_{\theta2}\mathbf{Q}_{2y} + M_{\varphi2}\mathbf{Q}_{2x}) \cdot {}^{N}\boldsymbol{\omega}_i^{Q_2} \\
&= (\mathbf{I}^{Q_1/Q_{1o}} \cdot {}^{N}\boldsymbol{\alpha}^{Q_1} + {}^{N}\boldsymbol{\omega}^{Q_1} \times \mathbf{I}^{Q_1/Q_{1o}} \cdot {}^{N}\boldsymbol{\omega}^{Q_1}) \cdot {}^{N}\boldsymbol{\omega}_i^{Q_1} + m\, {}^{N}\mathbf{a}^{Q_{1o}} \cdot {}^{N}\mathbf{v}_i^{Q_{1o}} \\
&+ (\mathbf{I}^{Q_2/Q_{2o}} \cdot {}^{N}\boldsymbol{\alpha}^{Q_2} + {}^{N}\boldsymbol{\omega}^{Q_2} \times \mathbf{I}^{Q_2/Q_{2o}} \cdot {}^{N}\boldsymbol{\omega}^{Q_2}) \cdot {}^{N}\boldsymbol{\omega}_i^{Q_2} + m\, {}^{N}\mathbf{a}^{Q_{2o}} \cdot {}^{N}\mathbf{v}_i^{Q_{2o}} \\
&+ (\mathbf{I}^{B/B_o} \cdot {}^{N}\boldsymbol{\alpha}^{B} + {}^{N}\boldsymbol{\omega}^{B} \times \mathbf{I}^{B/B_o} \cdot {}^{N}\boldsymbol{\omega}^{B}) \cdot {}^{N}\boldsymbol{\omega}_i^{B} + m_b\, {}^{N}\mathbf{a}^{B_o} \cdot {}^{N}\mathbf{v}_i^{B_o}
\end{aligned} \tag{6.60}$$

The nonlinear equations (6.60) of motions for the model shown in Figure 6.28 are complicated. Therefore, a simplified model is essential for dynamic analysis and oscillation suppression.

6.3.2 Model Reduction

6.3.2.1 Planar Model

When the dual quadrotors are assumed to undergo planar motions, a planar model can be derived from the full dynamic model. In the planar model, the thrust forces, f_1 and f_2, and thrust moments, $M_{\theta1}$ and $M_{\theta2}$, are inputs to the planar model, while outputs are quadrotor displacements x_1, z_1, x_2, and z_2, attitudes, θ_1 and θ_2, swing angles, α_{1y} and α_{2y}, and slope angle, γ.

The velocities of the mass center of the quadrotors, $\mathbf{Q}_1$ and $\mathbf{Q}_2$, and the load, $\mathbf{B}$, in the planar model are:

$$\begin{cases}
{}^{N}\mathbf{vp}^{Q_{1o}} = \dot{x}_1\mathbf{N}_x + \dot{z}_1\mathbf{N}_z \\
{}^{N}\mathbf{vp}^{Q_{2o}} = \dot{x}_2\mathbf{N}_x + \dot{z}_2\mathbf{N}_z \\
{}^{N}\mathbf{vp}^{B_o} = \dot{x}_1\mathbf{N}_x + \dot{z}_1\mathbf{N}_z + L\dot{\theta}_1\mathbf{Q}_{1x} + l(\dot{\theta}_1 + \dot{\alpha}_{1y})\mathbf{C}_{1x} - 0.5l_b\dot{\gamma}\mathbf{B}_z
\end{cases} \tag{6.61}$$

The angular velocities of the quadrotors, $\mathbf{Q}_1$ and $\mathbf{Q}_2$, and the load, $\mathbf{B}$, in the planar model are:

$$\begin{cases}
{}^{N}\boldsymbol{\omega}\mathbf{p}^{Q_1} = \dot{\theta}_1 C_{\theta_1}\mathbf{Q}_{1y} \\
{}^{N}\boldsymbol{\omega}\mathbf{p}^{Q_2} = \dot{\theta}_2 C_{\theta_2}\mathbf{Q}_{2y} \\
{}^{N}\boldsymbol{\omega}\mathbf{p}^{B} = \dot{\gamma}\mathbf{B}_y
\end{cases} \tag{6.62}$$

The generalized speeds, up_i ($i = 1, 2, \ldots, 9$), in the planar model are:

$$\begin{cases}
up_1 = \dot{x}_1, \quad up_2 = \dot{z}_1, \quad up_3 = \dot{\theta}_1, \quad up_4 = \dot{x}_2, \quad up_5 = \dot{z}_2, \\
up_6 = \dot{\theta}_2, \quad up_7 = \dot{\alpha}_{1y}, \quad up_8 = \dot{\alpha}_{2y}, \quad up_9 = \dot{\gamma}.
\end{cases} \tag{6.63}$$

The planar dynamic model is given by:

$$
\begin{aligned}
&(mg\mathbf{N}_z - f_1\mathbf{Q}_{1z}) \cdot {}^N\mathbf{vp}_i^{Q_{1o}} + (mg\mathbf{N}_z - f_2\mathbf{Q}_{2z}) \cdot {}^N\mathbf{vp}_i^{Q_{2o}} + m_b g\mathbf{N}_z \cdot {}^N\mathbf{vp}_i^{B_o} \\
&+ M_{\theta 1}\mathbf{Q}_{1y} \cdot {}^N\boldsymbol{\omega}\mathbf{p}_i^{Q_1} + M_{\theta 2}\mathbf{Q}_{2y} \cdot {}^N\boldsymbol{\omega}\mathbf{p}_i^{Q_2} \\
&= I_{yy}\mathbf{Q}_{1y}\mathbf{Q}_{1y} \cdot {}^N\boldsymbol{\alpha}\mathbf{p}^{Q_1} \cdot {}^N\boldsymbol{\omega}\mathbf{p}_i^{Q_1} + I_{yy}\mathbf{Q}_{2y}\mathbf{Q}_{2y} \cdot {}^N\boldsymbol{\alpha}\mathbf{p}^{Q_2} \cdot {}^N\boldsymbol{\omega}\mathbf{p}_i^{Q_2} + m\,{}^N\mathbf{ap}^{Q_{1o}} \cdot {}^N\mathbf{vp}_i^{Q_{1o}} \\
&+ \frac{1}{12}m_b l_b^2\mathbf{B}_y\mathbf{B}_y \cdot {}^N\boldsymbol{\alpha}\mathbf{p}^B \cdot {}^N\boldsymbol{\omega}\mathbf{p}_i^B + m\,{}^N\mathbf{ap}^{Q_{2o}} \cdot {}^N\mathbf{vp}_i^{Q_{2o}} + m_b\,{}^N\mathbf{ap}^{B_o} \cdot {}^N\mathbf{vp}_i^{B_o}
\end{aligned}
\tag{6.64}
$$

where ${}^N\mathbf{vp}_i^{Q_{1o}}$, ${}^N\mathbf{vp}_i^{Q_{2o}}$, and ${}^N\mathbf{vp}_i^{B_o}$ denote the derivative of Equations (6.61) with respect to generalized speed, up_i, ${}^N\boldsymbol{\omega}\mathbf{p}_i^{Q_1}$, ${}^N\boldsymbol{\omega}\mathbf{p}_i^{Q_2}$, and ${}^N\boldsymbol{\omega}\mathbf{p}_i^B$ derive from the derivative of Equation (6.62) with respect to generalized speed, up_i, ${}^N\mathbf{ap}^{Q_{1o}}$, ${}^N\mathbf{ap}^{Q_{2o}}$, and ${}^N\mathbf{ap}^{B_o}$ derive from the derivative of Equation (6.61) with respect to time, ${}^N\boldsymbol{\alpha}\mathbf{p}^{Q_1}$, ${}^N\boldsymbol{\alpha}\mathbf{p}^{Q_2}$, and ${}^N\boldsymbol{\alpha}\mathbf{p}^B$ denote the derivative of Equation (6.62) with respect to time.

6.3.2.2 Planar Near-Hover Model

The dual quadrotors are assumed to undergo planar and near-hover motions. Then a planar near-hover model can be derived from the planar model (6.64). In the planar near-hover model, the thrust moments, $M_{\theta 1}$ and $M_{\theta 2}$, are inputs to the model, while outputs are quadrotor attitudes, θ_1 and θ_2, swing angles, α_{1y} and α_{2y}, and slope angle, γ.

The velocity of the load, $\mathbf{B}$, in the planar near-hover model is:

$$
{}^N\mathbf{vph}^{B_o} = L\dot{\theta}_1\mathbf{Q}_{1x} + l(\dot{\theta}_1 + \dot{\alpha}_{1y})\mathbf{C}_{1x} - 0.5l_b\dot{\gamma}\mathbf{B}_z \tag{6.65}
$$

The angular velocities of the quadrotors, $\mathbf{Q}_1$ and $\mathbf{Q}_2$, and the load, $\mathbf{B}$, in the planar near-hover model are:

$$
\begin{cases}
{}^N\boldsymbol{\omega}\mathbf{ph}^{Q_1} = \dot{\theta}_1\mathbf{Q}_{1y} \\
{}^N\boldsymbol{\omega}\mathbf{ph}^{Q_2} = \dot{\theta}_2\mathbf{Q}_{2y} \\
{}^N\boldsymbol{\omega}\mathbf{ph}^{B} = \dot{\gamma}\mathbf{B}_y
\end{cases}
\tag{6.66}
$$

The generalized speeds, uph_i ($i = 1, 2, \ldots, 5$), in the planar near-hover model are:

$$
uph_1 = \dot{\theta}_1,\ uph_2 = \dot{\theta}_2,\ uph_3 = \dot{\alpha}_{1y},\ uph_4 = \dot{\alpha}_{2y},\ uph_5 = \dot{\gamma} \tag{6.67}
$$

The planar near-hover dynamic model is given by:

$$
\begin{aligned}
& m_b g \mathbf{N}_z \cdot {}^N\mathbf{vph}_i^{B_o} + M_{\theta 1}\mathbf{Q}_{1y} \cdot {}^N\boldsymbol{\omega}\mathbf{ph}_i^{Q_1} + M_{\theta 2}\mathbf{Q}_{2y} \cdot {}^N\boldsymbol{\omega}\mathbf{ph}_i^{Q_2} \\
& = I_{yy}\mathbf{Q}_{1y}\mathbf{Q}_{1y} \cdot {}^N\boldsymbol{\alpha}\mathbf{ph}^{Q_1} \cdot {}^N\boldsymbol{\omega}\mathbf{ph}_i^{Q_1} + I_{yy}\mathbf{Q}_{2y}\mathbf{Q}_{2y} \cdot {}^N\boldsymbol{\alpha}\mathbf{ph}^{Q_2} \cdot {}^N\boldsymbol{\omega}\mathbf{p}_i^{Q_2} \\
& + \frac{1}{12} m_b l_b^2 \mathbf{B}_y\mathbf{B}_y \cdot {}^N\boldsymbol{\alpha}\mathbf{ph}^B \cdot {}^N\boldsymbol{\omega}\mathbf{ph}_i^B
\end{aligned}
\tag{6.68}
$$

where ${}^N\mathbf{vph}_i^{B_o}$ denotes the derivative of Equation (6.65) with respect to generalized speed, uph_i, ${}^N\boldsymbol{\omega}\mathbf{ph}_i^{Q_1}$, ${}^N\boldsymbol{\omega}\mathbf{ph}_i^{Q_2}$, and ${}^N\boldsymbol{\omega}\mathbf{ph}_i^B$ derive from the derivative of Equation (6.66) with respect to generalized speed, uph_i, ${}^N\boldsymbol{\alpha}\mathbf{p}^{Q_1}$, ${}^N\boldsymbol{\alpha}\mathbf{p}^{Q_2}$, and ${}^N\boldsymbol{\alpha}\mathbf{p}^B$ denote the derivative of Equation (6.66) with respect to time.

When the distance between two quadrotors is assumed to be equal to the bar length, the load slope can be ignored because of the six-bar linkage. Oscillations of the quadrotor's attitude and load swing near the equilibria are assumed to be small. From the planar near-hover model (6.68), a linearized model can be derived based on those abovementioned assumptions:

$$
\begin{cases}
I_{yy}\ddot{\theta}_1 - 0.5 m_b g L \alpha_{1y} - M_{\theta 1} = 0 \\
I_{yy}\ddot{\theta}_2 - 0.5 m_b g L \alpha_{2y} - M_{\theta 2} = 0 \\
2lI_{yy}\ddot{\alpha}_{1y} + I_{yy} g\theta_1 + I_{yy} g\theta_2 + [I_{yy} g + m_b g L(l + L)]\alpha_{1y} + I_{yy} g\alpha_{2y} + 2(l + L)M_{\theta 1} \\
\quad = 0 \\
2lI_{yy}\ddot{\alpha}_{2y} + I_{yy} g\theta_1 + I_{yy} g\theta_2 + I_{yy} g\alpha_{1y} + [I_{yy} g + m_b g L(l + L)]\alpha_{2y} + 2(l + L)M_{\theta 2} \\
\quad = 0
\end{cases}
\tag{6.69}
$$

6.3.3 Control of Dual Quadrotors Slung Loads

Figure 6.29 shows the hybrid control architecture of dual quadrotors slung loads, which includes a smoother, a prescribed model, and a tracking controller. The smoother is the prefilter, which should be added before the prescribed model and

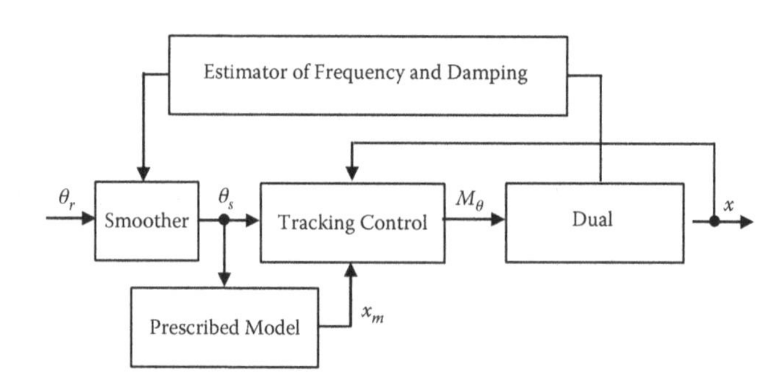

FIGURE 6.29 Hybrid control architecture of dual quadrotors slung loads.

tracking controller for decreasing the rise time of oscillation suppression. The operator's commands, θ_r, filter through the smoother to create modified commands, θ_s. The prescribed model is the feedforward controller, in which the column matrix of modified command, θ_s, is the input and output is the column matrix of model state, x_m. The tracking controller is the feedback controller, in which the column matrix of system state, x, is applied to regulate following error, e, in a closed loop and generate the column matrix of thrust moment, M_θ. An estimator of design frequency and damping ratio is installed for the smoother by using system parameters.

A discontinuous piecewise (DP) smoother will be used in Figure 6.29 for oscillation suppression:

$$c(\tau) = \begin{cases} \frac{\zeta_m \omega_m e^{-\zeta_m \omega_m \tau}}{(1+K)(1-K^h)}, & 0 \le \tau \le hT \\ 0, & hT < \tau < T \\ \frac{\zeta_m \omega_m e^{-\zeta_m \omega_m \tau}}{(1+K)(1-K^h)}, & T \le \tau \le (1+h)T \\ 0, & \text{others} \end{cases} \tag{6.70}$$

where

$$K = e^{\frac{-\pi\zeta_m}{\sqrt{1-\zeta_m^2}}} \tag{6.71}$$

$$T = \frac{\pi}{\omega_m \sqrt{1-\zeta_m^2}} \tag{6.72}$$

ζ_m is the design damping ratio and h is the modified factor. The modified factor satisfies $0 < h < 1$. The rise time of the DP smoother satisfies $(1 + h)T$.

The prescribed model is described as:

$$\begin{cases} \ddot{\theta}_{m1} + 2\zeta_p \sqrt{\frac{g}{l}}\dot{\theta}_{m1} + \frac{g}{l}\theta_{m1} = \frac{g}{l}\theta_{s1} \\ \ddot{\alpha}_{m1} + 2\zeta_p \sqrt{\frac{g}{l}}\dot{\alpha}_{m1} + \frac{g}{l}\alpha_{m1} = -\frac{g}{l}\theta_{s1} \\ \ddot{\theta}_{m2} + 2\zeta_p \sqrt{\frac{g}{l}}\dot{\theta}_{m2} + \frac{g}{l}\theta_{m2} = \frac{g}{l}\theta_{s2} \\ \ddot{\alpha}_{m2} + 2\zeta_m \sqrt{\frac{g}{l}}\dot{\alpha}_{m2} + \frac{g}{l}\alpha_{m2} = -\frac{g}{l}\theta_{s2} \end{cases} \tag{6.73}$$

where θ_{s1} and θ_{s2} are the modified commands and are the input to the prescribed model; θ_{m1}, α_{m1}, θ_{m2}, and α_{m2} are the outputs of the prescribed model; and ζ_p is the damping ratio of the prescribed model. A column matrix, which is referred to as $x_m = [\theta_{m1}, \dot{\theta}_{m1}\alpha_{m1}, \dot{\alpha}_{m1}, \theta_{m2}, \dot{\theta}_{m2}\alpha_{m2}, \dot{\alpha}_{m2}]^T$, is the output of the prescribed model. Another column matrix, which is $x = [\theta_1,\ \dot{\theta}_1,\ \alpha_{1y},\ \dot{\alpha}_{1y},\ \theta_2,\ \dot{\theta}_2,\ \alpha_{2y},\ \dot{\alpha}_{2y}]^T$, is the output of the quadrotors slung loads.

The asymptotic tracking controller forces the system state, x, to follow the model state, x_m. The asymptotic tracking control controller is expressed as:

$$\begin{cases} M_{\theta 1} = \frac{g}{l} I_{yy}\theta_{s1} - \frac{g}{l} I_{yy}\theta_1 - 2\zeta_p\sqrt{\frac{g}{l}} I_{yy}\dot{\theta}_1 - \frac{m_b g L}{2I_{yy}}\alpha_{1y} + \Lambda \\ M_{\theta 2} = \frac{g}{l} I_{yy}\theta_{s2} - \frac{g}{l} I_{yy}\theta_2 - 2\zeta_p\sqrt{\frac{g}{l}} I_{yy}\dot{\theta}_2 - \frac{m_b g L}{2I_{yy}}\alpha_{2y} + \Lambda \end{cases} \tag{6.74}$$

where

$$\begin{aligned} \Lambda = {} & \frac{l}{2g}\Big[(\theta_{m1} - \theta_1) + (\theta_{m2} - \theta_2) - (\alpha_{m1} - \alpha_{1y}) - (\alpha_{m2} - \alpha_{2y})\Big] \\ & + \frac{(g+l)\sqrt{l}}{4\zeta_p g\sqrt{g}}\Big[(\dot{\theta}_{m1} - \dot{\theta}_1) + (\dot{\theta}_{m2} - \dot{\theta}_2) - (\dot{\alpha}_{m1} - \dot{\alpha}_{1y}) - (\dot{\alpha}_{m2} - \dot{\alpha}_{2y})\Big] \end{aligned} \tag{6.75}$$

A Lyapunov function is defined as:

$$V(x) = e^T P e = (x - x_m)^T P (x - x_m) \tag{6.76}$$

where e is the column matrix of the tracking error between the system state, x, and the model state, x_m, and satisfies $e = x - x_m$. The square matrix, P, of order 8 defined in Equation (6.76) satisfies:

$$P = \begin{bmatrix} W & 0 & 0 & 0 \\ 0 & W & 0 & 0 \\ 0 & 0 & W & 0 \\ 0 & 0 & 0 & W \end{bmatrix} \tag{6.77}$$

where W is a square matrix of order 2 and is denoted by:

$$W = \begin{bmatrix} \frac{(4\zeta_p^2 l + g + l)}{4\zeta_p\sqrt{gl}} & \frac{l}{2g} \\ \frac{l}{2g} & \frac{(g+l)\sqrt{l}}{4\zeta_p g\sqrt{g}} \end{bmatrix} \tag{6.78}$$

Resulting from Equations (6.76)–(6.78), the Lyapunov function (6.76) is positive. The derivative of the Lyapunov function (6.76) with respect to time is given by:

$$\dot{V}(x) = e^T(A_m^T P + P A_m)e + 2e^T P[(A_m - A)x + B_m\theta_s - BM_\theta] \tag{6.79}$$

where

$$A_m = \begin{bmatrix} 0 & 1 & 0 & 0 & 0 & 0 & 0 & 0 \\ \frac{-g}{l} & \frac{-2\zeta_p\sqrt{g}}{\sqrt{l}} & 0 & 0 & 0 & 0 & 0 & 0 \\ 0 & 0 & 0 & 1 & 0 & 0 & 0 & 0 \\ 0 & 0 & \frac{-g}{l} & \frac{-2\zeta_p\sqrt{g}}{\sqrt{l}} & 0 & 0 & 0 & 0 \\ 0 & 0 & 0 & 0 & 0 & 1 & 0 & 0 \\ 0 & 0 & 0 & 0 & \frac{-g}{l} & \frac{-2\zeta_p\sqrt{g}}{\sqrt{l}} & 0 & 0 \\ 0 & 0 & 0 & 0 & 0 & 0 & 0 & 1 \\ 0 & 0 & 0 & 0 & 0 & 0 & \frac{-g}{l} & \frac{-2\zeta_p\sqrt{g}}{\sqrt{l}} \end{bmatrix} \tag{6.80}$$

$$B_m = \begin{bmatrix} 0 & g/l & 0 & -g/l & 0 & 0 & 0 & 0 \\ 0 & 0 & 0 & 0 & 0 & g/l & 0 & -g/l \end{bmatrix}^T \tag{6.81}$$

$$A = \begin{bmatrix} 0 & 1 & 0 & 0 & 0 & 0 & 0 & 0 \\ 0 & 0 & \frac{m_b gL}{2I_{yy}} & 0 & 0 & 0 & 0 & 0 \\ 0 & 0 & 0 & 1 & 0 & 0 & 0 & 0 \\ \frac{-g}{2l} & 0 & \frac{-g}{2l} - \frac{m_b gL}{2I_{yy}} & 0 & \frac{-g}{2l} & 0 & \frac{-g}{2l} & 0 \\ 0 & 0 & 0 & 0 & 0 & 1 & 0 & 0 \\ 0 & 0 & 0 & 0 & 0 & 0 & \frac{m_b gL}{2I_{yy}} & 0 \\ 0 & 0 & 0 & 0 & 0 & 0 & 0 & 1 \\ \frac{-g}{2l} & 0 & \frac{-g}{2l} & 0 & \frac{-g}{2l} & 0 & \frac{-g}{2l} - \frac{m_b gL}{2I_{yy}} & 0 \end{bmatrix} \tag{6.82}$$

$$B = \begin{bmatrix} 0 & 1/I_{yy} & 0 & -1/I_{yy} & 0 & 0 & 0 & 0 \\ 0 & 0 & 0 & 0 & 0 & 1/I_{yy} & 0 & -1/I_{yy} \end{bmatrix}^T \tag{6.83}$$

$$\theta_s = [\theta_{s1}, \theta_{s2}]^T \tag{6.84}$$

$$M_\theta = [M_{\theta 1}, M_{\theta 2}]^T \tag{6.85}$$

Substituting Equations (6.80)–(6.83) into the right side of Equation (6.79) gives rise to:

$$\dot{V}(x) = -e^T e - \frac{2\Lambda^2}{I_{yy}} \tag{6.86}$$

The right side of Equation (6.86) is negative such that the closed-loop control system including tracking controller (6.74) and linearized model (6.69) is asymptotically stable. By measuring the angular displacement and angular velocity of the quadrotor attitude and load swing, the tracking control law forces the error, e, to be zero asymptotically.

6.3.4 Computational-Dynamics Analyses

Substituting Equation (6.74) into the linearized model (6.69) gives rise to the three frequencies and damping ratios of controlled systems. Three frequencies and damping ratios are dependent on the moments of inertia, I_{yy}, link length, L, cable length, l, load mass, m_b, and damping ratio, ζ_p, of the prescribed model. The first-mode damping ratio is near zero, while the last two are approximately damping ratio of the prescribed model.

When the damping ratio of the prescribed model is designed to be much larger than zero, oscillations of the last two modes will be suppressed quickly. Under this condition, only the first-mode oscillations should be considered for the smoother. Therefore, the first-mode frequency and damping ratio of the controlled system will be used as the design frequency, ω_m, and damping ratio, ζ_m, for the smoother.

The nonlinear dynamics shown in Equation (6.60) are too complicated to analytically design tracking controller (6.74) and smoother (6.70). Therefore, the linearized model (6.69) was used to design tracking controller and smoother. The design including smoother (6.70), prescribed model (6.73), and tracking controller (6.74) will be applied on the nonlinear dynamics (6.60) to analyze the behavior of controlled dynamics in this section.

Figure 6.30 shows the simulated response of an initial condition. The initial vehicle attitude is zero, while the initial swing angle is non-zero value. The initial

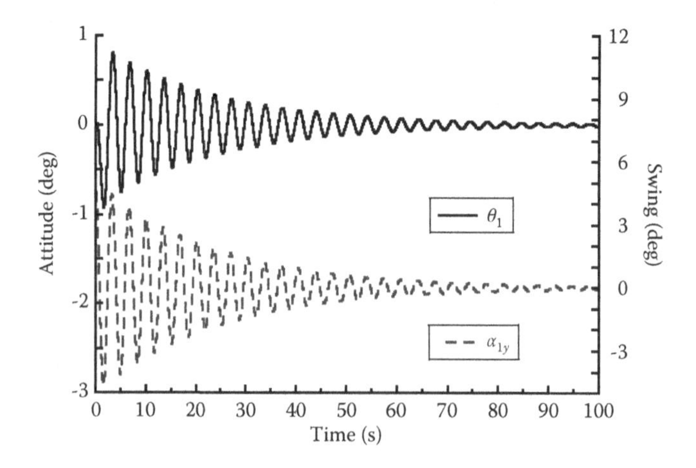

FIGURE 6.30 Simulated response to a non-zero initial condition.

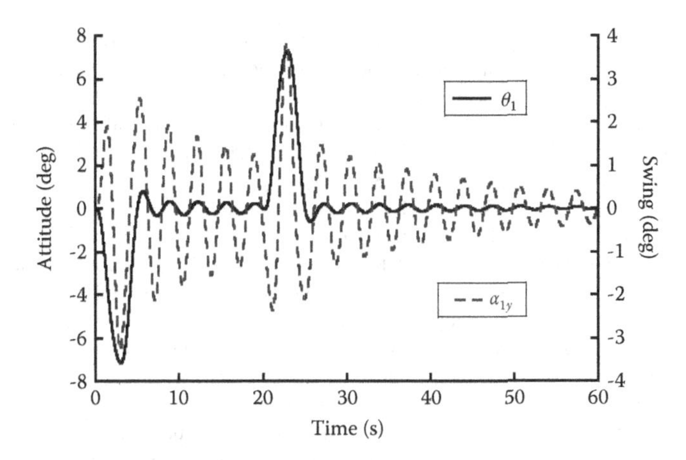

FIGURE 6.31 Response of vehicle attitude and load swing.

swing will result in oscillations of vehicle attitude. The vibrational frequencies are predicted accurately using the abovementioned method. The tracking controller (6.74) reduces the oscillations of vehicle attitude, and also caused vibration reduction of load swing.

Figure 6.31 shows the simulated response to a flight command. The fight commands caused oscillations of vehicle attitude and load swing. The abovementioned method for predicting frequency is effective to estimate the oscillational frequencies in Figure 6.30. The DP smoother and tracking controller suppressed the oscillations of vehicle attitude and load swing. Therefore, the combined control method is effective.

REFERENCES

[1] F.A. Leban, J. Díaz-Gonzalez, G.G. Parker, W.F. Zhao, Inverse kinematic control of a dual crane system experiencing base motion, *IEEE Transactions on Control Systems Technology*, 23(1) (2015) 331–339.

[2] A.V. Perig, A.N. Stadnik, A.A. Kostikov, S.V. Podlesny, Research into 2D dynamics and control of small oscillations of a cross-beam during transportation by two overhead cranes, *Shock and Vibration*, 2017 (2017) 9605657.

[3] J. Vaughan, J. Yoo, N. Knight, W. Singhose, Dynamics and control of multiple cranes with a connected payload, *19th International Congress on Sound and Vibration*, Vilnius, Lithuania, July 2012, pp. 1–8.

[4] J. Vaughan, J. Yoo, W. Singhose, Using approximate multi-crane frequencies for input shaper design, *12th International Conference on Control, Automation and Systems*, Jeju, Korea, 2012, pp. 639–644.

[5] E. Maleki, W. Singhose, J. Hawke, J. Vaughan, Dynamic response of a dual-hoist bridge crane, *Proceedings of the ASME 2013 Dynamic Systems and Control Conference*, Palo Alto, California, USA, 2013, pp. 1–8.

[6] J. Vaughan, J. Yoo, N. Knight, W. Singhose, Multi-input shaping control for multi-hoist cranes, *American Control Conference*, Washington DC, USA, 2013, pp. 3449–3454.

[7] A.S. Miller, P. Sarvepalli, W. Singhose, Dynamics and control of dual-hoist cranes moving triangular payloads, *Proceedings of the ASME Dynamic Systems and Control Conference*, San Antonio, Texas, USA, 2014, pp. 1–9.

[8] C. Rhee, A.S. Miller, W. Singhose, Operator testing on dual-hoist cranes moving triangular payloads, *Proceedings of the ASME Dynamic Systems and Control Conference*, Columbus, Ohio, USA, 2015, pp. 1–8.

[9] B. Lu, Y. Fang, N. Sun, Modeling and nonlinear coordination control for an underactuated dual overhead crane system, *Automatica*, 91 (2018) 244–255.

[10] M.Z. Rafat, M.S. Wheatland, T.R. Bedding, Dynamics of a double pendulum with distributed mass, *American Journal of Physics*, 77(3) (2009) 216–223.

[11] X. Xie, J. Huang, Z. Liang, Vibration reduction for flexible systems by command smoothing, *Mechanical Systems and Signal Processing*, 39(1) (2013) 461–470.

[12] J. Huang, X. Xie, Z. Liang, Control of bridge cranes with distributed-mass payload dynamics, *IEEE/ASME Transactions on Mechatronics*, 20(1) (2015) 481–486.

[13] L. Ramli, Z. Mohamed, H.I. Jaafar, A neural network-based input shaping for swing suppression of an overhead crane under payload hoisting and mass variations, *Mechanical Systems and Signal Processing*, 107 (2018) 484–501.

[14] L.A. Tuan, H.M. Cuong, P.V. Trieu, et al., Adaptive neural network sliding mode control of shipboard container cranes considering actuator backlash, *Mechanical Systems and Signal Processing*, 112 (2018) 233–250.

[15] H.I. Jaafar, Z. Mohamed, M.A. Shamsudin, et al., Model reference command shaping for vibration control of multimode flexible systems with application to a double-pendulum overhead crane, *Mechanical Systems and Signal Processing*, 115 (2019) 677–695.

[16] V.D. La, K.T. Nguyen, Combination of input shaping and radial spring-damper to reduce tridirectional vibration of crane payload, *Mechanical Systems and Signal Processing*, 116 (2019) 310–321.

[17] M. Ahmad, R. Ismail, M. Ramli, N. Hambali, Comparative assessment of feed-forward schemes with NCTF for sway and trajectory control of a DPTOC, *International Conference on Intelligent and Advanced* Systems, Manila, Philippines, 2010, pp. 1–6.

[18] Z. Masoud, K. Alhazza, E. Abu-Nada, M. Majeed, A hybrid command-shaper for double-pendulum overhead cranes, *Journal of Vibration and Control*, 20(1) (2014) 24–37.

[19] Z. Masoud, K. Alhazza, A smooth multimode waveform command shaping control with selectable command length, *Journal of Sound and Vibration*, 397 (2017) 1–16.

[20] L. Golovin, S. Palis, Robust control for active damping of elastic gantry crane vibrations, *Mechanical Systems and Signal Processing*, 121 (2019) 264–278.

[21] H. Chen, Y. Fang, N. Sun, An adaptive tracking control method with swing suppression for 4-DOF tower crane systems, *Mechanical Systems and Signal Processing*, 123 (2019) 426–442.

[22] M. Maghsoudi, L. Ramli, S. Sudin, Z. Mohamed, A. Husain, H. Wahid, Improved unity magnitude input shaping scheme for sway control of an underactuated 3D overhead crane with hoisting, *Mechanical Systems and Signal Processing*, 123 (2019) 466–482.

7 Flexible Link Manipulators

The study of flexible link manipulators has been actuated by the demands of lightweight robotic systems and space applications. However, flexible link manipulators suffer from undesired oscillations caused by operator-commanded motions. The detrimental effects degrade positioning accuracy, capable operating speeds, and safety. Therefore, it is essential to control unwanted oscillations in flexible link manipulators.

7.1 LINEAR-OSCILLATOR DYNAMICS

Broad attention has been attracted on modeling and dynamics of flexible link manipulators [1–8]. Dynamical analysis indicates that single-link flexible manipulator includes an infinite number of vibration modes. The first one is the fundamental mode, but high modes might have some effects. Therefore, designing a control system for reducing vibrations of all modes is essential.

Hundreds of papers reported vibration control for flexible link manipulators. The control schemes can be broken into two categories: feedback control and open-loop control. The feedback control strategies use measurements of the flexible link to control vibrations in a closed loop. The feedback control methods include the proportional-integral-derivative control [9,10], delayed feedback control [11], positive position feedback control [12], linear quadratic regulator [13], adaptive control [14,15], sliding-mode control [16,17], fuzzy control [18,19], and artificial-neural-networks-based control [20,21]. The open-loop control strategies filter inputs to produce a desirable motion that results in minimal vibrations. The open-loop control methods include the optimal trajectory planning [22–24], and input shaping [25–28]. More literature on dynamics and control of flexible link manipulators might be found from the review [29–33].

Many works have been directed at controlling multi-mode vibrations for single-link flexible manipulators. However, measuring high-mode vibrations is an obstacle toward the feedback-control application. Meanwhile, vibrations of high modes, whose frequencies lie outside of the bandwidth of the actuator or sensor, are challenging to be controlled by the feedback controller. The input shaper is designed for each mode of the flexible link independently, and then convolving them together. However, high-mode frequencies are challenging to measure or estimate. Additionally, convolving an infinite number of input shapers is impossible. Thus, previously presented control methods are focused on reducing vibrations for the first few modes.

DOI: 10.1201/9781003247210-7

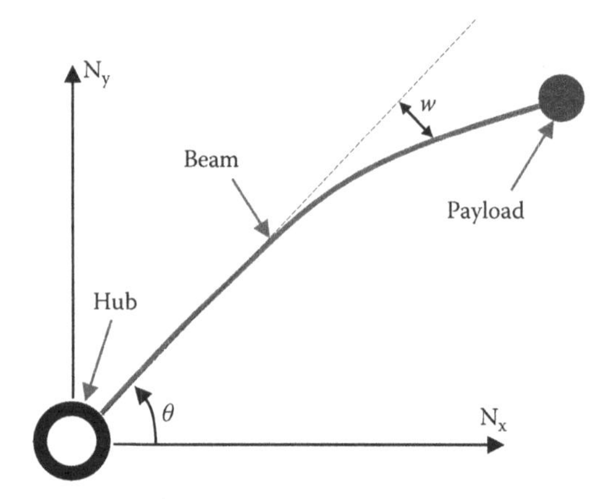

FIGURE 7.1 Model of a single-link flexible manipulator.

7.1.1 Modeling

A schematic representation of a single-link flexible manipulator is shown in Figure 7.1. The hub inputs an angular displacement, θ. A flexible beam of length, l_b, is attached to the hub, and supports a payload of mass, m_p. Ideally, responses of the payload and beam track the angular displacement, θ. However, such flexible structure often causes unwanted vibrations of the payload and beam.

The hub acceleration is the input to the model. The output is the deflection, $w(x, t)$, of a point along the beam at a distance, x, from the hub. It is assumed that the motion of the beam and payload does not affect the motion of the hub because of the large mechanical impedance in the drive system. The force equation of the motion is given by:

$$\rho dx\frac{\partial^2 w(x,\ t)}{\partial t^2} + \rho dx \cdot x\ddot{\theta}(t) + dV(x,\ t) = 0 \tag{7.1}$$

where ρ is the linear mass density of the beam, dx is an infinitely small change in the x, $w(x, t)$ is the deflection of a point along the beam at a distance, x, from the hub, $V(x, t)$ is the shearing force in the beam, and the differential $dV(x, t)$ represents an infinitely small change in the $V(x, t)$. The moment equation of the motion is given by:

$$dM(x,\ t) - V(x,\ t)\cdot dx - dV(x,\ t)\cdot dx - \rho dx\cdot x\ddot{\theta}(t)\cdot\frac{1}{2}dx = 0 \tag{7.2}$$

where $M(x, t)$ is the bending moment in the beam and the differential $dM(x, t)$ represents an infinitely small change in the bending moment, $M(x, t)$. From the theory of the beam bending, the relationship between bending moment and beam deflection satisfies:

$$M(x,\ t) = EI\frac{\partial^2 w(x,\ t)}{\partial x^2} \tag{7.3}$$

where I is the moment of inertia of the beam cross-section and E is Young's modulus. Substituting Equations (7.2) and (7.3) into (7.1), and ignoring terms involving high powers in dx produces:

$$EI\frac{\partial^4 w(x,t)}{\partial x^4} + \rho\frac{\partial^2 w(x,t)}{\partial t^2} = -\rho \cdot x\ddot{\theta}(t) \tag{7.4}$$

The boundary conditions are given by:

$$w(x,t)|_{x=0} = 0 \tag{7.5}$$

$$\left.\frac{\partial w(x,t)}{\partial x}\right|_{x=0} = 0 \tag{7.6}$$

$$\left.\frac{\partial^2 w(x,t)}{\partial x^2}\right|_{x=l_b} = 0 \tag{7.7}$$

$$\left.\frac{\partial}{\partial x}\left[EI\frac{\partial^2 w(x,t)}{\partial x^2}\right]\right|_{x=l_b} = m_p\left.\frac{\partial^2 w(x,t)}{\partial t^2}\right|_{x=l_b} \tag{7.8}$$

Using the mode superposition method, the deflection of the beam can be assumed as:

$$w(x,t) = \sum_{k=1}^{+\infty} \varphi_k(x) \cdot q_k(t) \tag{7.9}$$

where $\varphi_k(x)$ is the mode shape of the kth mode, and $q_k(t)$ is the time-dependent function of the kth vibration mode. The mode shape satisfies the following equations:

$$\varphi_k(x)|_{x=0} = 0 \tag{7.10}$$

$$\left.\frac{\partial \varphi_k(x)}{\partial x}\right|_{x=0} = 0 \tag{7.11}$$

$$\left.\frac{\partial^2 \varphi_k(x)}{\partial x^2}\right|_{x=l_b} = 0 \tag{7.12}$$

$$EI\left.\frac{\partial^3 \varphi_k(x)}{\partial x^3}\right|_{x=l_b} + m_p\omega_k^2\varphi_k(x)|_{x=l_b} = 0 \tag{7.13}$$

where ω_k is the natural frequency of the kth mode. Solving Equations (7.10)–(7.13) yields the frequency ω_k, and the mode shape $\varphi_k(x)$:

$$\omega_k = (\beta_k l_b)^2\sqrt{\frac{EI}{\rho l_b^4}} \tag{7.14}$$

$$\cos(\beta_k l_b)\cosh(\beta_k l_b) + 1 = h \cdot \beta_k l_b[\sin(\beta_k l_b)\cosh(\beta_k l_b) - \cos(\beta_k l_b)\sinh(\beta_k l_b)] \tag{7.15}$$

$$\varphi_k(x) = C\phi_k(x) \tag{7.16}$$

$$\phi_k(x) = \sin(\beta_k x) - \sinh(\beta_k x) - r\cos(\beta_k x) + r\cosh(\beta_k x) \tag{7.17}$$

$$r = \frac{\sin(\beta_k l_b) + \sinh(\beta_k l_b)}{\cos(\beta_k l_b) + \cosh(\beta_k l_b)} \tag{7.18}$$

where C is a constant and h is the ratio of the payload mass to the beam mass. The closed-form solution for the coefficients, β_k and r, is challenging, but the numerical solution can be derived. Substituting Equation (7.9) into (7.4) obtains:

$$EI\sum_{k=1}^{+\infty}\left[\frac{\partial^4\varphi_k(x)}{\partial x^4}q_k(t)\right] + \rho\sum_{k=1}^{+\infty}\left[\varphi_k(x)\frac{\partial^2 q_k(t)}{\partial t^2}\right] = -\rho x\ddot{\theta}(t) \tag{7.19}$$

Multiplying by $\varphi_k(x)$ and integrating over the length of the beam yield:

$$\frac{\partial^2 q_k(t)}{\partial t^2} + \omega_k^2 q_k(t) = -\frac{\gamma_k}{C\alpha_k}\ddot{\theta}(t) \tag{7.20}$$

where

$$\gamma_k = \int_0^{l_b} x\phi_k(x)dx \tag{7.21}$$

$$\alpha_k = \int_0^{l_b} \phi_k(x)\phi_k(x)dx \tag{7.22}$$

Equation (7.20) could be changed with the inclusion of proportional damping:

$$\frac{\partial^2 q_k(t)}{\partial t^2} + 2\zeta_k\omega_k\frac{\partial q_k(t)}{\partial t} + \omega_k^2 q_k(t) = -\frac{\gamma_k}{C\alpha_k}\ddot{\theta}(t) \tag{7.23}$$

where ζ_k is the damping ratio of the kth mode. The Laplace transform of the beam deflection at the distance, x, resulting from Equation (7.23) is:

$$w_x(s) = \sum_{k=1}^{+\infty}\frac{-\phi_k(x)\cdot\gamma_k}{\alpha_k\cdot(s^2 + 2\zeta_k\omega_k s + \omega_k^2)}\cdot\ddot{\theta}(s) \tag{7.24}$$

The model (7.24) includes an infinite number of vibrational modes. The deflection of the beam is a sum of response for each of an infinite number of vibration modes.

A trapezoidal velocity profile (bang-coast-bang acceleration) is applied to drive the hub. The Young's modulus, moment of inertia, linear mass density of the beam, damping ratio, maximum driving velocity, and maximum driving acceleration in the simulation were 2.06×10^5 MPa, 3.449 mm^4, 0.3143 kg/m, 0.06, 10°/s, and 100°/s^2, respectively. When the hub moves, the maximum peak-to-peak deflection is referred to as the transient deflection. After the hub stops, the maximum peak-to-peak deflection is defined as the residual amplitude. A simulated response for a driving distance of 30° is shown in Figure 7.2 when the hub moves between 0 s and 3 s. The transient deflection is 31.2 mm, while the residual amplitude is 46.1 mm.

Residual amplitudes of the first mode and high modes for varying normalized distance, x/l_b, are shown in Figure 7.3. The beam length, driving distance, and mass ratio were fixed at 95 cm, 19°, and 0.5, respectively. The residual amplitude of the first mode increases as the normalized distance increases. As the normalized distance changes, a peak and a trough occur in the high mode, and the peak occurs near

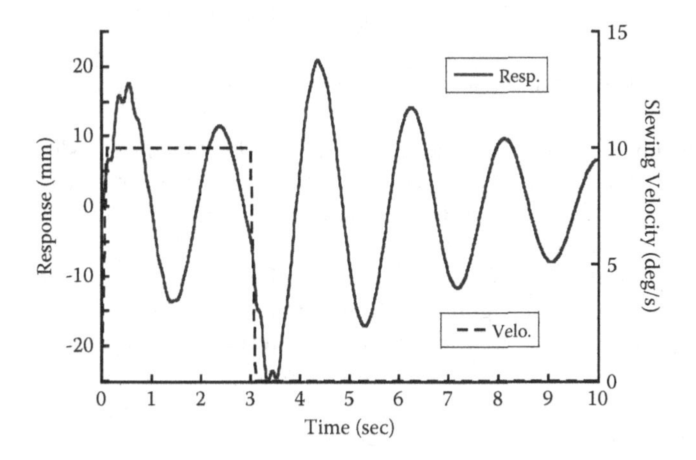

FIGURE 7.2 Simulated response for a driving distance of 30°.

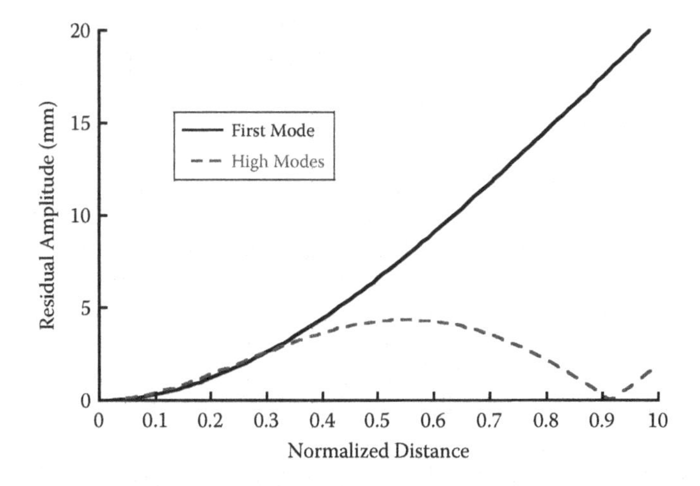

FIGURE 7.3 Residual amplitudes against normalized distance.

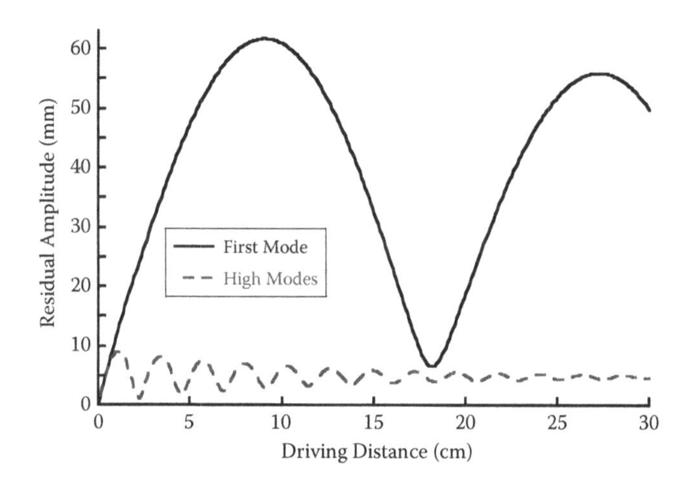

FIGURE 7.4 Residual amplitudes of the deflection against driving distance.

the middle point of the beam. The residual amplitude of the first mode at the middle point of the beam was 9.84 mm, while the corresponding magnitude of high modes was 4.43 mm. When the measurement point locates near the middle point of beam, the high modes cause large vibrations.

Residual amplitudes of the first mode and high modes for various driving distances are shown in Figure 7.4. The normalized distance, beam length, and mass ratio were fixed at 0.5, 95 cm, and 0.5, respectively. Peaks and troughs in the first mode and high modes exist as the driving distance changes. When vibrations caused by the acceleration and deceleration were in phase, peaks arose. When vibrations caused by the acceleration and deceleration were out of phase, troughs occurred. Peaks and troughs were spaced farther apart for the first mode. This is because the first mode had the low natural frequency. The average residual amplitudes of the first mode and high modes were 32.9 mm and 4.3 mm, respectively. The ratio of the average amplitude of high modes to that of the first mode was 13.1%. Therefore, the theoretical finding is that high modes have some effects in some specified cases.

Residual amplitudes of the first mode and high modes for varying beam lengths are shown in Figure 7.5. The normalized distance, driving distance, and mass ratio were fixed at 0.5, 19°, and 0.5, respectively. Peaks and troughs arose due to the interference between vibrations caused by the acceleration and deceleration. Residual vibrations of the first mode and high modes increase with increasing the beam length. The high modes cause large vibrations for long beams.

Residual amplitudes of the first mode and high modes for various mass ratios are shown in Figure 7.6. The normalized distance, driving distance, and beam length were fixed at 0.5, 19°, and 95 cm, respectively. Residual amplitudes of the first mode decreased as the mass ratio increased before 0.5. After this point, residual vibrations increased. Residual vibrations of high modes increase slightly as the mass ratio increases. Therefore, the mass ratio does not have large effects on the high-mode dynamics.

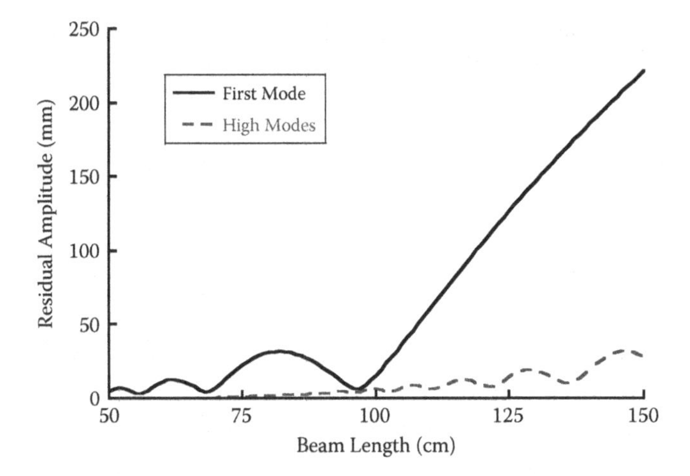

FIGURE 7.5 Residual amplitudes of the deflection against beam length.

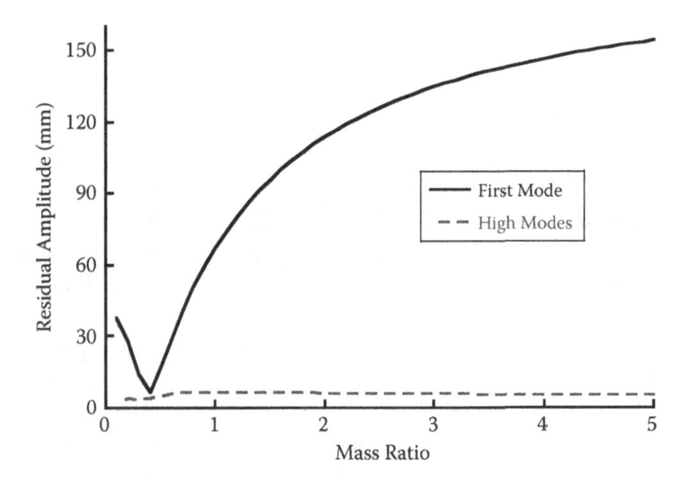

FIGURE 7.6 Residual amplitudes of the deflection against mass ratio.

Simulated result indicates that high modes have large effects when the middle point of the beam is chosen for measurement and the beam length is large. Therefore, it is essential to design a control system that can effectively reduce vibrations induced by total modes.

7.1.2 Computational Dynamics

Vibrations of the beam and payload caused by the commanded motions can be suppressed by smoothing the driving commands. The driving command filters through a four-pieces smoother to create a smoothed command, which moves the hub toward the desired state with minimum vibrations. The four-pieces smoother inherent in the limited response of the hub results in a limited response to high frequencies. The four-pieces smoother will be used to suppress infinite mode vibrations in flexible single-link manipulators.

$$u(\tau) = \begin{cases} M(1+\sigma)\tau e^{-2\zeta_m \omega_m \tau}, & 0 \le \tau \le 0.5T_m \\ M[(1+\sigma+\sigma K - K)T_m + (2K - 2K\sigma - \sigma - 1)\tau]e^{-2\zeta_m \omega_m \tau}, & \\ \quad 0.5T_m < \tau \le T_m & \\ M[K(3 - K\sigma - 3\sigma - K)T_m + K(K + \sigma K + 2\sigma - 2)\tau]e^{-2\zeta_m \omega_m \tau}, & \\ \quad T_m < \tau \le 1.5T_m & \\ M[2K^2(1+\sigma)T_m - K^2(1+\sigma)\tau]e^{-2\zeta_m \omega_m \tau}, & 1.5T_m < \tau \le 2T_m \end{cases} \tag{7.25}$$

where M, K, T_m, and σ are coefficients. T_m, K, and M are given by:

$$M = \zeta_m^2 \omega_m^2 / (1 - K^{-1})^2 \tag{7.26}$$

$$K = e^{2\pi\zeta_m / \sqrt{1-\zeta_m^2}} \tag{7.27}$$

$$T_m = 2\pi \Big/ \left(\omega_m \sqrt{1 - \zeta_m^2} \right) \tag{7.28}$$

The four-pieces smoother is a function of design frequency and damping ratio. However, the natural frequency might not be known accurately under certain conditions, then estimating how the modeling error in the frequency translates into the percent residual amplitude is important. The 5% insensitivity of the continuous function varies from 0.81 to infinity. In addition, changes in damping have little effect on the vibrations with the four-pieces smoother. This conclusion is fortunate because it is generally challenging to estimate the damping ratio accurately under real conditions.

The four-pieces smoother is designed to control first-mode vibrations of the single-link flexible manipulators, and then it will also suppress high-mode oscillations. Therefore, the high-mode frequency does not need to be estimated for the design of the four-pieces smoother. This conclusion is also fortunate because measurement of the first-mode frequency is generally easier than that of the high-mode frequencies.

The natural frequencies of the flexible link manipulators depend on the beam length and mass ratio. The dynamical analyses indicate that the variations of the high-mode frequencies are larger than that of the first-mode frequency. Thus, a vibration controller for infinite mode systems should provide more robustness to changes in the high-mode frequencies. The fundamental compromise is that an increase in robustness must be traded off against an increase in rise time.

The beam length and mass ratio can be easily measured for estimating the first-mode frequency. Then, the first-mode frequency and damping ratio can be used to design the four-pieces smoother. By using the first-mode frequency, the four-pieces smoother could reduce the vibrations of total modes. However, the first mode would

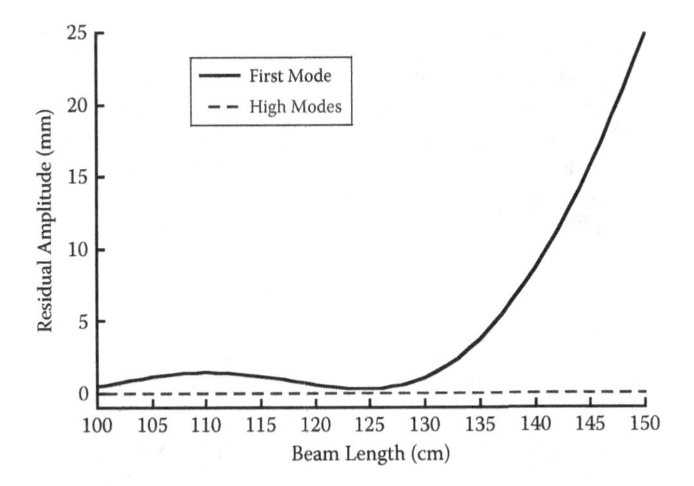

FIGURE 7.7 Simulated residual amplitudes against beam length.

have the longest vibration period. The rise time of the four-pieces smoother is twice times as long as the vibration period of the first mode, which is a drawback because of slowing down the flexible link manipulators. However, the maximum original driving velocity of the hub might increase because the presented function benefits vibration reduction. The higher-speed commands are modified by the continuous function to create the smoothed commands, which would move the hub faster.

The four-pieces smoother exploits the first-mode frequency. However, the first-mode frequency may not be known or change over time in some specified cases. Thus, this section will study the numerical verification of the four-pieces smoother for modeling errors in the first-mode frequency.

Simulated residual amplitudes for varying beam lengths are shown in Figure 7.7. The normalized distance, x/l_b, driving distance, and mass ratio were fixed at 0.5, 19°, and 0.5, respectively. The four-pieces smoother was designed when the modeled length of the beam was 125 cm. The real beam length ranged from 100 cm to 150 cm. The smoother limited first-mode vibrations to near zero before the 125 cm because of the low-passing effect of the smoother. As the beam length increased from 125 cm, residual amplitudes with the smoother increased sharply. This is because the smoother has relatively narrow insensitivity range at the low frequency. The smoother suppresses high-mode vibrations to near-zero values for all the beam length shown in Figure 7.7. This is also because the smoother has low-pass filtering effect.

7.1.3 Experimental Investigation

Experiments were performed on a Quanser testing apparatus shown in Figure 7.8. A motor with encoder drove the hub, to which a flexible beam is mounted. A motion control card connects amplifier to a MATLAB® script in the personal computer. The length, breadth, and thickness of the beam were 950 mm, 39 mm, and 1.02 mm, respectively. A tennis ball with the mass of 121 g, which was served as the payload, connected with the beam. Vibrations of the payload have few effects on the motion

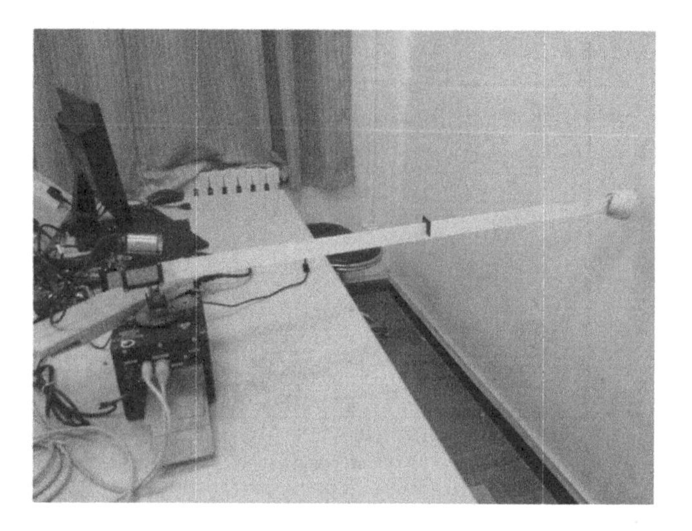

FIGURE 7.8 A single-link flexible manipulator.

of the hub because many metal blocks are mounted to the hub. The first-mode natural frequency and damping ratio in experiments were 3.35 rad/s and 0.06, respectively. The displacement of a black marker on the beam was recorded by a camera, which was also mounted to the hub. The marker in experiments was set to the middle point of the beam. Then mid-point responses based on the black marker are the experimental result.

Experimental control architecture is shown in Figure 7.9. The operator produced a bang-coast-bang acceleration (trapezoidal velocity profile) via the control interface. The original command is sent to a four-pieces smoother, which was described in Chapter 2. Then the smoother generates a driving command for rotating the hub in the flexible manipulator. The smoother is designed using the first-mode frequency in the flexible manipulator. The design frequency, ω_m, was estimated by using mass ratio, h, and beam length, l_b.

The first set of experiments investigated the effect of varying driving distances. Figure 7.10 shows simulated and experimental results of residual amplitudes. Without the controller, peaks and troughs occurred as the driving distance changed. The four-pieces smoother was designed when the design frequency and damping ratio were 3.35 rad/s and 0.06, respectively. The controlled results were limited to low levels. Experimental values were worse than simulated results because there was a small

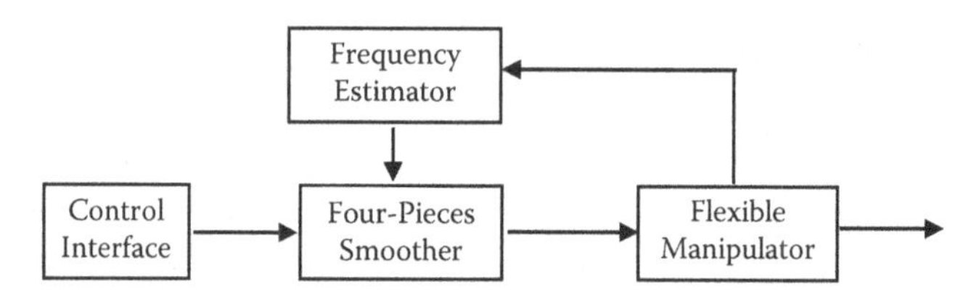

FIGURE 7.9 Control architecture.

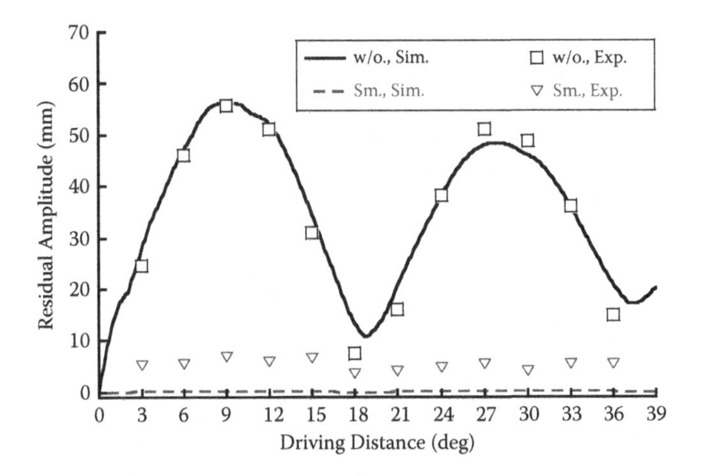

FIGURE 7.10 Residual amplitudes for various driving distances.

modeling error in the design frequency and uncertainty in the overall system. However, experimental data follow the general shape as the simulated curve.

The second set of experiments investigated the effect of various modeling errors in the design frequency. The experimental responses to a small modeling error in the design frequency are shown in Figure 7.11. Without the controller, transient and residual vibrations had a response with amplitudes of 30.3 mm and 38.2 mm, respectively. The residual amplitude was larger than the transient deflection because vibrations caused by the acceleration and deceleration were in phase. The four-pieces smoother was designed when the design frequency was held constant at 3.35 rad/s. The transient deflection and residual amplitude were 8.7 mm and 4.6 mm, respectively. The four-pieces smoother dramatically reduced vibrations in the case of small modeling errors.

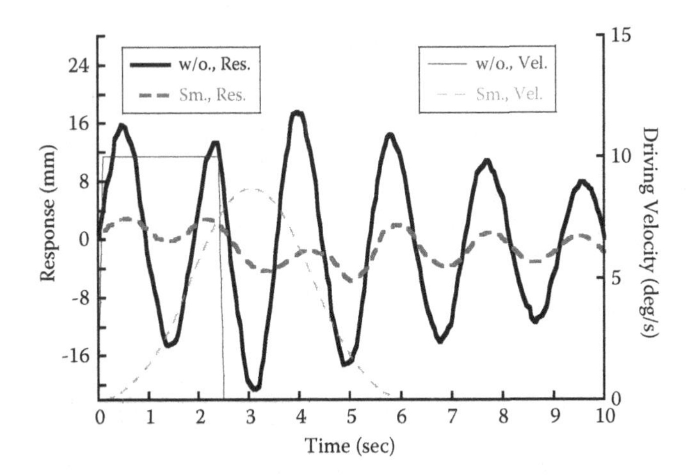

FIGURE 7.11 Responses to a small modeling error in the frequency.

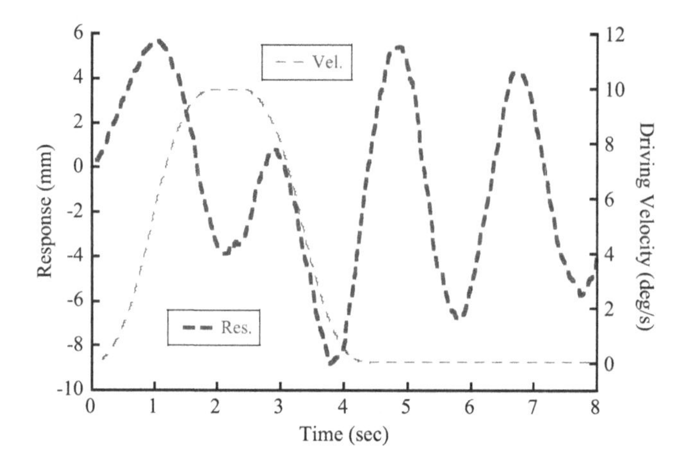

FIGURE 7.12 Responses to a negative modeling error in the frequency.

Experimental responses for the design frequency of 6.7 rad/s are shown in Figure 7.12. Note that the design frequency of 6.7 rad/s is corresponding to the normalized frequency of 0.5. The transient deflection with the smoother was 14.4 mm, while the residual amplitude was 12.2 mm. The residual amplitude with the design frequency of 6.7 rad/s are relatively larger than that with the design frequency of 3.35 rad/s. This is because the four-pieces smoother is less insensitive at the low frequency.

Experimental responses when the design frequency was fixed at 1.675 rad/s (corresponding to the normalized frequency of 2) are shown in Figure 7.13. The transient deflection and residual amplitude were 5.7 mm and 1.7 mm, respectively. The residual amplitude was limited to a low level in this case because of the low-pass filtering effect.

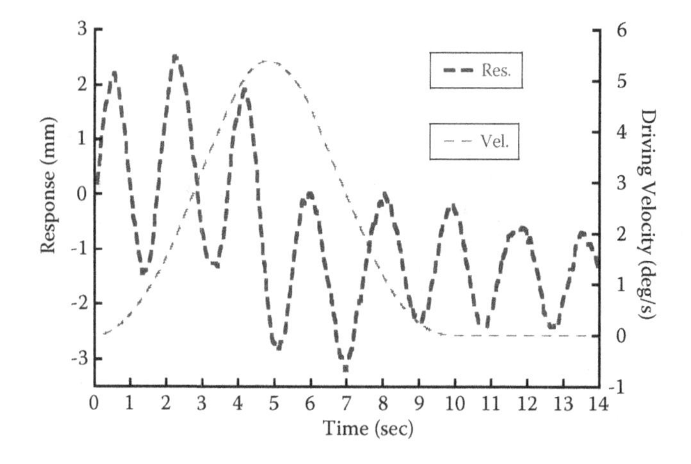

FIGURE 7.13 Responses to a positive modeling error in the frequency.

Experimental results in Figures 7.11–7.13 indicate that the four-pieces smoother provides good robustness to modeling errors in the design frequency. The experimental findings demonstrated that the four-pieces smoother is effective and robust for varying driving motions over a wide range of modeling errors in the flexible link manipulator.

7.2 DUFFING-OSCILLATOR DYNAMICS

Duffing oscillators are widely applied in many types of mechanical systems, such as pendulums, beams, cables, and nonlinear vibration isolators. Many methods have been reported to control the Duffing oscillator, such as time-delayed feedback control [34–41], combined linear-plus-nonlinear feedback control [42], state feedback control [43], optimal polynomial control [44], and sliding mode control [45]. Chen et al. proposed two-step and three-step shaped-command methods for controlling Duffing oscillators [46]. The magnitude and time of the shaped commands were derived using an optimization program. The effectiveness of the two- and three-step shaped-command methods was verified by simulating a fixed-fixed beam. However, the computational load for optimizing shaped commands on-the-fly is an obstacle in the application of control methods discussed in [46]. Moreover, accurately sensing vibrations is a barrier in the application of feedback controllers. Thus, it is highly desirable to control Duffing oscillators without measuring vibrations and performing heavy computational tasks.

7.2.1 MODELING

The input is the angular acceleration of the hub from the motor to the manipulator. The output is the deflection, $w(x, t)$, of a point on the link at a distance, x, from the hub. The movement of the flexible link and payload would not affect the motion of the hub due to a large inertia of the hub. The payload is assumed to be a point mass at the tip. The force equation of motion of a small element at distance x on the link can be expressed as:

$$\rho dx\frac{\partial^2 w(x,t)}{\partial t^2} + \rho x dx\frac{d^2\theta}{dt^2} + dV(x,t) = 0 \tag{7.29}$$

where the mass density of the uniform link denotes ρ, dx represents a small change of x, the shear force acting on the cross-section of the uniform link denotes $V(x, t)$, and $dV(x, t)$ denotes a small change of $V(x, t)$. The corresponding moment equation of motion of the small element can be written as:

$$dM(x,t) - V(x,t)dx - dV(x,t)dx - 0.5\rho x\frac{d^2\theta}{dt^2}(dx)^2 = 0 \tag{7.30}$$

where $M(x, t)$ denotes the bending moment acting at the distance of x on the link and $dM(x, t)$ represents the corresponding small change. From the mechanics of materials (flexure formula), the relationship between the bending moment and beam deflection is given by:

$$M(x,t) = EI \cdot \frac{\frac{\partial^2 w(x,t)}{\partial x^2}}{\left[1 + \left(\frac{\partial w(x,t)}{\partial x}\right)^2\right]^{\frac{3}{2}}} \tag{7.31}$$

where I is the moment of inertia of the cross-section of the beam and E is Young's modulus. Expanding the right-hand term of Equation (7.31) into Taylor series, and ignoring high-order terms in the resultant equation yield:

$$M(x,t) = EI \cdot \frac{\partial^2 w(x,t)}{\partial x^2} \cdot \left[1 - 1.5\left(\frac{\partial w(x,t)}{\partial x}\right)^2\right] \tag{7.32}$$

Substituting Equation (7.32) into (7.30) and then substituting into (7.29), and ignoring terms involving high-order powers in dx yield:

$$EI\frac{\partial^4 w(x,t)}{\partial x^4} - 1.5EI\frac{\partial^2\left(\frac{\partial^2 w(x,t)}{\partial x^2}\left(\frac{\partial w(x,t)}{\partial x}\right)^2\right)}{\partial x^2} + \rho\frac{\partial^2 w(x,t)}{\partial t^2} = -\rho x \cdot a(t) \tag{7.33}$$

The boundary conditions are:

$$w(x,t)|_{x=0} = 0 \tag{7.34}$$

$$\left.\frac{\partial w(x,t)}{\partial x}\right|_{x=0} = 0 \tag{7.35}$$

$$\left.\frac{\partial^2 w(x,t)}{\partial x^2}\right|_{x=l_b} = 0 \tag{7.36}$$

$$\left.\frac{\partial}{\partial x}\left[EI\frac{\partial^2 w(x,t)}{\partial x^2}\right]\right|_{x=l_b} = m_p\left.\frac{\partial^2 w(x,t)}{\partial t^2}\right|_{x=l_b} \tag{7.37}$$

The condition (7.34) indicates that the hub base does not experience any deflection. The condition (7.35) specifies that the derivative of the deflection function at the hub base is zero. The condition (7.36) shows that there is no bending moment acting at the free end of the beam. The condition (7.37) states that no shear force acts at the free end of the beam.

Using the mode superposition method, the deflection of the beam can be expressed as a linear combination of normal modes:

$$w(x,t) = \sum_{k=1}^{+\infty} \phi_k(x) \cdot q_k(t) \tag{7.38}$$

The linear frequency ω_k and the mode shape $\phi_k(x)$ can be derived by solving resultant equations (7.34)–(7.38):

$$\omega_k = \beta_k^2 \cdot \sqrt{\frac{EI}{\rho}} \tag{7.39}$$

$$\cos(\beta_k l_b)\cosh(\beta_k l_b) + 1 = c \cdot \beta_k l_b[\sin(\beta_k l_b)\cosh(\beta_k l_b) - \cos(\beta_k l_b)\sinh(\beta_k l_b)] \tag{7.40}$$

$$\phi_k(x) = \sin(\beta_k x) - \sinh(\beta_k x) - r\cos(\beta_k x) + r\cosh(\beta_k x) \tag{7.41}$$

$$r = \frac{\sin(\beta_k l_b) + \sinh(\beta_k l_b)}{\cos(\beta_k l_b) + \cosh(\beta_k l_b)} \tag{7.42}$$

where ω_k is the linear frequency of the kth mode and h is the ratio of the payload mass to the beam mass. The numerical solution for the coefficient, β_k, will be sought by using the mass ratio, h, and beam length, l_b.

Substituting Equation (7.38) into (7.33) and ignoring the modal coupling effects between different normal modes result in:

$$\rho \sum_{k=1}^{+\infty}\left[\phi_k \frac{d^2 q_k}{dt^2}\right] + EI \sum_{k=1}^{+\infty}\left[\frac{d^4\phi_k}{dx^4} q_k\right] - 1.5EI \sum_{k=1}^{+\infty}\left[\frac{d^2\left(\frac{d^2\phi_k}{dx^2}\left(\frac{d\phi_k}{dx}\right)^2\right)}{dx^2} q_k^3\right] = -\rho x \cdot a(t) \tag{7.43}$$

Multiplying by ϕ_k on both sides and integrating over the length of the beam $(0 \le x \le l_b)$ yield:

$$\frac{d^2 q_k}{dt^2} + \omega_k^2 q_k + e_k \omega_k^2 q_k^3 = -\gamma_k \cdot a(t); \; \omega_k > 0 \tag{7.44}$$

where ω_k is the linearized natural frequency of the kth mode and e_k is the nonlinear stiffness parameter of the kth mode. The nonlinear stiffness, e_k, and coefficient, γ_k, are determined by:

$$e_k = \frac{-1.5\int_0^{l_b} \frac{d^2\left(\frac{d^2\phi_k}{dx^2}\left(\frac{d\phi_k}{dx}\right)^2\right)}{dx^2}\phi_k dx}{\int_0^{l_b} \frac{d^4\phi_k}{dx^4}\phi_k dx} \tag{7.45}$$

$$\gamma_k = \frac{\int_0^{l_b} x\phi_k dx}{\int_0^{l_b} \phi_k \phi_k dx} \tag{7.46}$$

Considering, including a proportional damping term, Equation (7.44) can be re-written as:

$$\frac{d^2 q_k}{dt^2} + 2\zeta_k \omega_k \frac{dq_k(t)}{dt} + \omega_k^2 q_k + e_k \omega_k^2 q_k^3 = -\gamma_k \cdot a(t); \; \omega_k > 0, \, \zeta_k \ge 0 \tag{7.47}$$

where ζ_k is the damping ratio of the kth mode. The model (7.47) includes an infinite number of uncoupled Duffing oscillators. Moreover, the sum of the response from each of infinite Duffing oscillators is the deflection of the beam.

Numerical simulations and experimental tests can validate the dynamics of the flexible manipulator and its corresponding equation of the motion. Numerical simulations are conducted on the first four modes of infinite Duffing oscillators for simplicity. A trapezoidal velocity profile (bang-coast-bang acceleration) drove the hub. The Young's modulus, moment of inertia, linear mass density of the beam, beam length, mass ratio, damping ratio, maximum slewing acceleration, and maximum slewing velocity used in numerical simulations are 2.06×10^5 MPa, 3.449 mm^4, 0.3143 kg/m, 95 cm, 0.5, 0.03, 200°/s^2, and 20°/s, respectively.

Figure 7.14 shows experimental and simulated responses to a trapezoidal velocity command for a slewing distance of 54°. The experimental data follow a similar shape pattern to the simulated curve. Figure 7.14 also gives the simulated response for the corresponding linear model, in which the nonlinear stiffness parameter is set to be $e_k = 0$. With the experimental result, the nonlinear model matches better than the corresponding linear model. The comparing results indicate the correctness of the nonlinear modeling. With the experimental vibration period, the vibration period for the nonlinear model is smaller than that for the linear model because the nonlinear stiffness parameter is positive in this case. The hub rotates between 0 seconds and 2.7 seconds in Figure 7.14. The experimental results of the transient deflection and residual amplitude are found to be 186.1 mm and 338.7 mm, respectively.

The nonlinear stiffness parameter of the first mode as a function of the mass ratio and beam length is shown in Figure 7.15. Increasing beam length decreases the magnitude of the nonlinear stiffness parameters. When the mass ratio is smaller than 0.07, the nonlinear stiffness parameter is negative, which corresponds to the softening spring type. The positive nonlinear stiffness parameter corresponds to the type of hardening spring. The nonlinear stiffness parameters increase as the mass

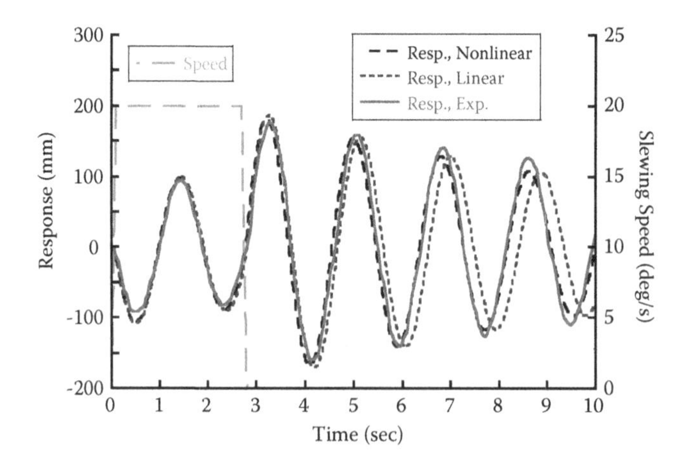

FIGURE 7.14 Experimental and simulated response.

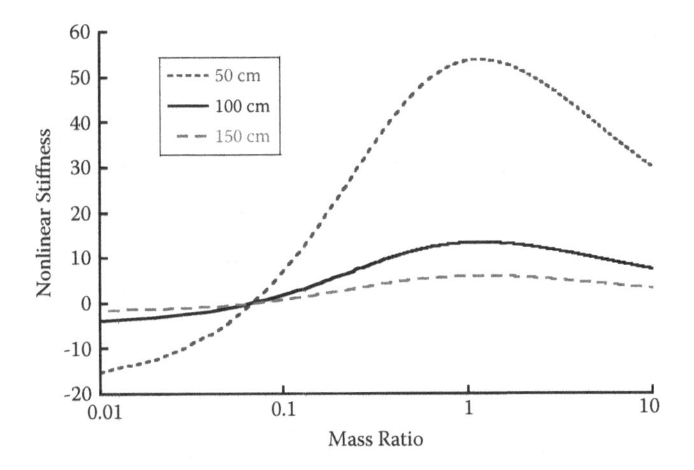

FIGURE 7.15 Nonlinear stiffness parameters of the first mode.

ratio increases before 1.16. After this critical mass ratio, the nonlinear stiffness parameters decrease with an increase in the mass ratio.

7.2.2 Experimental Investigation

Experiments are conducted on a test rig shown in Figure 7.16. A flexible beam is mounted to a hub that is driven by a motor with encoder. The control hardware consists of a personal computer for program development and user interface, and a motion control card connecting amplifier to the personal computer. The length, width, and thickness of the beam are 950 mm, 39 mm, and 1.02 mm, respectively. A tennis ball served as the payload, which is attached to the free end of the beam. The mass of the ball is 121 g. A number of heavy metal blocks are mounted to the hub such that vibrations of the payload have negligible effects on the motion of the hub. A camera is mounted to the hub to record deflections of the payload. The

FIGURE 7.16 Test bench of a flexible single-link manipulator.

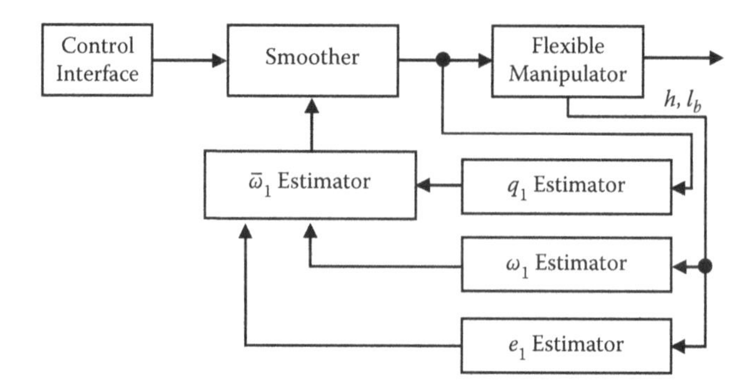

FIGURE 7.17 Control architecture.

maximum slewing velocity and maximum slewing acceleration are 20°/s and 200°/s^2, respectively. Experimental measurement for the first-mode linear frequency and damping ratio is found to be approximately 3.35 rad/s and 0.03, respectively.

The experimental control architecture is shown in Figure 7.17. A bang-coast-bang acceleration (trapezoidal velocity profile) is modified by the SD function or MD smoother to produce a slewing acceleration, a, for rotating the hub. The first-mode linear frequency, ω_1, is estimated using the mass ratio, h, and beam length, l_b. Solving Equations (7.45) and (7.47) yields the nonlinear stiffness parameter, e_1, and the time-dependent function, q_1, of the first mode. Then the nonlinear frequency of the first mode is then estimated for the design of the SD function and MD smoother. The damping ratio of the first mode was fixed at 0.03 in the experiment.

The SD function is given by:

$$u(\tau) = \begin{bmatrix} A_i \\ \tau_i \end{bmatrix} = \begin{bmatrix} \frac{1}{(1+K+K^3+K^4)} & \frac{K+K^3}{(1+K+K^3+K^4)} & \frac{K^4}{(1+K+K^3+K^4)} \\ 0 & 0.5T & T \end{bmatrix} \tag{7.48}$$

where

$$K = e^{\left(-\pi\zeta_k/\sqrt{1-\zeta_k^2}\right)} \tag{7.49}$$

$$T = \frac{2\pi}{\bar{\omega}_k \cdot \sqrt{1-\zeta_k^2}} \tag{7.50}$$

The MD smoother is given by:

$$u(\tau) = \begin{cases} M \cdot \tau e^{-2\zeta_k \bar{\omega}_k \tau}, \quad 0 \le \tau \le 0.5T \\ (-0.5K^{-1} + 1 - 0.5K) M \cdot T e^{-2\zeta_k \bar{\omega}_k \tau} + (K^{-1} - 1 + K) M \\ \quad \cdot \tau e^{-2\zeta_k \bar{\omega}_k \tau}, \ 0.5T \le \tau \le T \\ (1.5K^{-1} - 1 + 1.5K) M \cdot T e^{-2\zeta_k \bar{\omega}_k \tau} + (-K^{-1} + 1 - K) M \\ \quad \cdot \tau e^{-2\zeta_k \bar{\omega}_k \tau}, \ T \le \tau \le 1.5T \\ 2M \cdot T e^{-2\zeta_k \bar{\omega}_k \tau} - 2M \cdot \tau e^{-2\zeta_k \bar{\omega}_k \tau}, \quad 1.5T \le \tau \le 2T \end{cases} \tag{7.51}$$

where

$$M = \frac{4\zeta_k^2 \overline{\omega}_k^2}{(1 + K + K^3 + K^4)(1 - 2K^2 + K^4)} \tag{7.52}$$

Both the SD function and MD smoother modify operated commands to produce shaped commands for slewing the manipulator. The comparison of experimental response under the SD function and MD smoother is shown in Figure 7.18. The transient deflections with the SD function and MD smoother are 60.4 mm and 34.8 mm, respectively. The corresponding residual amplitudes with the SD function and MD smoother are 14.1 mm and 3.7 mm, respectively. The MD smoother reduces more oscillations than the SD function. Both the SD function and MD smoother control Duffing oscillator dynamics to a very low level.

The second set of experiments investigates the effect of varying slewing distances. Figure 7.19 shows simulated and experimental results of the residual amplitude when the slewing distance is ranged from 27° to 66°. Without the controller, peaks and troughs occur as the slewing distance is changed. The experimental data match the corresponding simulated curve very well. The SD function and MD smoother were designed when the modeled linear frequency and damping ratio are 3.35 rad/s and 0.03, respectively. The SD function and MD smoother both represent a significant reduction of vibrations for all cases. The experimental data with both the SD function and MD smoother are slightly higher than the simulated curve because of modeling errors in the design frequency and uncertainty in the overall system.

The last set of experiments is aimed at investigating the effect of various modeling errors in the design frequency. The experimental response to a small modeling error in the design frequency is shown in Figure 7.20. Without the controller, the transient and residual vibrations are found to have amplitudes of 186.1 mm and 338.7 mm, respectively. The residual amplitude is larger than the transient deflection. This is because vibrations caused by the acceleration and deceleration are in phase. Both the SD

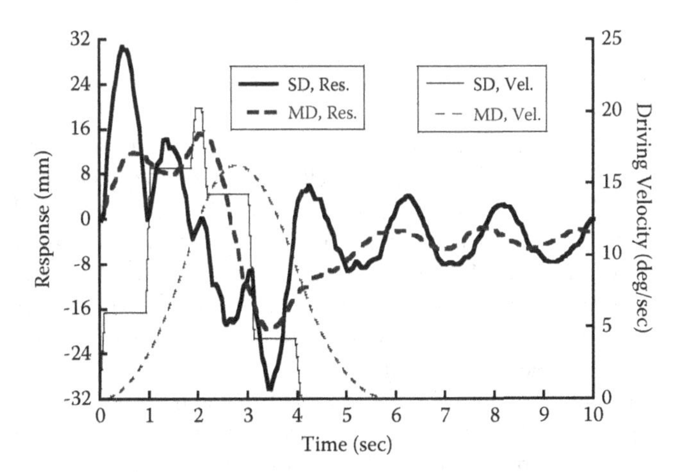

FIGURE 7.18 Experimental results at a slewing distance.

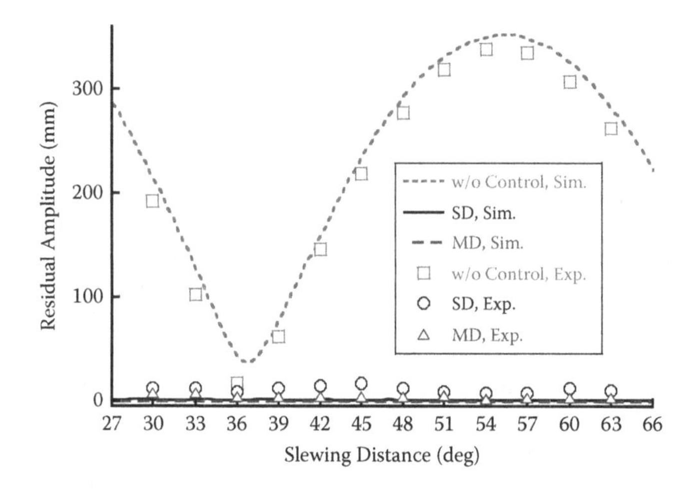

FIGURE 7.19 Experimental results at various slewing distances.

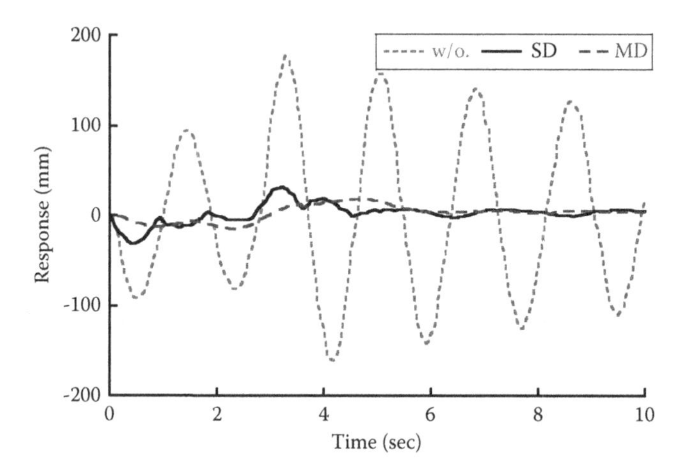

FIGURE 7.20 Response to a small modeling error in the frequency.

function and MD smoother were designed when the designed linear frequency was held constant at 3.35 rad/s. The transient deflections with the SD function and MD smoother are 62.9 mm and 32.5 mm, respectively. Meanwhile, the residual amplitudes with the SD function and MD smoother are 8.2 mm and 2.1 mm, respectively. Both the SD function and MD smoother suppressed dramatically vibrations in the case of small modeling error. Nevertheless, residual vibrations with the MD smoother are relatively smaller. This is because vibrations of the high-mode Duffing oscillators cannot be controlled under the SD function.

The experimental response for the modeled linear frequency of 4.79 rad/s is shown in Figure 7.21. Note that the modeled frequency of 4.79 rad/s is corresponding to –30% error. Transient deflections with the SD function and MD smoother are 156.2 mm and 58.8 mm, respectively. Meanwhile, residual amplitudes

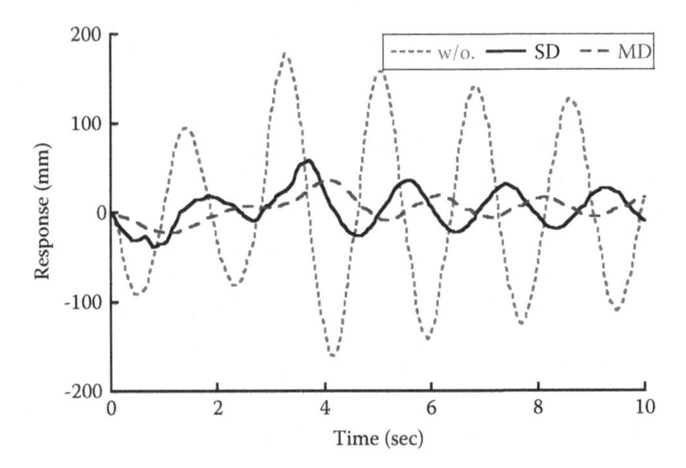

FIGURE 7.21 Response to a negative modeling error in the frequency.

with the SD function and MD smoother are 62.3 mm and 27.4 mm, respectively. Because the MD smoother is less sensitive to negative errors, the residual amplitude with the SD function is larger than that of the MD smoother.

The experimental response is shown in Figure 7.22 when the design frequency is fixed at 2.58 rad/s, which is corresponding to +30% error. Transient deflections with the SD function and MD smoother are 85.5 mm and 34.9 mm, respectively. Meanwhile, residual amplitudes with the SD function and MD smoother are 34.8 mm and 12.7 mm, respectively. The MD-smoothed vibrations are suppressed to a lower level because of a wide range of insensitivity. The experimental findings demonstrate that both the SD function and MD smoother are effective to reduce vibrations of the flexible link manipulator having Duffing oscillator dynamics, and have good robustness in the frequency.

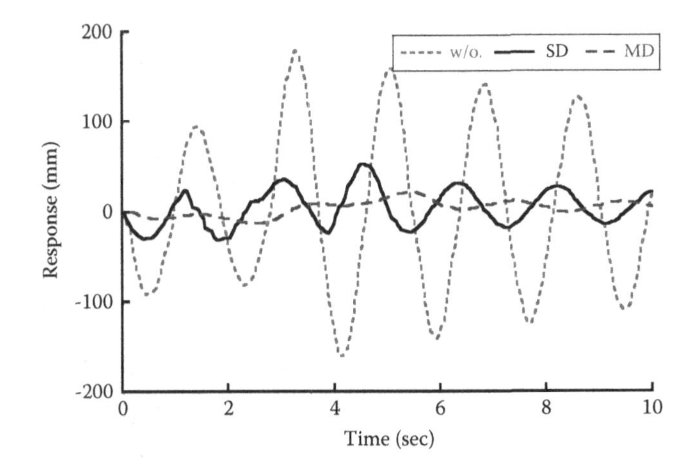

FIGURE 7.22 Response to a positive modeling error in the frequency.

7.3 COUPLED DUFFING-OSCILLATOR DYNAMICS

Broad attention has been attracted on dynamics and control of flexible link manipulators. The structural nonlinearities were unnoticed for simplicity in most of the existing studies, and the flexible links were considered as a linear model with infinite vibrational modes. Recently the dynamic modeling of flexible link was developed from the nonlinear dynamics perspective. The nonlinear model exhibits relatively large-amplitude vibrations and complicated dynamics. Definitely, the large bending deflection of the flexible link may demonstrate Duffing oscillator dynamics. Each vibrational mode may act as a Duffing oscillator, then the coupling effect and dynamic interaction naturally arise among different vibrational modes. Therefore, coupled Duffing oscillator dynamics is critical to describe the nonlinear oscillations of flexible link manipulators. However, most work did not consider coupled Duffing oscillators in the dynamic modeling.

7.3.1 MODELING

A flexible link of length, l_b, connects a payload of mass, m_p, to a hub. Then the rotating hub moves the link to a prescribed angular displacement, θ. The movement of the link and payload tracks the specified angular displacement of the flexible link in an ideal and desired situation. However, the motion-induced vibrations during the operation can deteriorate the manipulator performance due to the flexible lightweight structure.

The input is the angular acceleration of the hub from the motor to the manipulator. The output is the deflection, $w(x, t)$, of a point on the link at a distance, x, from the hub. The movement of the flexible link and payload would not affect the motion of the hub due to a large inertia of the hub. The payload is assumed to be a point mass at the tip. Figure 7.23 shows a free-body diagram of a small element on the link. The force equation of motion of a small element at distance x on the link can be expressed as:

$$\rho dx\frac{\partial^2 w(x, t)}{\partial t^2} + \rho x dx\frac{d^2\theta}{dt^2} + dV(x, t) = 0 \tag{7.53}$$

where the mass density of the uniform link denotes ρ, dx represents a small change of x, the shear force acting on the cross-section of the uniform link denotes $V(x, t)$, and $dV(x, t)$ denotes a small change of $V(x, t)$. The corresponding moment equation of motion of the small element can be written as:

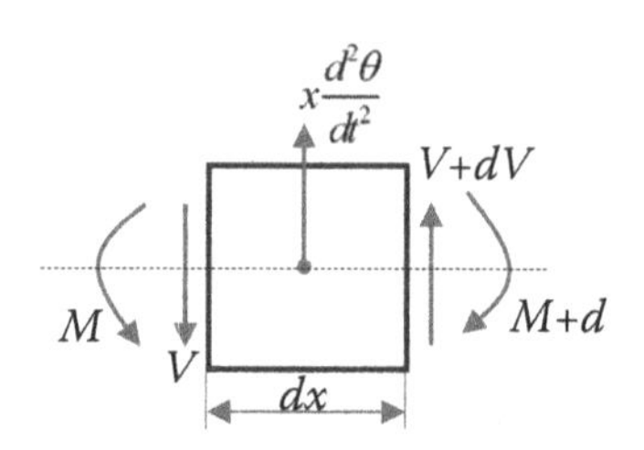

FIGURE 7.23 Free-body diagram of a small element.

$$dM(x,t) - V(x,t)dx - dV(x,t)dx - 0.5\rho x\frac{d^2\theta}{dt^2}(dx)^2 = 0 \tag{7.54}$$

where $M(x, t)$ denotes the bending moment acting at the distance of x on the link, and $dM(x, t)$ represents the corresponding small change. From the bending theory for the flexible link beam, the relationship between the link deflection and bending moment at the distance of x is given by:

$$M(x,t) = EI\frac{\frac{\partial^2 w(x,t)}{\partial x^2}}{\left[1 + \left(\frac{\partial w(x,t)}{\partial x}\right)^2\right]^{1.5}} \tag{7.55}$$

where the Young's modulus denotes E and the moment of inertia of the cross-section of the link represents I. Expanding Equation (7.55) and then neglecting the high-order power terms derive:

$$M(x,t) = EI\frac{\partial^2 w(x,t)}{\partial x^2}\left[1 - 1.5\left(\frac{\partial w(x,t)}{\partial x}\right)^2\right] \tag{7.56}$$

Substituting Equations (7.54) and (7.56) into (7.53) and neglecting the high-order power terms of dx yield:

$$EI\frac{\partial^4 w(x,t)}{\partial x^4} - 1.5EI\frac{\partial^2\left(\frac{\partial^2 w(x,t)}{\partial x^2}\left(\frac{\partial w(x,t)}{\partial x}\right)^2\right)}{\partial x^2} + \rho\frac{\partial^2 w(x,t)}{\partial t^2} = -\rho x\frac{d^2\theta}{dt^2} \tag{7.57}$$

The boundary conditions of the link at free and fixed ends are:

$$w(x,t)|_{x=0} = 0 \tag{7.58}$$

$$\left.\frac{\partial w(x,t)}{\partial x}\right|_{x=0} = 0 \tag{7.59}$$

$$\left.\frac{\partial^2 w(x,t)}{\partial x^2}\right|_{x=l_b} = 0 \tag{7.60}$$

$$\frac{\partial}{\partial x}\left[EI\frac{\partial^2 w(x,t)}{\partial x^2}\right]\Bigg|_{x=l_b} = m_p\left.\frac{\partial^2 w(x,t)}{\partial t^2}\right|_{x=l_b} \tag{7.61}$$

By using the mode superposition method, the deflection of the link is assumed as:

$$w(x,t) = \sum_{k=1}^{+\infty}\phi_k(x)q_k(t) \tag{7.62}$$

where the time-dependent function of the kth vibrational mode denotes $q_k(t)$ and the corresponding mode shape is $\phi_k(x)$. Substituting Equation (7.62) into the boundary conditions (7.58)–(7.61) yields the mode shape $\phi_k(x)$ and linear frequency ω_k:

$$\omega_k = \beta_k^2 \sqrt{\frac{EI}{\rho}} \tag{7.63}$$

$$\cos(\beta_k l_b)\cosh(\beta_k l_b) + 1 = c\beta_k l_b [\sin(\beta_k l_b)\cosh(\beta_k l_b) - \cos(\beta_k l_b)\sinh(\beta_k l_b)] \tag{7.64}$$

$$\phi_k(x) = \sin(\beta_k x) - \sinh(\beta_k x) - r\cos(\beta_k x) + r\cosh(\beta_k x) \tag{7.65}$$

$$r = \frac{\sin(\beta_k l_b) + \sinh(\beta_k l_b)}{\cos(\beta_k l_b) + \cosh(\beta_k l_b)} \tag{7.66}$$

where the mass ratio of the payload to link denotes c. A closed-form expression for coefficient β_k cannot be derived from Equation 7.64, but numerical solutions can be obtained for the link length, l_b, and mass ratio, c.

The link length, l_b, and mass ratio, c, have a large influence on the linear frequency, ω_k. Increasing the link length and mass ratio decreases the linear frequency. In addition, the frequency ratio of the second-mode linear frequency to the first-mode linear frequency is also dependent on the link length and mass ratio. The smallest frequency ratio is 6.267 for all the possible combinations of different mass ratios and link lengths. Therefore, the first two vibrational modes are not under any internal resonances for the flexible link manipulators considered.

Substituting Equation (7.62) into (7.57) and only considering the modal coupling of the first two modes result in:

$$\begin{aligned} &EI\left[q_1\frac{d^4\phi_1}{dx^4} + q_2\frac{d^4\phi_2}{dx^4}\right] + \rho\left[\phi_1\frac{d^2q_1}{dt^2} + \phi_2\frac{d^2q_2}{dt^2}\right] \\ &- 1.5EI\frac{d^2\left[\left(q_1\frac{d^2\phi_1}{dx^2} + q_2\frac{d^2\phi_2}{dx^2}\right)\left(q_1\frac{d^2\phi_1}{dx^2} + q_2\frac{d^2\phi_2}{dx^2}\right)^2\right]}{dx^2} = -\rho x\frac{d^2\theta}{dt^2} \end{aligned} \tag{7.67}$$

The modal coupling effects between the first two modes are included in Equation (7.67). Multiplying both sides of Equation (7.67) by $\phi_k(x)$, and then integrating over the length of the link ($0 \le x \le l_b$) yields:

$$\begin{aligned} &EIq_1\int_0^{l_b}\phi_1\frac{d^4\phi_1}{dx^4}dx + \rho\frac{d^2q_1}{dt^2}\int_0^{l_b}\phi_1^2 dx \\ &- 1.5EI\int_0^{l_b}\phi_1\frac{d^2\left[\left(q_1\frac{d^2\phi_1}{dx^2} + q_2\frac{d^2\phi_2}{dx^2}\right)\left(q_1\frac{d\phi_1}{dx} + q_2\frac{d\phi_2}{dx}\right)^2\right]}{dx^2}dx = -\rho\frac{d^2\theta}{dt^2}\int_0^{l_b}x\phi_1 dx \end{aligned} \tag{7.68}$$

$$\begin{aligned} &EIq_2\int_0^{l_b}\phi_2\frac{d^4\phi_2}{dx^4}dx + \rho\frac{d^2q_2}{dt^2}\int_0^{l_b}\phi_2^2 dx \\ &- 1.5EI\int_0^{l_b}\phi_2\frac{d^2\left[\left(q_1\frac{d^2\phi_1}{dx^2} + q_2\frac{d^2\phi_2}{dx^2}\right)\left(q_1\frac{d\phi_1}{dx} + q_2\frac{d\phi_2}{dx}\right)^2\right]}{dx^2}dx = -\rho\frac{d^2\theta}{dt^2}\int_0^{l_b}x\phi_2 dx \end{aligned} \tag{7.69}$$

Then dividing Equation (7.68) on both sides by $\rho \int_0^{l_b} \phi_1^2(x)dx$ and substituting Equation (7.63) into the resultant equation yield:

$$\frac{d^2q_1}{dt^2} + \omega_1^2 q_1 - \frac{1.5EI\int_0^{l_b}\phi_1 \frac{d^2\left[\left(q_1\frac{d^2\phi_1}{dx^2}+q_2\frac{d^2\phi_2}{dx^2}\right)\left(q_1\frac{d\phi_1}{dx}+q_2\frac{d\phi_2}{dx}\right)^2\right]}{dx^2}dx}{\rho\int_0^{l_b}\phi_1^2 dx} = -\gamma_1\frac{d^2\theta}{dt^2} \quad (7.70)$$

Similarly, dividing the both sides of Equation (7.69) by $\rho \int_0^{l_b} \phi_2^2(x)dx$ and substituting Equation (7.63) into the resultant equation result in:

$$\frac{d^2q_2}{dt^2} + \omega_2^2 q_2 - \frac{1.5EI\int_0^{l_b}\phi_2 \frac{d^2\left[\left(q_1\frac{d^2\phi_1}{dx^2}+q_2\frac{d^2\phi_2}{dx^2}\right)\left(q_1\frac{d\phi_1}{dx}+q_2\frac{d\phi_2}{dx}\right)^2\right]}{dx^2}dx}{\rho\int_0^{l_b}\phi_2^2 dx} = -\gamma_2\frac{d^2\theta}{dt^2} \quad (7.71)$$

where

$$\gamma_1 = \frac{\int_0^{l_b} x\phi_1 dx}{\int_0^{l_b}\phi_1^2 dx} \quad (7.72)$$

$$\gamma_2 = \frac{\int_0^{l_b} x\phi_2 dx}{\int_0^{l_b}\phi_2^2 dx} \quad (7.73)$$

Including the proportional damping into Equations (7.70) and (7.71) gives rise to a two degree-of-freedom nonlinear system with cubic nonlinearities [47]:

$$\begin{cases} \frac{d^2q_1}{dt^2} + 2\zeta_1\omega_1\frac{dq_1}{dt} + \omega_1^2 q_1 \\ \quad +[b_{11}q_1^3 + b_{12}q_1q_2^2 + b_{13}q_1^2q_2 + b_{14}q_2^3] = -\gamma_1\frac{d^2\theta}{dt^2} \\ \frac{d^2q_2}{dt^2} + 2\zeta_2\omega_2\frac{dq_2}{dt} + \omega_2^2 q_2 \\ \quad +[b_{21}q_1^3 + b_{22}q_1q_2^2 + b_{23}q_1^2q_2 + b_{24}q_2^3] = -\gamma_2\frac{d^2\theta}{dt^2} \end{cases} \quad (7.74)$$

where ζ_1 is the damping ratio of the first mode, and ζ_2 is the damping ratio of the second mode. The nonlinear stiffness coefficients, b_{kj}, are given by:

$$b_{11} = \frac{-1.5EI}{\rho}\frac{\int_0^{l_b}\phi_1\frac{d^2\left[\frac{d^2\phi_1}{dx^2}\left(\frac{d\phi_1}{dx}\right)^2\right]}{dx^2}dx}{\int_0^{l_b}\phi_1^2 dx} \quad (7.75)$$

$$b_{12} = \frac{-1.5EI}{\rho} \frac{\int_0^{l_b} \phi_1 \frac{d^2\left[\frac{d^2\phi_1}{dx^2}\left(\frac{d\phi_2}{dx}\right)^2 + 2\frac{d^2\phi_2}{dx^2}\frac{d\phi_1}{dx}\frac{d\phi_2}{dx}\right]}{dx^2} dx}{\int_0^{l_b} \phi_1^2 dx} \tag{7.76}$$

$$b_{13} = \frac{-1.5EI}{\rho} \frac{\int_0^{l_b} \phi_1 \frac{d^2\left[\frac{d^2\phi_2}{dx^2}\left(\frac{d\phi_1}{dx}\right)^2 + 2\frac{d^2\phi_1}{dx^2}\frac{d\phi_1}{dx}\frac{d\phi_2}{dx}\right]}{dx^2} dx}{\int_0^{l_b} \phi_1^2 dx} \tag{7.77}$$

$$b_{14} = \frac{-1.5EI}{\rho} \frac{\int_0^{l_b} \phi_1 \frac{d^2\left[\frac{d^2\phi_2}{dx^2}\left(\frac{d\phi_2}{dx}\right)^2\right]}{dx^2} dx}{\int_0^{l_b} \phi_1^2 dx} \tag{7.78}$$

$$b_{21} = \frac{-1.5EI}{\rho} \frac{\int_0^{l_b} \phi_2(x) \frac{d^2\left[\frac{d^2\phi_1}{dx^2}\left(\frac{d\phi_1}{dx}\right)^2\right]}{dx^2} dx}{\int_0^{l_b} \phi_2^2 dx} \tag{7.79}$$

$$b_{22} = \frac{-1.5EI}{\rho} \frac{\int_0^{l_b} \phi_2 \frac{d^2\left[\frac{d^2\phi_1}{dx^2}\left(\frac{d\phi_2}{dx}\right)^2 + 2\frac{d^2\phi_2}{dx^2}\frac{d\phi_1}{dx}\frac{d\phi_2}{dx}\right]}{dx^2} dx}{\int_0^{l_b} \phi_2^2 dx} \tag{7.80}$$

$$b_{23} = \frac{-1.5EI}{\rho} \frac{\int_0^{l_b} \phi_2 \frac{d^2\left[\frac{d^2\phi_2}{dx^2}\left(\frac{d\phi_1}{dx}\right)^2 + 2\frac{d^2\phi_1}{dx^2}\frac{d\phi_1}{dx}\frac{d\phi_2}{dx}\right]}{dx^2} dx}{\int_0^{l_b} \phi_2^2 dx} \tag{7.81}$$

$$b_{24} = \frac{-1.5EI}{\rho} \frac{\int_0^{l_b} \phi_2 \frac{d^2\left[\frac{d^2\phi_2}{dx^2}\left(\frac{d\phi_2}{dx}\right)^2\right]}{dx^2} dx}{\int_0^{l_b} \phi_2^2 dx} \tag{7.82}$$

The vibration of the flexible link manipulators can be mathematically represented by an infinite number of coupled Duffing oscillators. The first two coupled Duffing oscillators shown in Equation (7.74) are fundamental to the overall system dynamics. The nonlinear stiffness coefficients depend on the link length and mass ratio. Decreasing the link length increases significantly the magnitude of the nonlinear stiffness coefficients, while the changes in the link length do not affect the sign of the nonlinear stiffness coefficients. The mass ratio can cause the changes in both the sign and value of the nonlinear stiffness coefficient. The negative coefficient means the softening spring type, and the positive stiffness coefficient represents the hardening spring type.

The experimental and simulated responses to a trapezoidal velocity command at a slewing distance of 54° are shown in Figure 7.24 to verify the coupled-oscillators model.

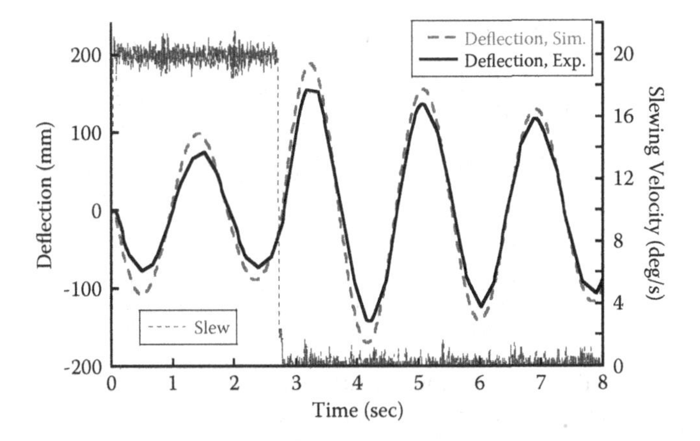

FIGURE 7.24 Experimental validation of the coupled-oscillator model.

The Young's modulus of the link material, the moment of inertia of the cross-section of the uniform link, the mass density of the link, the link length, the mass ratio, the maximum slewing velocity and acceleration are fixed at 2.06 × 105 MPa, 3.449 mm^4, 0.3143 kg/m, 95 cm, 0.5, 20°/s, and 200°/s^2, respectively. The hub starts accelerating at 0 second, and the hub decelerates after 2.7 seconds. The hub rotates at the maximum slewing velocity of 20°/s between its acceleration and deceleration. In order to achieve a good agreement between the simulated and experimental responses, a damping ratio of 0.03 is included in the model. Then simulated curve with the linear damping creates a similar shape to the experimental data.

When the hub moves, the maximum peak-to-peak deflection is referred to as the transient deflection. After the hub stops moving, the maximum peak-to-peak deflection is defined as the residual amplitude. The experimental transient deflection is 153.3 mm, while the experimental residual amplitude is 296.7 mm. The amplitude of residual vibrations is large because the vibrations produced by decelerating the hub are in phase with those generated by accelerating the hub.

By using the method of multiple scales and Lagrange's method of variation of parameters, the approximate solutions to the model (7.74) resulting from the acceleration of the hub can be written as:

$$
\begin{aligned}
q_1(t) \approx & \int_{\tau=0}^{+\infty} C_{11}\frac{d^2\theta}{dt^2}e^{-\zeta_1\Omega_1(t-\tau)}\sin\left[\Omega_1\sqrt{1-\zeta_1^2}\,(t-\tau)+\varphi_1\right]d\tau \\
& + \int_{\tau=0}^{+\infty} C_{12}\frac{d^2\theta}{dt^2}e^{-3\zeta_1\Omega_1(t-\tau)}\sin\left[3\Omega_1\sqrt{1-\zeta_1^2}\,(t-\tau)+\varphi_2\right]d\tau \\
& + \int_{\tau=0}^{+\infty} C_{13}\frac{d^2\theta}{dt^2}e^{-(\zeta_2\Omega_2-2\zeta_1\Omega_1)(t-\tau)}\sin\left[\left(\Omega_2\sqrt{1-\zeta_2^2}-2\Omega_1\sqrt{1-\zeta_1^2}\right)\right. \\
& \left.(t-\tau)+\varphi_3\right]d\tau \\
& + \int_{\tau=0}^{+\infty} C_{14}\frac{d^2\theta}{dt^2}e^{-\zeta_2\Omega_2(t-\tau)}\sin\left[\Omega_2\sqrt{1-\zeta_2^2}\,(t-\tau)+\varphi_4\right]d\tau + \ldots
\end{aligned}
\tag{7.83}
$$

$$q_2(t) \approx \int_{\tau=0}^{+\infty} C_{21}\frac{d^2\theta}{dt^2}e^{-\zeta_1\Omega_1(t-\tau)}\sin\left[\Omega_1\sqrt{1-\zeta_1^2}\,(t-\tau)+\varphi_1\right]d\tau$$
$$+\int_{\tau=0}^{+\infty} C_{22}\frac{d^2\theta}{dt^2}e^{-3\zeta_1\Omega_1(t-\tau)}\sin\left[3\Omega_1\sqrt{1-\zeta_1^2}\,(t-\tau)+\varphi_2\right]d\tau$$
$$+\int_{\tau=0}^{+\infty} C_{23}\frac{d^2\theta}{dt^2}e^{-(\zeta_2\Omega_2-2\zeta_1\Omega_1)(t-\tau)}\sin\left[\left(\Omega_2\sqrt{1-\zeta_2^2}-2\Omega_1\sqrt{1-\zeta_1^2}\right)(t-\tau)+\varphi_3\right]d\tau \quad (7.84)$$
$$+\int_{\tau=0}^{+\infty} C_{24}\frac{d^2\theta}{dt^2}e^{-\zeta_2\Omega_2(t-\tau)}\sin\left[\Omega_2\sqrt{1-\zeta_2^2}\,(t-\tau)+\varphi_4\right]d\tau+\ldots$$

where C_{kj} is the vibration-contribution function and φ_j is the phase function. The nonlinear frequencies of the first two modes are:

$$\Omega_1 = \omega_1 + (1.5b_{11}A_1^2 + b_{12}A_2^2)/\omega_1 \quad (7.85)$$

$$\Omega_2 = \omega_2 + (b_{23}A_1^2 + 1.5b_{24}A_2^2)/\omega_2 \quad (7.86)$$

where A_1 is the peak value of the displacement q_1, and A_2 is the peak value of the displacement q_2. The nonlinear frequency, Ω_k, is dependent on linear natural frequency, ω_k, nonlinear stiffness coefficient, b_{kj}, and the vibration amplitudes. In the case of hardening spring type, the nonlinear frequency increases as the vibration amplitudes increase. In the case of softening spring type, the nonlinear frequency decreases as the vibration amplitudes increase.

7.3.2 Vibration Control

The frequencies in the coupled-oscillator model (7.74) include the first-mode frequency Ω_1, its triple frequency $3\Omega_1$ due to cubic nonlinearities, the coupled-mode frequency (Ω_2–$2\Omega_1$), second-mode frequency Ω_2, and the frequencies of some combinations of two frequencies Ω_1 and Ω_2. Therefore, the control method should be effective to attenuate the oscillations at the abovementioned frequencies.

In the vibration response of the coupled Duffing oscillators, the first two frequencies are Ω_1 and $3\Omega_1$. Arbitrary commands are convolved with three-impulses SD function given in Section 2.5.5 to produce the acceleration for slewing the hub. The shaped commands suppress the vibration components at the frequencies Ω_1 and $3\Omega_1$ to the minimal value. The three-impulses SD function in this case is given by:

$$\begin{bmatrix} A_k \\ \tau_k \end{bmatrix} = \begin{bmatrix} \frac{1}{[1+K+K^3+K^4]} & \frac{K+K^3}{[1+K+K^3+K^4]} & \frac{K^4}{[1+K+K^3+K^4]} \\ 0 & \pi/\left(\Omega_1\sqrt{1-\zeta_1^2}\right) & 2\pi/\left(\Omega_1\sqrt{1-\zeta_1^2}\right) \end{bmatrix} \quad (7.87)$$

where

$$K = e^{\left(-\pi\zeta_1/\sqrt{1-\zeta_1^2}\right)} \quad (7.88)$$

The frequency sensitivity curve for the three-impulses SD function is shown in Figure 7.25 when the first-mode linear frequency, damping ratio, and vibration amplitude are fixed at 3.36 rad/s, 0.03, and 0, respectively. The SD function can eliminate the residual vibrations at the frequency Ω_1 and its triple frequency $3\Omega_1$. In the case of the small damping ratio, the vibration frequency can deviate ±15% from the designed value, and the residual vibration can still be limited below the 5% level.

The third frequency in the vibration response is $(\Omega_2 - 2\Omega_1)$. The third frequency is induced by the coupling of the first and second vibrational modes. Arbitrary commands are convolved with a four-pieces smoother, and then slew the hub. Minimum vibrations at the coupled-mode frequency $(\Omega_2 - 2\Omega_1)$ and other higher frequencies will produce. The four-pieces smoother is given by:

$$f(\tau) = \begin{cases} Q\tau e^{-2W\tau}, \ 0 \le \tau \le 0.5T_3 \\ Q(T_3 - P^{-1}T_3 - \tau + 2P^{-1}\tau)e^{-2W\tau}, \ 0.5T_3 \le \tau \le T_3 \\ Q(3P^{-1}T_3 - P^{-2}T_3 - 2P^{-1}\tau + P^{-2}\tau)e^{-2W\tau}, \ T_3 \le \tau \le 1.5T_3 \\ Q(2P^{-2}T_3 - P^{-2}\tau)e^{-2W\tau}, \ 1.5T_3 \le \tau \le 2T_3 \end{cases} \tag{7.89}$$

where

$$W = \zeta_2\Omega_2 - 2\zeta_1\Omega_1 \tag{7.90}$$

$$P = e^{-\pi\left(\zeta_2\Omega_2 - 2\zeta_1\Omega_1\right)/\left(\Omega_2\cdot\sqrt{1-\zeta_2^2} - 2\Omega_1\cdot\sqrt{1-\zeta_1^2}\right)} \tag{7.91}$$

$$T_3 = 2\pi/\left(\Omega_2\sqrt{1-\zeta_2^2} - 2\Omega_1\sqrt{1-\zeta_1^2}\right) \tag{7.92}$$

$$Q = \frac{4\left(\Omega_2\zeta_2 - 2\Omega_1\zeta_1\right)^2}{[1 + P - P^2 - P^3]^2} \tag{7.93}$$

The frequency sensitivity for the four-pieces smoother is shown in Figure 7.25 when the nonlinear frequencies Ω_1 and Ω_2 are set to be 3.36 rad/s and 28.16 rad/s, respectively. The residual amplitudes at the coupled-mode frequency of $(\Omega_2 - 2\Omega_1)$ is limited to zero, while their derivatives with respect to frequency is also zero. The four-pieces smoother shows the low-pass filtering effect to reduce the vibrations at other higher frequencies.

Figure 7.25 also shows the frequency sensitivity curve for the combination of the SD function and four-pieces smoother. The combined control method suppresses the components of the residual vibration at the first two frequencies, the coupled-mode frequency, and other higher-mode frequencies. Moreover, the combined scheme also exhibits a wide range of frequency insensitivity, and provides a robust effect to the changes in the frequency for various system parameters.

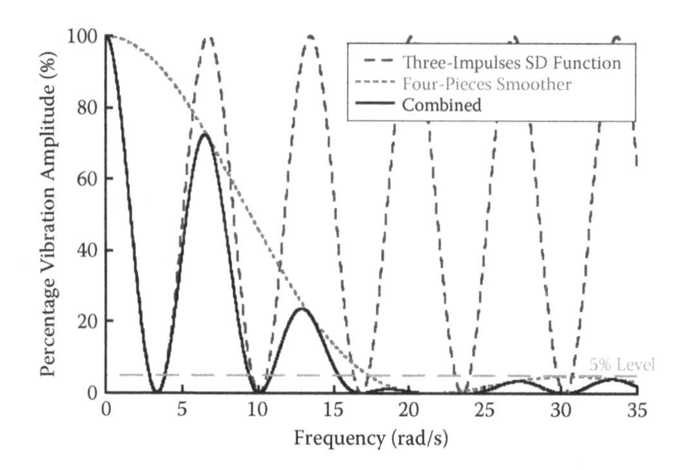

FIGURE 7.25 Frequency sensitive curves.

7.3.3 Experimental Investigation

Experimental investigations are performed on the testbed shown in Figure 7.16. The length of the uniform link is 950 mm, and the width and thickness of the link are 39 mm and 1.02 mm, respectively. The mass of the tennis ball is 121 g. The baseline command, which is a trapezoidal velocity profile, filters through the combined SD function and four-pieces smoother to create the slewing command to drive the hub.

The transient deflection and residual amplitude of the payload induced by various slewing distances are shown in Figures 7.26 and 7.27, respectively. The maximum slewing velocity and acceleration are fixed at 20°/s and 200°/s^2, respectively. In the absence of control, the peaks in the transient deflection are created because the vibrations caused by acceleration and deceleration of the hub are in phase. As the slewing distance increases, the residual amplitudes also contain peaks and troughs because the vibrations induced by acceleration and deceleration are

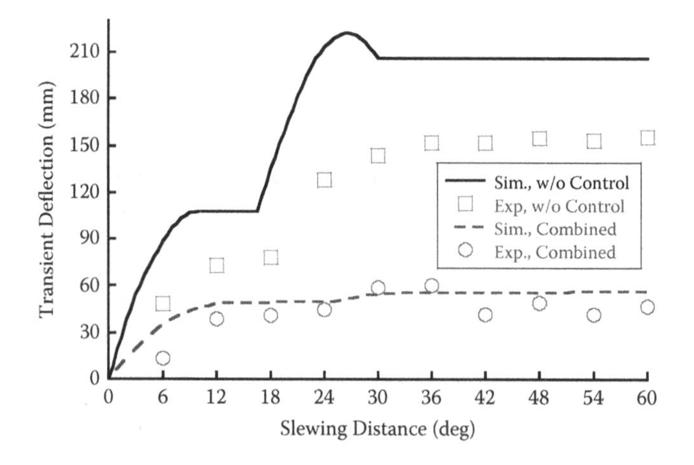

FIGURE 7.26 Transient deflection at different slewing distances.

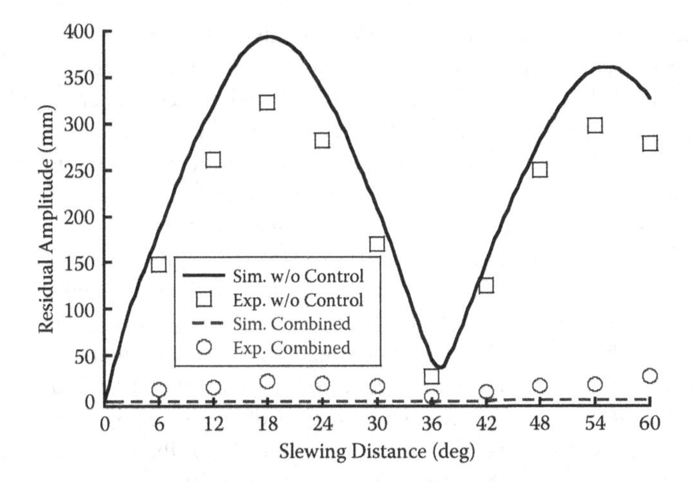

FIGURE 7.27 Residual amplitude at different slewing distances.

sometimes in phase, and sometimes out of phase. As the slewing distance increases, the damping effect causes a decrease in the peak magnitude.

In order to design the combined control scheme, the nonlinear first-mode frequency, Ω_1, second-mode frequency, Ω_2, and the damping ratio are 3.36 rad/s, 28.16 rad/s, and 0.03, respectively. When the combined control scheme is implemented, the transient and residual vibrations are significantly suppressed for all the slewing distances. The slight differences between the simulated and experimental results are caused by small modeling errors in the system. Overall, the experimental data match the corresponding simulated curves very well. Therefore, the experimental results validate the dynamic modeling and the effectiveness of the combined control architecture.

The experimental response under different modeling errors of frequency is shown in Figure 7.28. The slewing distance of the hub is set to be 54°. The uncontrolled

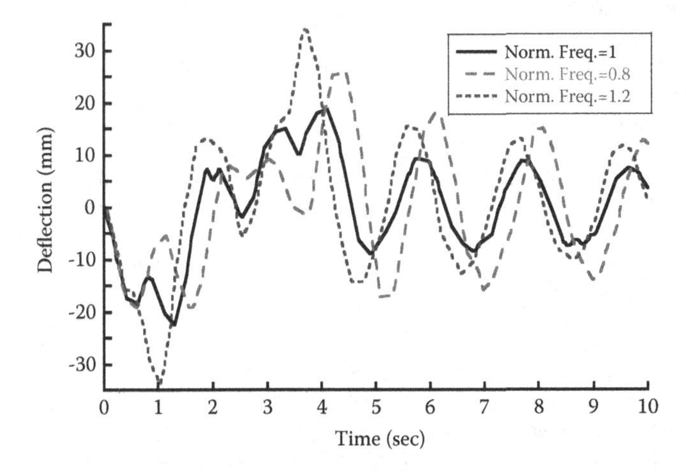

FIGURE 7.28 Experimental response under different frequency modeling errors.

transient deflection is 153.26 mm, and the amplitude of uncontrolled residual vibration is 296.7 mm. When the nonlinear frequencies, Ω_1 and Ω_2, are held constant at 3.36 rad/s and 28.16 rad/s, the transient deflection and the residual amplitude are 41.4 and 17.6 mm, respectively. The normalized frequency is defined as the ratio of the modeled frequency to the real frequency to study the effects of modeling errors in the frequency. In the case of the nonlinear frequencies, Ω_1 and Ω_2, of 3.36 rad/s and 28.16 rad/s, the normalized frequency is approximately 1.0.

In the case of the normalized frequency of 0.8, the nonlinear frequencies Ω_1 and Ω_2 are 2.69 rad/s and 22.53 rad/s, respectively. The transient deflection and the residual amplitude are 45.13 and 34.08 mm, respectively. Note that the rise time for the normalized frequency of 1.0 is shorter than that for the normalized frequency of 0.8. In the case of the normalized frequency of 1.2, the nonlinear frequencies Ω_1 and Ω_2 are 4.03 rad/s and 33.79 rad/s. The transient deflection is 67.75 mm, while the residual amplitude becomes 25.52 mm. The rise time for the normalized frequency of 1.0 is longer than that for the normalized frequency of 1.2. As the frequency modeling error increases, the transient and residual vibrations increase. However, when the frequency error deviates ±20% from the real value, the motion-induced vibrations can be limited below a low level.

REFERENCES

[1] W. Book, Recursive Lagrangian dynamics of flexible manipulator arms, *The International Journal of Robotics Research*, 3 (3) (1984) 87–101.

[2] P.B. Usoro, R. Nadira, S.S. Mahil, A finite element/Lagrange approach to modeling lightweight flexible manipulators, *Journal of Dynamic Systems, Measurement, and Control*, 108 (3) (1986) 198–205.

[3] J.Z. Sasiadek, R. Srinivasan, Dynamic modeling and adaptive control of a single-link flexible manipulator, *Journal of Guidance, Control, and Dynamics*, 12 (6) (1989) 838–844.

[4] T. Yoshikawa, K. Hosoda, Modeling of flexible manipulators using virtual rigid links and passive joints, *The International Journal of Robotics Research*, 15 (3) (1996) 290–299.

[5] D.J. Wagg, S.A. Neild, *Nonlinear Vibration with Control*, Springer-Verlag, 2009.

[6] S.M. Kim, Lumped element modeling of a flexible manipulator system, *IEEE/ASME Transactions on Mechatronics*, 20 (2) (2015) 967–974.

[7] A. Walsh, J.R. Forbes, Modeling and control of flexible telescoping manipulators, *IEEE Transactions on Robotics*, 31 (4) (2015) 936–947.

[8] K. Alipour, P. Zarafshan, A. Ebrahimi, Dynamics modeling and attitude control of a flexible space system with active stabilizers, *Nonlinear Dynamics*, 84 (4) (2016) 2535–2545.

[9] S. Choura, A.S. Yigit, Control of a two-link rigid-flexible manipulator with a moving payload mass, *Journal of Sound and Vibration*, 243 (5) (2001) 883–897.

[10] Z. Mohamed, M. Khairudin, A.R. Husain, et al., Linear matrix inequality-based robust proportional derivative control of a two-link flexible manipulator, *Journal of Vibration and Control*, 22 (5) (2016) 1244–1256.

[11] A. Jnifene, Active vibration control of flexible structures using delayed position feedback, *Systems & Control Letters*, 56 (3) (2007) 215–222.

[12] J. Shan, H. Liu, D. Sun, Slewing and vibration control of a single-link flexible manipulator by positive position feedback, *Mechatronics*, 15 (2005) 487–503.

[13] M. Baroudi, M. Saad, W. Ghie, State-feedback and linear quadratic regulator applied to a single-link flexible manipulator, *IEEE International Conference on Robotics and Biomimetics*, Guilin, China, 2009, pp. 1381–1386.

[14] J.H. Yang, F.L. Lian, L.C. Fu, Nonlinear adaptive control for flexible-link manipulators, *IEEE Transactions on Robotics and Automation*, 13 (1) (1997) 140–148.

[15] V. Feliu, K.S. Rattan, H.B. Brown, Adaptive control of a single-link flexible manipulator, *IEEE Control Systems Magazine*, 10 (2) (1990) 29–33.

[16] K.S. Yeung, Y.P. Chen, Sliding-mode controller design of a single-link flexible manipulator under gravity, *International Journal of Control*, 52 (1) (1990) 101–117.

[17] S.B. Choi, C.C. Cheong, H.C. Shin, Sliding mode control of vibration in a single-link flexible arm with parameter variations, *Journal of Sound and Vibration*, 179 (5) (1995) 737–748.

[18] L. Tian, C. Collins, Adaptive neuro-fuzzy control of a flexible manipulator, *Mechatronics*, 15 (10) (2005) 1305–1320.

[19] J. Lin, F.L. Lewis, Fuzzy controller for flexible-link robot arm by reduced-order techniques, *IEE Proceedings-Control Theory and Applications*, 149 (3) (2002) 177–187.

[20] M. Isogai, F. Arai, T. Fukuda, Modeling and vibration control with neural network for flexible multi-link structures, *IEEE International Conference on Robotics and Automation*, Detroit, MI, USA, 1999, pp. 1096–1101.

[21] M. Tinkir, Ü Önen, M. Kalyoncu, Modelling of neurofuzzy control of a flexible link, *Proceedings of the Institution of Mechanical Engineers, Part I: Journal of Systems and Control Engineering*, 224 (5) (2010) 529–543.

[22] A. Mohri, P.K. Sarkar, M. Yamamoto, An efficient motion planning of flexible manipulator along specified path, *IEEE International Conference on Robotics and Automation*, Leuven, Belgium, 1998, pp. 1104–1109.

[23] A. Abe, Trajectory planning for residual vibration suppression of a two-link rigid-flexible manipulator considering large deformation, *Mechanism and Machine Theory*, 44 (9) (2009) 1627–1639.

[24] H. Kojima, T. Kibe, Optimal trajectory planning of a two-link flexible robot arm based on genetic algorithm for residual vibration reduction, *IEEE/RSJ International Conference on Intelligent Robots and Systems*, Maui, Hawaii, USA, 2001, 4, pp. 2276–2281.

[25] S. Rhim, W.J. Book, Adaptive time-delay command shaping filter for flexible manipulator control, *IEEE/ASME Transactions on Mechatronics*, 9 (4) (2004) 619–626.

[26] J. Shan, H.T. Liu, D. Sun, Modified input shaping for a rotating single-link flexible manipulator, *Journal of Sound and Vibration*, 285 (1) (2005) 187–207.

[27] S. Kapucu, N. Yildirim, H. Yavuz, et al., Suppression of residual vibration of a translating-swinging load by a flexible manipulator, *Mechatronics*, 18 (3) (2008) 121–128.

[28] Q. Zhang, J.K. Mills, W.L. Cleghorn, et al., Dynamic model and input shaping control of a flexible link parallel manipulator considering the exact boundary conditions, *Robotica*, 33 (6) (2015) 1201–1230.

[29] M. Benosman, G. Le Vey, Control of flexible manipulators: a survey, *Robotica*, 22 (5) (2004) 533–545.

[30] S. Dwivedy, P. Eberhard, Dynamic analysis of flexible manipulators, a literature review, *Mechanism and machine theory*, 41 (7) (2006) 749–777.

[31] H. Rahimi, M. Nazemizadeh, Dynamic analysis and intelligent control techniques for flexible manipulators: a review, *Advanced Robotics*, 28 (2) (2014) 63–76.

[32] C.T. Kiang, A. Spowage, C.K. Yoong, Review of control and sensor system of flexible manipulator, *Journal of Intelligent & Robotic Systems*, 77 (1) (2015) 187–213.

[33] M. Sayahkarajy, Z. Mohamed, A. Faudzi, Review of modelling and control of flexible-link manipulators, *Proceedings of the Institution of Mechanical Engineers, Part I: Journal of Systems and Control Engineering*, 230 (8) (2016) 861–873.

[34] H. Hu, E.H. Dowell, L.N. Virgin, Resonances of a harmonically forced Duffing oscillator with time delay state feedback, *Nonlinear Dynamics*, 15 (4) (1998) 311–327.

[35] J.C. Ji, Nonresonant Hopf bifurcations of a controlled van der Pol-Duffing oscillator, *Journal of Sound and Vibration*, 297 (1) (2006) 183–199.

[36] X. Li, J.C. Ji, C.H. Hansen, et al., The response of a Duffing-van der Pol oscillator under delayed feedback control, *Journal of Sound and Vibration*, 291 (3) (2006) 644–655.

[37] Y. Jin, H. Hu, Dynamics of a Duffing oscillator with two time delays in feedback control under narrow-band random excitation, *The Transactions of the ASME - Journal of Computational and Nonlinear Dynamics*, 3 (2) (2008) 021205.

[38] J.C. Ji, N. Zhang, Additive resonances of a controlled van der Pol-Duffing oscillator, *Journal of Sound and Vibration*, 315 (1) (2008) 22–33.

[39] C. Feng, W. Zhu, Asymptotic Lyapunov stability with probability one of Duffing oscillator subject to time-delayed feedback control and bounded noise excitation, *Acta Mechanica*, 208 (1–2) (2009) 55–62.

[40] M.S. Siewe, C. Tchawoua, S. Rajasekar, Parametric resonance in the Rayleigh-Duffing oscillator with time-delayed feedback, *Communications in Nonlinear Science and Numerical Simulation*, 17 (11) (2012) 4485–4493.

[41] Y. Wang, F. Li, Dynamical properties of Duffing-van der Pol oscillator subject to both external and parametric excitations with time delayed feedback control, *Journal of Vibration and Control*, 21 (2) (2015) 371–387.

[42] H. Yabuno, Bifurcation control of parametrically excited Duffing system by a combined linear-plus-nonlinear feedback control, *Nonlinear Dynamics*, 12 (3) (1997) 263–274.

[43] M. Ghandchi-Tehrani, L. Wilmshurst, S. Elliott, Bifurcation control of a Duffing oscillator using pole placement, *Journal of Vibration and Control*, 21 (14) (2015) 2838–2851.

[44] Y. Peng, J. Li, Exceedance probability criterion based stochastic optimal polynomial control of Duffing oscillators, *International Journal of Non-Linear Mechanics*, 46 (2) (2011) 457–469.

[45] F. Khadra, Super-twisting control of the Duffing-Holmes chaotic system, *International Journal of Modern Nonlinear Theory and Application*, 5 (4) (2016) 160–170.

[46] K.S. Chen, T.S. Yang, K. Ou, et al., Design of command shapers for residual vibration suppression in Duffing nonlinear systems, *Mechatronics*, 19 (2) (2009) 184–198.

[47] J. Huang, JC Ji, Vibration control of coupled Duffing oscillators in flexible single-link manipulators, *Journal of Vibration and Control*, 27 (17-18) (2021) 2058–2068.

8 Liquid Sloshing

Motions of a free liquid surface inside its container are defined as sloshing. The interaction of the sloshing dynamics with the container has unwanted effects on the overall system. Those impacts cause detrimental problems in many industrial applications ranging from packing engineering to space vehicles [1–3]. Therefore, there exists a need for a control system that can effectively reduce sloshing for safe operations.

8.1 PLANAR LINEAR SLOSHING

Numerous scientists have worked to provide solutions to challenging problems posed by the sloshing dynamics. Absorbers or baffles are applied to dissipate sloshing energy [4–6]. However, the passive technique adds additional weight and complexity to the whole system. Active control method is a further solution for sloshing suppression. Venugopal and Bernstein have been proposed two active controllers. The first one used surface pressure, whereas the second controller applied a flap actuator to the surface of the fluid. Effectiveness of the controller was demonstrated by simulations [7]. The feedback control schemes measure the sloshing motion to control the container in a closed loop, such as proportional-integral-derivative control [8], sliding-mode control [9–12], H_∞ control [13,14], and Lyapunov-based feedback control [15,16]. Kurode et al. presented a nonlinear sliding-mode controller for sloshing suppression. Effectiveness of the controller was experimentally verified on a moving container [10]. Reyhanoglu and Hervas designed a Lyapunov-based feedback controller to suppress sloshing for a Prismatic-Prismatic-Revolute robot. Effectiveness of the control law was demonstrated in simulations [15]. Grundelius and Bernhardsson used optimal control and iterative learning control to reduce sloshing in the packing industry [17]. Yano et al. applied hybrid shape approach to control sloshing. Effectiveness of the controller was demonstrated in experiments [18–20]. Gandhi and Duggal stabilized sloshing in a cylindrical tank by using translational excitation as the control input. The design of the controller was based on a Lyapunov approach, and implementation of the controller applied force feedback [21].

Open-loop techniques modify inputs to create prescribed motions that induce minimal sloshing including infinite impulse response filter [22], acceleration compensation [23], and input shaper [24–29]. Feddema et al. presented an infinite impulse response filter to modify the acceleration profile to produce slosh-free motions. Experiments on a FANUC S-800 robot moving a hemispherical container of water verified the performance of the method [22]. Chen et al. reduced sloshing using acceleration compensation. Experimental results on a KUKA-KR16 industrial manipulator demonstrated the effectiveness [23]. Pridgen et al. presented a two-mode specified-insensitivity input shaper for slosh suppression. Experiments on a

DOI: 10.1201/9781003247210-8

moving container verified the effectiveness [25]. Additionally, significant works on the modeling of sloshing [30–33] and construction of experimental test rigs [34,35] have been reported.

8.1.1 Modeling

A schematic representation of sloshing in a moving container is shown in Figure 8.1. A fluid contains in a planar tank of length, $2a$. The surface of the fluid at rest is at a height, h, from the bottom of the tank. The fluid surface elevation measured from the undisturbed free surface denotes η. The acceleration of the container is $C(t)$. The following assumptions are employed to simplify the modeling of sloshing:

1. The tank is rigid and flow field is considered to be irrotational.
2. The fluid is assumed to be incompressible and nonviscous homogeneous.
3. Displacement and velocity of the liquid-free surface are assumed to be small.

For the irrotational flow motion, the velocity of the fluid in the container satisfies:

$$\nu = \nu_0 + \nabla\phi \tag{8.1}$$

where ∇ is the gradient operator, ν_0 is the velocity of the tank, and ϕ is the perturbed velocity potential function. The boundary value problem is summarized as follows:

$$\nabla^2\phi = 0 \tag{8.2}$$

$$\left.\frac{\partial\phi}{\partial x}\right|_{x=0;2a} = 0 \tag{8.3}$$

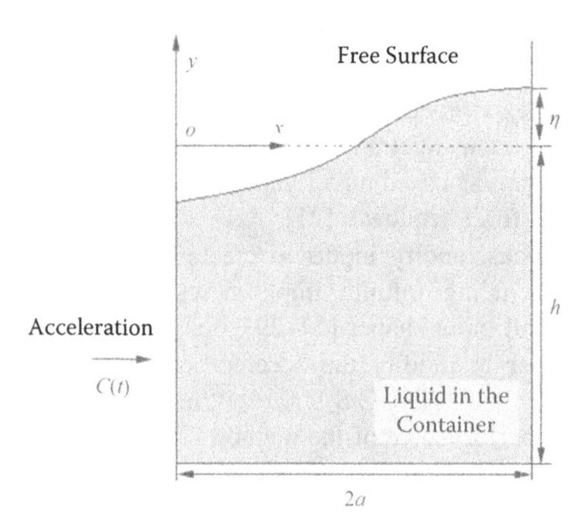

FIGURE 8.1 Liquid slosh in a moving container.

$$\left.\frac{\partial \phi}{\partial y}\right|_{y=-h} = 0 \tag{8.4}$$

$$\frac{\partial \phi}{\partial y} = \frac{\partial \eta}{\partial t},\ y = \eta(x, t) \tag{8.5}$$

$$\frac{\partial \phi}{\partial t} + g\eta + C(t)x = 0,\ y = \eta(x, t) \tag{8.6}$$

where g is the gravitational constant. The perturbed velocity potential function, ϕ, and surface elevation, η, satisfy:

$$\phi(x, y, t) = \sum_k \varphi_k(x, y)\dot{q}_k(t) \tag{8.7}$$

$$\eta(x, y, t) = \sum_k H_k(x, y)q_k(t) \tag{8.8}$$

where $q_k(t)$ is the time-dependent function, and $\varphi_k(x, y)$ and $H_k(x, y)$ are corresponding spatial functions. Spatial functions are the solution of following equations:

$$\nabla^2 \varphi_k = 0 \tag{8.9}$$

$$\left.\frac{\partial \varphi_k}{\partial x}\right|_{x=0;2a} = 0 \tag{8.10}$$

$$\left.\frac{\partial \varphi_k}{\partial y}\right|_{y=-h} = 0 \tag{8.11}$$

$$\frac{\partial \varphi_k}{\partial y} = H_k = \frac{\omega_k^2 \varphi_k}{g},\ y = \eta(x, t) \tag{8.12}$$

Solving Equations (8.9)–(8.12) yields the natural frequency, ω_k, of sloshing modes, and spatial functions, φ_k and H_k:

$$\omega_k^2 = g\frac{k\pi}{2a}\tanh\left(\frac{k\pi}{2a}h\right) \tag{8.13}$$

$$\varphi_k = \cos\left(\frac{k\pi x}{2a}\right)\cosh\left(k\pi\frac{y+h}{2a}\right) \tag{8.14}$$

$$H_k = \frac{\omega_k^2}{g}\varphi_k \tag{8.15}$$

Equation (8.13) can be used to predict the sloshing frequency. Substituting Equations (8.7) and (8.8) into (8.6), then multiplying by φ_k, and integrating over the free surface ($0 \le x \le 2a$) yield:

$$u_k \ddot{q}_k(t) + u_k \omega_k^2 q_k(t) + \alpha_k C(t) = 0 \tag{8.16}$$

where

$$u_k = \rho \int_0^{2a} H_k \varphi_k dx \tag{8.17}$$

$$\alpha_k = \rho \int_0^{2a} x H_k dx \tag{8.18}$$

ρ is the density of the liquid. When k is even, α_k is limited to zero. Only odd sloshing modes can be excited by the horizontal acceleration $C(t)$. Equations (8.16) with the inclusion of proportional damping could be changed [36]:

$$\ddot{q}_k(t) + 2\zeta_k \omega_k \dot{q}_k(t) + \omega_k^2 q_k(t) + \frac{\alpha_k}{u_k} C(t) = 0 \tag{8.19}$$

where ζ_k is the damping ratio of sloshing modes. The damping ratio is a function of the Galilei number and container geometry. The damping ratio has been developed theoretically to be approximately 0.01 for the water. Therefore, the damping ratio for each of sloshing modes is assumed to be 0.01. Substituting Equation (8.15) into (8.8) produces the surface elevation at the measurement point:

$$\eta(x, 0, t) = \sum_k H_k(x, 0) q_k(t) = \sum_{k=odd} \left[\frac{\omega_k^2}{g} \varphi_k q_k(t) \right] \tag{8.20}$$

Resulting from Equations (8.19) and (8.20), the transfer function of the surface elevation at the rightmost edge of the container is:

$$\eta_{x=2a,y=0}(s) = \sum_{k=odd} \left[\frac{8a}{gk^2\pi^2} \cdot \frac{-\omega_k^2}{\left(s^2 + 2\zeta_k \omega_k s + \omega_k^2\right)} C(s) \right] \tag{8.21}$$

Then substituting Equations (8.14) and (8.19) into (8.7) yields the transfer function of the perturbed velocity potential function at the rightmost edge of the container:

$$\phi_{x=2a,y=0}(s) = \sum_{k=odd} \left[\frac{8a}{k^2\pi^2} \cdot \frac{-s}{\left(s^2 + 2\zeta_k \omega_k s + \omega_k^2\right)} C(s) \right] \tag{8.22}$$

The models (8.21) and (8.22) include an infinite number of sloshing modes. A sum of response for each of an infinite number of sloshing modes is the total system

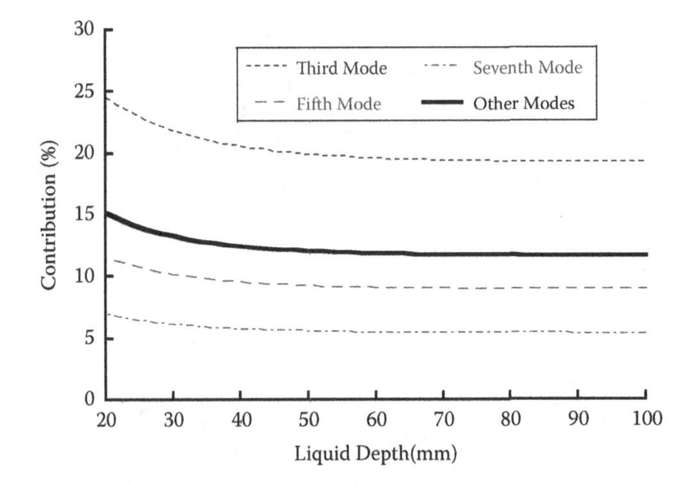

FIGURE 8.2 Relative amplitude contribution of high modes.

response. The ratio of vibration amplitude of the impulse response from the high mode to that from the first mode is defined as the relative amplitude contribution, which is applied to evaluate the effect of high sloshing modes.

The relative amplitude contribution of high modes for a large range of liquid depths is shown in Figure 8.2. The relative amplitude contribution decreases as the liquid depth increases before 50 mm, and then changes slightly as the liquid depth increases after this point. The average relative amplitude contribution of the third mode, fifth mode, and seventh mode are 19.8%, 9.2%, and 5.6%, respectively. Meanwhile, the average relative amplitude contribution of the sum of other six higher modes is 12.0%. Thus, theoretical results indicate that high sloshing modes have some impacts on the system dynamics. Therefore, reducing effectively sloshing induced by total modes is essential.

8.1.2 Computational Dynamics

Numerical verification of the two-pieces smoother over a large range of motions and liquid depths is presented in this section. A model including first four modes is used in the simulation. A trapezoidal-velocity profile drove the container. The container length, maximum velocity, and acceleration are 92 mm, 0.2 m/s, and 2 m/s^2, respectively. There are two stages in the system response. The peak-to-peak deflection when the container is moving is referred to as the transient deflection. The peak-to-peak deflection when the container is stopped is defined as the residual amplitude.

The transient deflection and residual amplitude of sloshing induced by various driving motions are shown in Figures 8.3 and 8.4. The liquid depth was set to 92 mm. Without the controller, peaks and troughs in the transient deflection and residual amplitude arise. This is because the transient deflection and residual amplitude are the result of the interference between the sloshing caused by the acceleration and deceleration.

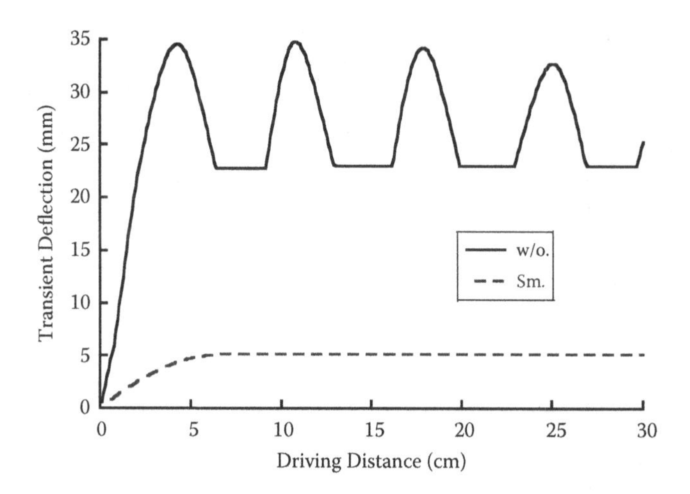

FIGURE 8.3 Transient vibrations against driving distance.

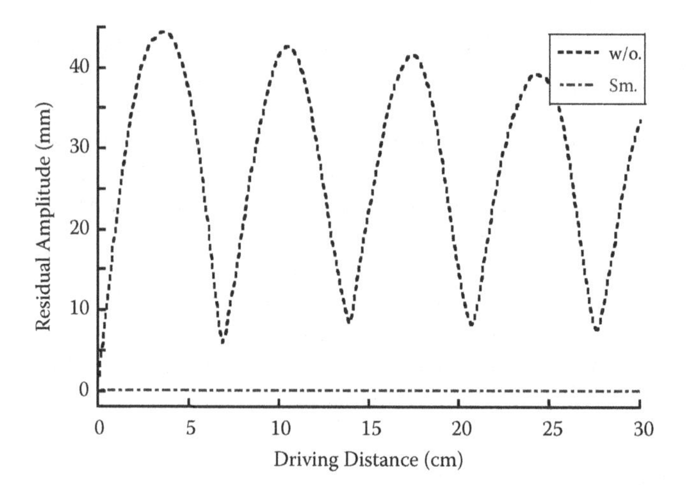

FIGURE 8.4 Residual vibrations against driving distance.

A two-pieces smoother is applied to suppress the infinite modes of the sloshing. The two-pieces smoother is described by:

$$u(\tau) = \begin{Bmatrix} \tau u_0 e^{-\zeta_k \omega_k \tau}, & 0 \le \tau \le T_d \\ (2T_d - \tau) u_0 e^{-\zeta_k \omega_k \tau}, & T_d < \tau \le 2T_d \\ 0 & \text{others} \end{Bmatrix} \tag{8.23}$$

where

$$u_0 = \frac{\zeta_k^2 \omega_k^2}{\left(1 - e^{-2\pi\zeta_k / \sqrt{1-\zeta_k^2}}\right)^2} \tag{8.24}$$

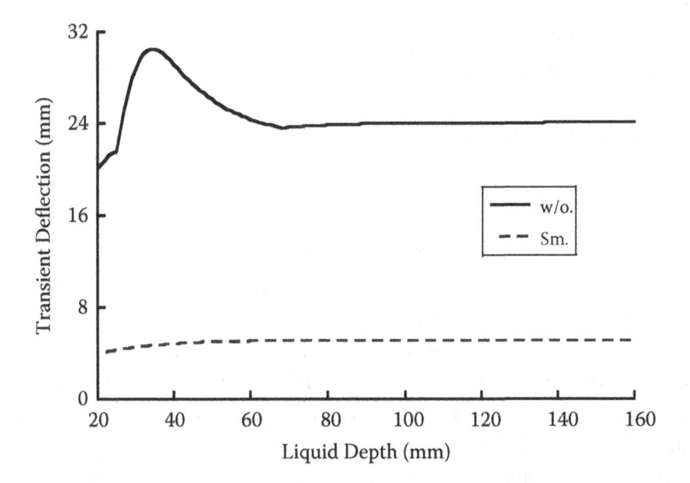

FIGURE 8.5 Transient vibrations against liquid depth.

The two-pieces smoother suppresses the transient deflection and residual amplitude by an average of 82.2% and 99.8%, respectively. The two-pieces smoother reduces the transient and residual sloshing to a low level for various driving motions.

The transient deflection and residual amplitude of sloshing for varying liquid depth are shown in Figures 8.5 and 8.6. The driving distance was fixed at 20 cm. Without the controller, the driving command causes the greatest transient deflection and residual amplitude at all liquid depths. The transient deflection and residual amplitude increase as the liquid depth increases before a shallow liquid depth. Then after this shallow depth, the transient deflection and residual amplitude decrease with increasing liquid depth. However, the transient deflection and residual amplitude change very little when the liquid depth increases after the container length. The two-pieces smoother suppresses the transient deflection and residual amplitude by an average of 78.8% and 99.8%,

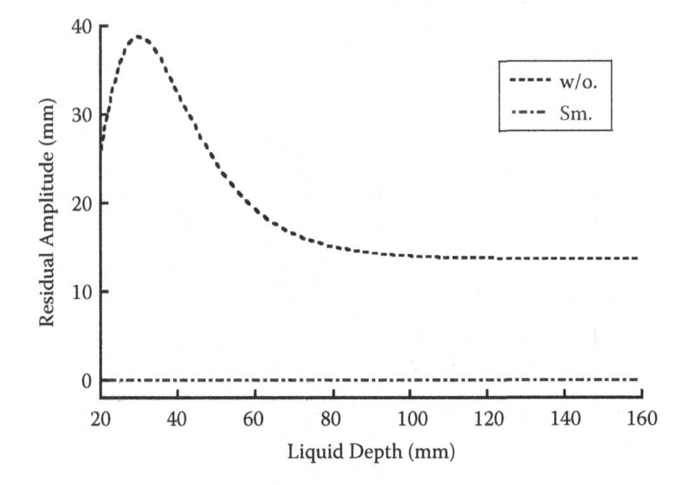

FIGURE 8.6 Residual vibrations against liquid depth.

respectively. Thus, the two-pieces smoother reduces the transient and residual sloshing to a low level over a wide range of liquid depths.

8.1.3 Experimental Verification

Figure 8.7 shows a container and camera mounted to a XY gantry, in which experiments were conducted. Panasonic AC servomotors with encoders drove the gantry. A DSP-based motion control card connects a personal computer to servo amplifier. The baseline command is sent to a Visual C++ program, which uses the two-pieces smoother algorithm, and produces the smoothed command for the drive. The sloshing frequencies were estimated by measuring the liquid depth via a plastic ruler. The travel of the gantry was 30 cm, and the size of the container was 92 mm × 92 mm × 180 mm. The gantry mounted a CMOS camera to record surface elevation at the rightmost edge of the container.

Experimental responses of the surface elevation at the rightmost edge of the container caused by a trapezoidal-velocity command is shown in Figure 8.8. The

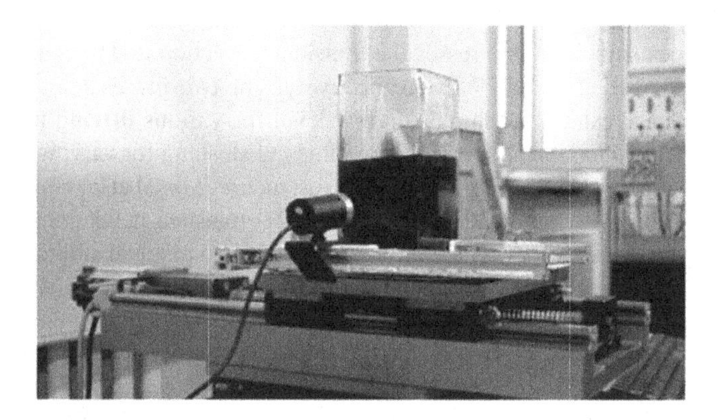

FIGURE 8.7 A moving liquid container.

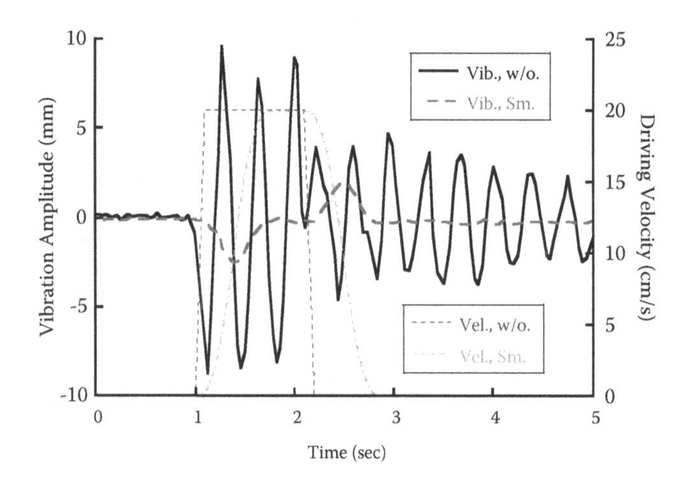

FIGURE 8.8 Experimental sloshing response to driving commands.

liquid depth and driving distance were set to 92 mm and 22 cm, respectively. The entire surface was level between 0 and 1 second. The container accelerated at 1 second, and decelerated at 2.1 seconds. Without the controller, the transient deflection and residual amplitude were 17.4 mm and 7.7 mm, respectively. The residual amplitude is smaller than the transient deflection because the sloshing induced by the acceleration is out of phase with that caused by the deceleration. With the two-pieces smoother, the transient deflection and residual amplitude were 4.3 mm and 0.2 mm, respectively. It is clear that the two-pieces smoother dramatically suppresses both transient and residual sloshing.

Two sets of experiments were performed to verify the dynamics and the effectiveness of the smoother on suppressing sloshing for variations of system parameters and operation conditions. The effect of variation in the driving distance is investigated in the first set of experiments. The liquid depth was set to 92 mm. The experimental transient deflection is shown in Figure 8.9. Without the controller, the transient deflection varied with changing driving distance because of the interference between the sloshing caused by the acceleration and deceleration. A smooth velocity profile was produced by the two-pieces smoother for moving the container. The smooth transitions between boundary conditions reduced the transient sloshing. The two-pieces smoother suppressed the transient deflection by an average of 78.3%.

Figure 8.10 shows experimental residual amplitude as a function of driving distance. Without the controller, peaks and troughs occurred as the driving distance varied. The two-pieces smoother with the notch and low-pass filtering effect suppressed the residual sloshing for infinite sloshing modes. The two-pieces smoother reduced the residual amplitude to <0.43 mm for all driving distances tested.

Effectiveness of the smoother at reducing sloshing from varying liquid depths was verified in another set of experiments. The transient deflection from these tests is shown in Figure 8.11. Without the controller, transient deflection varied slightly with changing liquid depth. The smooth velocity profile, which was produced by the

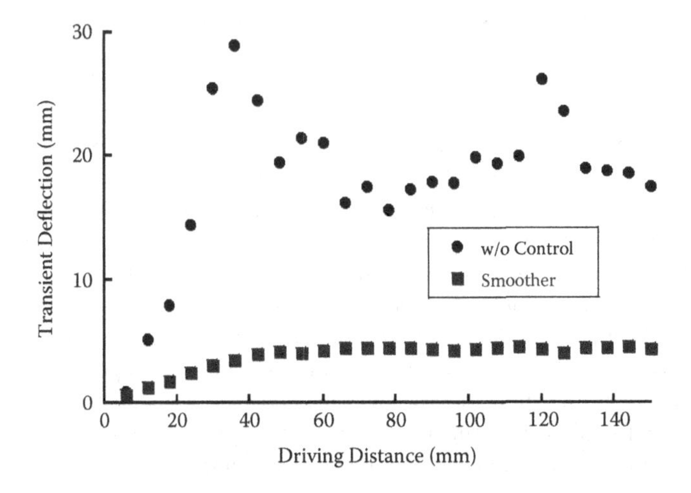

FIGURE 8.9 Experimental transient deflection induced by driving motions.

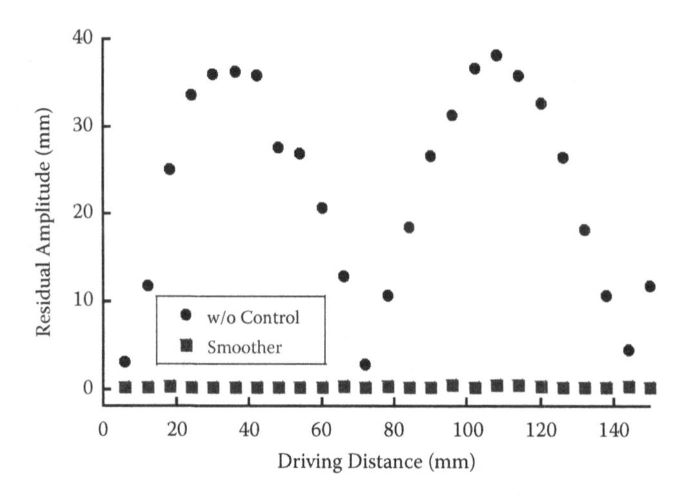

FIGURE 8.10 Experimental residual amplitude induced by driving motions.

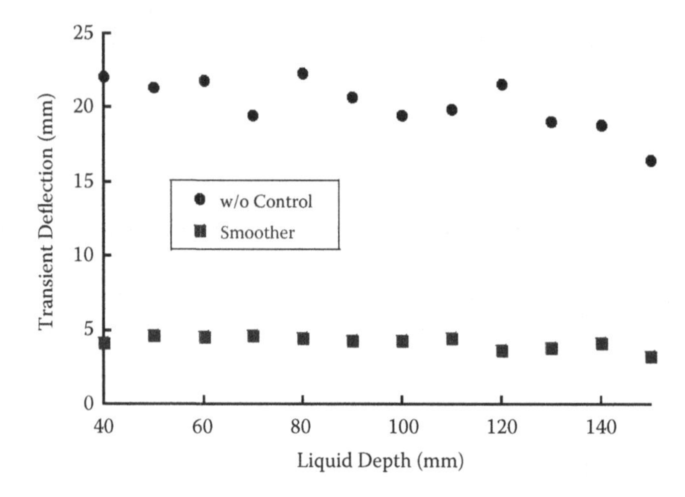

FIGURE 8.11 Experimental transient deflection for varying liquid depths.

two-pieces smoother, reduced the transient sloshing. The two-pieces smoother reduced the transient sloshing by an average of 79.1%.

Figure 8.12 shows the residual sloshing for varying liquid depths. Without the controller, the residual amplitude decreased as liquid depth increased. The two-pieces smoother reduced the residual amplitude to <0.32 mm for all the depths tested. The notch and low-pass filtering effect of the two-pieces smoother suppressed the residual sloshing for infinite sloshing modes. Thus, the two-pieces smoother was effective for all liquid depths. These experiments demonstrated that the two-pieces smoother can effectively eliminate the transient and residual sloshing induced by various combinations of driving motions and system parameters.

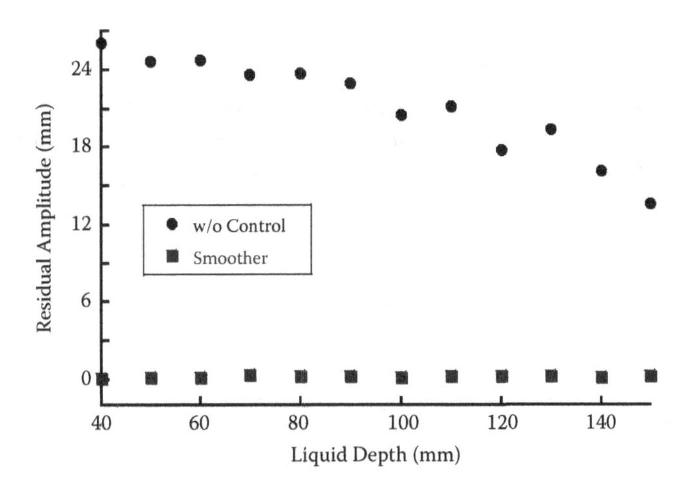

FIGURE 8.12 Experimental residual amplitude for varying liquid depths.

8.2 THREE-DIMENSIONAL LINEAR SLOSHING

8.2.1 Modeling

A schematic representation of the three-dimensional sloshing in a moving rectangular liquid container is shown in Figure 8.13. A liquid fluid contains in a rectangular tank of length a and width b. Fluid surface at rest is at height, h, from the bottom of the tank. The fluid surface elevation measured from the undisturbed free surface denotes η. Motions of the container include the linear movement of the container along the directions, X' and Y', in the inertial Newtonian frame. The acceleration, $C(t)$, of the container and steering angle, α, between the acceleration and the X' direction are defined as input variables. The moving coordinates *oxyz* are fixed to the tank, and the *oxy* plane coincides with the undisturbed free surface. Axes of the moving coordinate *oxyz* system are parallel to the inertial Newtonian frame $O'X'Y'Z'$.

The fluid is incompressible, inviscid, and irrotational. The displacement and velocity of the liquid-free surface are assumed to be small. The container is rigid and impermeable. For irrotational flow motions, the absolute velocity of the fluid in the container satisfies:

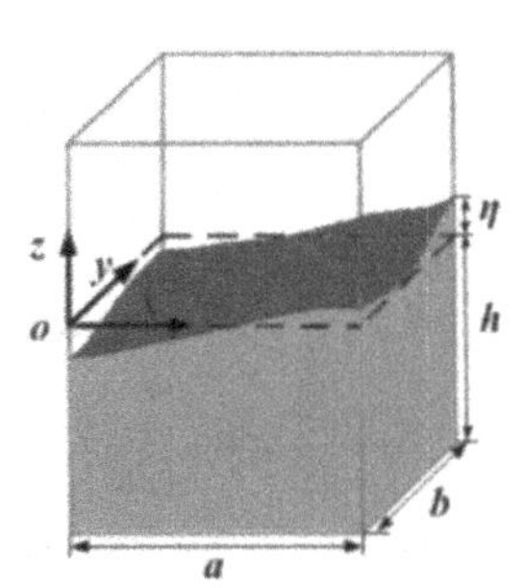

FIGURE 8.13 A three-dimensional sloshing model in a rectangular container.

$$v = v_0 + \nabla\phi \tag{8.25}$$

where $\nabla$is the gradient operator, v_0 is the velocity of the tank, and ϕ is the perturbed velocity potential function. The boundary value problem in terms of the perturbed velocity potential in the moving coordinate system is given by:

$$\nabla^2\phi = 0 \tag{8.26}$$

$$\left.\frac{\partial\phi}{\partial x}\right|_{x=0,a} = 0 \tag{8.27}$$

$$\left.\frac{\partial\phi}{\partial y}\right|_{y=0,b} = 0 \tag{8.28}$$

$$\left.\frac{\partial\phi}{\partial z}\right|_{z=-h} = 0 \tag{8.29}$$

$$\frac{\partial\phi}{\partial z} = \frac{\partial\eta}{\partial t},\ z = \eta(x, y, t) \tag{8.30}$$

$$\frac{\partial\phi}{\partial t} + g\eta + xC(t)\cos\alpha + yC(t)\sin\alpha = 0,\ z = \eta(x, y, t) \tag{8.31}$$

The perturbed velocity potential function, ϕ, and surface elevation, η, satisfy:

$$\phi(x, y, z, t) = \sum_{ij} \varphi_{ij}(x, y, z)\dot{q}_{ij}(t) \tag{8.32}$$

$$\eta(x, y, z, t) = \sum_{ij} H_{ij}(x, y, z)q_{ij}(t) \tag{8.33}$$

where $\varphi_{ij}(x, y, z)$ and $H_{ij}(x, y, z)$ are spatial functions, i and j are nonnegative integers, and $q_{ij}(t)$ is the time-dependent function. Spatial functions are the solution of following equations:

$$\nabla^2\varphi_{ij} = 0 \tag{8.34}$$

$$\left.\frac{\partial\varphi_{ij}}{\partial x}\right|_{x=0,a} = 0 \tag{8.35}$$

$$\left.\frac{\partial\varphi_{ij}}{\partial y}\right|_{y=0,b} = 0 \tag{8.36}$$

$$\left.\frac{\partial\varphi_{ij}}{\partial z}\right|_{z=-h} = 0 \tag{8.37}$$

$$\frac{\partial \varphi_{ij}}{\partial z} = H_{ij} = \frac{\omega_{ij}^2 \varphi_{ij}}{g}, \; z = \eta(x, y, t) \tag{8.38}$$

Solving Equations (8.34)–(8.38) yields the natural frequencies, ω_{ij}, and spatial functions, $\varphi_{ij}(x, y, z)$ and $H_{ij}(x, y, z)$:

$$\omega_{ij}^2 = g\pi\sqrt{\left(\frac{i}{a}\right)^2 + \left(\frac{j}{b}\right)^2} \cdot \tanh\left[\pi h\sqrt{\left(\frac{i}{a}\right)^2 + \left(\frac{j}{b}\right)^2}\right] \tag{8.39}$$

$$\varphi_{ij} = \cos\left(\frac{i\pi x}{a}\right)\cos\left(\frac{j\pi y}{b}\right) \cdot \cosh\left[\pi(z + h)\sqrt{\left(\frac{i}{a}\right)^2 + \left(\frac{j}{b}\right)^2}\right] \tag{8.40}$$

$$H_{ij} = \frac{\omega_{ij}^2 \varphi_{ij}}{g} \tag{8.41}$$

where ω_{i0} is the frequency of the transverse mode, ω_{0j} is the frequency of the longitudinal mode, and the set $\{\omega_{ij}, i \neq 0 \,\&\, j \neq 0\}$ represents the frequency of the mixed mode. The natural frequency of the sloshing mode is dependent on the container size and liquid depth. Substituting Equations (8.32) and (8.33) into (8.31), then multiplying by φ_{ij}, and integrating over the free surface ($0 \le x \le a$; $0 \le y \le b$) yield:

$$\lambda_{ij}\ddot{q}_{ij}(t) + \lambda_{ij}\omega_{ij}^2 q_{ij}(t) + \gamma_{ij}C(t)\cos\alpha + \beta_{ij}C(t)\sin\alpha = 0 \tag{8.42}$$

where

$$\lambda_{ij} = \rho\int_0^a\int_0^b H_{ij}\varphi_{ij}\,dxdy \tag{8.43}$$

$$\gamma_{ij} = \rho\int_0^a\int_0^b xH_{ij}\,dxdy \tag{8.44}$$

$$\beta_{ij} = \rho\int_0^a\int_0^b yH_{ij}\,dxdy \tag{8.45}$$

From Equations (8.44) and (8.45), coefficients γ_{ij} and β_{ij} are given by:

$$\gamma_{ij} = \begin{cases} \dfrac{-2a^2\rho\omega_{ij}^2\cosh(i\pi h/a)}{g\pi^2 i^2}, & i = odd, \;\; j = 0 \\ 0, & \text{others} \end{cases} \tag{8.46}$$

$$\beta_{ij} = \begin{cases} \dfrac{-2b^2\rho\omega_{ij}^2\cosh(j\pi h/b)}{g\pi^2 j^2}, & i = 0, \;\; j = odd \\ 0, & \text{others} \end{cases} \tag{8.47}$$

From Equations (8.46) and (8.47), the transverse modal response cannot be excited by the acceleration in the y direction, the acceleration in the x direction cannot excite the longitudinal modal response, and accelerations in both the x direction and y direction cannot excite the mixed modal response. Equation (8.42) with the inclusion of the proportional damping could be changed:

$$\ddot{q}_{i0}(t) + 2\zeta_{i0}\omega_{i0}\dot{q}_{i0}(t) + \omega_{i0}^2 q_{i0}(t) + \frac{\gamma_{i0}}{\lambda_{i0}}C(t)\cos\alpha = 0,\ i = odd \tag{8.48}$$

$$\ddot{q}_{0j}(t) + 2\zeta_{0j}\omega_{0j}\dot{q}_{0j}(t) + \omega_{0j}^2 q_{0j}(t) + \frac{\beta_{0j}}{\lambda_{0j}}C(t)\sin\alpha = 0,\ j = odd \tag{8.49}$$

where ζ_{i0} is the damping ratio of the transverse mode and ζ_{0j} is the damping ratio of the longitudinal mode. The damping ratio is also 0.01 for the water. By substituting Equation (8.41) into (8.33), the surface elevation at the measurement point is:

$$\begin{aligned}\eta(x, y, 0, t) &= \sum_{i=odd,j=0} H_{ij}(x, y, 0)q_{ij}(t) + \sum_{i=0,j=odd} H_{ij}(x, y, 0)q_{ij}(t)\\ &= \sum_{i=odd} \frac{\omega_{i0}^2\varphi_{i0}}{g}q_{i0}(t) + \sum_{j=odd}\frac{\omega_{0j}^2\varphi_{0j}}{g}q_{0j}(t)\end{aligned} \tag{8.50}$$

The transfer function of the surface elevation at the diagonal corner of the container resulting from Equations (8.48)–(8.50) is given by:

$$\begin{aligned}\eta_{x=a,y=b,z=0}(s) = &\sum_{i=odd}\frac{-4a\omega_{i0}^2\cdot C(s)\cos\alpha}{g\pi^2 i^2\left(s^2 + 2\zeta_{i0}\omega_{i0}s + \omega_{i0}^2\right)}\\ &+ \sum_{j=odd}\frac{-4b\omega_{0j}^2\cdot C(s)\sin\alpha}{g\pi^2 j^2\left(s^2 + 2\zeta_{0j}\omega_{0j}s + \omega_{0j}^2\right)}\end{aligned} \tag{8.51}$$

The transfer function of the perturbed velocity potential function at the diagonal corner of the container can be derived by substituting Equations (8.40), (8.48), (8.49) into (8.32):

$$\begin{aligned}\phi_{x=a,y=b,z=0}(s) = &\sum_{i=odd}\frac{-4as\cdot C(s)\cos\alpha}{\pi^2 i^2\left(s^2 + 2\zeta_{i0}\omega_{i0}s + \omega_{i0}^2\right)}\\ &+ \sum_{j=odd}\frac{-4bs\cdot C(s)\sin\alpha}{\pi^2 j^2\left(s^2 + 2\zeta_{0j}\omega_{0j}s + \omega_{0j}^2\right)}\end{aligned} \tag{8.52}$$

The sloshing resulting from accelerations along two directions is independent. Both the surface elevation and the perturbed velocity potential are the sum of the response for each of an infinite number of sloshing modes along two directions. The amplitude of surface elevation response from an impulse command at time zero resulting from Equation (8.51) is:

$$A_{x=a,y=b,z=0} = \sum_{i=odd} \frac{-4a\omega_{i0}\cos\alpha}{g\pi^2 i^2\sqrt{1-\zeta_{i0}^2}} + \sum_{j=odd} \frac{-4b\omega_{0j}\sin\alpha}{g\pi^2 j^2\sqrt{1-\zeta_{0j}^2}} \tag{8.53}$$

The ratio of the vibration amplitude of the impulse response from the high mode to that from the first mode is used to identify the amplitude contribution of each sloshing mode to the whole system. While the ratio is called relative amplitude contribution, c_{ij}, Equation (8.53) can be changed:

$$A_{x=a,y=b,z=0} = \frac{-4a\omega_{10}\cos\alpha}{g\pi^2\sqrt{1-\zeta_{10}^2}} \cdot \sum_{i=odd} c_{i0} + \frac{-4b\omega_{01}\sin\alpha}{g\pi^2\sqrt{1-\zeta_{01}^2}} \cdot \sum_{j=odd} c_{0j} \tag{8.54}$$

where ω_{10} and ω_{01} are the frequency for the first mode along the transverse and longitudinal direction, ζ_{10} and ζ_{01} are the damping ratio for the first mode along the transverse and longitudinal direction. The variables c_{i0} and c_{0j} are given by:

$$c_{i0} = \frac{\omega_{i0}\sqrt{1-\zeta_{10}^2}}{i^2\omega_{10}\sqrt{1-\zeta_{i0}^2}}, \quad i = odd \tag{8.55}$$

$$c_{0j} = \frac{\omega_{0j}\sqrt{1-\zeta_{01}^2}}{j^2\omega_{01}\sqrt{1-\zeta_{0j}^2}}, \quad j = odd \tag{8.56}$$

Thus, multiplying the sum of the relative amplitude contribution by the vibration amplitude from the first mode produces the amplitude of surface elevation responses.

The relative amplitude contribution of each sloshing modes with a liquid depth of 90 mm is illustrated in Figure 8.14. The container length, container width, maximum

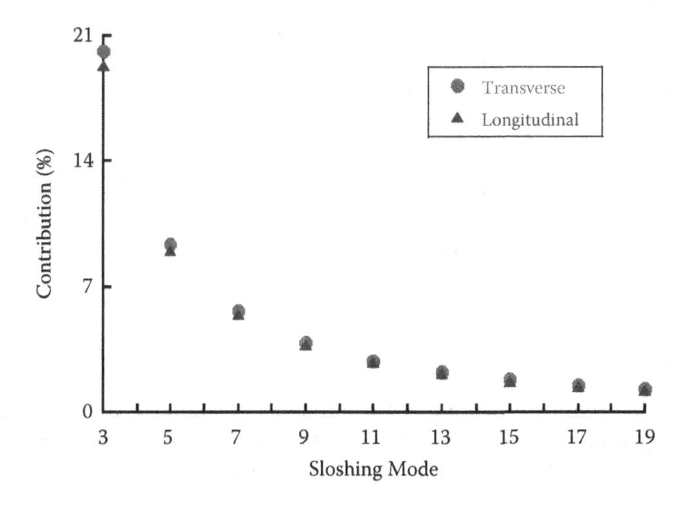

FIGURE 8.14 Relative amplitude contributions.

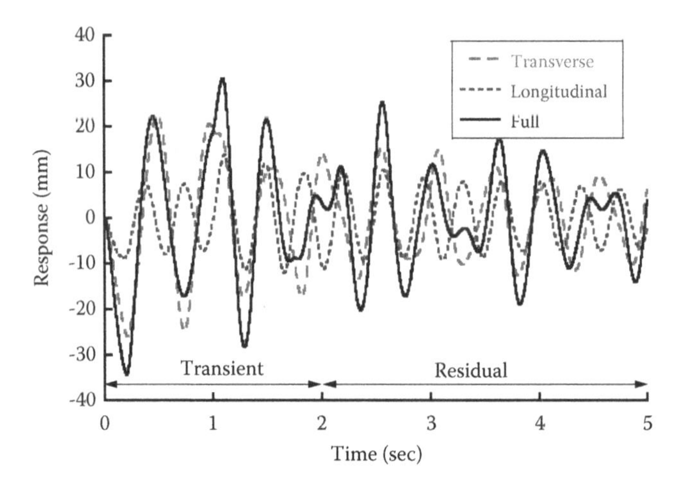

FIGURE 8.15 Sloshing responses resulting from two directions.

velocity, and acceleration are 182 mm, 102 mm, 0.2 m/s, and 2 m/s^2, respectively. The relative amplitude contribution of third transverse and longitudinal modes are 20.1%, and 19.3%, respectively. Meanwhile, the relative amplitude contribution of fifth transverse and longitudinal modes are 9.4%, and 9.0%, respectively. In addition, the relative amplitude contribution of seventh transverse and longitudinal modes are 5.6%, and 5.4%, respectively. Increasing sloshing modes decreases relative amplitude contribution. However, the sum of the relative amplitude contribution from the ninth transverse mode to the nineteenth transverse mode is 13.5%, and that from the longitudinal mode is 13.0%. High sloshing modes have some effects on the overall system dynamics. Therefore, suppressing sloshing induced by total modes is essential.

The simulated response from first ten modes along two directions is shown in Figure 8.15. The liquid depth, driving angle, and resultant driving distance were fixed at 90 mm, 30°, and 40 cm, respectively. The transient deflection and residual amplitude of full modes are 60.8 mm, and 45.6 mm, respectively. The transient deflection and residual amplitude of the transverse mode are 48.1 mm, and 29.0 mm, respectively. Meanwhile, the transient deflection and residual amplitude of the longitudinal mode are 25.8 mm, and 21.6 mm, respectively. The sum of the response along each of two directions is the whole system response. Thus, two fundamental modes have significant effects on the sloshing dynamics.

8.2.2 Computational Dynamics

Numerical robustness verification of the three-pieces smoother over a large range of motions and liquid depths is presented in this section. Using the first ten modes in each of two directions, simulations were conducted. The three-pieces smoother is given by:

$$u(\tau) = \begin{cases} \mu(e^{-r\zeta_m\omega_m\tau} - e^{-p\zeta_m\omega_m\tau}), & 0 \le \tau \le (T_m/p) \\ \mu e^{-r\zeta_m\omega_m\tau}(1-\delta), & (T_m/p) < \tau < (T_m/r) \\ \mu(\sigma e^{-p\zeta_m\omega_m\tau} - \delta e^{-r\zeta_m\omega_m\tau}), & (T_m/r) \le \tau \le (T_m/p + T_m/r) \\ 0, & \text{others} \end{cases} \quad (8.57)$$

where p and r are coefficients, and

$$\delta = e^{2\pi(r/p-1)\zeta_m/\sqrt{1-\zeta_m^2}} \quad (8.58)$$

$$\sigma = e^{2\pi(p/r-1)\zeta_m/\sqrt{1-\zeta_m^2}} \quad (8.59)$$

$$\mu = \frac{pr\zeta_m\omega_m}{(p-r)\left(1 - e^{-2\pi\zeta_m/\sqrt{1-\zeta_m^2}}\right)^2} \quad (8.60)$$

$$T_m = 2\pi/\left(\omega_m\sqrt{1-\zeta_m^2}\right) \quad (8.61)$$

The transient deflection and residual amplitude induced by varying resultant driving distances are shown in Figure 8.16. The liquid depth and driving angle were fixed at 90 mm and 30°, respectively. Without the controller, the transient deflection increases as the driving distance increases before 5.8 cm. Peaks and troughs arise in the transient deflection when the sloshing induced by the acceleration and deceleration is in phase or out of phase. The residual amplitude is also the result of the interference between the sloshing caused by the acceleration and deceleration. The smoother attenuated the transient deflection and residual amplitude by an average of 83.1%, and 96.2%, respectively. The smoother created a smooth

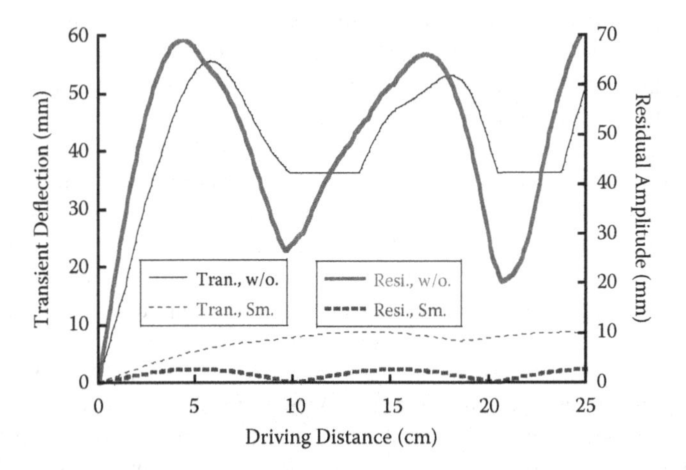

FIGURE 8.16 Transient and residual sloshing against driving distance.

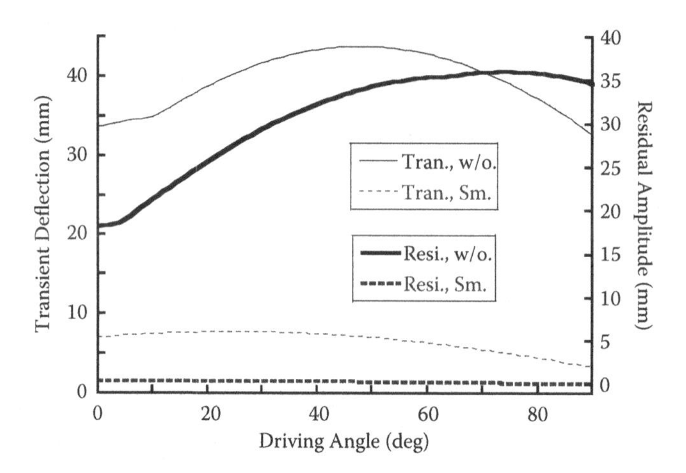

FIGURE 8.17 Transient and residual sloshing against driving angle.

velocity profile between boundary conditions to move the container. The smooth transitions reduced the transient sloshing. The smoother suppressed the residual sloshing to a low level because the smoother is more insensitive between two modified frequencies.

The transient deflection and residual amplitude for various driving angles are shown in Figure 8.17. The resultant driving distance and liquid depth were fixed at 20 cm and 90 mm, respectively. The sloshing dynamic response is the sum of the transverse modal response and longitudinal modal response. In the case of zero driving angle, only transverse modal responses exist. Meanwhile, the container is moved longitudinally for a driving angle of 90°. Without the controller, a maximum value in the transient deflection appears near the driving angle of 48° because of the interference between the transverse and longitudinal modal response. In addition, an extreme in the residual amplitude occurs near the driving angle of 75°. With the smoother, the transient deflection and residual amplitude were suppressed by an average of 83.5% and 98.9%, respectively. Simulations demonstrated that the smoother has good performance at reducing the transient and residual sloshing caused by the excitation in the random direction.

The liquid depth may not be known in many cases. Therefore, it becomes important for the controller to have insensitivity to variations in the liquid depth. Figure 8.18 shows the transient deflection and residual amplitude as a function of liquid depth. The resultant driving distance and driving angle were fixed at 20 cm and 30°, respectively. Without the controller, container motions induce large sloshing. Both the transient deflection and residual amplitude decrease with increasing liquid depth, then they vary slightly as the liquid depth increases after the container width. The smoother was designed for a liquid depth of 90 mm. The transient deflection and residual amplitude were suppressed by an average of 82.4% and 98.1%, respectively. The residual amplitudes are limited to near-zero values for all liquid depths because the smoother provides more insensitivity to changes in the frequency for a wide range of liquid depths.

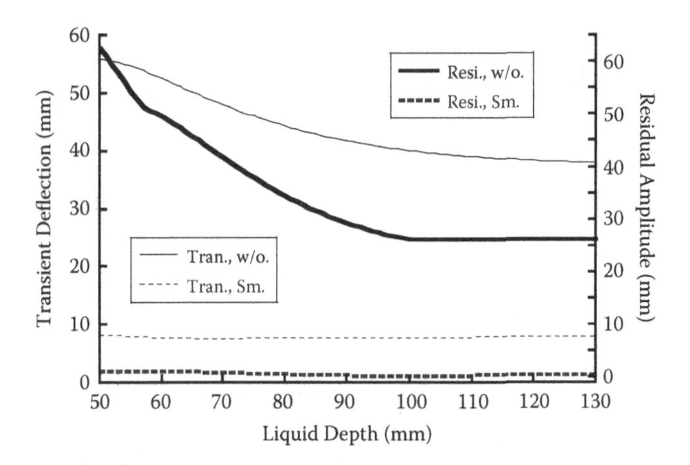

FIGURE 8.18 Transient and residual sloshing against liquid depth.

8.2.3 Experimental Verifications

Experiments were conducted on a testing apparatus shown in Figure 8.19. An XY gantry mounted a liquid container. The gantry is driven transversely and longitudinally by Panasonic AC servomotors with encoders. A DSP-based motion control card connects servo amplifiers to a Visual C++ program in a personal computer. The original command is sent to the program, which applies the three-pieces smoother algorithm and produces a modified command for the drive. The size of the container was set to 182 mm × 102 mm × 200 mm. The gantry also mounts a CMOS camera for recording the surface elevation at the diagonal corner of the container.

The experimental response of the surface elevation at the diagonal corner of the container caused by the trapezoidal-velocity command is shown in Figure 8.20. The resultant distance, driving angle, and liquid depth were fixed at 20 cm, 30°, and 90 mm, respectively. The container was still between 0 and 1 second. The container accelerated at 1 second, and decelerated 1.0 second later. Without the controller, the

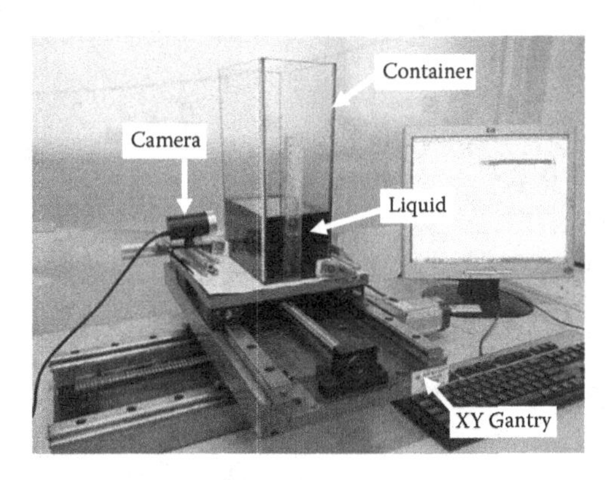

FIGURE 8.19 Experimental testing apparatus.

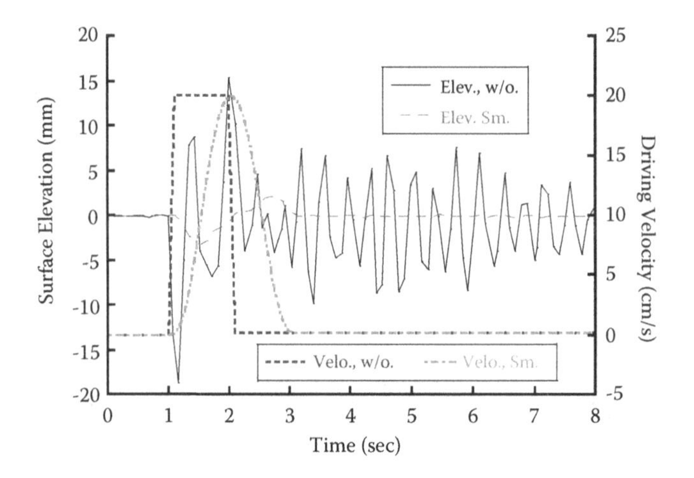

FIGURE 8.20 Experimental sloshing response to a driving command.

experimental transient deflection and residual amplitude were 34.0 mm and 17.4 mm, respectively. The residual amplitude was smaller than the transient deflection. This is because the sloshing caused by the deceleration attenuated that induced by the acceleration. With the three-pieces smoother, the experimental transient deflection and residual amplitude were 5.4 mm and 0.4 mm, respectively. Experimental results demonstrated that the smoother can suppress the transient and residual sloshing.

One set of experiments was performed to verify the effectiveness of the smoother on suppressing sloshing for variations of the liquid depth. The container was driven at a distance of 20 cm while the driving angle was set to 30°. Simulated and experimental results of the transient and residual sloshing from those tests are shown in Figures 8.21 and 8.22. Without the controller, both the

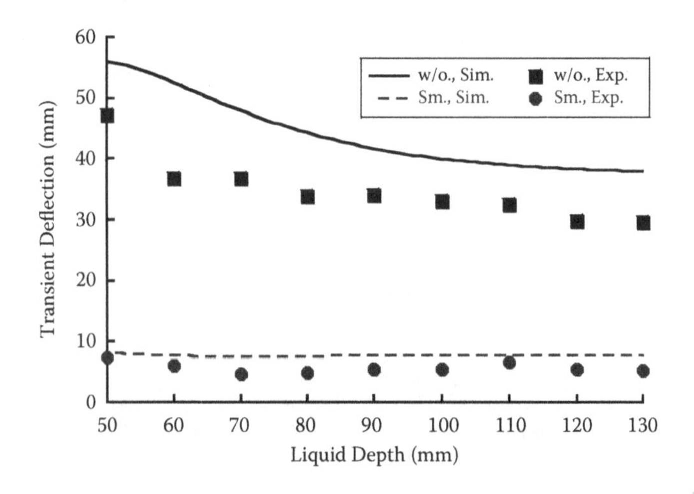

FIGURE 8.21 Experimental transient deflection for various liquid depths.

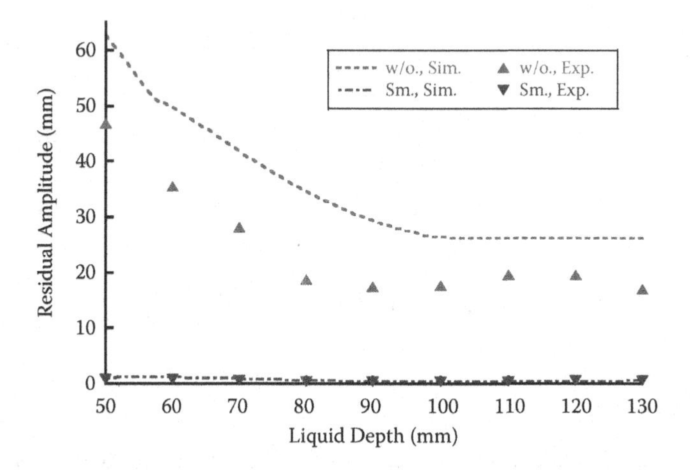

FIGURE 8.22 Experimental residual amplitude for various liquid depths.

transient and residual sloshing decreased slowly as the liquid depth increased. The simulated results were worse than the experimental data because the liquid was assumed to be inviscid and the surface tension was ignored. The smoother was designed for a liquid depth of 90 mm. The smoother suppressed the experimental transient sloshing by an average of 83.7%, and reduced the residual sloshing to <0.92 mm for all depths tested. The smoother has good performance to reduce the transient sloshing because the smooth velocity profile produced by the smoother benefits reduction of the transient sloshing. Meanwhile, the smoother limited the residual sloshing to a low level. This is because the smoother is more insensitive to changes in the sloshing frequency. Those experiments verified that the smoother can effectively reduce the three-dimensional sloshing for a large range of various system parameters as it was predicted by the simulation.

8.3 PLANAR NONLINEAR SLOSHING

Significant attentions for modeling of the sloshing have been focused on the equivalent mechanical model or linear sloshing model. The planar surface without rotation of its nodal diameter (for the case of the cylindrical container) or nodal line (for the rectangular container) was kept in the equivalent mechanical model or linear sloshing model. Furthermore, the free surface of the nonlinear sloshing exhibits non-planar motions with rotation of its nodal diameter or nodal line. This nonlinear sloshing dynamics will present large amplitude oscillations. Faltinsen et al. proposed the modeling of the nonlinear sloshing [37]. Nevertheless, coefficients presented by Faltinsen are only expressed in a non-expanded fashion for three-dimensional nonlinear slosh. Meanwhile, much of work for slosh suppression has been directed at controlling the sloshing based on the equivalent mechanical model or linear sloshing model. However, no effects have been directed at controlling the nonlinear sloshing dynamics.

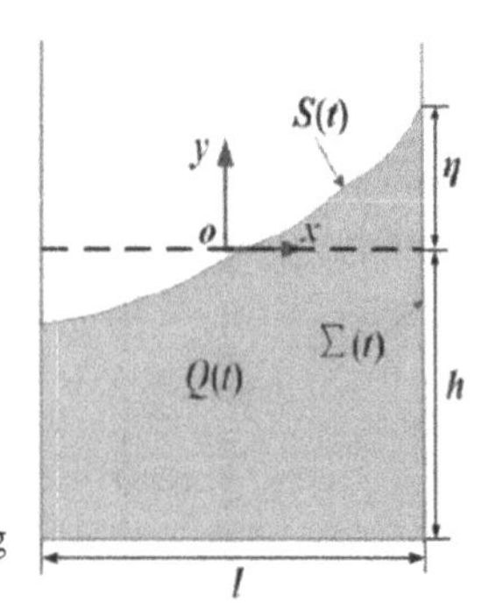

FIGURE 8.23 Planar nonlinear slosh model in a moving container.

8.3.1 MODELING

A moving liquid container of length, l, partly filled by an inviscid incompressible fluid with a depth, h, is shown in Figure 8.23. The flow is also assumed to be irrotational and the container is rigid. Let $O'X'Y'$ be the inertial coordinates and oxy be the moving coordinates fixed to the container. Note that the axes of the two coordinates are parallel to each other. The origin of the moving coordinates is in the mean free surface at the centerplane of the tank. The container was driven along the X' direction with the velocity of $v_0(t)$. $\eta(x, t)$ denotes the liquid-free surface elevation measured from the undisturbed free surface.

For the irrotational flow, the absolute velocity of the liquid can be written as:

$$\nu = \nabla\phi \tag{8.62}$$

where ϕ is the absolute velocity potential and ∇ is the gradient operator. The free boundary value problem can be described in the moving coordinates:

$$\left.\begin{array}{l}\nabla^2\phi = 0 \text{ in } Q(t),\ \frac{\partial\phi}{\partial\upsilon} = \nu_0\cdot\upsilon \text{ on } \Sigma(t),\ \frac{\partial\phi}{\partial\upsilon} = \nu_0\cdot\upsilon - \frac{\xi_t}{|\nabla\xi|}\ on\ S(t),\\ \frac{\partial\phi}{\partial t} + \frac{1}{2}(\nabla\phi)^2 - \nabla\phi\cdot\nu_0 - g\cdot r' = 0 \text{ on } S(t),\ \int_{Q(t)} dQ = const.\end{array}\right\} \tag{8.63}$$

where $Q(t)$ is the fluid volume bounded by the free surface $S(t)$ and the wetted tank surface $\Sigma(t)$, υ is the outer normal to the boundary of $Q(t)$, r' is the radius-vector of a point of the body-fluid system with respect to O', and g is the gravitational constant. The equation of the free surface is $\xi(x, y, t) = y - \eta(x, t) = 0$ and the normal velocity component on the free surface $S(t)$ in the moving coordinates can be given by:

$$-\frac{\partial\xi/\partial t}{|\nabla\xi|} = \frac{\partial\eta/\partial t}{\sqrt{1 + (\partial\eta/\partial x)^2}} \tag{8.64}$$

To get a unique solution of the free boundary problem shown in Equation (8.63), the flow is also assumed to start from rest and the surface elevation is zero at the initial time. Based on the Bate-Luke variational principle, the boundary value problem given by (8.63) can be described by examining the necessary conditions for the extrema of the following function:

$$W = \int_{t_1}^{t_2} L dt \tag{8.65}$$

where the Lagrangian, L, is the pressure integral. The pressure on the free surface is equal to a constant, p_0, and then the hydrodynamic pressure, p, in $Q(t)$ can be obtained by:

$$\frac{\partial \phi}{\partial t} + \frac{1}{2}(\nabla \phi)^2 - \nabla \phi \cdot \nu_0 + g \cdot r' + \frac{p - p_0}{\rho} = 0, \ in\ Q(t) \tag{8.66}$$

where ρ is the density of the liquid and the Lagrangian, L, is given by:

$$L = \int_{Q(t)} (p - p_0) dQ = -\rho \int_{Q(t)} \left[\frac{\partial \phi}{\partial t} + \frac{1}{2}(\nabla \phi)^2 - \nabla \phi \cdot \nu_0 + g \cdot r' \right] dQ \tag{8.67}$$

In addition, the test functions satisfy:

$$\delta\phi(x, y, t_1) = 0, \ \delta\phi(x, y, t_2) = 0; \ \delta\xi(x, y, t_1) = 0, \ \delta\xi(x, y, t_2) = 0. \tag{8.68}$$

The variations of Equation (8.65) can be written as:

$$\delta W = \rho \int_{t_1}^{t_2} \left\{ \begin{array}{l} \int_{s_0} \left[\frac{\partial \phi}{\partial t} + \frac{1}{2}(\nabla \phi)^2 - \nabla \phi \cdot \nu_0 + g \cdot r' \right]_{y=\eta} \delta \eta dS \\ + \int_{s_0} \left[\frac{\partial \phi}{\partial \upsilon} - \nu_0 \cdot \upsilon + \frac{\xi_t}{|\nabla \xi|} \right]_{y=\eta} \delta \phi dS \\ + \int_{\Sigma(t)} \left[\frac{\partial \phi}{\partial v} - \nu_0 \cdot \upsilon \right] \delta \phi dS - \int_{Q(t)} \nabla^2 \phi \delta \phi dV \end{array} \right\} dt = 0 \tag{8.69}$$

and the absolute velocity potential can be expressed as:

$$\phi(x, y, t) = \boldsymbol{\nu}_0(t) \cdot \boldsymbol{r} + \varphi(x, y, t) \tag{8.70}$$

where $\varphi(x, y, t)$ is the relative velocity potential and r is the radius-vector with respect to the origin O'. The surface elevation and the relative velocity potential can be expressed as:

$$\eta(x, t) = \sum_{i=1}^{\infty} \beta_i(t) \eta_i(x) \tag{8.71}$$

$$\varphi(x, y, t) = \sum_{n=1}^{\infty} R_n(t)\varphi_n(x, y) \tag{8.72}$$

where the surface modes, $\eta_i(x)$, and the domain modes, $\varphi_n(x, y)$, are the solutions of the linear eigenvalue problem.

$$\left.\begin{array}{l} \nabla^2\varphi_i = 0,\ (-l/2 < x < l/2,\ -h < y < 0) \\ \frac{\partial\varphi_i}{\partial x}|_{x=-l/2,l/2} = 0,\ \frac{\partial\varphi_i}{\partial y}|_{y=-h} = 0,\ \frac{\partial\varphi_i}{\partial y}|_{y=0} = \lambda_i\varphi_i,\ \eta_i(x) = \varphi_i(x, 0) \end{array}\right\} \tag{8.73}$$

Solving Equation (8.73) yields the eigenvalues, λ_i, and surface modes, $\eta_i(x)$, and domain modes, $\varphi_i(x, y)$:

$$\left.\begin{array}{l} \lambda_i = \frac{\pi i}{l}\tanh\left(\frac{\pi i}{l}h\right),\ \eta_i(x) = \cos\left(\frac{\pi i}{l}(x + l/2)\right), \\ \varphi_i(x, y) = \eta_i(x)\frac{\cosh((\pi i/l)(y + h))}{\cosh((\pi i/l)h)} \end{array}\right\} \tag{8.74}$$

Substituting Equations (8.70)–(8.72) into (8.69) yields:

$$\delta W = \int_{t_1}^{t_2} \left\{ \begin{array}{l} \Sigma_n A_n \delta\dot{R}_n + \Sigma_n \Sigma_k A_{nk} R_k \delta R_n \\ + \Sigma_i \left[\Sigma_n \dot{R}_n \frac{\partial A_n}{\partial\beta_i} + \frac{1}{2}\Sigma_n \Sigma_k R_n R_k \frac{\partial A_{nk}}{\partial\beta_i} + \dot{v}_O \frac{\partial l_1}{\partial\beta_i} - g\frac{\partial l_3}{\partial\beta_i} \right] \delta\beta_i \end{array} \right\} dt \tag{8.75}$$

where

$$A_n = \rho\int_{Q(t)} \varphi_n dx,\ A_{nk} = \rho\int_{Q(t)} \nabla\varphi_n \cdot \nabla\varphi_k dx \tag{8.76}$$

$$l_1 = \rho\int_{Q(t)} x\eta_i dx,\ l_3 = \rho\int_{Q(t)} \eta_i^2 dx \tag{8.77}$$

Using the initial condition of the test functions and integrating by parts in Equation (8.75), the following infinite system can be obtained:

$$\dot{A}_n - \sum_k A_{nk} R_k = 0,\ n = 1, 2, \ldots \tag{8.78}$$

$$\sum_n \dot{R}_n \frac{\partial A_n}{\partial\beta_i} + \frac{1}{2}\sum_n\sum_k R_n R_k \frac{\partial A_{nk}}{\partial\beta_i} + \dot{v}_0(t)\frac{\partial l_1}{\partial\beta_i} - g\frac{\partial l_3}{\partial\beta_i} = 0,\ i = 1, 2, \ldots \tag{8.79}$$

The infinite-dimensional mode system can be detuned to a finite-dimensional mode system by the asymptotic relation.

$$O(\beta_1) = \varepsilon^{1/3},\ O(\beta_2) = \varepsilon^{2/3},\ O(\beta_3) = \varepsilon \tag{8.80}$$

where ε is the small parameter. Higher orders than ε will be neglected in the nonlinear equations. By inserting Equation (8.74) into (8.76) and (8.77), A_n and A_{nk} can be expanded in Taylor series by β_i. Furthermore, R_n can also be expressed correct to three-order polynomials in β_i by inserting A_n and A_{nk} into Equation (8.78). Substituting A_n, A_{nk}, and R_n into Equation (8.79), and forcing all the terms correct to $O(\varepsilon)$, the nonlinear slosh equations can be yielded:

$$\begin{aligned} &\ddot{\beta}_1 + \omega_1^2\beta_1 + \left(E_1 + \frac{2E_0}{E_1}\right)\left(\ddot{\beta}_1\beta_2 + \dot{\beta}_1\dot{\beta}_2\right) + \left(E_1 - \frac{2E_0}{E_2}\right)\ddot{\beta}_2\beta_1 \\ &+ \left(\frac{8E_0^2}{E_1E_2} - 2E_0\right)\left(\ddot{\beta}_1\beta_1^2 + \dot{\beta}_1^2\beta_1\right) - \frac{8E_1 l}{\pi^2}\dot{v}_0(t) = 0 \end{aligned} \tag{8.81}$$

$$\ddot{\beta}_2 + \omega_2^2\beta_2 + \left(2E_2 - \frac{4E_0}{E_1}\right)\ddot{\beta}_1\beta_1 - \left(\frac{4E_0}{E_1} + \frac{E_2(2E_0 - E_1^2)}{E_1^2}\right)\dot{\beta}_1^2 = 0 \tag{8.82}$$

$$\begin{aligned} &\ddot{\beta}_3 + \omega_3^2\beta_3 + \left(3E_3 - \frac{6E_0}{E_1}\right)\ddot{\beta}_1\beta_2 + \left(-\frac{9E_0E_3}{E_1} - 3E_0 + \frac{24E_0^2}{E_1E_2}\right)\ddot{\beta}_1\beta_1^2 \\ &+ \left(3E_3 - \frac{6E_0}{E_2}\right)\ddot{\beta}_2\beta_1 + \left(3E_3 - \frac{6E_0}{E_1} - \frac{6E_0}{E_2} - \frac{6E_0E_3}{E_1E_2}\right)\dot{\beta}_1\dot{\beta}_2 \\ &+ \left(-\frac{24E_0E_3}{E_1} - 6E_0 + \frac{48E_0^2}{E_1E_2} + \frac{24E_0^2E_3}{E_1^2E_2}\right)\dot{\beta}_1^2\beta_1 - \frac{8lE_3}{3\pi^2}\dot{v}_0(t) = 0 \end{aligned} \tag{8.83}$$

$$R_1 = \frac{\dot{\beta}_1}{2E_1} + \frac{E_0}{E_1^2}\dot{\beta}_1\beta_2 - \frac{E_0}{E_1E_2}\dot{\beta}_2\beta_1 + \frac{E_0}{E_1}\left(-\frac{1}{2} + \frac{4E_0}{E_1E_2}\right)\dot{\beta}_1\beta_1^2 \tag{8.84}$$

$$R_2 = \frac{1}{4E_2}\left(\dot{\beta}_2 - \frac{4E_0}{E_1}\dot{\beta}_1\beta_1\right) \tag{8.85}$$

$$R_3 = \frac{\dot{\beta}_3}{6E_3} - \frac{E_0}{E_1E_3}\dot{\beta}_1\beta_2 - \frac{E_0}{E_2E_3}\dot{\beta}_2\beta_1 + \left(-\frac{E_0}{2E_3} + \frac{4E_0^2}{E_1E_2E_3}\right)\dot{\beta}_1\beta_1^2 \tag{8.86}$$

where

$$E_0 = \frac{1}{8}\left(\frac{\pi}{l}\right)^2;\ E_i = \frac{\pi}{2l}\tanh\left(\frac{\pi i}{l}h\right),\ i \geq 1 \tag{8.87}$$

and ω_i is the natural frequency of ith sloshing mode. It is given by:

$$\omega_i = \sqrt{2giE_i} \tag{8.88}$$

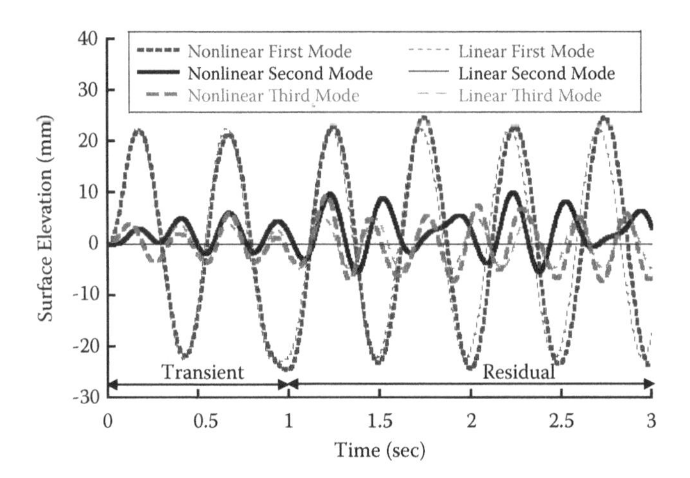

FIGURE 8.24 Simulated slosh response for the nonlinear and linear models.

Figure 8.24 shows the simulated responses of the first three modes for both the nonlinear and linear slosh models when the driving distance, container length, and liquid depth were fixed at 22.5 cm, 182 mm, and 120 mm, respectively. The original command used to drive the container is a trapezoidal-velocity profile (bang-coast-bang acceleration). An acceleration pulse causes a rise in the velocity until the maximum velocity is reached. The container then moves at its maximum velocity until a deceleration pulse act. This causes a decrease in the velocity until zero.

The slosh responses of the first and third modes for the nonlinear model match that for the linear model very well. However, there is a huge difference for the second mode. The second-mode response for the linear model is limited to zero because the container motion cannot excite the even modes for the linear slosh model. Meanwhile, the second-mode response for the nonlinear model exhibits relatively large amplitude oscillations.

There are two stages in the system response. The transient stage is defined as the time frame when the container is in motion. The maximum peak-to-peak deflection during the transient stage is referred to as the transient deflection. The residual stage is defined as the time frame when the container is stopped. The maximum peak-to-peak deflection during the residual stage is defined as the residual amplitude. Maximum driving velocity and acceleration were 0.25 m/s and 2.5 m/s^2, respectively. The damping ratio was zero in the simulations. Note that the duration of the simulation in this section is 5 seconds and the slosh measurement position is chosen at the area of the free surface near the left side of the container.

The simulated residual amplitudes of the first three sloshing modes for the nonlinear model are shown in Figure 8.25. While the average residual amplitude of the first mode and the third mode are 51.8 mm and 22.1 mm, respectively, that of the second mode reaches 18.5 mm. The maximum peak-to-peak amplitudes of the velocity potential from the first three nonlinear modes are also given in Figure 8.26. The amplitudes of the velocity potential for the second-mode slosh are larger than that for the third-mode slosh when the liquid depths are shallow. Simulated results

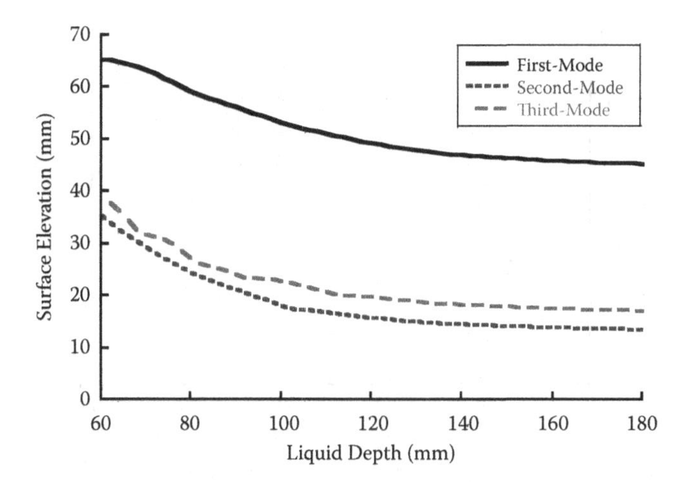

FIGURE 8.25 Simulated residual amplitude of surface elevation for the nonlinear model.

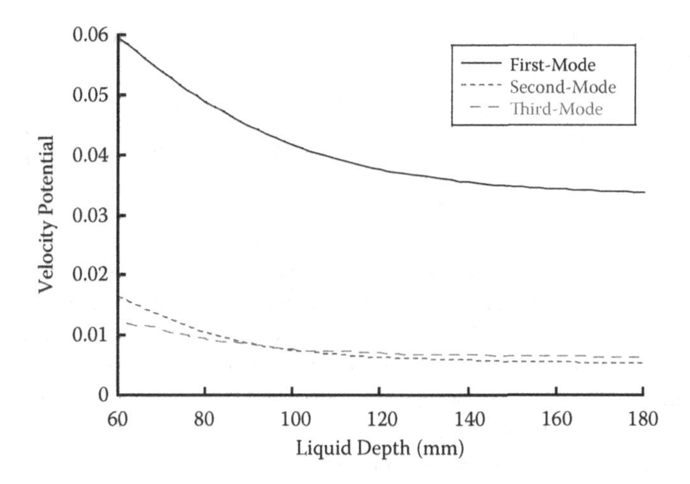

FIGURE 8.26 Simulated residual amplitude of velocity potential for the nonlinear model.

indicate that the odd modes have large impacts on the weakly nonlinear sloshing dynamics, but the even modes may also have some effects, which is significantly different from the linear slosh. Thus, there is a need for a control system that can effectively reduce weakly nonlinear slosh induced by both odd and even modes.

The higher modes than three, which are neglected in nonlinear model, can be described in the linear model. Therefore, the slosh model in this section includes the first three nonlinear modes and other linear higher modes. The sloshing response is the sum of the response for the total modes. Simulations were conducted using the first ten modes model in this section.

Two one-piece smoothers with the first-mode frequency and the second-mode frequency are convolved together to create a combined smoother. The combined

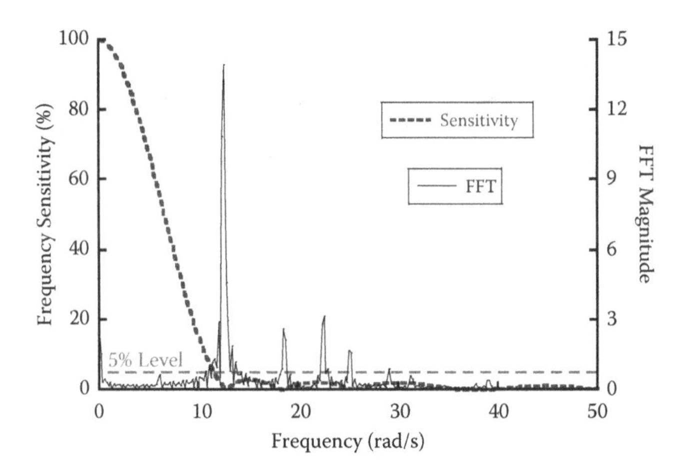

FIGURE 8.27 FFT magnitude of the slosh response and frequency sensitivity curve for the combined smoother.

smoother provides notch and low-pass filtering effect. The one-piece smoother is described by:

$$u(\tau)=\begin{Bmatrix} \zeta\omega e^{-\zeta\omega\tau}\Big/\left(1-e^{-2\pi\zeta/\sqrt{1-\zeta^2}}\right), & 0\le\tau\le 2\pi\Big/\left(\omega\sqrt{1-\zeta^2}\right) \\ 0 & \text{others} \end{Bmatrix} \tag{8.89}$$

To better identify the slosh suppression, the fast Fourier transform (FFT) of a slosh response is shown in Figure 8.27 when the driving distance, container length, and liquid depth were fixed at 22.5 cm, 182 mm, and 120 mm, respectively. The estimates of the sloshing frequencies from Equation (8.65) are revealed by the peaks of the FFT magnitude at 12.8 rad/s, 18.4 rad/s, and 22.5 rad/s, etc. Using the frequencies of the first two sloshing modes (12.8 rad/s and 18.4 rad/s) as design frequencies, the combined smoother would suppress slosh induced by total modes. Figure 8.27 also shows the frequency-sensitive curve for the combined smoother. The frequency insensitivity, defined as the range of each curve that lies below 5% of percentage residual amplitude, provides a quantitative measure of robustness. It is clear that the combined smoother produces a low-pass filtering effect. The 5% insensitivity of the combined smoother ranges from 11.55 rad/s to infinity. Thus, the presented combined smoother could suppress a wide range of sloshing frequencies.

To better evaluate slosh suppression for high modes, the maximum peak-to-peak amplitude of the velocity potential from the first ten modes is given in Table 8.1. Without the controller, the amplitude of the velocity potential for the first nonlinear mode is larger than that for the others. Thus, the first mode is fundamental for sloshing dynamics. The maximum peak-to-peak amplitude of the second nonlinear mode is relatively large. Thus, it cannot be ignored. The velocity potential for the case of the fourth, the sixth, the eighth, and the tenth mode are limited to zero.

TABLE 8.1
Velocity Potential From the First Ten Modes

Mode	Without Control	Combined Smoother
1	3.76×10^{-2}	2.68×10^{-5}
2	6.35×10^{-3}	6.91×10^{-5}
3	7.06×10^{-3}	8.83×10^{-5}
4	0	0
5	1.01×10^{-3}	2.07×10^{-5}
6	0	0
7	1.84×10^{-4}	1.25×10^{-6}
8	0	0
9	4.15×10^{-4}	3.30×10^{-7}
10	0	0

This is because the container motion cannot trigger the even modes for the linear slosh. The velocity potential resulting from the combined smoother is also given in Table 8.1. The combined smoother reduces the velocity potential for the total modes below a very low level.

8.3.2 Computational Dynamics

In many cases, the liquid depth and tank length may not be known accurately and the driving distance often changes. Then it becomes important for the combined smoother to have insensitivity to the variations in the system parameters and to provide good performance over a large range of container motions.

Figures 8.28 and 8.29 show the transient deflection and residual amplitude of the surface elevation induced by varying driving distances when the container length

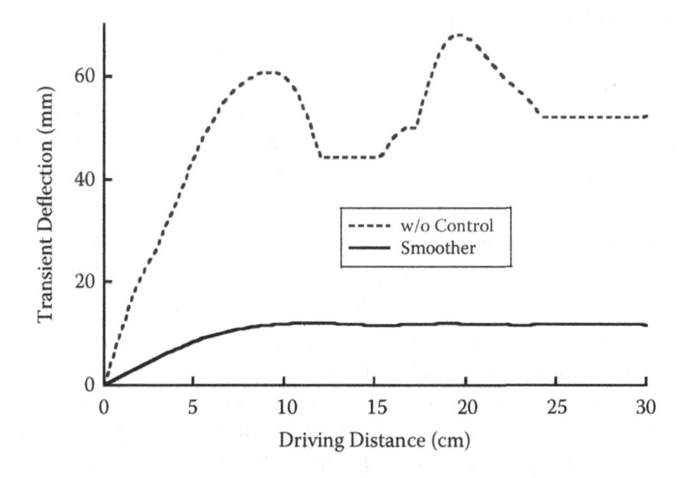

FIGURE 8.28 Transient amplitude against driving distance.

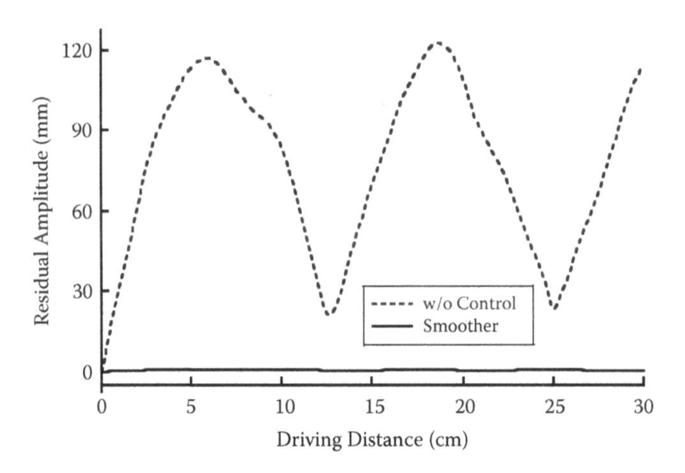

FIGURE 8.29 Residual amplitude against driving distance.

and liquid depth were fixed at 182 mm and 120 mm, respectively. The uncontrolled transient deflection increases with increasing driving distance before 9.5 cm because the amount of transient deflection depends on the size of the acceleration pulse and the duration of the transient stage. After this point, the transient deflection is dependent on the interference between the slosh caused by the acceleration and deceleration. When the vibration induced by the deceleration is in phase with that caused by the acceleration, the peaks will arise in the uncontrolled transient deflection. When the vibrations caused by the acceleration and deceleration are out of phase, the troughs occur. The combined smoother attenuated the transient deflection by an average of 78.7%. This is because the smoother produces a smooth velocity profile to move the container. The smooth transitions between boundary conditions reduce the transient slosh. The uncontrolled residual amplitude is also the result of the interference between the slosh caused by the acceleration and deceleration. The combined smoother eliminated the residual amplitude by an average of 99.3% because the modeled sloshing frequencies were correct.

Figures 8.30 and 8.31 show the transient deflection and residual amplitude of the surface elevation as a function of liquid depth, h, when the combined smoother was designed for a liquid depth of 120 mm. The driving distance and container length were fixed at 22.5 cm and 182 mm in this case. Without the controller, the container motion results in large slosh. The uncontrolled transient deflection changes slightly with increasing liquid depth. The uncontrolled residual amplitude decreases sharply as the liquid depth increases. The combined smoother suppressed the transient deflection and residual amplitude by an average of 79.6% and 98.7%, respectively. The transient deflection with the combined smoother keeps low values because of the transient-vibration constraint of the smoother. The residual amplitudes with the combined smoother are limited below a low level because the combined smoother can provide more insensitivity to the changes in the frequency.

Figures 8.32 and 8.33 show the transient deflection and residual amplitude of the surface elevation for various container lengths when the driving distance and liquid

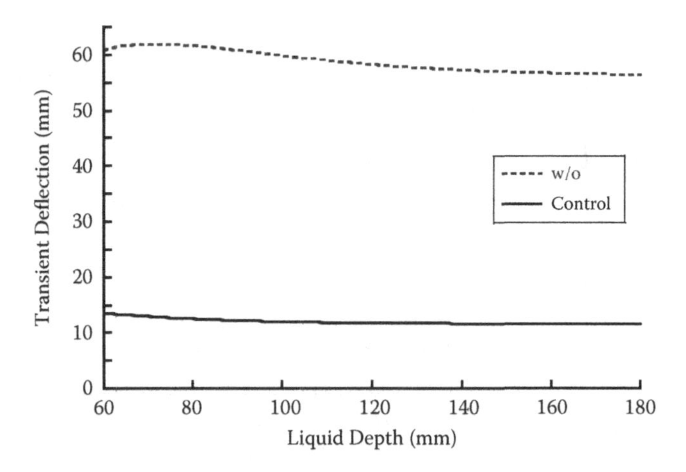

FIGURE 8.30 Transient amplitude against liquid depth.

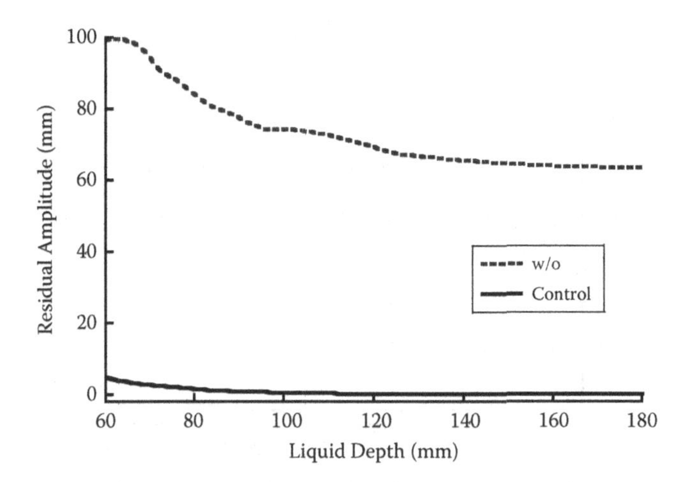

FIGURE 8.31 Residual amplitude against liquid depth.

depth were fixed at 22.5 cm and 120 mm in this case. The transient deflection without the controller decreases before 106 mm, and then increases as the container length increases. There is a local maximum in the uncontrolled transient deflection for the container length of 239 mm. The uncontrolled residual amplitude decreases before 150 mm, then increases with increasing the container length. The trough arises at the tank length of 150 mm because the vibrations caused by the container motion are out of phase. The combined smoother was designed for a container length of 182 mm. The transient deflection with the smoother increases with increasing the container length. The combined smoother attenuated the transient deflection by an average of 78.5%. This is because the smoother limits the transient vibrations to low level. In addition, the combined smoother eliminated that by an average of 95.7%. This is also because the combined smoother can provide more

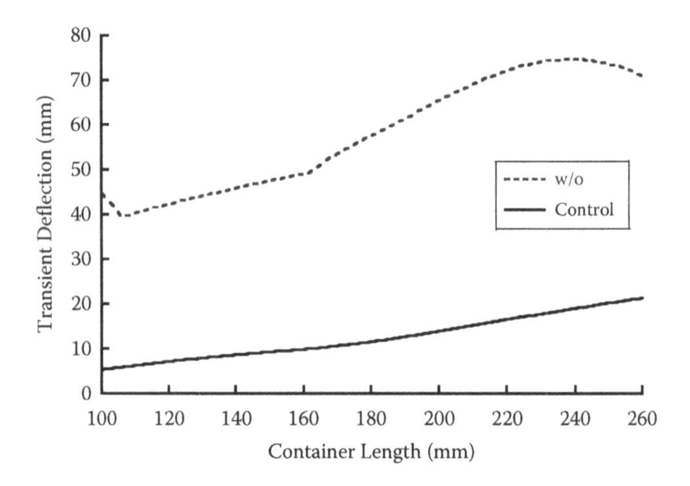

FIGURE 8.32 Transient amplitude against container length.

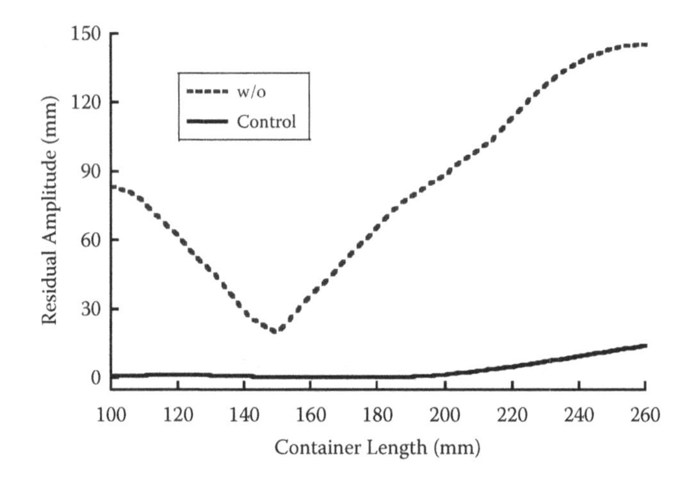

FIGURE 8.33 Residual amplitude against container length.

insensitivity to the changes in the frequency. The abovementioned simulations demonstrated that the smoother can robustly suppress both the transient and residual slosh for varying working conditions and system parameters.

8.3.3 Experimental Results

Experiments were also conducted on a testing apparatus shown in Figure 8.19. Figure 8.34 shows the experimental response of surface elevation at the left side of the container caused by the trapezoidal-velocity commands. The tank was driven for a distance of 22.5 cm and the liquid depth was fixed at 120 mm. The surface was level and not moving between 0 and 1 second. The tank accelerated at 1 second,

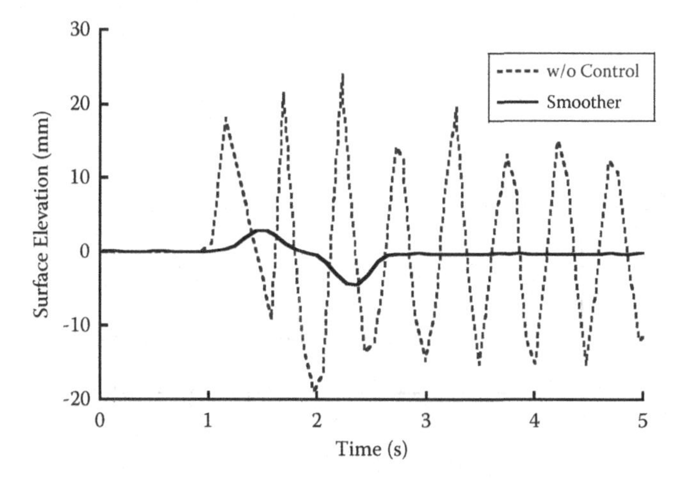

FIGURE 8.34 Experimental response to driving commands.

which resulted in slosh, and the tank decelerated at 0.9 second later, which resulted in additional slosh. Without the controller, the transient and residual slosh had a response with amplitudes of 40.3 mm and 38.7 mm, respectively. The uncontrolled residual amplitude was smaller than the uncontrolled transient deflection because the slosh caused by the acceleration was out of phase with that induced by the deceleration. The effect of the smoother is also shown in Figure 8.34. Experimental results show that the transient deflection and residual amplitude with the combined smoother were 7.4 mm and 0.2 mm, respectively. It demonstrated that the smoother can reduce the transient and residual slosh. This is because the combined smoother suppresses a wider range of slosh frequencies.

To verify the dynamics and effectiveness of the combined smoother on suppressing transient and residual slosh for variations of container motions and system parameters, two sets of experiments were performed. The first experiment investigated the effects of changes in the driving distance when the liquid depth was fixed at 120 mm. Figures 8.35 and 8.36 show the experimental results of transient deflection and residual amplitude from these tests. Without the controller, both the transient deflection and residual amplitude varied from driving distances because of the interference between the slosh caused by the acceleration and deceleration. The combined smoother attenuated the transient deflection by an average of 79.0%. The transient deflection with the combined smoother was limited to a lower level. In addition, the combined smoother eliminated the residual amplitude by an average of 99.3%. The combined smoother suppressed the residual slosh to a lower level because it has better frequency insensitivity.

Another experiment was conducted to verify the robustness of the combined smoother at suppressing transient and residual slosh for various liquid depths, h. The driving distance was fixed at 22.5 cm. The liquid depth ranged from 60 mm to 180 mm while the combined smoother was designed for a liquid depth of 120 mm. Figures 8.37 and 8.38 show experimental transient deflection and residual amplitude from these tests. Similar to the simulated results, the uncontrolled transient

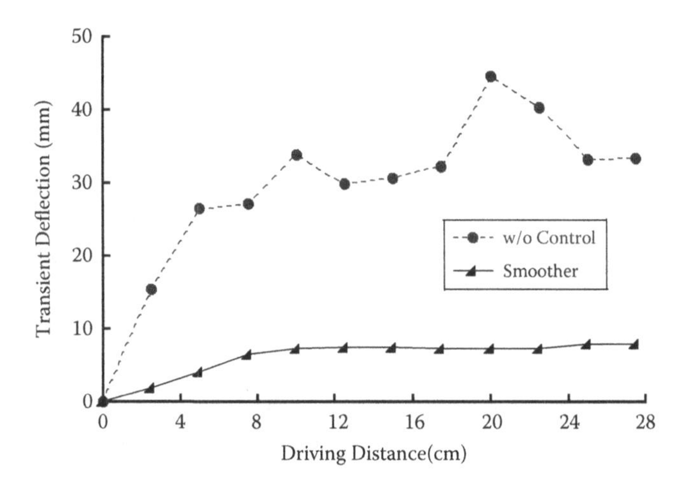

FIGURE 8.35 Transient amplitude induced by driving motions.

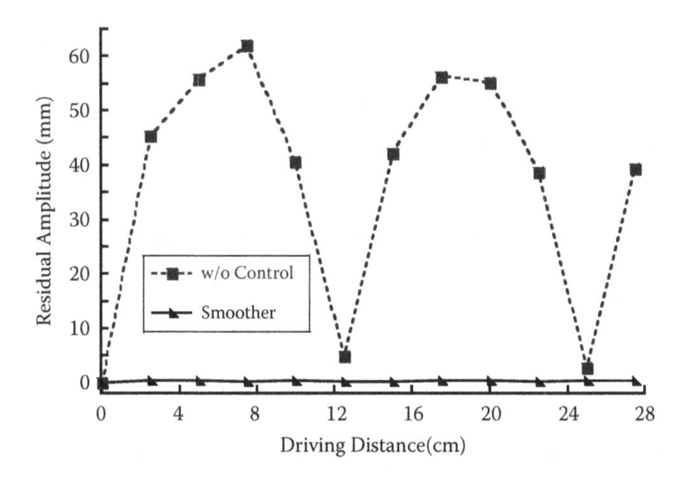

FIGURE 8.36 Transient and residual amplitude induced by driving motions.

deflection varied slightly as the liquid depth increased, and the uncontrolled residual amplitude decreased with increasing the liquid depth. Experimental results show that the combined smoother attenuated the transient deflection by an average of 78.9%. The combined smoother provided better performance at reducing transient slosh. Meanwhile, the combined smoother suppressed the residual amplitude to <0.67 mm for all depths tested. The combined smoother limited the residual slosh to a lower level because the combined smoother is more insensitive to changes in sloshing frequency.

Those experiments verified that the presented combined smoother can effectively suppress weakly nonlinear slosh for a wide range of various system parameters and motion conditions as it was predicted by the simulations.

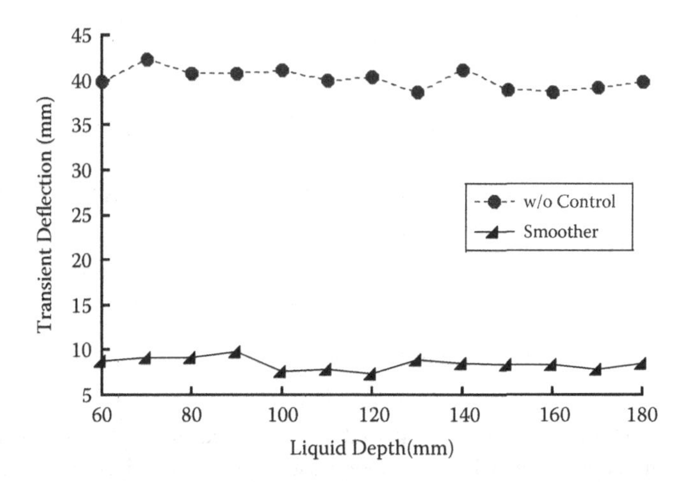

FIGURE 8.37 Transient amplitude for various liquid depths.

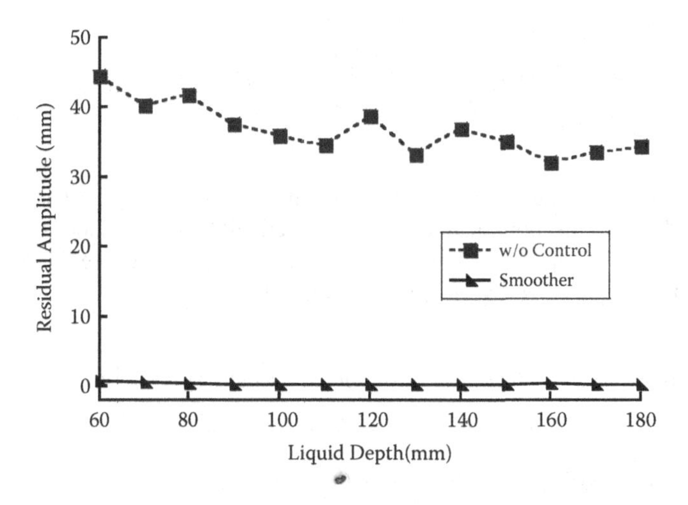

FIGURE 8.38 Residual amplitude for various liquid depths.

8.4 THREE-DIMENSIONAL NONLINEAR SLOSHING

8.4.1 Modeling

A schematic representation of three-dimensional nonlinear sloshing in a moving liquid container is shown in Figure 8.39. A rectangular tank of length, a, and breadth, b, contains an inviscid incompressible fluid. The depth of fluid is denoted as h. The container is rigid, and the sloshing does not affect the motions of the container.

Let $O'X'Y'Z'$ be the inertial coordinates, while $oxyz$ be the moving coordinates. The two coordinates are parallel to each other. The mean free surface at the centerplane of the tank is the origin of the moving coordinates. The liquid free surface elevation measured from the undisturbed free surface is defined as $\eta(x, y, z, t)$. Planar excitations

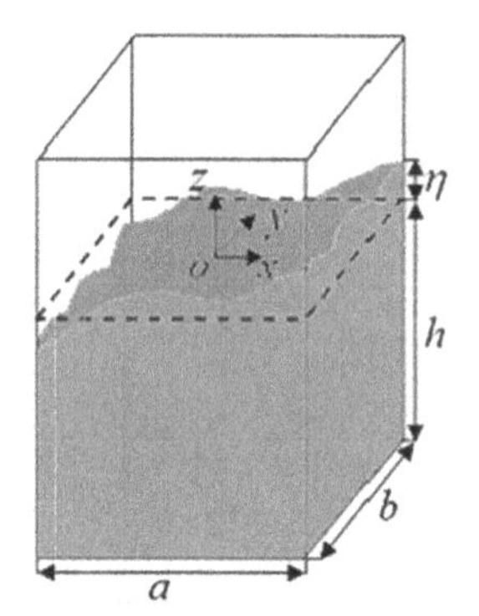

FIGURE 8.39 A slosh model in a rectangular tank.

drive the container. The motion is divided into the linear movement of the container along two directions, X' and Y'. The motion can also be defined as the velocity, $v_0(t)$, of the container and steering angle, α, between the velocity and the X' direction.

The equation of the surface elevation for the nonlinear sloshing is given by:

$$\eta(x, y, 0, t) = a\sum_i \sum_j \cos\left[\pi i\left(\frac{x}{a} + 0.5\right)\right] \cdot \cos\left[\pi j\left(\frac{ry}{a} + 0.5\right)\right] \cdot \beta_{i,j}(t), \quad i + j \neq 0 \tag{8.90}$$

where $\beta_{i,j}(t)$ is the time-dependent function and r is the ratio of the tank length to tank breadth. In this nonlinear sloshing model, higher-order modes will be neglected. Then, the time-dependent function, $\beta_{i,j}(t)$, in [37] including the proportional damping terms are given by Huang and Zhao [38]:

$$\begin{aligned}
&\ddot{\beta}_{1,0} + 2\zeta_{1,0}\omega_{1,0}\dot{\beta}_{1,0} + \omega_{1,0}^2\beta_{1,0} + d_{1,1}\left(\ddot{\beta}_{1,0}\beta_{2,0} + \dot{\beta}_{1,0}\dot{\beta}_{2,0}\right) + d_{1,2}\left(\ddot{\beta}_{1,0}\beta_{1,0}^2 + \dot{\beta}_{1,0}^2\beta_{1,0}\right) \\
&\quad + d_{1,3}\ddot{\beta}_{2,0}\beta_{1,0} + d_{1,4}\ddot{\beta}_{1,0}\beta_{0,1}^2 + d_{1,5}\ddot{\beta}\ddot{\beta}_{0,1}\beta_{1,1} + d_{1,6}\beta_{1,0}\beta_{0,1}\ddot{\beta}_{0,1} + d_{1,7}\beta_{0,1}\ddot{\beta}_{1,1} \\
&\quad + d_{1,8}\dot{\beta}_{1,0}\beta_{0,1}\dot{\beta}_{0,1} + d_{1,9}\beta_{1,0}\dot{\beta}_{0,1}^{\;2} + d_{1,10}\dot{\beta}_{0,1}\dot{\beta}_{1,1} + d_{1,11}\dot{v}_0(t)\cos(\alpha) = 0
\end{aligned} \tag{8.91}$$

$$\begin{aligned}
&\ddot{\beta}_{0,1} + 2\zeta_{0,1}\omega_{0,1}\dot{\beta}_{0,1} + \omega_{0,1}^2\beta_{0,1} + d_{2,1}\left(\ddot{\beta}_{0,1}\beta_{0,2} + \dot{\beta}_{0,1}\dot{\beta}_{0,2}\right) + d_{2,2}\left(\ddot{\beta}_{0,1}\beta_{0,1}^2 + \dot{\beta}_{0,1}^2\beta_{0,1}\right) \\
&\quad + d_{2,3}\ddot{\beta}_{0,2}\beta_{0,1} + d_{2,4}\ddot{\beta}_{0,1}\beta_{1,0}^2 + d_{2,5}\ddot{\beta}_{1,0}\beta_{1,1} + d_{2,6}\beta_{1,0}\beta_{0,1}\ddot{\beta}_{1,0} + d_{2,7}\beta_{0,1}\dot{\beta}_{1,0}^2 \\
&\quad + d_{2,8}\beta_{1,0}\ddot{\beta}_{1,1} + d_{2,9}\dot{\beta}_{1,0}\dot{\beta}_{0,1}\beta_{1,0} + d_{2,10}\dot{\beta}_{1,0}\dot{\beta}_{1,1} + d_{2,11}\dot{v}_0(t)\sin(\alpha) = 0
\end{aligned} \tag{8.92}$$

$$\ddot{\beta}_{2,0} + 2\zeta_{2,0}\omega_{2,0}\dot{\beta}_{2,0} + \omega_{2,0}^2\beta_{2,0} + d_{3,1}\ddot{\beta}_{1,0}\beta_{1,0} + d_{3,2}\dot{\beta}_{1,0}^2 = 0 \tag{8.93}$$

$$\ddot{\beta}_{0,2} + 2\zeta_{0,2}\omega_{0,2}\dot{\beta}_{0,2} + \omega_{0,2}^2\beta_{0,2} + d_{4,1}\ddot{\beta}_{0,1}\beta_{0,1} + d_{4,2}\dot{\beta}_{0,1}^2 = 0 \tag{8.94}$$

$$\ddot{\beta}_{1,1} + 2\zeta_{1,1}\omega_{1,1}\dot{\beta}_{1,1} + \omega_{1,1}^{2}\beta_{1,1} + d_{5,1}\ddot{\beta}_{1,0}\beta_{0,1} + d_{5,2}\ddot{\beta}_{0,1}\beta_{1,0} + d_{5,3}\dot{\beta}_{1,0}\dot{\beta}_{0,1} = 0 \tag{8.95}$$

$$\begin{aligned}\ddot{\beta}_{2,1} &+ 2\zeta_{2,1}\omega_{2,1}\dot{\beta}_{2,1} + \omega_{2,1}^{2}\beta_{2,1} + d_{6,1}\beta_{1,0}\beta_{0,1}\ddot{\beta}_{1,0} + d_{6,2}\ddot{\beta}_{1,0}\beta_{1,1} + d_{6,3}\beta_{0,1}\dot{\beta}_{1,0}^{2} \\ &+ d_{6,4}\ddot{\beta}_{0,1}\beta_{2,0} + d_{6,5}\dot{\beta}_{1,0}\dot{\beta}_{0,1}\beta_{1,0} + d_{6,6}\ddot{\beta}_{0,1}\beta_{1,0}^{2} + d_{6,7}\dot{\beta}_{1,0}\dot{\beta}_{1,1} \\ &+ d_{6,8}\beta_{0,1}\ddot{\beta}_{2,0} + d_{6,9}\beta_{1,0}\ddot{\beta}_{1,1} + d_{6,10}\dot{\beta}_{0,1}\dot{\beta}_{2,0} = 0\end{aligned} \tag{8.96}$$

$$\begin{aligned}\ddot{\beta}_{1,2} &+ 2\zeta_{1,2}\omega_{1,2}\dot{\beta}_{1,2} + \omega_{1,2}^{2}\beta_{1,2} + d_{7,1}\beta_{1,0}\beta_{0,1}\ddot{\beta}_{0,1} + d_{7,2}\ddot{\beta}_{0,1}\beta_{1,1} + d_{7,3}\dot{\beta}_{1,0}\dot{\beta}_{0,1}\beta_{0,1} \\ &+ d_{7,4}\ddot{\beta}_{1,0}\beta_{0,1}^{2} + d_{7,5}\ddot{\beta}_{1,0}\beta_{0,2} + d_{7,6}\beta_{0,1}\ddot{\beta}_{1,1} + d_{7,7}\beta_{1,0}\ddot{\beta}_{0,2} + d_{7,8}\beta_{1,0}\dot{\beta}_{0,1}^{2} \\ &+ d_{7,9}\dot{\beta}_{0,1}\dot{\beta}_{1,1} + d_{7,10}\dot{\beta}_{1,0}\dot{\beta}_{0,2} = 0\end{aligned} \tag{8.97}$$

$$\begin{aligned}\ddot{\beta}_{3,0} &+ 2\zeta_{3,0}\omega_{3,0}\dot{\beta}_{3,0} + \omega_{3,0}^{2}\beta_{3,0} + d_{8,1}\ddot{\beta}_{1,0}\beta_{2,0} + d_{8,2}\ddot{\beta}_{1,0}\beta_{1,0}^{2} + d_{8,3}\beta_{1,0}\ddot{\beta}_{2,0} \\ &+ d_{8,4}\dot{\beta}_{1,0}^{\;2}\beta_{1,0} + d_{8,5}\dot{\beta}_{1,0}\dot{\beta}_{2,0} + d_{8,6}\dot{v}_{0}(t)\cos(\alpha) = 0\end{aligned} \tag{8.98}$$

$$\begin{aligned}\ddot{\beta}_{0,3} &+ 2\zeta_{0,3}\omega_{0,3}\dot{\beta}_{0,3} + \omega_{0,3}^{2}\beta_{0,3} + d_{9,1}\ddot{\beta}_{0,1}\beta_{0,2} + d_{9,2}\dot{\beta}_{0,1}\dot{\beta}_{0,2} + d_{9,3}\beta_{0,1}\ddot{\beta}_{0,2} \\ &+ d_{9,4}\ddot{\beta}_{0,1}\beta_{0,1}^{2} + d_{9,5}\dot{\beta}_{0,1}^{\;2}\beta_{0,1} + d_{9,6}\dot{v}_{0}(t)\sin(\alpha) = 0\end{aligned} \tag{8.99}$$

where $\zeta_{i,j}$ is the damping ratio of $(i, j)^{th}$ sloshing mode, $\omega_{i,j}$ is the natural frequency of $(i, j)^{th}$ sloshing mode. The coefficient $d_{i,j}$ is given by:

$$d_{1,1} = \frac{1}{2}E_{1,0} + \frac{\pi^2}{2E_{1,0}} \tag{8.100}$$

$$d_{1,2} = \frac{\pi^4}{E_{1,0}E_{2,0}} - \frac{\pi^2}{4} \tag{8.101}$$

$$d_{1,3} = \frac{1}{2}E_{1,0} - \frac{\pi^2}{E_{2,0}} \tag{8.102}$$

$$d_{1,4} = \frac{\pi^4}{2E_{1,0}E_{1,1}} - \frac{\pi^2}{2} \tag{8.103}$$

$$d_{1,5} = \frac{1}{2}E_{1,0} \tag{8.104}$$

$$d_{1,6} = \frac{\pi^4 r^2}{2E_{0,1}E_{1,1}} \tag{8.105}$$

$$d_{1,7} = \frac{1}{2}E_{1,0} - \frac{\pi^2}{2E_{1,1}} \tag{8.106}$$

$$d_{1,8} = \frac{\pi^4}{E_{1,0}E_{1,1}} - \pi^2 \tag{8.107}$$

$$d_{1,9} = \frac{\pi^4 r^2}{2E_{0,1}E_{1,1}} + \frac{\pi^2 r^2 E_{1,0}}{2E_{0,1}} - \frac{\pi^4 r^4 E_{1,0}}{2E_{0,1}^2 E_{1,1}} \tag{8.108}$$

$$d_{1,10} = \frac{1}{2}E_{1,0} - \frac{\pi^2}{2E_{1,1}} + \frac{\pi^2 r^2 E_{1,0}}{2E_{0,1}E_{1,1}} \tag{8.109}$$

$$d_{1,11} = -\frac{4E_{1,0}}{\pi^2 a} \tag{8.110}$$

$$d_{2,1} = \frac{1}{2}E_{0,1} + \frac{\pi^2 r^2}{2E_{0,1}} \tag{8.111}$$

$$d_{2,2} = \frac{\pi^4 r^4}{E_{0,1}E_{0,2}} - \frac{\pi^2 r^2}{4} \tag{8.112}$$

$$d_{2,3} = \frac{1}{2}E_{0,1} - \frac{\pi^2 r^2}{E_{0,2}} \tag{8.113}$$

$$d_{2,4} = \frac{\pi^4 r^4}{2E_{0,1}E_{1,1}} - \frac{\pi^2 r^2}{2} \tag{8.114}$$

$$d_{2,5} = \frac{1}{2}E_{0,1} \tag{8.115}$$

$$d_{2,6} = \frac{\pi^4 r^2}{2E_{1,0}E_{1,1}} \tag{8.116}$$

$$d_{2,7} = \frac{\pi^4 r^2}{2E_{1,0}E_{1,1}} + \frac{\pi^2 E_{0,1}}{2E_{1,0}} - \frac{\pi^4 E_{0,1}}{2E_{1,0}^2 E_{1,1}} \tag{8.117}$$

$$d_{2,8} = \frac{1}{2}E_{0,1} - \frac{\pi^2 r^2}{2E_{1,1}} \tag{8.118}$$

$$d_{2,9} = \frac{\pi^4 r^4}{E_{0,1}E_{1,1}} - \pi^2 r^2 \tag{8.119}$$

$$d_{2,10} = \frac{1}{2}E_{0,1} + \frac{\pi^2 E_{0,1}}{2E_{1,0}E_{1,1}} - \frac{\pi^2 r^2}{2E_{1,1}} \tag{8.120}$$

$$d_{2,11} = -\frac{4E_{0,1}}{r\pi^2 a} \tag{8.121}$$

$$d_{3,1} = \frac{1}{2}E_{2,0} - \frac{\pi^2}{E_{1,0}} \tag{8.122}$$

$$d_{3,2} = \frac{1}{4}E_{2,0} - \frac{\pi^2}{E_{1,0}} - \frac{\pi^2 E_{2,0}}{4E_{1,0}{}^2} \tag{8.123}$$

$$d_{4,1} = \frac{1}{2}E_{0,2} - \frac{\pi^2 r^2}{E_{0,1}} \tag{8.124}$$

$$d_{4,2} = \frac{1}{4}E_{0,2} - \frac{\pi^2 r^2}{E_{0,1}} - \frac{\pi^2 r^2 E_{0,2}}{4E_{0,1}{}^2} \tag{8.125}$$

$$d_{5,1} = E_{1,1} - \frac{\pi^2}{E_{1,0}} \tag{8.126}$$

$$d_{5,2} = E_{1,1} - \frac{\pi^2 r^2}{E_{0,1}} \tag{8.127}$$

$$d_{5,3} = E_{1,1} - \frac{\pi^2}{E_{1,0}} - \frac{\pi^2 r^2}{E_{0,1}} \tag{8.128}$$

$$d_{6,1} = \frac{\pi^4}{E_{1,0}E_{1,1}} - \pi^2 - \frac{\pi^2 E_{2,1}}{E_{1,0}} + \frac{4\pi^4}{E_{1,0}E_{2,0}} + \frac{\pi^4 r^2}{2E_{1,0}E_{1,1}} \tag{8.129}$$

$$d_{6,2} = \frac{1}{2}E_{2,1} - \frac{\pi^2}{E_{1,0}} \tag{8.130}$$

$$d_{6,3} = \frac{\pi^4 E_{2,1}}{2E_{1,0}^2 E_{1,1}} - \pi^2 - \frac{3\pi^2 E_{2,1}}{2E_{1,0}} + \frac{4\pi^4}{E_{1,0}E_{2,0}} + \frac{\pi^4 r^2}{2E_{1,0}E_{1,1}} + \frac{\pi^4}{E_{1,0}E_{1,1}} \tag{8.131}$$

$$d_{6,4} = E_{2,1} - \frac{\pi^2 r^2}{E_{0,1}} \tag{8.132}$$

$$\begin{aligned} d_{6,5} = & \frac{\pi^4 r^4}{E_{0,1}E_{1,1}} + \frac{2\pi^4 r^2}{E_{0,1}E_{1,1}} + \frac{4\pi^4}{E_{1,0}E_{2,0}} + \frac{\pi^4 r^2}{2E_{1,0}E_{1,1}} + \frac{\pi^4}{E_{1,0}E_{1,1}} \\ & + \frac{\pi^4 r^4 E_{2,1}}{2E_{1,0}E_{0,1}E_{1,1}} - \frac{\pi^2 r^2 E_{2,1}}{2E_{0,1}} - \frac{\pi^2 E_{2,1}}{E_{1,0}} - \frac{\pi^2 r^2}{2} - \pi^2 \end{aligned} \tag{8.133}$$

$$d_{6,6} = \frac{\pi^4 r^2}{E_{0,1}E_{1,1}} - \frac{\pi^2 r^2}{4} - \frac{\pi^2 r^2 E_{2,1}}{4E_{0,1}} + \frac{\pi^4 r^2}{2E_{0,1}E_{1,1}} \tag{8.134}$$

$$d_{6,7} = \frac{1}{2}E_{2,1} - \frac{\pi^2}{E_{1,0}} - \frac{\pi^2 r^2}{2E_{1,1}} - \frac{\pi^2}{E_{1,1}} - \frac{\pi^2 E_{2,1}}{2E_{1,0}E_{1,1}} \tag{8.135}$$

$$d_{6,8} = E_{2,1} - \frac{4\pi^2}{E_{2,0}} \tag{8.136}$$

$$d_{6,9} = \frac{1}{2}E_{2,1} - \frac{\pi^2 r^2}{2E_{1,1}} - \frac{\pi^2}{E_{1,1}} \tag{8.137}$$

$$d_{6,10} = E_{2,1} - \frac{\pi^2 r^2}{\mathrm{E}_{0,1}} - \frac{4\pi^2}{E_{2,0}} \tag{8.138}$$

$$d_{7,1} = \frac{4\pi^4 r^4}{E_{0,1}E_{0,2}} + \frac{\pi^4 r^2}{2E_{0,1}E_{1,1}} + \frac{\pi^4 r^4}{E_{0,1}E_{1,1}} - \pi^2 r^2 - \frac{\pi^2 r^2 E_{1,2}}{E_{0,1}} \tag{8.139}$$

$$d_{7,2} = \frac{1}{2}E_{1,2} - \frac{\pi^2 r^2}{E_{0,1}} \tag{8.140}$$

$$\begin{aligned} d_{7,3} = & \frac{\pi^4}{E_{1,0}E_{1,1}} + \frac{2\pi^4 r^2}{E_{1,0}E_{1,1}} + \frac{4\pi^4 r^4}{E_{0,1}E_{0,2}} + \frac{\pi^4 r^2}{2E_{0,1}E_{1,1}} + \frac{\pi^4 r^4}{E_{0,1}E_{1,1}} \\ & + \frac{\pi^4 r^2 E_{1,2}}{2E_{0,1}E_{1,0}E_{1,1}} - \frac{\pi^2 E_{1,2}}{2E_{1,0}} - \frac{\pi^2 r^2 E_{1,2}}{E_{0,1}} - \frac{\pi^2}{2} - \pi^2 r^2 \end{aligned} \tag{8.141}$$

$$d_{7,4} = \frac{\pi^4}{2E_{1,0}E_{1,1}} + \frac{\pi^4 r^2}{E_{1,0}E_{1,1}} - \frac{\pi^2}{4} - \frac{\pi^2 E_{1,2}}{4E_{1,0}} \tag{8.142}$$

$$d_{7,5} = E_{1,2} - \frac{\pi^2}{E_{1,0}} \tag{8.143}$$

$$d_{7,6} = \frac{1}{2}E_{1,2} - \frac{\pi^2}{2E_{1,1}} - \frac{\pi^2 r^2}{E_{1,1}} \tag{8.144}$$

$$d_{7,7} = E_{1,2} - \frac{4\pi^2 r^2}{E_{0,2}} \tag{8.145}$$

$$d_{7,8} = \frac{4\pi^4 r^4}{E_{0,1}E_{0,2}} + \frac{\pi^4 r^2}{2E_{0,1}E_{1,1}} + \frac{\pi^4 r^4}{E_{0,1}E_{1,1}} + \frac{\pi^4 r^4 E_{1,2}}{2E_{0,1}^2 E_{1,1}} - \pi^2 r^2 - \frac{3\pi^2 r^2 E_{1,2}}{2E_{0,1}} \tag{8.146}$$

$$d_{7,9} = \frac{1}{2}E_{1,2} - \frac{\pi^2 r^2}{E_{0,1}} - \frac{\pi^2}{2E_{1,1}} - \frac{\pi^2 r^2}{E_{1,1}} - \frac{\pi^2 r^2 E_{1,2}}{2E_{0,1}E_{1,1}} \tag{8.147}$$

$$d_{7,10} = E_{1,2} - \frac{\pi^2}{E_{1,0}} - \frac{4\pi^2 r^2}{E_{0,2}} \tag{8.148}$$

$$d_{8,1} = \frac{1}{2}E_{3,0} - \frac{3\pi^2}{2E_{1,0}} \tag{8.149}$$

$$d_{8,2} = \frac{3\pi^4}{E_{1,0}E_{2,0}} - \frac{3\pi^2}{8} - \frac{3\pi^2 E_{3,0}}{8E_{1,0}} \tag{8.150}$$

$$d_{8,3} = \frac{1}{2}E_{3,0} - \frac{3\pi^2}{E_{2,0}} \tag{8.151}$$

$$d_{8,4} = \frac{6\pi^4}{E_{1,0}E_{2,0}} + \frac{\pi^4 E_{3,0}}{E_{1,0}^2 E_{2,0}} - \frac{3\pi^2}{4} - \frac{\pi^2 E_{3,0}}{E_{1,0}} \tag{8.152}$$

$$d_{8,5} = \frac{1}{2}E_{3,0} - \frac{3\pi^2}{2E_{1,0}} - \frac{3\pi^2}{E_{2,0}} - \frac{\pi^2 E_{3,0}}{E_{1,0}E_{2,0}} \tag{8.153}$$

$$d_{8,6} = -\frac{4E_{3,0}}{9\pi^2 a} \tag{8.154}$$

$$d_{9,1} = \frac{1}{2}E_{0,3} - \frac{3\pi^2 r^2}{2E_{0,1}} \tag{8.155}$$

$$d_{9,2} = \frac{1}{2}E_{0,3} - \frac{3\pi^2 r^2}{2E_{0,1}} - \frac{3\pi^2 r^2}{E_{0,2}} - \frac{\pi^2 r^2 E_{0,3}}{E_{0,1}E_{0,2}} \tag{8.156}$$

$$d_{9,3} = \frac{1}{2}E_{0,3} - \frac{3\pi^2 r^2}{E_{0,2}} \tag{8.157}$$

$$d_{9,4} = \frac{3\pi^4 r^4}{E_{0,1}E_{0,2}} - \frac{3\pi^2 r^2}{8} - \frac{3\pi^2 r^2 E_{0,3}}{8E_{0,1}} \tag{8.158}$$

$$d_{9,5} = \frac{\pi^4 r^4 E_{0,3}}{E_{0,1}{}^2 E_{0,2}} - \frac{3\pi^2 r^2}{4} - \frac{\pi^2 r^2 E_{0,3}}{E_{0,1}} + \frac{6\pi^4 r^4}{E_{0,1}E_{0,2}} \tag{8.159}$$

$$d_{9,6} = -\frac{4E_{0,3}}{9r\pi^2 a} \tag{8.160}$$

where a is tank length, r is the ratio of the tank length to tank breadth, h is liquid depth from the bottom of the tank to the fluid surface at rest, and $E_{i,j}$ is given by:

$$E_{i,j} = \pi\sqrt{i^2 + (rj)^2} \cdot \tanh\left(\frac{\pi h}{a}\sqrt{i^2 + (rj)^2}\right) \tag{8.161}$$

The $(i, 0)$ denotes the transverse mode, $(0, j)$ denotes the longitudinal mode, and the set $(i \neq 0, j \neq 0)$ is the mixed mode. The natural frequency of the sloshing mode is described as:

$$\omega_{i,j}^2 = \frac{g\pi}{a}\sqrt{i^2 + (rj)^2} \cdot \tanh\left(\frac{\pi h}{a}\sqrt{i^2 + (rj)^2}\right) \tag{8.162}$$

The sloshing frequency is dependent on the liquid depth and tank size. However, the tank size provides a larger influence on the sloshing frequency. A trapezoidal-velocity profile (bang-coast-bang acceleration) is applied to move the container. The maximum driving velocity and maximum driving acceleration of the container in the simulation were set to 25 cm/s and 2.5 m/s^2, respectively. The sloshing measurement position ($-0.5a$, $-0.5b$) was selected at the area of the free surface near the left diagonal corner of the container.

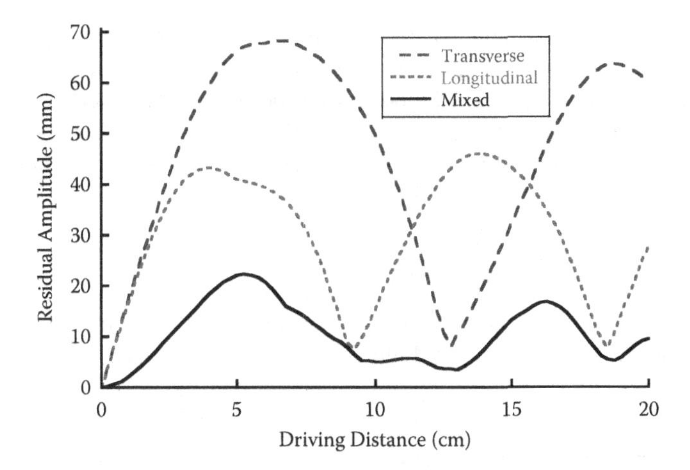

FIGURE 8.40 Residual amplitudes against driving distance.

Simulated residual amplitudes induced by various driving distances are shown in Figure 8.40. The driving angle, liquid depth, tank length, and breadth were set to 45°, 90 mm, 182 mm, and 102 mm, respectively. As the driving distance changes, peaks and troughs appear in all of the transverse, longitudinal, and mixed modes. Peaks and troughs arise when the sloshing caused by the acceleration and deceleration are in phase or out of phase. The locations of the peak and trough vary because of difference of the natural frequency of the transverse, longitudinal, and mixed modes. Due to a huge difference between the nonlinear sloshing model and the linear model, container motions cannot excite the mixed mode for the linear sloshing model. However, the mixed-mode response for the nonlinear model will exhibit relatively large amplitude oscillations.

Residual amplitudes induced by various tank lengths are shown in Figure 8.41. The driving distance, steering angle, tank breadth, and liquid depth were set to

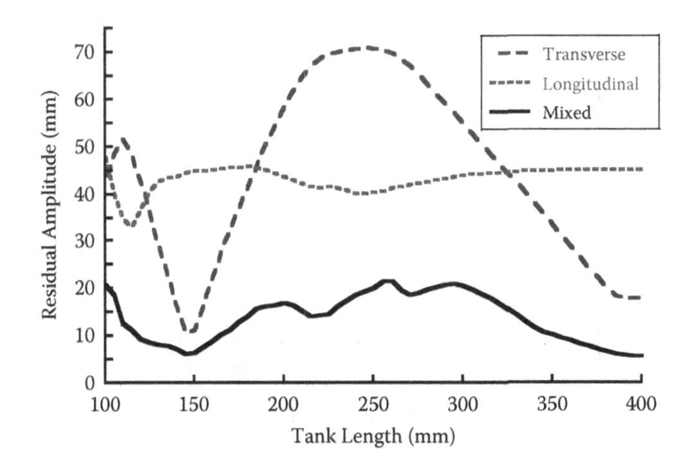

FIGURE 8.41 Residual amplitudes against container length.

22.5 cm, 45°, 102 mm, and 90 mm, respectively. The residual sloshing in the longitudinal mode varied slightly because of constant tank breadth. Peaks and troughs for transverse and mixed modes occur because of the interference between the sloshing caused by the acceleration and deceleration. Therefore, both the transverse and longitudinal modes are fundamental, while mixed modes still have some effects on the sloshing dynamics.

8.4.2 Computational Dynamics

Simulated residual amplitudes with the four-pieces smoother induced by various driving distances are shown in Figure 8.42. The four-pieces smoother is described by:

$$u(\tau) = \begin{cases} M\mu\tau e^{-2\zeta_m\omega_m\tau}, & 0 \le \tau \le 0.5T_m \\ M(\mu T_m - 2KT_m + \mu KT_m + 4K\tau - \mu\tau - 2\mu K\tau)e^{-2\zeta_m\omega_m\tau}, & 0.5T_m < \tau \le T_m \\ M(6KT_m - 3\mu KT_m - \mu K^2T_m + \mu K^2\tau - 4K\tau + 2\mu K\tau)e^{-2\zeta_m\omega_m\tau}, & T_m < \tau \le 1.5T_m \\ M(2\mu K^2T_m - \mu K^2\tau)e^{-2\zeta_m\omega_m\tau}, & 1.5T_m < \tau \le 2T_m \end{cases} \tag{8.163}$$

where M, K, T_m, and μ are coefficients. M, K, and T_m are given by:

$$M = \zeta_m^2\omega_m^2/\left(1 - e^{-2\pi\zeta_m/\sqrt{1-\zeta_m^2}}\right)^2 \tag{8.164}$$

$$K = e^{2\pi\zeta_m/\sqrt{1-\zeta_m^2}} \tag{8.165}$$

$$T_m = 2\pi/\left(\omega_m\sqrt{1-\zeta_m^2}\right) \tag{8.166}$$

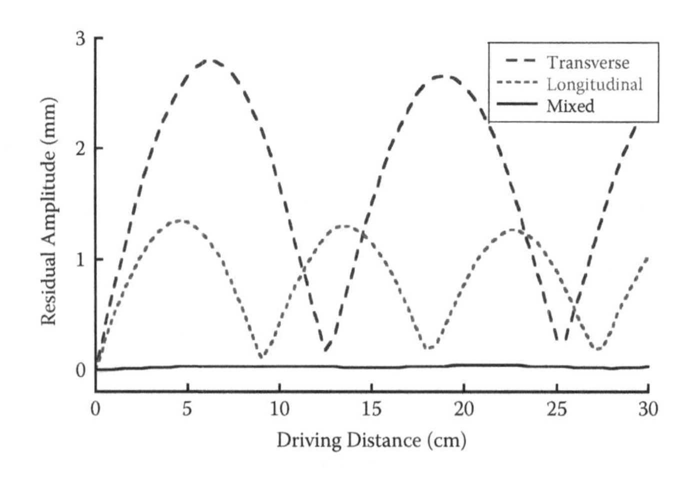

FIGURE 8.42 Residual amplitudes for various driving distances.

The smoother profile is a piecewise continuous curve with four pieces, each of which has continuous transitions between boundary conditions.

The steering angle, liquid depth, tank length, and breadth were set to 45°, 90 mm, 182 mm, and 102 mm, respectively. As the driving distance changes, peaks and troughs occur in all of the transverse, longitudinal, and mixed modes. This is because the sloshing caused by the acceleration and deceleration are sometimes in phase and sometimes out of phase. The transverse-mode sloshing was not zero because residual oscillations were designed below the tolerable level for the four-pieces smoother. The dynamics of both the longitudinal mode and mixed mode are similar to that of the transverse mode. With the four-pieces smoother, the residual sloshing of the transverse, longitudinal, and mixed modes was suppressed by an average of 96%, 97%, and 99%, respectively. The four-pieces smoother suppressed the residual sloshing to be a low level for a wide range of driving motions.

Simulated residual amplitudes with the four-pieces smoother induced by various tank lengths ranged from 102 mm to 500 mm are shown in Figure 8.43. The driving distance, steering angle, tank breadth, and liquid depth were fixed at 22.5 cm, 45°, 102 mm, and 90 mm, respectively. The smoother was designed for the tank length of 300 mm, which is corresponding to the design frequency of 8.7 rad/s. The tank length of 102 mm and 500 mm correspond to the frequency of 17.3 rad/s and 5.6 rad/s, respectively. For the case of tank length of 300 mm, the four-pieces smoother limited the sloshing of the transverse, longitudinal, and mixed modes to a lower level. For the case of tank length of 102 mm, the four-pieces smoother eliminated the sloshing of the transverse, longitudinal, and mixed modes to <0.1 mm. For the case of 500 mm tank length, the residual amplitude in the transverse mode was large because the four-pieces smoother gets less insensitive at lower frequencies. Therefore, the four-pieces smoother can effectively reduce the total sloshing modes for a large range of working conditions and system parameters.

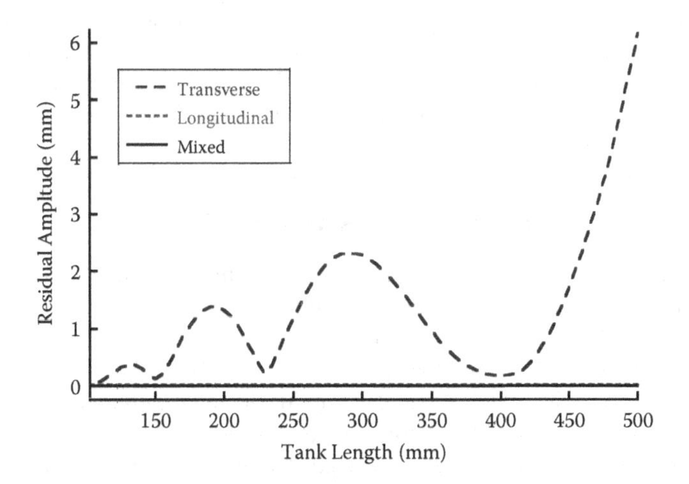

FIGURE 8.43 Residual amplitudes for various container lengths.

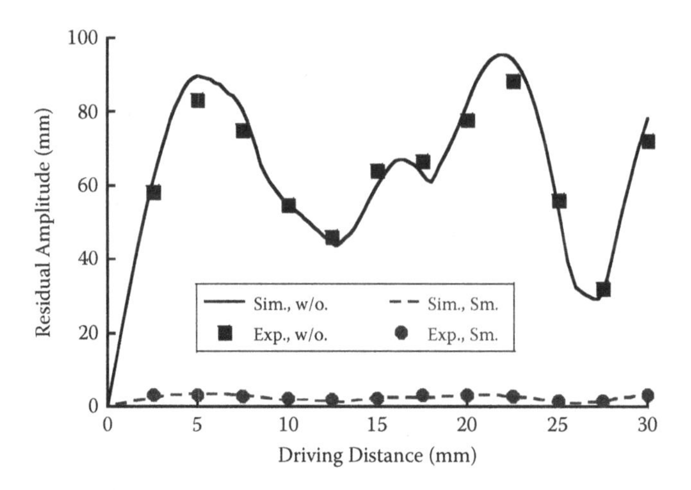

FIGURE 8.44 Experimental results induced by driving distances.

8.4.3 Experimental Results

Experiments were also conducted on a testing apparatus shown in Figure 8.19. The liquid depth and steering angle were set to 90 mm and 45°, respectively. A camera was also mounted to the gantry for recording the surface elevation at the left diagonal corner of the container. One set of experiments was performed to verify the sloshing dynamics and the effectiveness of the smoother on suppressing nonlinear sloshing. Figure 8.44 shows residual amplitudes of the surface elevation from those tests. As the driving distance increased, experimental residual amplitudes changed. Peaks and troughs occurred because the sloshing caused by the acceleration and deceleration was sometimes in phase and sometimes out of phase. Experimental results match the general trend as simulated curves. Thus, experimental results verified the simulated sloshing dynamics.

The effect of the smoother is also shown in Figure 8.44. Residual amplitudes with the smoother were <3.3 mm for all cases because they were limited to the tolerable level. Experimental data also follow the same general shape as simulated curves. The experiments clearly verified that the four-pieces smoother can effectively eliminate the nonlinear sloshing.

8.5 PENDULUM-SLOSHING DYNAMICS

Liquid containers suspended by cables are commonly applied in industrial applications. In the casting industry, molten metal is transported by bridge cranes. In the case of firefighting services, helicopters transport water containers using cables. In the space station, liquid containers are captured by SPHERES robots via tethers. However, undesirable cable swing and liquid sloshing caused by the operator-commanded motions degrade effectiveness and safety.

A schematic representation of a suspended liquid container undergoing planar motions is shown in Figure 8.45. A trolley slides in the *OX* direction, and two

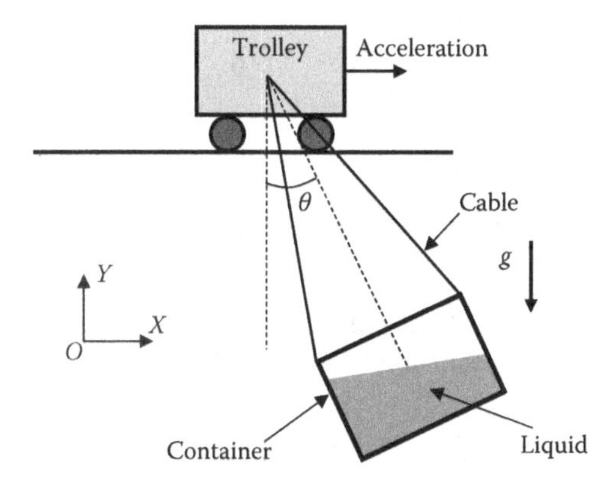

FIGURE 8.45 Planar model of a suspended container.

massless and inelastic cables of length, l, suspend below the trolley. The cables support a liquid rectangular container of mass, m, height, H, and length, $2r$, which contains a fluid. The liquid surface at rest is at a height, h, from the bottom of the container. The mass of the liquid is M. Rotating the swing angle, θ, of the cables about the vertical direction converts the inertial coordinates OXY to the moving Cartesian coordinates oxy fixed to the container. η denotes the fluid surface elevation measured from the undisturbed free surface in the oy direction. The acceleration, a, of trolley is the input to the model, and the swing angle, θ, of the cables, and surface elevation, η, of the liquid are outputs to the model.

The motion of the trolley is assumed to be unaffected by motions of the pendulum and sloshing due to the high-ratio geared drives. The cable length does not change during the motions. The following assumptions are also applied to simplify the model: (1) the flow is irrotational; (2) the container is rigid; (3) the fluid is nonviscous homogeneous; (4) the fluid is incompressible; and (5) displacement and velocity of the liquid free surface are small.

8.5.1 Sloshing Modeling

The velocity of the fluid in the container is:

$$v = v_0 + \nabla\varphi \tag{8.167}$$

where v_0 is the velocity of the container, ∇ is the gradient operator, and φ is the perturbed velocity potential function. The boundary values in terms of the perturbed velocity potential is given by:

$$\frac{\partial^2\varphi}{\partial x^2} + \frac{\partial^2\varphi}{\partial y^2} = 0 \tag{8.168}$$

$$\left.\frac{\partial\varphi}{\partial x}\right|_{x=\pm r} = 0 \tag{8.169}$$

$$\left.\frac{\partial \varphi}{\partial y}\right|_{y=-h} = 0 \tag{8.170}$$

$$\left.\frac{\partial \varphi}{\partial y}\right|_{y=\eta} = \frac{\partial \eta}{\partial t} \tag{8.171}$$

At the fluid-free surface, the fluid pressure is zero. Ignoring the high-order terms and employing Bernoulli's equation derive:

$$\begin{aligned}&\left.\frac{\partial \varphi}{\partial t}\right|_{y=\eta} + \eta(g\cos\theta - a\sin\theta)\\ &\quad + x\left[g\sin\theta - a\cos\theta + \ddot{\theta}\left(\sqrt{l^2 - r^2} + H - 0.5h\right)\right] = 0\end{aligned} \tag{8.172}$$

where the gravitational constant is g. The perturbed velocity potential and surface elevation can be written as:

$$\varphi(x, y, t) = \sum_k \phi_k(x, y)\cdot\dot{q}_k(t) \tag{8.173}$$

$$\eta(x, y, t) = \sum_k \sigma_k(x, y)\cdot q_k(t) \tag{8.174}$$

where k is the positive integer, q_k is the time-dependent function of the kth vibrational mode, and φ_k and σ_k are the corresponding spatial functions. Substituting Equations (8.173) and (8.174) into (8.168)–(8.172) yields:

$$\frac{\partial^2 \phi_k}{\partial x^2} + \frac{\partial^2 \phi_k}{\partial y^2} = 0 \tag{8.175}$$

$$\left.\frac{\partial \phi_k}{\partial x}\right|_{x=\pm r} = 0 \tag{8.176}$$

$$\left.\frac{\partial \phi_k}{\partial y}\right|_{y=-h} = 0 \tag{8.177}$$

$$\left.\frac{\partial \phi_k}{\partial y}\right|_{y=\eta} = \sigma_k = \frac{\omega_k^2 \phi_k}{g\cos\theta - a\sin\theta} \tag{8.178}$$

Solving Equations (8.175)–(8.178) yields the natural frequency, ω_k, and spatial function, φ_k:

$$\omega_k^2 = (g\cos\theta - a\sin\theta)\cdot\frac{k\pi}{2r}\tanh\left(k\pi\frac{h}{2r}\right) \tag{8.179}$$

$$\phi_k(x, y) = \cos\left[k\pi\left(\frac{x}{2r} + \frac{1}{2}\right)\right] \cdot \cosh\left[k\pi\frac{(y+h)}{2r}\right] \tag{8.180}$$

The liquid depth, h, container length, $2r$, swing angle, θ, and trolley acceleration, a, have a large influence on the sloshing frequencies (8.179). As the swing angle and trolley acceleration increase, the sloshing frequencies decreases. The complex nonlinear behavior occurs in the sloshing frequencies because trolley acceleration has some effect on the sloshing frequencies. Substituting Equations (8.173) and (8.174) into (8.172), and multiplying by spatial function, φ_k, then integrating over the free surface ($-r \le x \le r$) yield:

$$\ddot{q}_k(t) + \omega_k^2 q_k(t) + \gamma_k \cdot \left[g \sin\theta - a\cos\theta + \ddot{\theta}\left(\sqrt{l^2 - r^2} + H - 0.5h\right)\right] = 0 \tag{8.181}$$

where

$$\gamma_k = \frac{\int_{-r}^{r} x\cdot\phi_k\cdot dx}{\int_{-r}^{r} \phi_k\cdot\phi_k\cdot dx} = \frac{4r(\cos k\pi - 1)}{k^2\pi^2 \cosh\left[k\pi\frac{(y+h)}{2r}\right]} \tag{8.182}$$

When the k is even, the coefficient, γ_k, in Equation (8.182) is zero. Both the perturbed velocity potential, φ, and surface elevation, η, are a sum of vibrational response for infinite sloshing modes.

8.5.2 Pendulum Modeling

The acceleration at the center of the container mass along the ox direction induced by the swing is $(\sqrt{l^2 - r^2} + 0.5H)\ddot{\theta}$. $a \cdot \cos\theta$ and $g \sin\theta$ are the acceleration at the center of the container mass along the ox direction caused by the trolley acceleration and the gravitation, respectively. Then the pendulum dynamics are described as:

$$m(\sqrt{l^2 - r^2} + 0.5H)\ddot{\theta} + mg\sin\theta - \Delta - ma \cdot \cos\theta = 0 \tag{8.183}$$

where the fluid force acting on the container sidewall denotes Δ. The fluid force can be written as:

$$\Delta = \int_{-h}^{\eta(r,t)} p(r, y)dy - \int_{-h}^{\eta(-r,t)} p(-r, y)dy \tag{8.184}$$

where

$$p(x, y) = -\rho\frac{\partial\varphi}{\partial t} - \rho y(g\cos\theta - a\sin\theta) - \rho x\left[g\sin\theta - a\cos\theta + \ddot{\theta}\left(\sqrt{l^2 - r^2} + H - 0.5h\right)\right] \tag{8.185}$$

where the area mass density of the liquid is ρ. The fluid force on the container sidewall can be obtained by substituting Equation (8.184) into (8.185):

$$\begin{aligned}\Delta = & -\tfrac{\rho}{2}(g\cos\theta - a\sin\theta)[\eta(r,t)^2 - \eta(-r,t)^2] \\ & - M\left[g\sin\theta - a\cos\theta + \ddot{\theta}\left(\sqrt{l^2-r^2} + H - 0.5h\right)\right] \\ & - \textstyle\sum_k [\cos(k\pi) - 1]\frac{2\rho r}{k\pi}\sinh\left(\frac{k\pi h}{2r}\right)\ddot{q}_k(t)\end{aligned} \tag{8.186}$$

Then the dynamic model of the pendulum can be obtained by substituting Equation (8.186) into (8.183):

$$\begin{aligned}&\left[M(\sqrt{l^2-r^2} + H - 0.5h) + m(\sqrt{l^2-r^2} + 0.5H)\right]\cdot\ddot{\theta} + (M+m)g\sin\theta \\ &- (M+m)a\cos\theta \\ &+ \frac{\rho}{2}(g\cos\theta - a\sin\theta)[\eta(r,t)^2 - \eta(-r,t)^2] \\ &+ \sum_k [\cos(k\pi) - 1]\frac{2\rho r}{k\pi}\sinh\left(\frac{k\pi h}{2r}\right)\ddot{q}_k(t) = 0\end{aligned} \tag{8.187}$$

8.5.3 Frequency Analyses

The equilibrium points of the swing angle, θ, and surface elevation, η, are zero resulting from the dynamic models (8.181) and (8.187). Assuming small oscillations around the equilibria and small acceleration of the trolley yield a linearized model. The linearized model can be used to calculate numerical solutions of the natural frequency of the coupled pendulum-sloshing dynamics.

The natural frequencies, Ω_i, of the first three modes of coupled pendulum-sloshing dynamics for various container lengths are shown in Figure 8.46. The tank mass, m, tank height, H, liquid height, h, and area mass density of the liquid, ρ, were

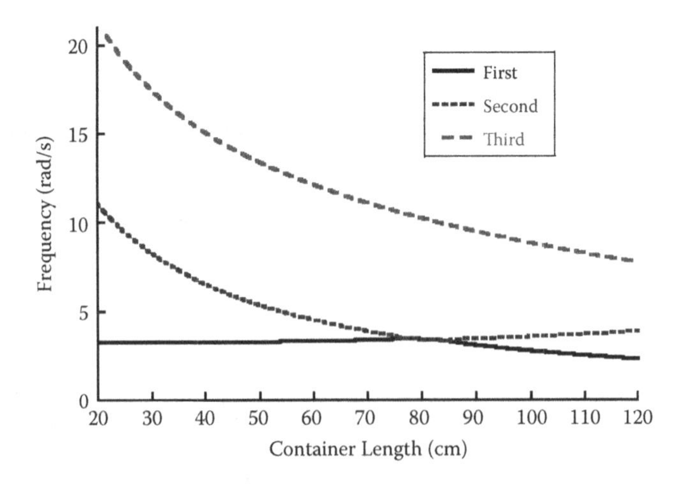

FIGURE 8.46 Frequencies of first three modes against the container length.

fixed at 1.395 kg, 28 cm, 14.5 cm, and 85 kg/m^2, respectively. The cable length, l, was 74 cm. As the container length increases, the first-mode frequency, Ω_1, changes slightly. As the container length increases before 76 cm, the second-mode frequency, Ω_2, decreases. As the container length increases from 76 cm, the second-mode frequency, Ω_2, increases slightly. The first-mode frequency, Ω_1, is more related with the pendulum motion when the container length is less than 76 cm. When the container length ranges from 76 cm to 84 cm, the second-mode frequency, Ω_2, approaches the first-mode frequency, Ω_1. The internal resonance might arise between the pendulum and first-mode sloshing in this case. The first-mode frequency, Ω_1, is more related with the first-mode sloshing motion when the container length is larger than 84 cm.

8.5.4 Oscillation Control

Oscillations of the coupled pendulum-sloshing dynamics in cable-suspended liquid containers are attenuated by smoothing operator's commands. A four-pieces smoother inherent in the limited motions of the trolley produces a limited response to oscillations of the coupled pendulum-sloshing dynamics. The four-pieces smoother is given by:

$$u(\tau) = \begin{cases} \tau \cdot Me^{-2\zeta\Omega_i\tau}, & 0 \le \tau \le 0.5T \\ (T - K^{-1}T - \tau + 2K^{-1}\tau)\cdot Me^{-2\zeta\Omega_i\tau}, & 0.5T \le \tau \le T \\ (3K^{-1}T - K^{-2}T - 2K^{-1}\tau + K^{-2}\tau)\cdot Me^{-2\zeta\Omega_i\tau}, & T \le \tau \le 1.5T \\ (2K^{-2}T - K^{-2}\tau)\cdot Me^{-2\zeta\Omega_i\tau}, & 1.5T < \tau \le 2T \end{cases} \tag{8.188}$$

where

$$K = e^{(-\pi\zeta/\sqrt{1-\zeta^2})} \tag{8.189}$$

$$M = 4\zeta^2\Omega_i^2/(1 + K - K^2 - K^3)^2 \tag{8.190}$$

$$T = 2\pi/(\Omega_i\sqrt{1 - \zeta^2}) \tag{8.191}$$

Figure 8.47 shows the oscillation control scheme. The operated commands generated via the control interface filter through the smoother to create modified commands.

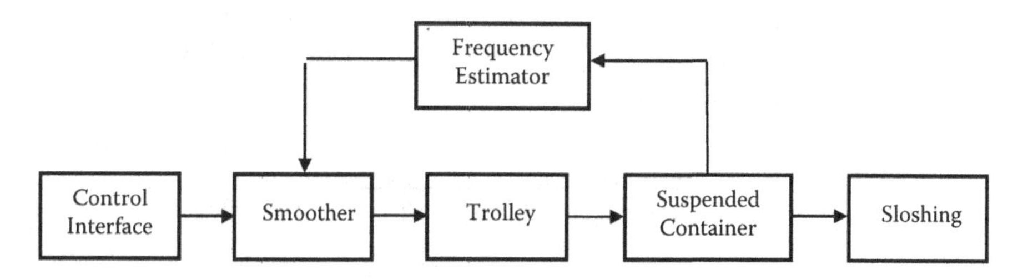

FIGURE 8.47 Oscillation suppression architecture.

The modified commands drove the trolley toward the desired state. The system parameters are used to calculate the natural frequencies, Ω_i. The smoother for oscillation suppression is designed using natural frequency and zero damping.

8.5.5 Computational Dynamics

The system parameters are applied to estimate the first-mode frequency, Ω_1, for the design of the four-pieces smoother. Nevertheless, it is challenging to measure the system parameters accurately. Therefore, it is essential to study the robustness to changes in the system parameters. The dynamic models (8.181) and (8.187) will be used in this section to perform the numerical verification of the robustness of smoother.

The trapezoidal-velocity profile is used to move the trolley as the original command. The maximum velocity of the trolley is 20 cm/s, while the acceleration of the trolley is 2 m/s^2. The transient deflection is referred to as the maximum peak-to-peak deflection when the trolley moves, while the residual amplitude is the maximum peak-to-peak deflection after the trolley stops.

The transient deflection and residual amplitude of the swing deflection and surface elevation induced by various tank lengths are shown in Figures 8.48 and 8.49. The cable length, l, and driving distance of the trolley were set to 74 cm and 60 cm, respectively. The uncontrolled transient deflection of the container swing deflection increased slowly as the container length increases. The uncontrolled residual amplitude of the container swing deflection changed slightly as the tank length is less than 80 cm. This is because the first-mode frequency changes slightly. The residual amplitude of the swing occurred a near-zero value at the tank length of 130 cm because oscillations caused by the trolley acceleration and deceleration are out of phase at this point. The design length of the container for the smoother was fixed at 75 cm, while the tank length varied from 10 cm to 140 cm.

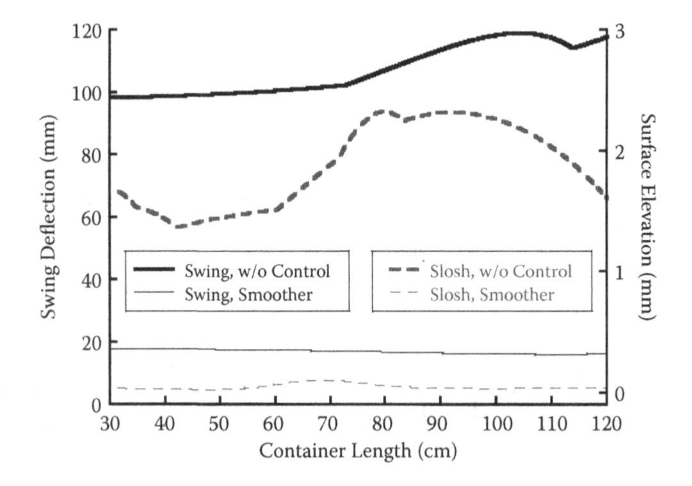

FIGURE 8.48 Transient deflection against container length.

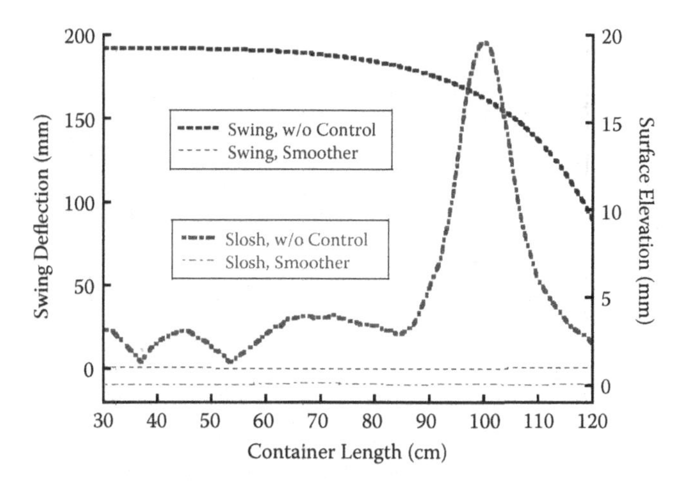

FIGURE 8.49 Residual amplitude against container length.

The smoother suppressed the transient deflection and residual amplitude of the swing deflection by an average of 84.3% and 99.6%, respectively.

The uncontrolled transient deflection of the surface elevation of the liquid varied slightly as the tank length changes. Peaks and troughs occurred in the residual amplitude of the surface elevation. Moreover, peaks and troughs are spaced farther apart as the container length increases. This is because the sloshing frequency decreases with increasing the tank length. The tank length is set to 75 cm for the design of the smoother when the tank length changed between 10 cm and 140 cm. The smoother reduced the transient deflection and residual amplitude of the surface elevation by an average of 97.7% and 99.2%, respectively.

The transient deflection and residual amplitude of the container swing deflection and surface elevation for various cable lengths are shown in Figures 8.50 and 8.51,

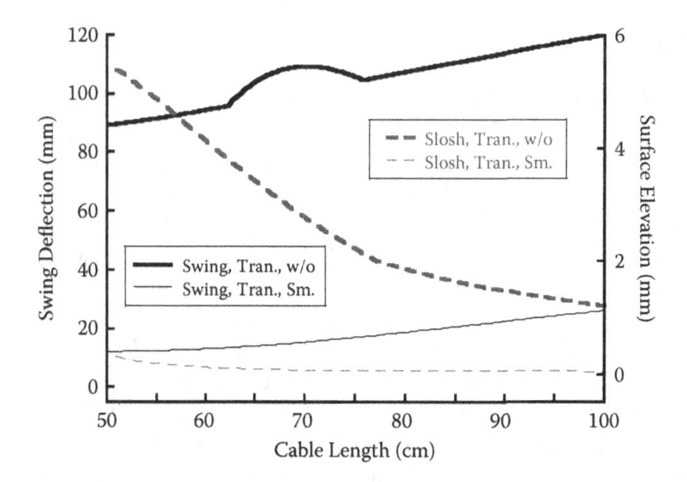

FIGURE 8.50 Transient deflection against cable length.

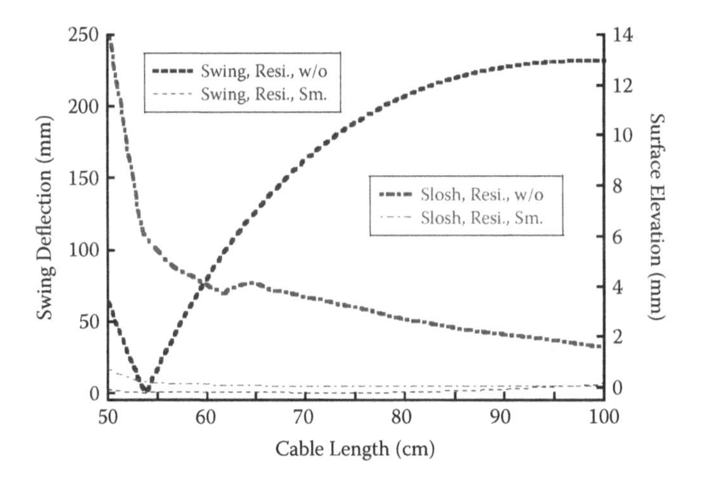

FIGURE 8.51 Residual amplitude against cable length.

respectively. The container length, $2r$, and the driving distance of the trolley were fixed at 80 cm and 60 cm, respectively. The uncontrolled transient deflection of the swing deflection increased slightly as the cable length increased. Peaks and troughs arose in the residual amplitude of the swing deflection. The cable length for the design of the smoother was fixed at 74 cm, while the cable length ranged from 45 cm to 105 cm. The smoother attenuated the transient deflection and residual amplitude of the container swing deflection by an average of 83.0% and 98.6%, respectively.

The uncontrolled transient deflection of the surface elevation of the liquid decreased as the cable length increased. The residual amplitude of the surface elevation exhibits a local maximum at the cable length of 48 cm. The cable length for the design of the smoother is 74 cm. The smoother suppressed the transient deflection and residual amplitude of the surface elevation by an average of 96.1% and 96.6%, respectively.

8.5.6 Experimental Verification

Figure 8.52 shows testbed of a small-scale industrial crane carrying a liquid container by four cables to perform experiments. A trolley was driven by a servomotor with encoder. The control hardware includes a host computer and a motion control card. The host computer is used for program development and user interface. The motion control card connects the computer to servo amplifiers. A MATLAB® script embedded in the host computer applies the control algorithm to filter through the operator's commands, which are baseline trapezoidal-velocity profiles. Smoothed commands are generated for driving the trolley. A liquid container was supported by four cables below the trolley. The cables were made of Dyneema fishing line. A camera, which recorded the displacement of the container, was attached to the trolley. Another camera, which recorded surface elevation at the rightmost edge of the container, was mounted on the container. The length, breadth, and height of the container were 8.5 cm, 8.5 cm, and 28 cm, respectively. The cable length, liquid depth, tank mass, and liquid mass were held constant at 74 cm, 14.5 cm, 1.395 kg, and 1.048 kg, respectively.

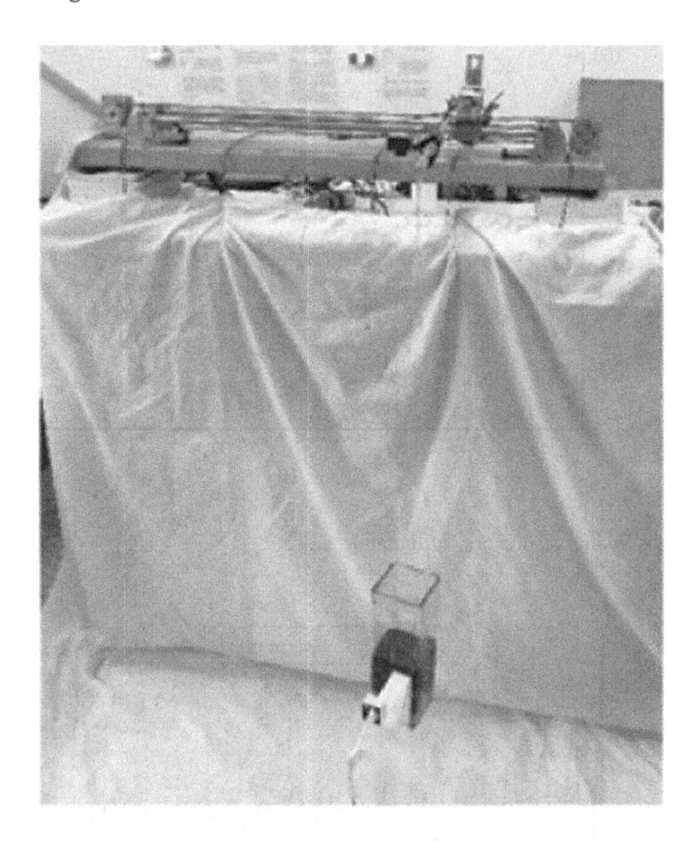

FIGURE 8.52 Testbed of a suspended liquid container.

Figures 8.53 and 8.54 show experimental and simulated responses of container swing deflection and surface elevation of the liquid to a non-zero initial condition to validate the nonlinear equations (8.181) and (8.187) of motion. The experimental data follow the similar trends to the simulated results. The one difference is that experimental magnitudes in the surface elevation of the liquid were smaller than the simulated amplitudes. This is because the real system had a small amount of damping, while the model was undamped. The swing frequency is similar to the first-mode frequency in Figure 8.46. This is because the contribution of the first mode in this case is fundamental to the swing deflection. The sloshing frequency is estimated approximately by the second-mode frequency in Figure 8.46. This is because the contribution of the second mode in this case is fundamental to the surface elevation of the liquid.

Figure 8.55 shows the uncontrolled experimental and simulated residual amplitudes for a wide range of driving motions to verify the dynamic behavior of the nonlinear model and the effectiveness of the smoother. The driving distance of the trolley varied from 0 to 45 cm. Peaks and troughs arose in the swing and sloshing as the distance varied when the uncontrolled trapezoidal commands moved the trolley. This is because oscillations of the swing and sloshing caused by the trolley acceleration and deceleration were sometimes in phase, and sometimes out of phase.

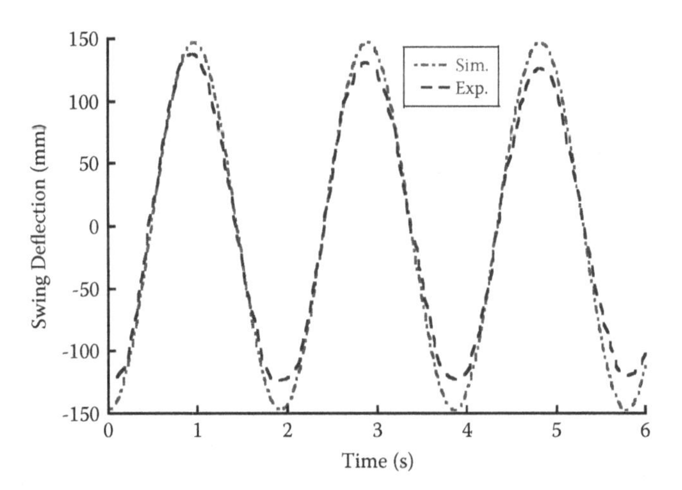

FIGURE 8.53 Swing verification of dynamic model.

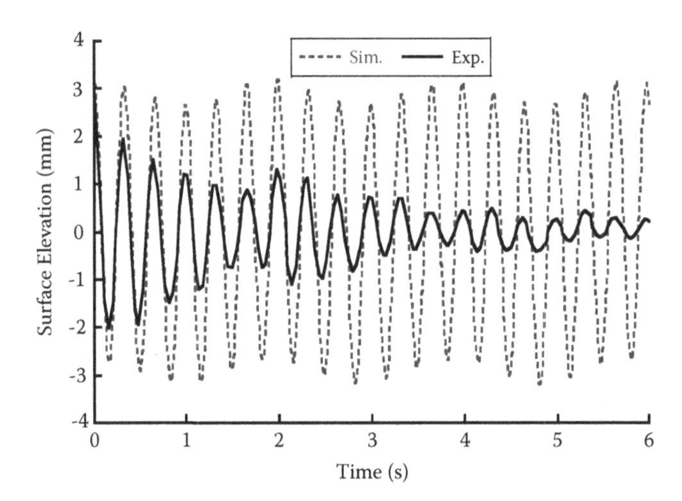

FIGURE 8.54 Twisting verification of dynamic model.

The uncontrolled simulated curve in the surface elevation of sloshing were larger than the experimental amplitudes because the model of the sloshing is undamped. However, experimental frequencies in the swing and sloshing follow the same shape as the simulated results very well. Hence, the experimental results verified the dynamic behavior of the nonlinear model.

Figure 8.56 shows the corresponding experimental and simulated residual amplitudes with the smoother. The design frequency of the smoother was set to 3.288 rad/s. The smoother suppressed the experimental amplitudes of the swing and sloshing by an average of 89.2% and 75.6%, respectively. The controlled results were nearly independent of the driving distance because the smoother attenuated most of swing and

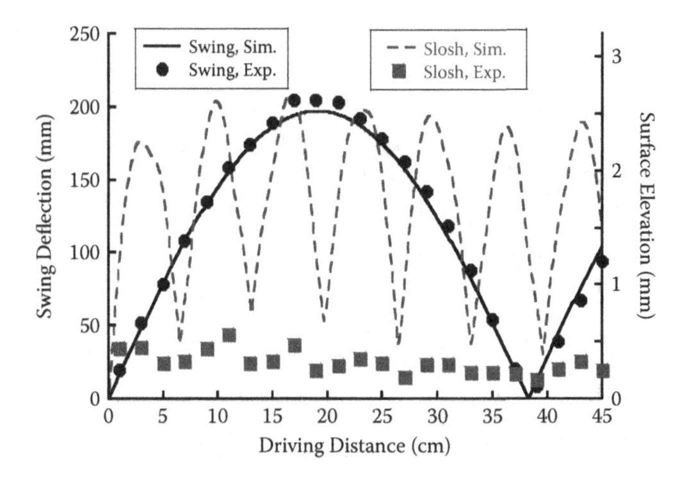

FIGURE 8.55 Uncontrolled residual amplitudes for various driving distances.

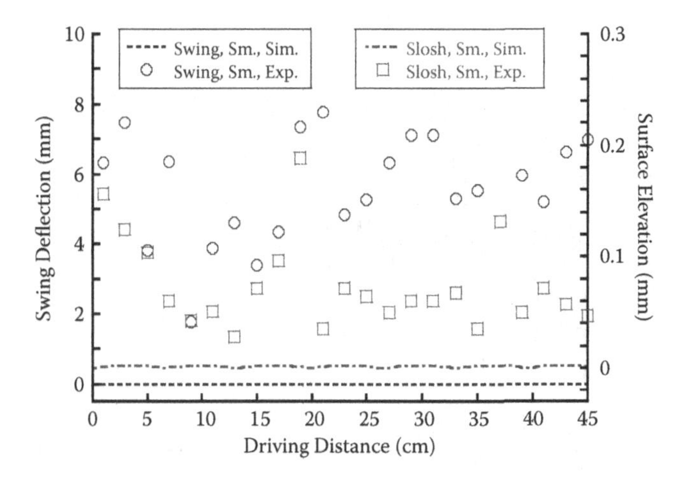

FIGURE 8.56 Controlled residual amplitudes for various driving distances.

sloshing. The controlled experimental data in the swing and sloshing were larger than the simulated results. This is because small modeling errors in the design frequency of the smoother arise.

Figures 8.57 and 8.58 show the experimental responses of the swing deflection and surface elevation resulting from different modeling errors in the frequency to examine the robustness of the smoother. In the case of a small modeling error in the frequency, the design frequency was 3.288 rad/s. The transient deflection and residual amplitude of the container swing deflection were limited to 4.2 mm and 3.9 mm, and the transient deflection and residual amplitude of the surface elevation were limited to 0.21 mm and 0.05 mm. In the case of negative error of –30%, the design frequency was set to 2.302 rad/s. The transient deflection and residual

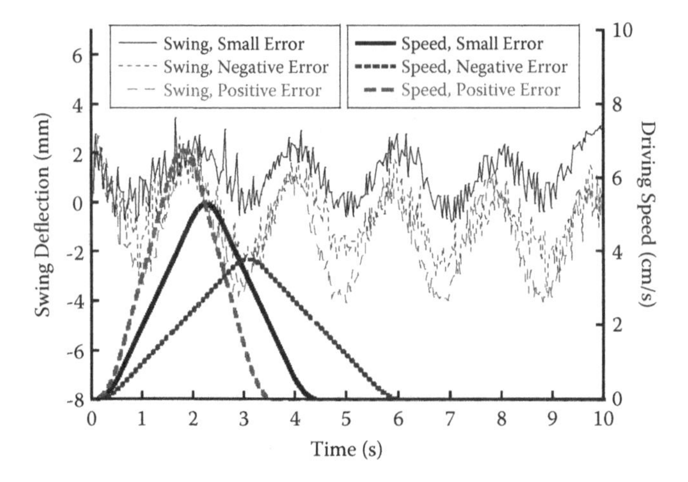

FIGURE 8.57 Experimental results of cable swing and driving speed to modeling errors in the frequency.

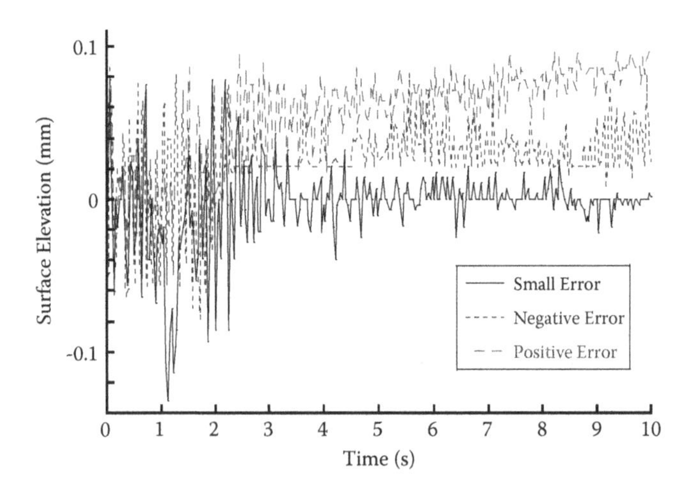

FIGURE 8.58 Experimental results of liquid sloshing to modeling errors in the frequency.

amplitude of the swing deflection were 6.1 mm and 4.9 mm. The transient deflection and residual amplitude of the surface elevation were 0.16 mm and 0.075 mm. In the case of positive error of +30%, the design frequency was fixed at 4.274 rad/s. the transient deflection and residual amplitude of the swing deflection were 6.5 mm and 5.4 mm, while the transient deflection and residual amplitude of the surface elevation were 0.13 mm and 0.078 mm. Comparison results indicate that oscillations of the cable swing and liquid sloshing increase as modeling errors in the design frequency increase. However, the controlled results are robust to modeling errors in the frequency.

8.6 LINK-SLOSHING DYNAMICS

8.6.1 Modeling

Figure 8.59 shows a flexible single-link manipulator transporting a liquid container. A hub slews a flexible link to a prescribed angular displacement, θ. The length and mass of the uniform link are denoted by l_B and m_B, respectively. A liquid tank is attached at the tip of the light-weight link. The length, breadth, height, and mass of the rectangular tank are denoted by l_C, b_C, h_C, and m_C, respectively. The mass of the liquid inside the partially filled tank is denoted by m_L. The liquid depth is, h_L, from the tank bottom to the fluid surface at rest, while the fluid surface elevation is denoted by η measured from the undisturbed free surface.

Let *OXYZ* be the inertial coordinates and *oxyz* be the moving Cartesian coordinates fixed at the tank. The origin of the moving coordinates exists in the mean free surface at the center plane of the tank, as shown in Figure 8.60. The input to the model shown in Figure 8.59 is the angular acceleration, $d^2\theta/dt^2$, of the hub. The outputs are the vibrational deflection, w, on the link and the surface elevation, η, of the fluid sloshing. It is assumed that the motion of the hub is unaffected by the movement of the link and fluid due to large inertia of the hub and small oscillations of the flexible link and fluid sloshing.

8.6.1.1 Modeling of Flexible Link

The force equation of the motion of a small element at the distance, r, on the flexible link is expressed by:

$$\rho_B dr \frac{\partial^2 w(r,t)}{\partial t^2} + dV(r,t) - \rho_B r \frac{d^2\theta}{dt^2} dr - f_L(r,t)dr = 0; \quad 0 \le r \le l_B \tag{8.192}$$

where r is the distance along the long axis of the link measured from the hub, $w(r, t)$ denotes the vibrational deflection at the distance, r, on the link, dr is a small variation of r, ρ_B is the linear mass density of the link, $V(r, t)$ is the shear force on the

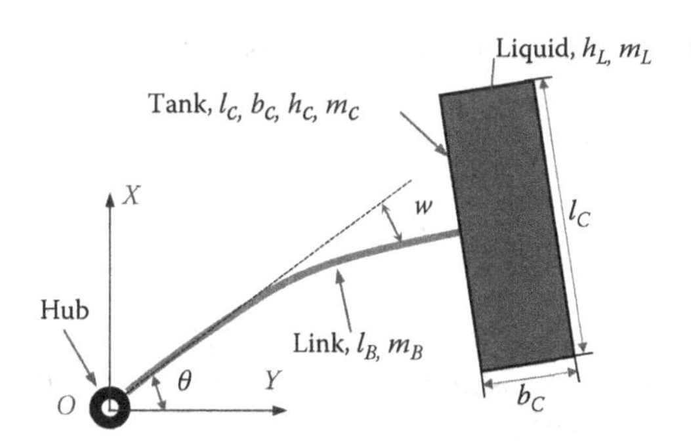

FIGURE 8.59 Schematic of a single-link flexible manipulator carrying a liquid tank viewed from above.

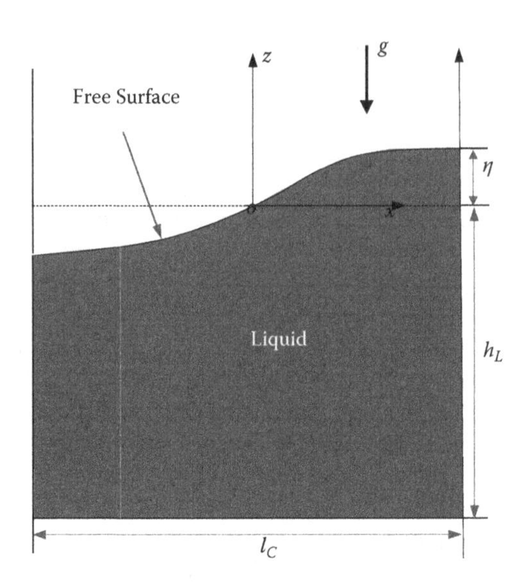

FIGURE 8.60 Liquid tank viewed from right.

cross-section of the link, $dV(r, t)$ is the small variation of $V(r, t)$, and $f_L(r, t)$ is the fluid force acting on the per unit length of the link. The force, $f_L(r, t)$, satisfies:

$$f_L(r, t) = \delta(r - l_B)F_L(t) \tag{8.193}$$

where $F_L(t)$ denotes the fluid force on the tank sidewall, and $\delta(r - l_B)$ denotes the unit impulse function. The function, $\delta(r - l_B)$, is unit at the distance, l_B, and zero otherwise. The moment equation of the motion of the small element is given by:

$$dM(r, t) - [V(r, t) + dV(r, t)]dr - \rho_B r\frac{d^2\theta}{dt^2}\frac{1}{2}dr^2 - f_L(r, t)dr\frac{1}{2}dr = 0 \tag{8.194}$$

where $M(r, t)$ is the bending moment on the cross-section of the link, and $dM(r, t)$ is the corresponding small variation. The linearized equation between the bending moment and link deflection is described by:

$$M(r, t) = EI_B\frac{\partial^2 w(r, t)}{\partial r^2} \tag{8.195}$$

where E is the Young's modulus of the link and I_B is the moment of inertia of the cross-section of the link. Substituting Equations (8.193) and (8.195) into (8.192) and ignoring the high-order power terms of dr yield:

$$EI_B\frac{\partial^4 w(r, t)}{\partial r^4} + \rho_B\frac{\partial^2 w(r, t)}{\partial t^2} = \rho_B r\frac{d^2\theta}{dt^2} + f_L(r, t) \tag{8.196}$$

The boundary conditions of the flexible link at the fixed end and tip are given by:

$$w(r, t)|_{r=0} = 0 \tag{8.197}$$

$$\left.\frac{\partial w(r, t)}{\partial r}\right|_{r=0} = 0 \tag{8.198}$$

$$\left.\frac{\partial}{\partial r}\left[EI_B\frac{\partial^2 w(r, t)}{\partial r^2}\right]\right|_{r=l_B} = (m_C + m_L)\left.\frac{\partial^2 w(r, t)}{\partial t^2}\right|_{r=l_B} \tag{8.199}$$

$$\left.EI_B\frac{\partial^2 w(r, t)}{\partial r^2}\right|_{r=l_B} = -I_C\left.\frac{\partial^2}{\partial t^2}\left(\frac{\partial w(r, t)}{\partial r}\right)\right|_{r=l_B} \tag{8.200}$$

where I_C is the moment of inertia of the container about the long axis of the link. By the mode superposition method, the link deflection can be assumed as:

$$w(r, t) = \sum_{k=1}^{+\infty} [\Theta_k(r)\mu_k(t)] \tag{8.201}$$

where μ_k is the time-dependent function of the *kth* vibrational mode of the link, and Θ_k is the corresponding mode shape. The natural frequency and mode shape can be obtained by substituting Equation (8.201) into (8.197)–(8.200):

$$\upsilon_k^2 = \beta_k^4\frac{EI_B}{\rho_B} \tag{8.202}$$

$$\Theta_k(r) = [\sin(\beta_k r) - \sinh(\beta_k r)] - \sigma_k[\cos(\beta_k r) - \cosh(\beta_k r)] \tag{8.203}$$

$$\sigma_k = \frac{EI_B\beta_k[\sin(\beta_k l_B) + \sinh(\beta_k l_B)] + I_C\upsilon_k^2[\cos(\beta_k l_B) - \cosh(\beta_k l_B)]}{EI_B\beta_k[\cos(\beta_k l_B) + \cosh(\beta_k l_B)] - I_C\upsilon_k^2[\sin(\beta_k l_B) + \sinh(\beta_k l_B)]} \tag{8.204}$$

$$1 + \cos(\beta_k l_B)\cosh(\beta_k l_B) = \frac{(m_C + m_L)\beta_k l_B}{m_B}[\sin(\beta_k l_B)\cosh(\beta_k l_B) - \cos(\beta_k l_B)\sinh(\beta_k l_B)] \tag{8.205}$$

where υ_k is the natural frequency of the kth vibrational mode of the link. Substituting Equation (8.201) into (8.196) and neglecting the modal coupling between different vibrational modes of the flexible link result in:

$$EI_B\sum_{k=1}^{+\infty}[\frac{\partial^4\Theta_k}{\partial r^4}\mu_k] + \rho_B\sum_{k=1}^{+\infty}[\Theta_k\frac{d^2\mu_k}{dt^2}] - \rho_B r\frac{d^2\theta}{dt^2} - f_L = 0 \tag{8.206}$$

Multiplying Equation (8.206) by the mode shape, Θ_k, and then integrating over the link length give rise to:

$$\frac{d^2\mu_k}{dt^2} + \upsilon_k^2\mu_k - \frac{\int_0^{l_B} r\Theta_k dr}{\int_0^{l_B} \Theta_k\Theta_k dr}\frac{d^2\theta}{dt^2} - \frac{\Theta_k(l_B)}{\rho_B\int_0^{l_B} \Theta_k\Theta_k dr}F_L = 0 \tag{8.207}$$

Including the proportional damping into Equation (8.207) results in the dynamic equation of the kth vibrational mode of the flexible link:

$$\frac{d^2\mu_k}{dt^2} + 2\varsigma_k\upsilon_k\frac{d\mu_k}{dt} + \upsilon_k^2\mu_k - \frac{\int_0^{l_B} r\Theta_k dr}{\int_0^{l_B} \Theta_k\Theta_k dr}\frac{d^2\theta}{dt^2} - \frac{\Theta_k(l_B)F_L}{\rho_B\int_0^{l_B} \Theta_k\Theta_k dr} = 0 \tag{8.208}$$

where ς_k is the damping ratio of the kth vibrational mode of the link.

8.6.1.2 Modeling of Fluid Sloshing

The fluid flow is assumed to be irrotational, incompressible, and nonviscous homogeneous. The velocity of the fluid flow in the tank is:

$$v_L = l_B\frac{d\theta}{dt} + \frac{\partial w(l_B, t)}{\partial t} + \nabla\varphi_L \tag{8.209}$$

where φ_L is the perturbed velocity potential and ∇ is the gradient operator. The first two terms on the right side of Equation (8.209) are the tank velocity along the ox direction. The first one is caused by the slewing angular speed, and the second one is induced by the deflection velocity of the flexible link. The boundary conditions of the velocity potential at the tank wall are:

$$\nabla^2\varphi_L = 0 \tag{8.210}$$

$$\left.\frac{\partial\varphi_L}{\partial x}\right|_{x=-0.5l_C;0.5l_C} = 0 \tag{8.211}$$

$$\left.\frac{\partial\varphi_L}{\partial z}\right|_{z=-h_L} = 0 \tag{8.212}$$

$$\left.\frac{\partial\varphi_L}{\partial z}\right|_{z=\eta} = \frac{\partial\eta}{\partial t} \tag{8.213}$$

Substituting Equation (8.209) into the Euler's fluid equation and neglecting the high-order power terms yield the fluid dynamics at the free surface:

$$\frac{\partial\varphi_L}{\partial t} + g\eta + l_B\frac{d^2\theta}{dt^2}x + \frac{\partial^2 w(l_B, t)}{\partial t^2}x = 0 \tag{8.214}$$

By the mode superposition method, the surface elevation and velocity potential can also be assumed as:

$$\eta(x,t) = \sum_{n=1}^{+\infty} [H_n(x) q_n(t)] \tag{8.215}$$

$$\varphi_L(x,z,t) = \sum_{n=1}^{+\infty} \left[\phi_n(x,z)\frac{dq_n(t)}{dt}\right] \tag{8.216}$$

where q_n is the time-dependent function of the nth vibrational mode of the fluid sloshing, and H_n and ϕ_n are the corresponding mode shape. The natural frequency, ω_n, and mode shapes, H_n and ϕ_n, of the sloshing can be obtained by substituting Equations (8.215) and (8.216) into (8.210)–(8.213):

$$\omega_n^2 = g\frac{n\pi}{l_C}\tanh\left(\frac{n\pi h_L}{l_C}\right) \tag{8.217}$$

$$\phi_n(x,z) = \cosh\left[n\pi\frac{(z+h_L)}{l_C}\right]\cos\left[n\pi\frac{(x+0.5l_C)}{l_C}\right] \tag{8.218}$$

$$H_n(x) = \frac{\omega_n^2\phi_n(x, z=0)}{g} \tag{8.219}$$

Multiplying Equation (8.214) by the mode shape, ϕ_n, and then integrating over the tank length yield:

$$\frac{d^2q_n}{dt^2} + \omega_n^2 q_n + \frac{2l_C[\cos(n\pi)-1]}{n^2\pi^2\cosh\left(n\pi\frac{h_L}{l_C}\right)}\left(l_B\frac{d^2\theta}{dt^2} + \frac{\partial^2 w(l_B,t)}{\partial t^2}\right) = 0 \tag{8.220}$$

Including the proportional damping into Equation (8.220) gives rise to the dynamic equation of the nth vibrational mode of the fluid sloshing:

$$\frac{d^2q_n}{dt^2} + 2\zeta_n\omega_n\frac{dq_n}{dt} + \omega_n^2 q_n + \frac{2l_C[\cos(n\pi)-1]}{n^2\pi^2\cosh\left(n\pi\frac{h_L}{l_C}\right)}\left(l_B\frac{d^2\theta}{dt^2} + \frac{\partial^2 w(l_B,t)}{\partial t^2}\right) = 0 \tag{8.221}$$

where ζ_n is the damping ratio of the nth vibrational mode of the sloshing.

The fluid force, F_L, acting on the tank sidewall is described by:

$$F_L(t) = \int_{-h_C}^{\eta(0.5l_C,t)} p(0.5l_C,z,t)dz - \int_{-h_C}^{\eta(-0.5l_C,t)} p(-0.5l_C,z,t)dz \tag{8.222}$$

where

$$p(x, z, t) = -\rho_L \frac{\partial \varphi_L}{\partial t} - \rho_L g z - \rho_L l_B \frac{d^2\theta}{dt^2} x - \rho_L \frac{\partial^2 w(l_B, t)}{\partial t^2} x \tag{8.223}$$

Substituting Equation (8.223) into (8.222) gives rise to the equation of the fluid force on the sidewall of the tank:

$$F_L(t) = -m_L \left[l_B \frac{d^2\theta}{dt^2} + \frac{\partial^2 w(l_B, t)}{\partial t^2} \right] - \sum_{n=1}^{+\infty} \left(\frac{l_C \rho_L [\cos(n\pi) - 1]}{n\pi} \sinh\left(n\pi \frac{h_L}{l_C} \right) \ddot{q}_n \right) \tag{8.224}$$

8.6.2 Dynamics

Equation (8.208) described the flexible-link dynamics and Equation (8.221) reported the liquid-sloshing dynamics. Furthermore, the fifth term in the left side of Equation (8.209) is the fluid force. Meanwhile, the link deflection has been included in Equation (8.221). The fluid-force term in Equation (8.208) and the link-deflection term in Equation (8.221) generate coupling oscillations between link and sloshing dynamics.

The link-sloshing dynamics indicate that both the flexible link and liquid sloshing have an infinite number of vibrational modes. Resulting from Equations (8.208) and (8.221), neglecting the vibrational modes higher than two in the link and sloshing yields a simplified model:

$$X_1 \begin{bmatrix} \ddot{\mu}_1 \\ \ddot{\mu}_2 \\ \ddot{q}_1 \\ \ddot{q}_2 \end{bmatrix} + X_2 \begin{bmatrix} \dot{\mu}_1 \\ \dot{\mu}_2 \\ \dot{q}_1 \\ \dot{q}_2 \end{bmatrix} + X_3 \begin{bmatrix} \mu_1 \\ \mu_2 \\ q_1 \\ q_2 \end{bmatrix} + X_4 \frac{d^2\theta}{dt^2} = 0 \tag{8.225}$$

where

$$X_1 = \begin{bmatrix} 1 + \dfrac{m_L \Theta_1(l_B)\Theta_1(l_B)}{\rho_B \int_0^{l_B} \Theta_1 \Theta_1 dr} & \dfrac{m_L \Theta_1(l_B)\Theta_2(l_B)}{\rho_B \int_0^{l_B} \Theta_1 \Theta_1 dr} & \dfrac{-2 l_C \rho_L \Theta_1(l_B)}{\pi \rho_B \int_0^{l_B} \Theta_1 \Theta_1 dr} \sinh\left(\dfrac{\pi h_L}{l_C}\right) & 0 \\ \dfrac{m_L \Theta_1(l_B)\Theta_2(l_B)}{\rho_B \int_0^{l_B} \Theta_2 \Theta_2 dr} & 1 + \dfrac{m_L \Theta_2(l_B)\Theta_2(l_B)}{\rho_B \int_0^{l_B} \Theta_2 \Theta_2 dr} & \dfrac{-2 l_C \rho_L \Theta_2(l_B)}{\pi \rho_B \int_0^{l_B} \Theta_2 \Theta_2 dr} \sinh\left(\dfrac{\pi h_L}{l_C}\right) & 0 \\ \dfrac{-4 l_C \Theta_1(l_B)}{\pi^2 \cosh\left(\frac{\pi h_L}{l_C}\right)} & \dfrac{-4 l_C \Theta_2(l_B)}{\pi^2 \cosh\left(\frac{\pi h_L}{l_C}\right)} & 1 & 0 \\ 0 & 0 & 0 & 1 \end{bmatrix} \tag{8.226}$$

$$X_2 = \begin{bmatrix} 2\varsigma_1 \upsilon_1 & 0 & 0 & 0 \\ 0 & 2\varsigma_2 \upsilon_2 & 0 & 0 \\ 0 & 0 & 2\zeta_1 \omega_1 & 0 \\ 0 & 0 & 0 & 2\zeta_2 \omega_2 \end{bmatrix} \tag{8.227}$$

$$X_3 = \begin{bmatrix} \upsilon_1{}^2 & 0 & 0 & 0 \\ 0 & \upsilon_2{}^2 & 0 & 0 \\ 0 & 0 & \omega_1{}^2 & 0 \\ 0 & 0 & 0 & \omega_2^2 \end{bmatrix} \tag{8.228}$$

$$X_4 = \begin{bmatrix} -\dfrac{\int_0^{l_B} r\Theta_1 dr}{\int_0^{l_B} \Theta_1\Theta_1 dr} + \dfrac{m_L l_B \Theta_1(l_B)}{\rho_B \int_0^{l_B} \Theta_1\Theta_1 dr} \\ -\dfrac{\int_0^{l_B} r\Theta_2 dr}{\int_0^{l_B} \Theta_2\Theta_2 dr} + \dfrac{m_L l_B \Theta_2(l_B)}{\rho_B \int_0^{l_B} \Theta_2\Theta_2 dr} \\ \dfrac{-4 l_C l_B}{\pi^2 \cosh\left(\frac{\pi h_L}{l_C}\right)} \\ 0 \end{bmatrix} \tag{8.229}$$

From the simplified model (8.225), the frequencies and damping ratios of the coupled system can be calculated:

$$|X_1\lambda^2 + X_2\lambda + X_3| = 0 \tag{8.230}$$

where

$$\lambda = -\xi\Omega \pm j\Omega\sqrt{1 - \xi^2} \tag{8.231}$$

Ω and ξ are the frequency and damping ratio of the coupled link-sloshing system. Both frequency and damping ratio are dependent on system parameters including Young's modulus of the link material, E, moment of inertia of the cross-section of the uniform link, I_B, link length, l_B, link mass, m_B, tank length, l_C, tank breadth, b_C, tank height, h_C, tank mass, m_C, liquid depth, h_L, and liquid mass, m_L.

Figure 8.61 shows the frequencies and damping ratios of the link-sloshing dynamics for various link lengths. The Young's modulus of the link material, E,

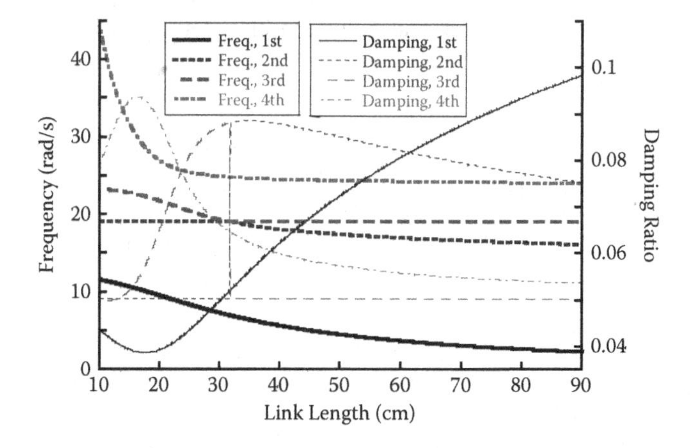

FIGURE 8.61 Frequencies and damping ratios for various link lengths.

moment of inertia of the cross section of the uniform link, I_B, link mass, m_B, tank length, l_C, tank breadth, b_C, tank height, h_C, tank mass, m_C, liquid depth, h_L, liquid mass, m_L, link damping ratio, ς_k, and liquid damping ratio, ζ_n, were fixed at 206 GPa, 12.692 mm^4, 174.9 g, 165 mm, 24 mm, 130 mm, 180.2 g, 60 mm, 237.6 g, 0.1949, and 0.0502, respectively. As the link length increased, the first-mode damping ratio decreased before 18 cm, and increased sharply after this point. Increasing link length decreased first-mode frequency slowly. The second- and third-mode damping ratios exchanged near the link length of 31.8 cm. This is because the second-mode frequency approached the third-mode frequency at this point, and an internal resonance might arise in this case. As the link length increased, the fourth-mode damping ratio increased before 16.4 cm, and then decreased quickly. The fourth-mode frequency decreased rapidly with increasing link length before 30 cm. After this point, the fourth-mode frequency would hold constant. In addition, the first mode of the link-sloshing system is more related with the motion of the flexible link. Before the link length of 31.8 cm, the second mode of the coupled system is more related with the motion of the liquid sloshing. After 31.8 cm, the second mode of the coupled system is more related with the motion of the flexible link.

When the link length is set to 10 cm, 45 cm, and 90 cm, the corresponding first frequency, υ_1, of the link resulting from Equation (8.202) is 135.4 rad/s, 13.6 rad/s, and 4.6 rad/s, respectively. Increasing link length decreases the link frequency. The first frequency, ω_1, of the sloshing resulting from Equation (8.217) is held constant at 12.3 rad/s in this case. Both the link frequency (8.202) and sloshing frequency (8.217) are significantly different with the coupled-system frequency given in Equation (8.230) and Figure 8.61. This is because the coupling effect between flexible link and liquid sloshing creates the modification between the frequencies of link and sloshing and coupled frequency.

8.6.3 Control

A discontinuous piecewise (DP) smoother can be used to suppress the coupled oscillations between flexible link and liquid sloshing defined in Equations (8.208) and (8.221):

$$c(\tau) = \begin{cases} \dfrac{\xi_m \Omega_m e^{-\xi_m \Omega_m \tau}}{(1+K)(1-K^h)}, & 0 \le \tau \le \dfrac{h\pi}{\Omega_m \sqrt{1-\xi_m^{\ 2}}} \\ 0, & \dfrac{h\pi}{\Omega_m \sqrt{1-\xi_m^{\ 2}}} < \tau \le \dfrac{\pi}{\Omega_m \sqrt{1-\xi_m^{\ 2}}} \\ \dfrac{\xi_m \Omega_m e^{-\xi_m \Omega_m \tau}}{(1+K)(1-K^h)}, & \dfrac{\pi}{\Omega_m \sqrt{1-\xi_m^{\ 2}}} \le \tau \le \dfrac{(1+h)\pi}{\Omega_m \sqrt{1-\xi_m^{\ 2}}} \\ 0, & \text{others} \end{cases} \tag{8.232}$$

where

$$K = e^{\frac{-\pi \xi_m}{\sqrt{1-\xi_m^{\ 2}}}} \tag{8.233}$$

ξ_m is the design damping ratio and h is the modified factor. The modified factor satisfies $0 \le h \le 1$. The rise time of the DP smoother (48) is $0.5(1 + h)$ times the vibration period. As the modified factor, h, increases, both the rise time and the frequency insensitivity increase, and the maximum transient vibrations decrease. The modified factor, h, is chosen to be 0.6 in this section for short rise time, wide range in the frequency insensitivity, and low amplitude of transient vibrations.

Multiple smoothers with different modes can be convoluted to generate a combined smoother. The combined smoother has multi-notch and low-pass filtering effect. The multi-notch filtering effect targets vibrations of the first few modes, while the low-pass filtering effect attenuates the amplitude of the high-mode vibrations. The control architecture for link-sloshing dynamics is shown in Figure 8.62. The slewing commands filter through double DP smoothers, which are designed for each of the first two modes. The smoothed commands will be applied to drive the flexible link manipulators moving liquid containers. Resulting from Equation (8.230), the system parameters are used to estimate the frequencies and damping ratios of the first two modes. The prediction of frequencies and damping ratios will be applied to design double DP smoothers.

Figure 8.63 shows the FFT of the simulated responses for identifying the oscillation attenuation. The Young's modulus of the link material, E, moment of

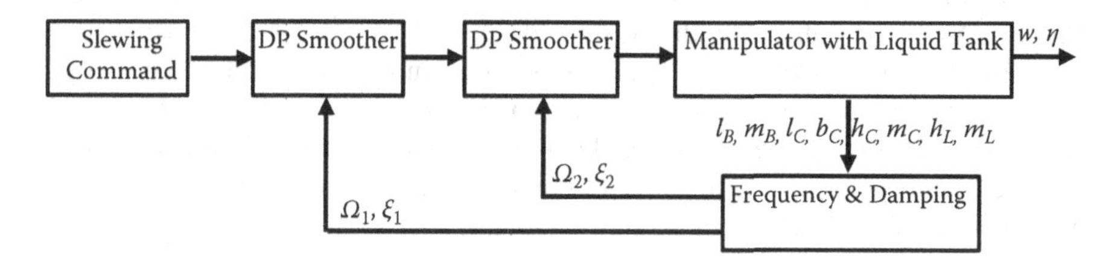

FIGURE 8.62 Control architecture.

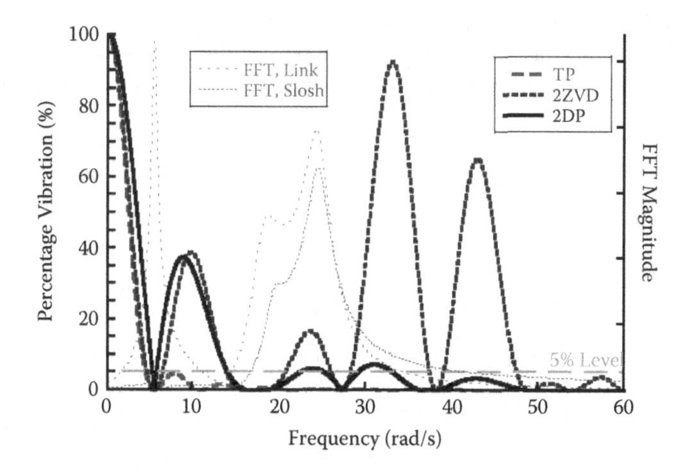

FIGURE 8.63 FFT magnitude of the link and sloshing responses and frequency insensitivity curves for two DP smoothers.

inertia of the cross-section of the uniform link, I_B, link length, l_B, link mass, m_B, tank length, l_C, tank breadth, b_C, tank height, h_C, tank mass, m_C, liquid depth, h_L, and liquid mass, m_L, were 206 GPa, 12.692 mm^4, 42 cm, 174.9 g, 165 mm, 130 mm, 24 mm, 180.2 g, 60 mm, and 237.6 g, respectively. Peaks of the FFT magnitude disclose that the first-, second-, third-, and fourth-mode frequencies of the link-sloshing dynamics are 5.4 rad/s, 18.0 rad/s, 19.1 rad/s, and 24.6 rad/s, respectively. Figure 8.63 also describes the frequency insensitivity curves for double DP smoothers. Double DP smoothers were designed using the first two modes. Then multi-notching filtering effect of double DP smoothers eliminates oscillations of the first two modes, and the low-passing filtering effect inherent in the double DP smoothers reduces oscillations of the high modes. Therefore, double DP smoothers attenuate oscillations for all modes.

A two-pieces (TP) smoother was reported to suppress sloshing. The frequency insensitivity of the TP smoother is also displayed in Figure 8.63. The TP smoother provides wider range in 5% frequency insensitivity than double DP smoothers because of a penalty in the rise time. The rise time of the TP smoother is twice times as long as the vibration period. Thus, the rise time of TP smoother is larger than that of the double DP smoothers.

A two-mode input shaper was also reported to reduce oscillations from the liquid sloshing and flexible structure. The combined shaper was zero vibration and derivative (ZVD) input shaper with the first two modes. The frequency-insensitivity curve for the double ZVD shaper was also provided in Figure 8.63. The multi-notch filtering effect in the double ZVD shaper also attenuated the oscillations of the first two modes. However, the double ZVD shaper cannot control vibrations at some specified frequencies of 33 rad/s and 43 rad/s. As the system parameters vary, FFT magnitude at those frequencies might be large.

8.6.4 Experiments

Experimental validation was performed on a flexible link manipulator carrying a rectangular container, as shown in Figure 8.64. A hub is driven by a motor with encoder. A flexible light-weight link is attached to the hub, and supports a

FIGURE 8.64 Testbed of a flexible link manipulator transporting a rectangular container.

rectangular tank. The tank is partially filled with water. The water tank is located at the tip of the light-weight link. The link connects metal blocks on the back for counterweight. A computer is applied for operator interface and implementation of the control methods for the drives. A motion control card connects the computer to the motor amplifier. One camera is installed on the hub for recording vibrations of the tank, which is the link deflection at the tip. Moreover, the other camera is mounted on the tank for measuring the surface elevation of the fluid at the left diagonal corner of the container.

The original slewing commands were trapezoidal velocity profiles, which would be sent to double DP smoothers to create modified commands for slewing the hub. A proportional-integral-derivative controller implanted in the computer forced the hub motions to follow the modified slewing profile. The link breadth, link height, tank length, tank breadth, tank height, tank mass, liquid depth, maximum slewing velocity, and acceleration in experiments were fixed at 31 mm, 1.7 mm, 165 mm, 24 mm, 130 mm, 180.2 g, 60 mm, 10°/s, and 100°/s^2, respectively. Experiments for each of the damping ratio, ς_k, of the link and damping ratio, ζ_n, of the sloshing were developed to be approximately 0.1949 and 0.0502, respectively. The link length may change by changing the connected position between the hub and link.

The first experiment was performed to validate the theoretical findings: (1) the dynamic model in Equations (8.208) and (8.221) is effective for predicting the experimental response; (2) the frequency-estimation equation (8.230) is also effective for forecasting the frequency of the experimental data. The link length was held constant at 42 cm. A trapezoidal velocity command was applied to rotate the hub for 30°. Figure 8.65 displays the experimental responses from these tests. The simulated curve resulting from the dynamic model closely matches the experimental data. The difference between simulated and experimental results comes from the damping nonlinearity in the experiment.

The first frequency, υ_1, of the uncoupled link in this case is 15.2 rad/s from Equation (8.202), and the first frequency, ω_1, of the uncoupled sloshing is 12.3 rad/s

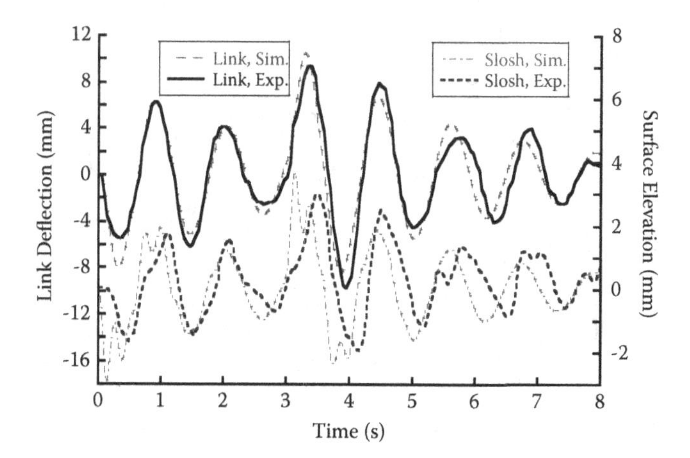

FIGURE 8.65 Experimental verification of dynamic model.

from Equation (8.217). Resulting from Equation (8.230), the estimation of the first-mode frequency, Ω_1, for the link-sloshing dynamics is 5.4 rad/s. The experimental responses of link deflection and surface elevation in Figure 8.65 show explicitly that the fundamental frequency is 5.4 rad/s approximately. Therefore, the experiment obviously validates the effectiveness of the frequency prediction defined in Equation (8.230). There is a very large change between the coupled frequency, Ω_1, of 5.4 rad/s, and uncoupled frequencies, which are link frequency, υ_1, of 15.2 rad/s, and sloshing frequency, ω_1, of 12.3 rad/s. The change was induced by the coupling effect between flexible link and fluid sloshing.

The transient deflection is defined as the peak-to-peak deflection when the hub moves. Meanwhile, the residual amplitude is referred to as the peak-to-peak deflection after the hub stops. The hub moves from 0 second to 3 seconds for a slewing displacement of 30° in Figure 8.65. Thus, the transient stage ranges from 0 second to 3 seconds, while the time frame after 3 seconds is residual stage. The experimental transient deflection and residual amplitude of the link deflection in Figure 8.65 were 12.46 mm and 19.11 mm, respectively. The experimental transient deflection and residual amplitude of the surface elevation of the liquid sloshing were 3.44 mm and 4.93 mm, respectively.

The second experiment was conducted to verify the dynamic behavior of the analytical model. Figure 8.66 displays the simulated and experimental results of the transient deflection and residual amplitude of the link and sloshing induced by slewing motions. The link length was also set to 42 cm. The trapezoidal velocity commands were used to slew the hub from 0° to 50°. As the slewing displacement varied, peaks and troughs arose in the residual amplitude of the link. This effect may be physically interpreted as the interference between vibrations induced by the acceleration and deceleration of the hub. When the interference was in phase, peaks arose. Troughs occurred when the interference was out of phase. In addition, the distance between peaks or troughs is corresponding to the first-mode frequency of the link-sloshing dynamics resulting from Equation (8.230).

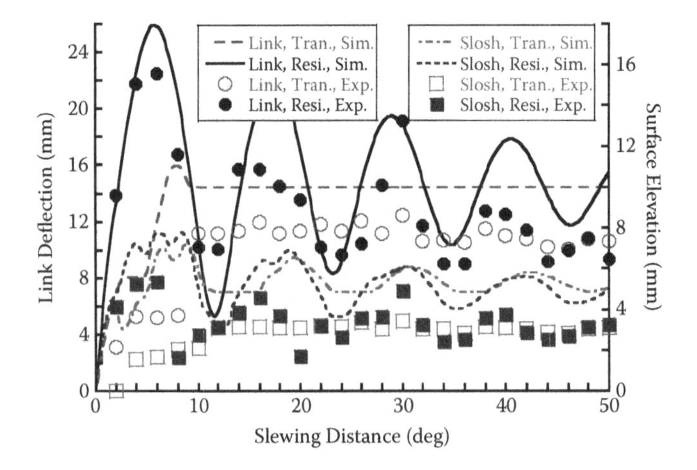

FIGURE 8.66 Experimental validation of the behavior of the dynamic model.

The residual amplitude of the surface elevation contains the low-frequency oscillations and high-frequency oscillations. The low frequency is approximately as same as the frequency in the residual amplitude of the link, and is corresponding to the first-mode frequency of coupled link-sloshing systems. The high frequency corresponds to the second-mode frequency of coupled systems.

The experimental frequency in residual amplitude of the link and sloshing closely matches that in the simulated curves. The experimental magnitudes in the link and sloshing were smaller than the simulated amplitudes because of proportional damping in the dynamic model. Moreover, the damping impacts produced a decrease in the peak values as the slewing distance increased.

Peaks in the transient deflection of the link and sloshing were also created by the interference between oscillations caused by the slewing acceleration and deceleration. The transient deflection of the sloshing exhibits more complicated dynamic behavior because of high modes. The experimental data in the transient deflection also follow the general shape of the simulated curves. The only difference is the simulated results are larger. This is also because the damping is assumed to be proportional in the model. Therefore, the experimental results validate the dynamic behavior of the model.

The third experiment was conducted to validate the effectiveness of the proposed active control method for oscillation attenuation, and compare with the previous methods. The link length was 42 cm. Resulting from Equation (8.230) and Figure 8.61, the first-mode frequency and damping ratio were calculated to be 5.4 rad/s and 0.0643, and the second-mode frequency and damping ratio were 18.0 rad/s and 0.0873. The frequency and damping ratio of the first two modes will be applied to design the double DP smoothers and double ZVD shaper. The frequency and damping ratio of the first mode will be used to design the TP smoother.

Experimental responses of flexible link and liquid sloshing are illustrated in Figure 8.67 for slewing an angular displacement of 16°. With the double DP smoothers, the transient deflections of link and sloshing were 4.11 mm and 1.26 mm,

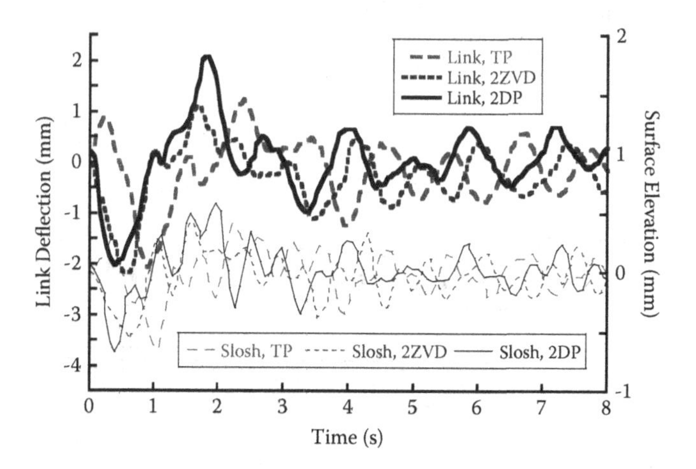

FIGURE 8.67 Experimental responses of link and sloshing to a slewing distance of 16°.

and the residual amplitudes of link and sloshing were 1.77 mm and 0.62 mm. When the TP smoother was implemented, the transient deflections of link and sloshing were 3.33 mm and 1.08 mm, and the residual amplitudes of link and sloshing were 1.81 mm and 0.56 mm. The transient deflections of link and sloshing with the double ZVD shaper were 3.30 mm and 0.99 mm, and the residual amplitudes of link and sloshing were 1.58 mm and 0.72 mm. The theoretical results in Figure 8.63 show that double ZVD shaper cannot control high-mode vibrations. Both the vibrations of link and sloshing in Figure 8.67 were suppressed to a low level by the double ZVD shaper because the contribution of high modes is weak in this case. The vibrational amplitudes of high modes may be large under some specified conditions.

The rise time of TP smoother, double ZVD shaper, and double DP smoothers in this case were 2.3122 seconds, 1.5067 seconds, and 1.2053 seconds, respectively. In order to achieve fast operation, the rise time should not be much longer than the first-mode vibrational period of the coupled link-sloshing system. Therefore, the double DP smoothers benefit the manipulator with liquid tank move quickly. The double DP smoothers attenuated residual vibrations of link and sloshing by 88.8% and 86.6%, respectively. Meanwhile, the double ZVD shaper reduced oscillations of link and sloshing by 89.9% and 84.4%, respectively. Instead, the rise time of the double DP smoothers decreases by 20% as compared with the double ZVD shaper. The vibration reduction of the previous methods comes at the cost of increased rise time. Finally, all of double DP smoothers, TP smoother, and double ZVD shaper can suppress the residual vibrations of link and sloshing below a low level, but the rise time of double DP smoothers is much shorter than that of previous two methods.

The fourth experiment was performed to verify the oscillation suppression for various slewing displacements. Both the simulated and experimental results with the double DP smoothers are given in Figure 8.68 when the link length was set to 42 cm. The transient deflection decreased slightly with increasing slewing displacements after 5° because of damping. The experimental transient deflections

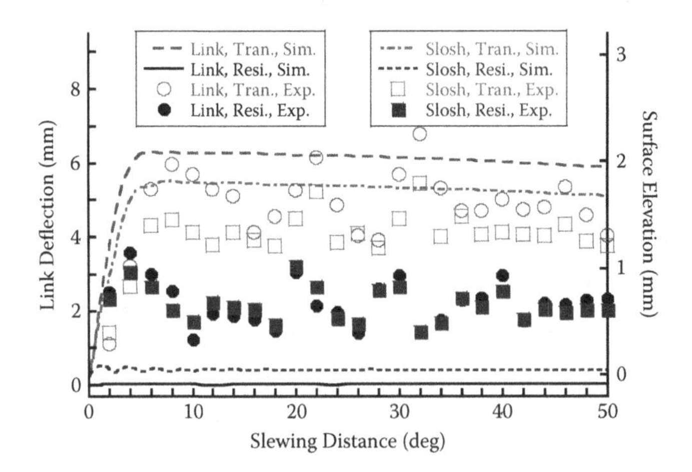

FIGURE 8.68 Transient and residual amplitudes of link and sloshing induced by slewing displacements.

were better than the simulated curves because of proportional damping ratio in the model. In addition, the residual amplitudes of the link and sloshing were nearly independent of the slewing displacements because most residual oscillations were suppressed. The experimental residual amplitudes were worse than the simulated results because of small modeling error in the frequency. The double DP smoothers attenuated experimental transient and residual amplitudes of the link by an average of 52.4% and 82.9% for all distances in Figure 8.68, and reduced those of the sloshing by an average of 53.0% and 80.5%. Therefore, the double DP smoothers can control the transient and residual vibrations below low level for various slewing displacements.

The fifth experiment was performed to verify the robustness of the double DP smoother to modeling error in the system parameters. The DP smoother depends on the frequency of the coupled system. System parameters have a large influence on the coupled frequency. Therefore, the control performance should be studied in the presence of the modeling errors in the system parameter.

Figure 8.69 shows the simulated and experimental results under different link lengths when the slewing displacement was fixed at 30°. In the absence of control, transient deflection of the link and sloshing increased with increasing the link length. Meanwhile, peaks and troughs in the residual amplitude arose as the link length varied. A trough occurred at the link length of 36 cm, while a peak arose at the link length of 44 cm. Peak and trough are also created by the interference between vibrations induced by the acceleration and deceleration of the hub.

When the double DP smoothers are implemented, the design length of the link in the smoother was held constant at 42 cm. The corresponding experimental results are illustrated in Figure 8.70. Increasing modeling errors in the link length increased residual vibrational amplitudes of link and sloshing. However, the double DP smoothers can control the residual oscillations below a low level for a wide range of

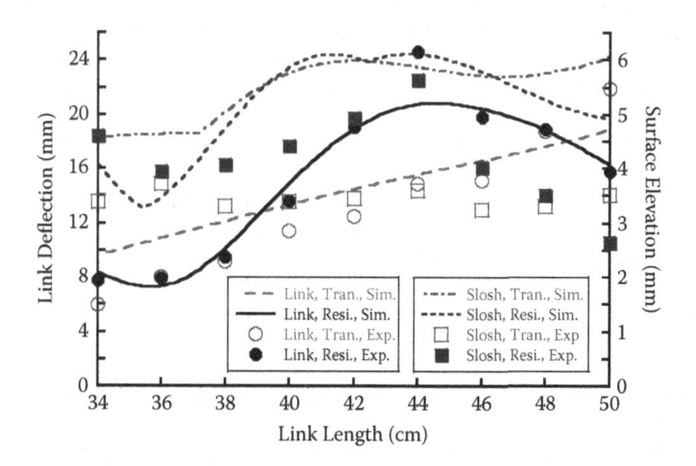

FIGURE 8.69 Uncontrolled transient and residual amplitudes of link and sloshing under different link lengths.

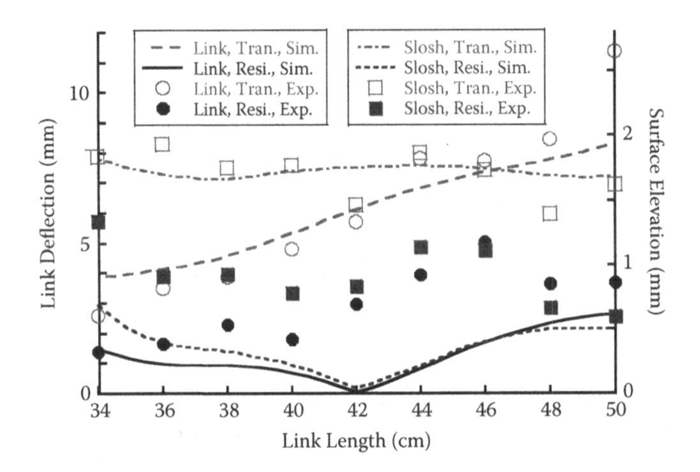

FIGURE 8.70 Controlled transient and residual amplitudes of link and sloshing under different link lengths.

various link lengths. Double DP smoothers attenuated the experimental residual amplitudes of link and sloshing by an average of 80.8% and 78.0%, respectively. The experimental data validated double DP smoothers have good robustness in the modeling error of the frequency and variation of the system parameters.

REFERENCES

[1] M. Grundelius, B. Bernhardsson, Control of liquid slosh in an industrial packaging machine, *IEEE International Conference on Control Applications*, Kohala Coast, HI, 1999, pp. 1654–1659.

[2] T. Acarman, Ü. Özgüner, Rollover prevention for heavy trucks using frequency shaped sliding mode control, *Vehicle System Dynamics*, 44 (10) (2006) 737–762.

[3] L. Perterson, E. Crawley, R. Hansman, Nonlinear fluid slosh coupled to the dynamics of spacecraft, *AIAA Journal*, 27 (9) (1989) 1230–1240.

[4] K. Muto, Y. Kasai, M. Nakahara, Experimental tests for suppression effects of water restraint plates on sloshing of a water pool, *Journal of Pressure Vessel Technology*, 110 (3) (1988) 240–246.

[5] V. Modi, S. Munshi, An efficient liquid sloshing damper for vibration control, *Journal of fluids and structures*, 12 (8) (1998) 1055–1071.

[6] K. Biswal, S. Bhattacharyya, P. Sinha, Dynamic characteristics of liquid filled rectangular tank with baffles, *Journal of the Institution of Engineers. India. Civil Engineering Division*, 84 (2003) 145–148.

[7] R. Venugopal, D. Bernstein, State space modeling and active control of slosh, *IEEE International Conference on Control Applications*, Dearborn, MI, 1996, pp. 1072–1077.

[8] H. Sira-Ramirez, M. Fliess, A flatness based generalized PI control approach to liquid sloshing regulation in a moving container, *American Control Conference*, Anchorage, AK, 2002, pp. 2909–2914.

[9] B. Bandyopadhyay, P. Gandhi, S. Kurode, Sliding mode observer based sliding mode controller for slosh-free motion through PID scheme, *IEEE Transactions on Industrial Electronics*, 56 (9) (2009) 3432–3442.

[10] S. Kurode, S. Spurgeon, B. Bandyopadhyay, et al., Sliding mode control for slosh-free motion using a nonlinear sliding surface, *IEEE/ASME Transactions on Mechatronics*, 18 (2) (2013) 714–724.
[11] S. Kurode, B. Bandyopadhyay, P. Gandhi, Sliding mode control for slosh-free motion of a container using partial feedback linearization, *International Workshop on Variable Structure Systems*, Antalya Turkey, 2008, pp. 367–372.
[12] H. Richter, Motion control of a container with slosh: constrained sliding mode approach, *Journal of Dynamic Systems, Measurement, and Control*, 132 (3) (2010) 031002.
[13] K. Yano, K. Terashima, Robust liquid container transfer control for complete sloshing suppression, *IEEE Transactions on Control Systems Technology*, 9 (3) (2001) 483–493.
[14] K. Terashima, G. Schmidt, Motion control of a cart-based container considering suppression of liquid oscillations, *IEEE International Symposium on Industrial Electronics*, Santiago, Chile, 1994, pp. 275–280.
[15] M. Reyhanoglu, J. Hervas, Nonlinear modeling and control of slosh in liquid container transfer via a PPR robot, *Communications in Nonlinear Science and Numerical Simulation*, 18 (6) (2013) 1481–1490.
[16] M. Reyhanoglun, J. Hervas, Nonlinear dynamics and control of space vehicles with multiple fuel slosh modes, *Control Engineering Practice*, 20(9) (2012) 912–918.
[17] M. Grundelius, B. Bernhardsson, Constrained iterative learning control of liquid slosh in an industrial packaging machine, *Proceedings of the 39th IEEE Conference on Decision and Control*, Sydney, Australia, 2000, pp. 4544–4549.
[18] K. Yano, T. Toda, K. Terashima, Sloshing suppression control of automatic pouring robot by hybrid shape approach, *Proceedings of the 40th IEEE Conference on Decision and Control*, Orlando, FL, 2001, pp. 1328–1333.
[19] K. Yano, K. Terashima, Sloshing suppression control of liquid transfer systems considering a 3-D transfer path, *IEEE/ASME Transactions on Mechatronics*, 20 (1) (2005) 8–16.
[20] Y. Noda, K. Yano, S. Horihata, et al., Sloshing suppression control during liquid container transfer involving dynamic tilting using Wigner distribution analysis, *43rd IEEE Conference on Decision and Control*, Atlantis, Bahamas, 2004, vol. 3, pp. 3045–3052.
[21] P. Gandhi, A. Duggal, Active stabilization of lateral and rotary slosh in cylindrical tanks, *IEEE International Conference on Industrial Technology*, Gippsland, VIC, 2009, pp. 1–6.
[22] J. Feddema, C. Dohrmann, G. Parker, et al., Control for slosh-free motion of an open container, *IEEE Control Systems*, 17 (1) (1997) 29–36.
[23] S. Chen, B. Hein, H. Worn, Using acceleration compensation to reduce liquid surface oscillation during a high speed transfer, *IEEE International Conference on Robotics and Automation*, Roma, Italy, 2007, pp. 2951–2956.
[24] K. Terashima, M. Hamaguchi, K. Yano, Modeling and input shaping control of liquid vibration for an automatic pouring system, *Proceedings of the 35th IEEE Conference on Decision and Control*, Kobe, Japan, 1996, pp. 4844–4850.
[25] B. Pridgen, K. Bai, W. Singhose, Shaping container motion for multimode and robust slosh suppression, *Journal of Spacecraft and Rockets*, 50 (2) (2013) 440–448.
[26] K. Terashima, K. Yano, Sloshing analysis and suppression control of tilting-type automatic pouring machine, *Control Engineering Practice*, 9 (6) (2001) 607–620.
[27] M. Hamaguchi, Y. Yoshida, T. Kihara, et al., Path design and trace control of a wheeled mobile robot to damp liquid sloshing in a cylindrical container, *IEEE International Conference on Mechatronics and Automation*, Niagara Falls, Ont. 2005, vol. 4, pp. 1959–1964.

[28] N. Qi, K. Dong, X. Wang, et al., Spacecraft propellant sloshing suppression using input shaping technique, *International Conference on Computer Modeling and Simulation*, Macau, China, 2009, pp. 162–166.
[29] A. Aboel-Hassan, M. Arafa, A. Nassef, Design and optimization of input shapers for liquid slosh suppression, *Journal of Sound and Vibration*, 320 (1–2) (2009) 1–15.
[30] H. Abramson, The dynamic behavior of liquids in moving containers, NASA SP-106, 1966.
[31] J. Roberts, E. Basurto, P. Chen, Slosh design handbook I, NASA TR CR-406, 1966.
[32] F. Dodge, The new dynamic behavior of liquids in moving containers, Southwest Research Institute Technical Report SP-106, San Antonio, TX, 2000.
[33] R. Ibrahim, V. Pilipchuk, T. Ikeda, Recent advances in liquid sloshing dynamics, *Applied Mechanics Reviews*, 54 (2) (2001) 133.
[34] P. Gandhi, K. Joshi, N. Ananthkrishnan, Design and development of a novel 2DOF actuation slosh rig, *Journal of Dynamic SystemsMeasurement, and Control*, 131 (1) (2009) 011006.
[35] P. Gandhi, J. Mohan, K. Joshi, et al., Development of 2DOF actuation slosh rig: a novel mechatronic system, *IEEE International Conference on Industrial Technology*, Bombay, India, 2006, pp. 1810–1815.
[36] Q. Zang, J. Huang, Z. Liang, Slosh suppression for infinite modes in a moving liquid container, *IEEE/ASME Transactions on Mechatronics*, 20 (1) (2015) 217–225.
[37] O. Faltinsen, O. Rognebakke, A. Timokha, Resonant three-dimensional nonlinear sloshing in a square-base basin, *Journal of Fluid Mechanics*, 487 (2003) 1–42.
[38] J. Huang, X. Zhao, Control of three-dimensional nonlinear slosh in moving rectangular containers, *The Transactions of the ASME - Journal of Dynamic Systems Measurement, and Control*, 140 (8) (2018) 081016–081018.

Index

Printed in the United States
by Baker & Taylor Publisher Services